Stochastic Differential Equations

Proceedings
of the
International Symposium
on
Stochastic Differential Equations

Kyoto, 1976

Edited by

Kiyosi Itô

A Wiley-Interscience Publication

JOHN WILEY & SONS

New York • Chichester • Brisbane • Toronto

PROCEEDINGS OF THE INTERNATIONAL SYMPOSIUM
ON STOCHASTIC DIFFERENTIAL EQUATIONS

Published in Japan by Kinokuniya Book-Store Co., Ltd.
17–7 3-chome, Shinjuku, Shinjuku-ku, Tokyo, 160-91 Japan

Published throughout the world except Japan by Wiley
Interscience, a division of John Wiley & Sons, Inc.
605 Third Avenue
New York N.Y. 10016 U.S.A.

Library of Congress Cataloging in Publication Data

International Symposium on Stochastic Differential Equations,
Kyoto University, 1976.
Proceedings of the International Symposium on Stochastic
Differential Equations, Kyoto, 1976.

"A Wiley-Interscience Publication."
1. Stochastic differential equations—Congresses.
I. Itō, Kiyoshi, 1915- II. Title.
QA274.23.I57 1976 515′.35 78-19655

ISBN 0-471-05375-9

Printed by Tokyo Press Co., Ltd. Tokyo, Japan

Printed in Japan

Preface

The International Symposium on Stochastic Differential Equations was held in the Research Institute for Mathematical Sciences at Kyoto University from July 9 through 14, 1976 under the auspices of the Institute, Japan Society for the Promotion of Science and the Mathematical Society of Japan. 151 mathematicians, 138 from Japan and 13 from abroad, participated in the Symposium and 27 invited speakers gave one hour lectures in session. On the fifth day of the symposium we made a bus tour for sightseeing around Kyoto to have foreign participants enjoy the scenery of this 1000 years old town.

We were given generous financial supports by Japan Society for the Promotion of Science, The Life Insurance Association of Japan, The Marine and Fire Insurance Association of Japan, Inc. and Taniguchi Foundation. In this connection we owe very much to Professor K. Yosida, Mr. T. Yamanouchi, Mr. S. Hishinuma and Professor Y. Akizuki. Professors H. Yoshizawa, G. Maruyama, M. Motoo, N. Ikeda, M. Nisio, H. Tanaka, S. Watanabe and H. Kunita were engaged in designing the Symposium as members of its Organizing Committee. Especially Ikeda and Watanabe did a hard job to carry out the Symposium and edit these proceedings. Many people in Kyoto University, especially Mr. T. Mitsui, eagerly helped us for the administrative and secretarial businesses of the Symposium. The Mathematical Society of Japan helped us and Mr. S. Okada, Kinokuniya Book-Store Co. Ltd., patiently cooperated with us for publication of these proceedings. We would like to express our sincere gratitude to all of those people and institutions.

In addition to the papers submitted to the Symposium we include in these proceedings an introduction to stochastic differential equations, which, we hope, will be useful to the reader.

May 1978, Kyoto

K. Itô

CONTENTS

Contents

Stochastic Differential Equations

Proc. of Intern. Symp. SDE
Kyoto 1976, pp. i–xxx

Introduction to Stochastic Differential Equations

Kiyosi ITÔ and Shinzo WATANABE

§ 1. Stochastic integrals

Stochastic integrals were first introduced by K. Itô [1] (see also [13]) in 1942 to rigorously formulate the stochastic differential equation that determines Kolmogorov's diffusion process [2]. J. L. Doob [3] pointed out the martingale character of stochastic integrals and suggested that a unified theory of stochastic integrals should be established in the framework of martingale theory. His program was accomplished by D. L. Fisk [4], P. Courrège [5], H. Kunita S. Watanabe [6] and P. Meyer [7]. Here we will give a brief account of this modern theory of stochastic integrals (*martingale integrals*).

We denote the basic probability space by $\Omega = (\Omega, \mathscr{F}_\Omega, P)$ and a generic point of Ω by ω. A stochastic process $X = \{X_t(\omega), 0 \leqq t < \infty\}$ is called *continuous* (or *left continuous*) if almost every function of X is continuous (or left continuous at $t > 0$ and right continuous at $t = 0$). A right continuous increasing family $\mathscr{F} = \{\mathscr{F}_t, 0 \leqq t < \infty\} \subset \mathscr{F}_\Omega$ is called a *reference family* if for every t $\mathscr{F}_t$ contains all P-null sets.

Take a reference family $\mathscr{F} = \{\mathscr{F}_t\}$ and fix it for the moment. A stochastic process $X = \{X_t\}$ is called *adapted* (to $\mathscr{F}$) if for every t $X_t(\omega)$ is $\mathscr{F}_t$-measurable in ω. Let $\mathscr{P} = \mathscr{P}(\mathscr{F})$ denote the least of all σ-algebras on $[0, \infty) \times \Omega$ with respect to which every left continuous adapted process is measurable in (t, ω). A stochastic process $X = \{X_t\}$ is called *predictable* (relative to $\mathscr{F}$) if $X_t(\omega)$ is $\mathscr{P}$-measurable in (t, ω). Every predictable process is adapted.

Let $\mathscr{M}_c^2 = \mathscr{M}_c^2(\mathscr{F})$ denote the family of all square integrable, continuous martingales (relative to $\mathscr{F}$). Let $\mathscr{S} = \mathscr{S}(\mathscr{F})$ denote the family of all square integrable, adapted, right continuous step processes with deterministic (= independent of ω) jump points.

Let $M \in \mathscr{M}_c^2$. Then $(M_t^2, t \leqq n)$, is a uniformly integrable submartingale for every n. The increasing part of the Doob-Meyer decomposition of $\{M_t^2\}$ is called the *quadratic variation* of M, $\langle M \rangle = \{\langle M \rangle_t, 0 \leqq t < \infty\}$ in notation. Let $L_{loc}^2 = L_{loc}^2(M) = L_{loc}^2(M, \mathscr{F})$ be the family of all predictable processes $X = \{X_t\}$ such that

$$(1.1) \qquad \|X\|_s^2 \equiv E\left(\int_0^s X_t^2 d\langle M\rangle_t\right) < \infty\,, \qquad 0 \leqq s < \infty\,.$$

We can check that $\mathscr{S} \subset L^2_{loc}$ and that for every $X \in L^2_{loc}$ there exists a sequence of processes $\{X^{(n)}\} \subset \mathscr{S}$ such that $\|X^{(n)} - X\|_s \to 0$ $(n \to \infty)$ for every s.

Now we will define the integral of the form

$$(1.2) \qquad I_t(X) = \int_0^t X_s dM_s\,, \qquad 0 \leqq t < \infty,\ M \in \mathscr{M}_c^2,\ X \in L^2_{loc}$$

and call it '*stochastic integral*'. The process $\{I_t(X), 0 \leqq t < \infty\}$ is denoted by $I(X)$. $I_t(X)$ is obviously linear in X. Let us first define $I_t(X)$ for $X \in \mathscr{S}$ in the obvious way. It is easy to check that $I(X) \in \mathscr{M}_c^2$ and that the following relation holds:

$$(1.3) \qquad \varepsilon^2 P\left\{\sup_{0 \leqq t \leqq s} |I_t(X)| \geqq \varepsilon\right\} \leqq E(I_s(X)^2) = \|X\|_s^2\,, \qquad \varepsilon > 0\,.$$

To define $I_t(X)$ for a general process $X \in L^2_{loc}$, take a sequence $\{X^{(n)}\} \subset \mathscr{S}$ such that

$$\|X^{(n)} - X\|_n < 4^{-n}/2\,.$$

Then

$$\begin{aligned}\|X^{(n+1)} - X^{(n)}\|_n &\leqq \|X^{(n+1)} - X\|_n + \|X^{(n)} - X\|_n \\ &\leqq \|X^{(n+1)} - X\|_{n+1} + \|X^{(n)} - X\|_n \\ &< 4^{-n}\,.\end{aligned}$$

Since $X^{(n+1)} - X^{(n)} \in \mathscr{S}$, we can apply (1.3) to obtain

$$P\left\{\sup_{0 \leqq t \leqq n} |I_t(X^{(n+1)}) - I_t(X^{(n)})| \geqq 2^{-n}\right\} \leqq 4^{-n}\,, \qquad n = 1, 2, \cdots\,.$$

Using the Borel-Cantelli lemma we can prove that $\{I_t(X^{(n)}), n = 1, 2, \cdots\}$ is convergent uniformly on every compact t-set a.s. Defining

$$I_t(X) = \lim_{n \to \infty} I_t(X^{(n)})\,,$$

we obtain a continuous process $I(X) = \{I_t(X)\}$. For s fixed and for $n > s$ we have

$$E((I_s(X^{(n+1)}) - I_s(X^{(n)}))^2) = \|X^{(n+1)} - X^{(n)}\|_s^2 < 4^{-2n}\,,$$

which implies that $\{I_s(X^{(n)}), n = 1, 2, \cdots\}$ converges to $I_s(X)$ in the $L^2(\Omega)$ norm. Now it is easy to check that $I(X) \in \mathcal{M}_c^2$ and that (1.3) holds for every $X \in L^2_{loc}$.

The stochastic integral $I(X) = I_t(X)$ for $X \in L^2_{loc}$ has the following properties:

(I_1) *$I: X \mapsto I(X)$ is a linear map from L^2_{loc} into $\mathcal{M}_c^2$,*

(I_2) (1.3) *holds for $X \in L^2_{loc}$,*

as we have already shown above. Hence

(I_3) *if $\|X^{(n)} - X\|_s \to 0$ for every s,*

$$\sup_{0 \leq t \leq t} |I_t(X^{(n)}) - I_t(X)| \to 0 \quad \text{i.p.} \qquad \textit{for every } s .$$

Using the uniqueness of the Doob-Meyer decomposition of submartingales we can prove that the *random inner product* $\langle M, N \rangle$ defined by

$$(1.4) \quad \langle M, N \rangle_t = \tfrac{1}{4}\{\langle M + N \rangle_t - \langle M - N \rangle_t\} , \qquad M, N \in \mathcal{M}_c^2$$

is bilinear and that

$$(1.5) \qquad \langle M, M \rangle_t = \langle M \rangle_t .$$

$\langle M, N \rangle$ is a continuous adapted process whose sample functions are locally bounded variation. We can use (I_2) to prove

(I_4) if $X \in L^2_{loc}(M)$ and $Y \in L^2_{loc}(N)$, then

$$E\left(\int_0^s X_t dM_t \int_0^s Y_t dN_t\right) = E\left(\int_0^s X_t Y_t d\langle M, N \rangle_t\right) .$$

Using the stopping argument, we will generalize the stochastic integral defined above. Let $\mathcal{M}_{c,loc} = \mathcal{M}_{c,loc}(\mathcal{F})$ denote the family of all continuous local martingales. Let $M \in \mathcal{M}_{c,loc}$. Then

$$\theta_n = \sup\{t : |M_t| \leq n\} , \qquad n = 1, 2, \cdots$$

are predictable stopping times (relative to $\mathcal{F}$) and $\theta_n \to \infty$ a.s. as $n \to \infty$. Let $M^{(n)} = \{M_t^{(n)}\}$ be the stopped process of M by θ_n for each n. Since $|M_t^{(n)}| \leq n$, $M^{(n)}$, $n = 1, 2, \cdots$ belong to $\mathcal{M}_c^2$ and

$$M_t = M_t^{(n)} = M_t^{(n+1)} = M_t^{(n+2)} = \cdots , \qquad t \leq \theta_n ,$$

so

$$\langle M^{(n)}\rangle_t = \langle M^{(n+1)}\rangle_t = \cdots , \qquad t \leqq \theta_n .$$

Keeping this in mind, we define the quadratic variation $\langle M\rangle$ by

$$\langle M\rangle_t = \langle M^{(n)}\rangle_t , \quad t \leqq \theta_n , \quad n = 1, 2, \cdots .$$

Let $\mathscr{L}^2_{loc} = \mathscr{L}^2_{loc}(M) = \mathscr{L}^2_{loc}(M, \mathscr{F})$ denote the family of all predictable processes X such that

$$(1.6) \qquad \int_0^s X_t^2 d\langle M\rangle_t < \infty \text{ a.s.} \qquad \text{for every } s .$$

Net $X \in \mathscr{L}^2_{loc}$. Then

$$\sigma_n = \sigma_{n,X} = \sup\left\{s : \int_0^s X_t^2 d\langle M\rangle_t \leqq n\right\} , \qquad n = 1, 2, \cdots$$

are predictable stopping times (relative to $\mathscr{F}$) and $\sigma_n \to \infty$ a.s. as $n \to \infty$. Also the processes

$$X^{(n)} = \{X_t^{(n)} \equiv 1_{[0,\sigma_n]}(t) X_t, \ 0 \leqq t < \infty\} , \qquad n = 1, 2, \cdots$$

belong to $L^2_{loc}(dM^{(n)})$ and

$$X_t = X_t^{(n)} = X_t^{(n+1)} = X_t^{(n+2)} = \cdots , \qquad t \leqq \sigma_n .$$

Keeping this in mind, we define the stochastic integral $I_s(X)$ by

$$I_s(X) \equiv \int_0^s X_t dM_t = \int_0^s X_t^{(n)} dM_t^{(n)} ,$$
$$s \leqq \theta_n \wedge \sigma_n , \qquad n = 1, 2, \cdots .$$

(Note that $\theta_n \wedge \sigma_n \to \infty$ as $n \to \infty$).

This generalized stochastic integral has the following properties.

(I_1') *$I : X \mapsto I(X)$ is a linear map from $\mathscr{L}^2_{loc}$ into $\mathscr{M}_{c,loc}$.*

(I_2') $$\underset{n\to\infty}{\text{l.i.p.}} \int_0^s (X_t^{(n)} - X_t)^2 d\langle M\rangle_t = 0$$

$$\Rightarrow \qquad \underset{n\to\infty}{\text{l.i.p.}} \sup_{0\leqq t\leqq s} |I_t(X^{(n)}) - I_t(X)| = 0 .$$

(I_3') *if $X = \{X_t\}$ is an adapted step process with deterministic jump times, then $X \in \mathscr{L}^2_{loc}$ and $I(X)$ is defined in the obvious way.*

(I_4') *if $X = \{X_t\}$ is an adapted continuous process, then $X \in \mathscr{L}^2_{loc}$ and*

$$\underset{n\to\infty}{\text{l.i.p.}}\left|I^s(X)-\sum_{i=1}^{\infty}X\Big(\frac{i-1}{n}\wedge s\Big)\Big(M\Big(\frac{i}{n}\wedge s\Big)-M\Big(\frac{i-1}{n}\wedge s\Big)\Big)\right|=0\,,$$

$$0\leqq s<\infty\,.$$

(I_1'), (I_2') and (I_3') follow from the definition. To prove (I_4'), take an approximating sequence of X:

$$(1.7)\qquad X_t^{(n)}=X\Big(\frac{i-1}{n}s\Big)\,,\qquad \frac{i-1}{n}s<t\leqq\frac{i}{n}s$$

$$i=1,2,\cdots \text{ (Replace '}<\text{' by '}\leqq\text{' for } i=1)\,.$$

Then these processes are adapted step processes with jump times in $\{is/n,\ i=1,2,\cdots\}$. By continuity of X we have

$$\int_0^s(X_t^{(n)}-X_t)^2dt\to 0\qquad\text{a.s.}\,.$$

Hence

$$I_s(X)=\underset{n\to\infty}{\text{l.i.p.}}\,I_s(X^{(n)})$$

which is nothing but the equation of (I_4') by virtue of (I_3').

Applying (I_4') to $X=M$ and noting

$$b^2-a^2=2a(b-a)+(b-a)^2\,,$$

we obtain

$$M_s^2-M_0^2=2I_s(M)+\underset{n\to\infty}{\text{l.i.p.}}\sum_{i=1}^{\infty}\Big(M\Big(\frac{i}{n}\wedge s\Big)-M\Big(\frac{i-1}{n}\wedge s\Big)\Big)^2$$

$$\text{a.s.}\,,$$

which implies that

$$(1.8)\qquad \langle M\rangle_s=\underset{n\to\infty}{\text{l.i.p.}}\sum_{i=1}^{\infty}\Big(M\Big(\frac{i}{n}\wedge s\Big)-M\Big(\frac{i-1}{n}\wedge s\Big)\Big)^2\,.$$

This formula justifies the name 'quadratic variation'.

The random inner product $\langle M,N\rangle$ can be defined for $M,N\in\mathcal{M}_{c,loc}$ similarly and we have

(I_5') $Y=I(X)$ if and only if $Y\in\mathcal{M}_{c,loc}$ and

$$\langle Y,N\rangle_s=\int_0^s Y_t d\langle M,N\rangle_t\qquad\text{for every } N\in\mathcal{M}_{c,loc}\,.$$

We can define stochastic integrals by this property and derive the other properties mentioned above from this definition.

Remark 1. Completion of stochastic integrals. Define a measure μ on the product space $[0, \infty) \times \Omega$ by setting

$$\mu(\Gamma) = E\left[\int_{t=0}^{\infty} l_\Gamma(t, \omega) d\langle M\rangle_t(\omega)\right]$$

for $\Gamma \in B[0, \infty) \times \mathscr{F}_\Omega$, $\mathscr{B}[0, \infty)$ being the topological σ-algebra, and then completing it in the usual way. Let $\overline{\mathscr{L}}^2_{loc}(M)$ denote the family of all processes Y such that $Y_t(\omega) = X_t(\omega)$ a.e. (μ) for some $X \in \mathscr{L}^2_{loc}(M)$ (depending on Y). By defining

$$\int_0^s Y_t dM_t = \int_0^s X_t dM_t$$

we can extend the stochastic integral onto $\overline{\mathscr{L}}^2_{loc}(M)$. If Y is adapted to $\mathscr{F}$, measurable with respect to $B[0, \infty) \times \mathscr{F}_\Omega$ and locally square integrable with respect to $d\langle M\rangle_t$ a.s., then Y belongs to the family $\overline{\mathscr{L}}^2_{loc}(M)$.

Remark 2. Reference families. The stochastic integral defined above appears to depend on the reference family $\mathscr{F}$ but in fact it is independent of $\mathscr{F}$. To prove this, suppose that $\mathscr{F}_1$ and $\mathscr{F}_2$ be two reference families and that

$$M \in \mathscr{M}_{c,loc}(\mathscr{F}_1) \cap \mathscr{M}_{c,loc}(\mathscr{F}_2) .$$

Then the stochastic integral relative to $\mathscr{F}_1$ coincides with that relative to $\mathscr{F}_2$ as we can easily see from the definition. Hence the stochastic integral

$$\int_0^s X_t dM_t$$

is well-defined for every process X in the union $\bigcup_{\mathscr{F}} \mathscr{L}^2_{loc}(M, \mathscr{F})$ where $\mathscr{F} = \{\mathscr{F}_t\}$ runs over all reference families with respect to which M is a continuous local martingale. See Itô's paper in these proceedings for related topics.

Remark 3. Time changes. Stochastic integrals are independent of stochastic time changes. Let $\tau = \{\tau(t)\}$ be a strictly increasing continuous process adapted to a given reference family $\mathscr{F} = \{\mathscr{F}_t\}$ such that $\tau(t) \to \infty$ as $t \to \infty$. Then $\mathscr{F}^\tau = \{\mathscr{F}_{\tau^{-1}(t)}\}$ is also a reference family. If $M = \{M_t\} \in \mathscr{M}_{c,loc}(\mathscr{F})$, then

$$M^\tau \equiv \{M_{\tau^{-1}(t)}\} \in \mathscr{M}_{c,loc}(\mathscr{F}^\tau)$$

and if $X \in \mathscr{L}^2_{loc}(M, \mathscr{F})$, then

$$X^\tau \equiv \{X_{\tau^{-1}(t)}\} \in \mathscr{L}^2_{loc}(M, \mathscr{F}^\tau)$$

and

$$\int_0^s X_t dM_t = \int_0^{\tau(s)} X_{\tau^{-1}(t)} dM_{\tau^{-1}(t)} .$$

Remark 4. *The classical stochastic integral* (*Itô integral*). A Wiener process $B = \{B_t\}$ is called a Wiener martingale relative to a given reference family $\mathscr{F} = \{\mathscr{F}_t\}$ if for every t the σ-algebra generated by the future increments $B_{t+u} - B_t$, $u \geqq 0$ is independent of $\mathscr{F}_t$. It is obvious that $B \in \mathscr{M}^2_c \subset \mathscr{M}_{c,loc}$ and

$$\langle B \rangle_t \equiv t ,$$

so

$$L^2_{loc}(B, \mathscr{F}) = \left\{X \text{ adapted, measurable} : E\left(\int_0^s X_t^2 dt\right) < \infty, \ \forall t\right\} ,$$

$$\mathscr{L}^2_{loc}(B, \mathscr{F}) = \left\{X \text{ adapted, measurable} : \int_0^s X_t^2 dt < \infty \text{ a.s.}, \ \forall t\right\} ,$$

and the stochastic integral

$$\int_0^s X_t dB_t , \qquad X \in \mathscr{L}^2_{loc}(B, \mathscr{F})$$

can be defined. This integral is the classical stochastic integral. Since M is a Wiener martingale if and only if

$$M \in \mathscr{M}_{c,loc} \quad \text{and} \quad \langle M \rangle_t = t ,$$

we can reduce the general stochastic integral to this special case by the time change with $\tau(t) \equiv \langle M \rangle_t$ in Remark 3.

Remark 5. *Quasimartingale integrals*. Let $Q = \{Q_t\}$ be a continuous local quasimartingale (or often called '*semimartingale*') relative to $\mathscr{F} = \{\mathscr{F}_t\}$ with the canonical decomposition

$$Q = M + A$$

where $M \in \mathscr{M}_{c,loc}$ and A is a locally bounded variation, right continuous process adapted to $\mathscr{F}$ with $A_0 \equiv 0$. The (quasimartingale) integral

$$\int_0^s X_t dQ_t$$

is defined to be the sum of

$$\int_0^s X_t dM_t \qquad \text{(martingale integral)}$$

and

$$\int_0^s X_t dA_t \qquad \text{(Stieltjes integral for each sample)}$$

whenever these two integrals are defined.

Remark 6. Stochastic integrals based on a discontinuous martingale (*or quasimartingale*). Thus far we have restricted ourselves to the case where the basic local martingale (or quasimartingale) is continuous. To extend this to the case where it may be discontinuous we need several delicate modifications in the arguments. See Kunita-Watanabe [6] and Meyer [7] for details.

§ 2. Stochastic differentials

In this section we will give a systematic account of stochastic differentials, see K. Itô [8].

We fix a reference family $\mathscr{F} = \{\mathscr{F}_t\}$ and consider the following families of processes:

$\mathscr{B}$ = the family of all locally bounded measurable processes adapted to $\mathscr{F}$

$\mathscr{Q}$ = the family of all continuous local quasimartingales relative to $\mathscr{F}$

$\mathscr{M} = \mathscr{M}_{c,loc}(\mathscr{F})$ = the family of all continuous local martingales

$\mathscr{A}$ = the family of all continuous locally bounded variation processes adapted to $\mathscr{F}$.

It is obvious that

$$\mathscr{B} \supset \mathscr{Q} \supset \mathscr{M} \cup \mathscr{A} \supset \mathscr{M} \cap \mathscr{A} \supset \boldsymbol{R}, \tag{2.1}$$

where the last inclusion relation is to be understood in the sense that every real number α represents a process identically equal to α. $\mathscr{B}$, $\mathscr{Q}$ and $\mathscr{A}$ are commutative rings but $\mathscr{M}$ is not.

With every $X \in \mathscr{Q}$ we associate a random interval function dX defined as follows:

(2.2) $$(dX)(I) = X_t - X_s \quad \text{for } I = (s, t] .$$

Let $d\mathcal{Q}$ denote the family $\{dX : X \in \mathcal{Q}\}$. Similarly we define $d\mathcal{M}$ and $d\mathcal{A}$. We introduce the following operations in $d\mathcal{Q}$.

A (Addition): $dX + dY = d(X + Y)$

M (Multiplication):

$$(\Phi \cdot dX)(s, t] = \int_s^t \Phi dX \text{ (stochastic integral)} , \qquad \Phi \in \mathcal{B} .$$

Then we have

(D_1) $d\mathcal{Q}$ with the operations A and M is a $\mathcal{B}$-module, both $d\mathcal{M}$ and $d\mathcal{A}$ are submodules of $d\mathcal{Q}$ and

$$d\mathcal{Q} = d\mathcal{M} \oplus d\mathcal{A} .$$

Let us introduce a third operation P in $d\mathcal{Q}$.

P (Product): $dX \cdot dY = d(XY) - X \cdot dY - Y \cdot dX$

It is easy to check that

(2.3) $$(dX \cdot dY)(s, t] = \underset{|\Delta| \to 0}{\text{l.i.p.}} \sum_{i=1}^{n} (X(t_i) - X(t_{i-1}))(Y(t_i) - Y(t_{i-1})) \quad \text{a.s.}$$

where $\Delta = (s = t_0 < t_1 < \cdots < t_n = t)$ is an arbitrary partition of $(s, t]$. Noting that the absolute variation process $|A| = \{|A|_t\}$ of $A \in \mathcal{A}$ belongs to $\mathcal{A}$ and that

$$\left| \sum_{i=1}^{n} (X(t_i) - X(t_{i-1}))(A(t_i) - A(t_{i-1})) \right| \leq \max_i |X(t_i) - X(t_{i-1})| \, (d|A|)(s, t] ,$$

we obtain

(2.4) $$dX \cdot dA = 0 \quad \text{whenever} \quad dX \in d\mathcal{Q} \quad \text{and} \quad dA \in d\mathcal{A} .$$

This implies that if $X = M_X + A_X$ and $Y = M_Y + A_Y$ are canonical decompositions of $X, Y \in \mathcal{Q}$, then

(2.5) $$dX \cdot dY = dM_X \cdot dM_Y = d\langle M_X, M_Y \rangle ,$$

where $\langle \; , \; \rangle$ denotes 'random inner product'. Keeping these facts in mind we can prove

(D$_2$) $d\mathcal{Q}$ with the operations A, M and P is a commutative algebra over $\mathcal{B}$ and $d\mathcal{Q}\cdot d\mathcal{Q} \subset d\mathcal{A}$, $d\mathcal{Q}\cdot d\mathcal{A} = 0$ and $d\mathcal{Q}\cdot d\mathcal{Q}\cdot d\mathcal{Q} = 0$, so $d\mathcal{A}$ is an ideal of the algebra $d\mathcal{Q}$.

The following fact corresponds to the chain rule in the ordinary calculus:

(D$_3$) (*Stochastic chain rule*). If $X^1, X^2, \cdots, X^n \in \mathcal{Q}$, then $Y = f(X^1, X^2, \cdots, X^n) \in \mathcal{Q}$ for every $f \in C^2$ and

$$df(X^1, X^2, \cdots, X^n) = \sum_{i=0}^{n} \partial_i f\cdot dX^i + \frac{1}{2}\sum_{i,j=1}^{n} \partial_i\partial_j f\cdot dX^i\cdot dX^j$$

where ∂_i denotes the partial derivative with respect to the i-th variable.

This formula can be written in the integral form as follows:

$$\begin{aligned} Y_t - Y_0 &= \sum_{i=1}^{n}\int_0^t \partial_i f\cdot dX^i + \frac{1}{2}\int_0^t \partial_i\partial_j f\cdot dX^i\cdot dX_j \\ &= \sum_{i=1}^{n}\left(\int_0^t \partial_i f\cdot dM^i + \int_0^t \partial_i f\cdot dA^i\right) + \frac{1}{2}\sum_{i,j=1}^{n}\int_0^t \partial_i\partial_i f\cdot d\langle M^i, M^j\rangle, \end{aligned}$$

where $X^i = M^i + A^i$ is the canonical decomposition of $X^i \in \mathcal{Q}$. Since both stochastic integrals and Lebesgue-Stieltjes integrals can be defined for measurable integrands with certain integrability conditions, we can imagine that our formula may hold for a more general function f, but we will not get into it.

Now we introduce a fourth operation M^s in $d\mathcal{Q}$:

M^s (*Symmetric multiplication*)

$$Y\circ dX = Y\cdot dX + \tfrac{1}{2}dY\cdot dX, \qquad Y \in \mathcal{Q}.$$

Since

$$\begin{aligned} \int_s^t Y\circ dX \equiv (Y\circ dX)(s,t] &= \int_s^t Y\cdot dX + \frac{1}{2}\int_s^t dY\cdot dX \\ &= \operatorname*{l.i.p.}_{|\Delta|\to 0} \sum_{i=1}^{n} \frac{1}{2}(Y_{t_{i-1}} + Y_{t_i})(X_{t_i} - X_{t_{i-1}}) \qquad \text{a.s.}, \end{aligned}$$

the symmetric multiplication corresponds to the so-called *Stratonovich integral* [8] or *Fisk integral* [4]. Since $d\mathcal{Q}\cdot d\mathcal{A} = 0$ and $(d\mathcal{Q})^3 = 0$, it is obvious that

(2.6) $X\circ dY = X\cdot dY$ (X or $Y \in \mathcal{A}$) and $(Z\circ dX)\cdot dY = Z\cdot dX\cdot dY$,

Corresponding to (D_1), (D_2) and (D_3) we have

(D_1^s) $d\mathscr{Q}$ with the operations A and M^s is a $\mathscr{Q}$-module.

$d\mathscr{A}$ is a submodule of $d\mathscr{Q}(A, M^s)$ but $d\mathscr{M}$ is not.

(D_2^s) $d\mathscr{Q}$ with the operations A, M^s and P is a commutative algebra over $\mathscr{Q}$ and

$$d\mathscr{Q}\cdot d\mathscr{Q} \subset d\mathscr{A}\,, \quad d\mathscr{Q}\cdot d\mathscr{A} = 0 \quad \text{and} \quad d\mathscr{Q}\cdot d\mathscr{Q}\cdot d\mathscr{Q} = 0\,,$$

so $d\mathscr{A}$ is an ideal of the ring $d\mathscr{Q}$.

(D_3^s) (*Symmetric stochastic chain rule*). If $X^1, X^2, \cdots, X^n \in \mathscr{Q}$, then

$$df(X^1, X^2, \cdots, X^n) = \sum_{i=1}^{n} \partial_i f \circ dX^i \qquad \text{for } Y = f(X^1, X^2, \cdots, X^n)$$

where $f \in C^\infty$.

These facts follow immediately from (D_1), (D_2), (D_3) and (2.6). For example, the symmetric stochastic chain rule can be proved as follows:

$$\begin{aligned}
&\sum_i \partial_i f \circ dX^i \\
&\quad = \sum_i \left(\partial_i f\cdot dX^i + \frac{1}{2} d(\partial_i f)\cdot dX^i\right) \\
&\quad = \sum_i \partial_i f\cdot dX^i \\
&\qquad + \frac{1}{2}\sum_i \left(\sum_j \partial_j\partial_i f\cdot dX^j + \frac{1}{2}\sum_{jk}\partial_j\partial_k\partial_i f\cdot dX^j\cdot dX^k\right)\cdot dX^i \\
&\quad = \sum_i \partial_i f\cdot dX^i + \frac{1}{2}\sum_{ij}\partial_i\partial_j f\cdot dX^i\cdot dX^j \qquad (\text{by } (d\mathscr{Q})^3 = 0) \\
&\quad = dY\,.
\end{aligned}$$

We can extend (D_3^s) to a more general f but we will not get into it. The importance of the symmetric stochastic chain rule lies in the fact that it takes the same form as in the ordinary calculus.

The stochastic chain rule (D_3) was first proved by K. Itô [10] for the special case

$$dX^i = a^i dt + \sum_{\alpha=1}^{m} b^i_\alpha dB^\alpha\,, \qquad i = 1, 2, \cdots, n\,.$$

$B^1, B^2, \cdots, B^m$ being independent Wiener martingales. The formula is as follows:

$$df(X^1, X^2, \cdots, X^n)$$

$$(2.7) \qquad = \sum_{i,\alpha} \partial_i f \cdot b^i_\alpha \cdot dB^\alpha + \left(\sum_i a_i \partial_i f + \frac{1}{2} \sum_{i,j} \partial_i \partial_j f \sum_\alpha b^i_\alpha b^j_\alpha\right) dt \, .$$

We can derive this immediately from (D_3) if we observe that

$$(2.8) \qquad dB^i \cdot dB^j = \delta^{ij} dt \qquad (\text{i.e. } \langle B^i, B^j \rangle = \delta^{ij} t) \, .$$

If $X^1, X^2, \cdots, X^n$ are martingales, the formula takes the form

$$df(X^1, X^2, \cdots, X^n) = \sum_i \partial_i f \cdot dX^i + \frac{1}{2} \sum_{i,j} \partial_i \partial_j f \cdot d\langle X^i, X^j \rangle$$

which was proved by Kunita-Watanabe [6].

The symmetric stochastic chain rule (D_3^s) was obtained by Fisk [4] in this general form. The special case related to Wiener processes was obtained by Stratonovich [9]. See also Wong [11] and McShane [12].

If both $\{X_t\}$ and $\{X_t^2 - t\}$ are martingales, then $\{X_t\}$ is a Wiener martingale. This fact is known as Lévy's theorem and rephrased in terms of stochastic differentials as follows: If $dX_t \in d\mathcal{M}$ and $(dX_t)^2$ $(\equiv d\langle X_t \rangle)$ $= dt$, then X_t is a Wiener martingale. Applying the stochastic chain rule to the real and imaginary parts of e^{izX_t}, we obtain, for $t > s$ (s: fixed),

$$d\, e^{iz(X_t - X_s)} = iz\, e^{iz(X_t - X_s)} dX_t - \frac{z^2}{2} e^{iz(X_t - X_s)} dt$$

$$d\, E(e^{iz(X_t - X_s)} \mid \mathcal{F}_s) = -\frac{z^2}{2} E(e^{i(X_t - X_s)} \mid \mathcal{F}_s) dt$$

and hence

$$E(e^{iz(X_t - X_s)} \mid \mathcal{F}_s) = \exp\left\{-\frac{(t-s)}{2} z^2\right\} ,$$

from which we can easily see that $\{X_t\}$ is a Wiener martingale. Such argument can be used to prove the n-dimensional version of the theorem: If $X^1, X^2, \cdots, X^n$ are martingales and $dX^i dX^j = \delta_{ij} dt$, then $X^1, X^2, \cdots,$ X^n are independent Wiener martingales. We can use this to derive the following fact due to H. P. McKean [13]. Let $B^1, B^2, \cdots, B^n$ be independent Wiener martingales. If

$$dX^i = \sum_{\alpha=1}^n \sigma^i_\alpha dB^\alpha \, , \qquad i = 1, 2, \cdots, n$$

where $(\sigma^i_j) = (\sigma^i_j(t, \omega))$ is an orthogonal $n \times n$-matrix whose component processes belong to $\mathscr{B}$. Then $X^1, X^2, \cdots, X^n$ are independent Wiener martingales. In fact

$$dX^i \in d\mathscr{M}\,, \qquad i = 1, 2, \cdots, n$$

and

$$dX^i dX^j = \sum_{\alpha,\beta} \sigma^i_\alpha \sigma^j_\beta dB^\alpha dB^\beta = \sum_\alpha \sigma^i_\alpha \sigma^j_\alpha dt = \delta_{ij} dt\,,$$

so $X^1, X^2, \cdots, X^n$ are independent Wiener martingales.

§ 3. Stochastic differential equations

Let $B = \{B_t, 0 \leqq t < \infty\}$ be a Wiener process and Ξ an arbitrary real random variable independent of B. Consider a stochastic differential equation:

$$\text{(3.1)} \qquad dX_t = a(t, X_t)dt + b(t, X_t)dB_t \quad \text{and} \quad X_0 = \Xi\,.$$

A process $X = \{X_t, 0 \leqq t < \infty\}$ is called a *solution* of the equation, if X is *non-anticipating*, i.e. for every t the past development $(B_s, X_s, s \leqq t)$ is independent of the future increments $(B_u - B_t, u \geqq t)$ and if X satisfies (3.1).

The equation is meaningful on every probability space (Ω, P) on which a Wiener process B and a real random variable Ξ independent of B are defined. A solution X of (3.1) on any such probability space is called *strict* if for every t X_t is measurable with respect to the σ-algebra generated by Ξ and $(B_s, s \leqq t)$. It is obvious that if there exists a strict solution on a probability space (Ω, P) chosen appropriately, then there exists a strict solution on every other probability space.

If there exists one and only one strict solution and if every solution of (3.1) is strict, then the equation is called *strictly regular*. If there exists at least one solution and if all solutions (which may be defined in different probability spaces) have the same probability law, then the equation is called *regular in law*. It is obvious that strict regularity implies regularity in law.

Suppose that

$$\text{(3.2)} \qquad dX_t = a(t+s, X_t)dt + b(t+s, X_t)dB(t)\,, \qquad X_0 = \Xi$$

is regular in law for every $s \geqq 0$ and for every Ξ. Then the solution of (1) defines a continuous Markov process. Using the stochastic chain rule, we obtain

$$df(x_t) = f'(X_t)b(t, X_t)dB_t + (f'(X_t)a(t, X_t) + \tfrac{1}{2}f''(X_t)b(t, X_t)^2)dt\,,$$

which implies that the Kolmogorov generator L_t of the diffusion, Kolmogorov [2], is given by

$$L_t = a(t, x)\frac{d}{dx} + \frac{1}{2}b(t, x)^2\frac{d^2}{dx^2}. \tag{3.3}$$

If $a(t, x)$ and $b(t, x)$ are continuous in t and satisfies the Lipschitz condition in x uniformly in t, i.e.

$$|a(t, x) - a(t, y)| + |b(t, x) - b(t, y)| \leqq K|x - y| \qquad (K\text{: constant}), \tag{3.4}$$

then the equation is strictly regular and the solution is obtained by the stochastic version of Picard's successive approximation method, K. Itô [14], [15]. More generally, if $a(t, x)$ and $b(t, x)$ are continuous in t and if

$$\begin{cases} |a(t, x)| + |b(t, x)| \leqq K(1 + x^2)^{1/2} & K\text{: Constant}, \\ |a(t, x) - a(t, y)| \leqq \kappa(|x - y|) & \int_{0+} \kappa^{-1}(\xi)d\xi = \infty, \\ |b(t, x) - b(t, y)| \leqq \rho(|x - y|) & \int_{0+} \rho^{-2}(\xi)d\xi = \infty, \end{cases} \tag{3.5}$$

then the equation (3.1) is strictly regular, as T. Yamada and S. Watanabe [16] showed by using their idea of *pathwise uniqueness*. The above condition is best possible in a sense, as we can imagine from the following example due to Girsanov [48]: The equation

$$dX_t = |X_t|^\beta dB_t \qquad (0 < \beta < \tfrac{1}{2}) \tag{3.6}$$

has infinitely many solutions.

Since the condition (3.4) is invariant under the time shift $t \to t + s$, the equation (3.2) is strictly regular, (hence regular in law) for every s, the solution of (1) defines a continuous Markov process with the Kolmogorov generator L_t given by (3.3) under the Lipschitz condition, K. Itô [14], [15]. Similarly for the case (3.5), T. Yamada and S. Watanabe [16].

The existence problem under the continuity of coefficients plus the first condition of (3.5) was solved by A. V. Skorohod [17]. He used the stochastic version of the Cauchy polygonal method. Let $X_n = X_t^n$, $n = 1, 2, \cdots$ be a sequence of approximate solutions. Noting that the joint processes (X^n, B), $n = 1, 2, \cdots$ are tight in probability law and appealing to Prohorov's theorem, Skorohod proved that the joint processes have at least one limit process in the sense of probability law and that the limit process satisfies the equation (3.1). But the solution thus obtained is not strict in general.

The following equation due to H. Tanaka [18] is regular in law but has no strict solution.

$$(3.7) \qquad dX_t = e(X_t)dB_t$$

where $e(x) = \pm 1$ according to whether $x \geqq 0$ or not. If we slightly modify e so that $e(x) = 1, 0$, or -1 according to whether $x > 0$, $=0$ or <0, then (3.7) has a strict solution $X_t \equiv 0$.

The equation

$$(3.8) \qquad dX_t = 3X_t^{1/3}dt + 3X_t^{2/3}dB_t\ , \qquad X_0 = 0$$

has infinitely many strict solutions:

$$X_t^a = \begin{cases} 0 & t \leqq \tau_a \equiv \inf\{s \geqq a : B_s = 0\} \\ B_t^3 & t \geqq \tau_a \end{cases}$$

where a is a constant $\in [0, \infty]$.

Analogously we can deal with a stochastic differential equation in several dimensions:

$$(3.1)' \quad dX_t^i = a^i(t, X_t)dt + \sum_{\alpha=1}^{n} b_\alpha^i(t, X_t)dB_t^\alpha\ , \qquad i = 1, 2, \cdots, m\ ,$$

where $B = (B_t^1, B_t^2, \cdots, B_t^n)$ is an n-dimensional Wiener process. Under the Lipschitz condition the solution defines a continuous Markov process whose Kolmogorov generator [2] is given by

$$(3.3)' \qquad L_t = \sum_{i=1}^{m} a^i \frac{\partial}{\partial x^i} + \frac{1}{2} \sum_{i,j=1}^{m} \left(\sum_{\alpha=1}^{n} b_\alpha^i b_\alpha^j \right) \frac{\partial^2}{\partial x^i \partial x^j}\ .$$

We can discuss the case where the solution process X_t takes values in a manifold, using the local coordinates and patching together the local solutions: See Itô [19], [20] Dynkin [21] and Mckean [13] for the case of smooth coefficients.

The non-Markovian case where the coefficients depend on the past development of $\{X(t)\}$ was discussed by Itô-Nisio [22]. See S. Watanabe [23] for the systematic treatment of this problem.

§ 4. Several topics related to stochastic differential equations

(a) *The martingale problem of Stroock-Varadhan*

When we use the stochastic differential equation (3.1) or (3.1)′ to construct a diffusion with a given generator L_t of (3.3) or (3.3)′, the Wiener process B of (3.1) appears only as a parametric process. Stroock and Varadhan [24] have formulated the equation (3.1) so that the parametric process B be eliminated. If X satisfies the equation (3.1), then we have

$$(4.1)\qquad f(X_t) - f(X_0) - \int_0^t L_s f(X_s)ds \equiv 0 \qquad (\mathrm{mod}\ \mathcal{M}_{c,loc})$$

for every twice continuously differentiable function f of compact support. Stroock and Varadhan called this equation the *martingale problem*. Under the condition that a and b are continuous and bound and that b is strictly elliptic they proved that there exists a solution unique in law and that the solution defines a diffusion whose Kolmogorov generator is L_t. Thus they have solved the long standing problem to construct a diffusion process whose generator is an elliptic differential operator with continuous coefficients; see also H. Tanaka [25]. The martingale formulation (4.1) is also convenient for the following purposes:

(i) To construct a diffusion with a given generator on a manifold, because it takes a form independent of local coordinates,

(ii) To prove the law-convergence of diffusions as their generators converge, or to prove that a given sequence of Markov chains converges to a certain diffusion, because (4.1) depends only on the probability law of the process X.

(b) *The support theorem of Stroock-Varadhan*

Because of the definition of the symmetric multiplication M^s in § 2 the following equations are equivalent:

$$(4.2)\qquad dX_t^i = a^i(t, X_t)dt + \sum_{\alpha=1}^n b_\alpha^i(t, X_t) \circ dB_t^\alpha\,, \qquad i = 1, 2, \cdots, m\,,$$

$$(4.3)\qquad dX_t^i = \tilde{a}^i(t, X_t)dt + \sum_{\alpha=1}^n b_\alpha^i(t, X_t)dB_t^\alpha\,, \qquad i = 1, 2, \cdots, m\,,$$

where

$$(4.4)\qquad \tilde{a}^i(t, x) = a^i(t, x) + \frac{1}{2}\sum_{\alpha=1}^n \sum_{j=1}^m b_\alpha^j(t, x)\partial_j b_\alpha^i(t, x) \qquad \left(\partial_j = \frac{\partial}{\partial x^j}\right).$$

Hence (4.2) defines a diffusion with generator

$$L_t = \sum_{i=1}^m \tilde{a}^i\partial_i + \frac{1}{2}\sum_{i,j=1}^m \sum_{\alpha=1}^n b_\alpha^i b_\alpha^j \partial_i\partial_j$$

$$= \sum_{i=1}^{m} a^i \partial_i + \frac{1}{2} \Big(\sum_{i=1}^{m} b_\alpha^i \partial_i \Big)^2$$

$$= X_0 + \frac{1}{2} \sum_{\alpha=1}^{n} X_\alpha^2$$

where

$$X_0 = \sum_{i=1}^{m} a^i(t, x) \partial_i , \quad X_\alpha = \sum_{i=1}^{m} b_\alpha^i(t, x) \partial_i , \quad \alpha = 1, 2, \cdots, n .$$

For simplicity we consider the case where the coefficients a^i and b_α^i are independent of t. The probability law P_x of the solution of (4.2) starting at $X_0 = x = (x^1, x^2, \cdots, x^m)$ is a Radon probability measure on the Polish space $C = C([0, \infty) \to \boldsymbol{R}^m)$ endowed with the topology of uniform convergence on compacts. Let $S(P_x)$ denote the topological support of P_x, i.e. the smallest closed subset of C that carries probability 1. Let S_x be a subset of C consisting of all solutions of the ordinary differential equations with the initial value $x = (x^1, x^2, \cdots, x^m)$:

$$(4.5) \qquad d\xi_t^i = a^i(\xi_t) + \sum_{\alpha=1}^{n} b_\alpha^i(\xi_t) d\varphi_t^\alpha , \qquad i = 1, 2, \cdots, m$$

where $\varphi^1, \varphi^2, \cdots, \varphi^n$ run over $C^\infty([0, \infty) \to \boldsymbol{R})$. The support theorem of Stroock-Varadhan [26] claims that *$S(P_x)$ coincides with the closure of S_x in C*.

As an immediate result of this theorem we have the following fact. If

$$(4.6) \qquad X_\alpha f = 0 , \quad \alpha = 0, 1, 2, \cdots, n , \quad f \in C^2(\boldsymbol{R}^m \to \boldsymbol{R}) ,$$

then $f(\xi_t) \equiv f(x)$ for $\xi \in S_x$ and therefore $f(X_t) \equiv f(x)$ a.s. for the solution X of (4.2). (Of course this follows also from the symmetric stochastic rule:

$$df(X_t) = \sum_{i=1}^{m} \partial_i f(X_t) \circ dX_t^i = X_0 f(X_t) dt + \sum_{\alpha=1}^{n} X_\alpha f(X_t) \circ dB_t^\alpha = 0 .)$$

This means that the solution of (3.1), i.e. the diffusion with generator $X_0 + 1/2 \sum_{\alpha=1}^{n} X_\alpha^2$ stays always on the surface $f(x) = \text{const.}$ whenever f satisfies (4.6).

If the Lie ring generated by $X_1, X_2, \cdots, X_n$ is m-dimensional at each point of M, then the set S_x is known to be dense in $C_x = \{\xi \in C : \xi(0) = x\}$, so $S(P_x) = C_x$ by the support theorem. See Kunita's paper in these proceedings for related topics.

(c) *Stochastic differential equations on the closed half-space*

Let D be the half n-space

$$\{x = (x_1, x_2, \cdots, x_n) : x_i \in R \ (i = 1, 2, \cdots, n-1),\ x_n > 0\}$$

and let $\bar{D}$ and ∂D be the closure and the boundary of D. Let $X = \{X_t\}$ be a time homogeneous diffusion on $\bar{D}$. Under a fairly general assumption $u(t, x) \equiv E(f(X_t) \mid X_0 = x)$ is determined by

$$\frac{\partial u}{\partial t} = L_D u \qquad \text{in } D \tag{4.7}$$

$$L_{\partial D} u = 0 \qquad \text{on } \partial D \tag{4.8}$$

and

$$u(0+, x) = f(x)\,, \tag{4.9}$$

where

$$L_D = \frac{1}{2} \sum_{i,j=1}^{n} a^{ij}(x) \frac{\partial^2}{\partial x^i \partial x^j}$$

and

$$L_{\partial D} = \frac{1}{2} \sum_{i,j=1}^{n-1} \alpha^{ij}(x) \frac{\partial^2}{\partial x^i \partial x^j} + \sum_{i=1}^{n-1} \beta^i(x) \frac{\partial}{\partial x^i} + \frac{\partial}{\partial x^n} - \rho(x) L_D\,.$$

In probabilistic language the operator L_D of (4.7) is the Kolmogorov generator determining the behavior of X in the interior D and the condition (4.8) (called *the Wentzell boundary condition* [27]) determines the behavior of X on the boundary ∂D. To construct such a process N. Ikeda [28] and S. Watanabe [29] used stochastic differential equations and Stroock-Varadhan [30] used the martingale problem.

(d) *Stochastic parallel displacement*

Let M be an m-dimensional Riemannian manifold with metric tensor $\{g_{ij}\}$ and $X = \{X_t\} \equiv (X_t^1, X_t^2, \cdots, X_t^m)$ be a diffusion process on M determined by a stochastic differential equation in § 3. K. Itô [31] defined the parallel displacement of a tangent vector V_0 at $X_0 \in M$ along the random curve $\{X_t\}$. Since the curve $\{X_t\}$ is not smooth in most interesting cases, we cannot use the Levi-Civita parallel displacement in its original sense. By connecting $X_{k/n}$ with $X_{(k+1)/n}$ by a geodesic curve for every k we obtain a piecewise smooth curve $\{X_t^{(n)}\}$, along which we can define the parallel displacement $V_t^{(n)}$ of V_0. $\{V_t^{(n)}\}$ is clearly a random curve on the tangent bundle. We can prove that, as $n \to \infty$, $\{V_t^{(n)}\}$ converges in the topology of uniform convergence on compacts i.p. The limit curve $\{V_t\}$ is called the parallel displacement of V_0 along $\{X_t\}$. Later Itô [32] noticed that V_t is determined by the following stochastic differential equation of Stratonovich:

$$dV_t^i = -\sum_{k,l=1}^{m} \Gamma_{lk}^i(X_t)V_t^k \circ dX_t^l \tag{4.10}$$

where $\{\Gamma_{lk}^i\}$ denotes the Riemannian connection on M. It should be noted that this equation takes the same form as in the Levi-Civita equation in differential geometry. Since the chain rule for the symmetric stochastic calculus takes the same form as in the ordinary calculus, the stochastic parallel displacement preserves inner products.

The stochastic parallel displacement of a tensor or of a frame can be defined similarly. Orthonormality of frames is preserved by stochastic parallel displacement.

(e) *Lifted diffusions of Malliavin*

Let M be an m-dimensional C^∞-manifold and A an elliptic differential operator:

$$A = \frac{1}{2}\sum_{i,j} a^{ij}(x)\frac{\partial^2}{\partial x^i \partial x^j} + \sum_i b^i(x)\frac{\partial}{\partial x^i}\,. \tag{4.11}$$

Since A is invariant under the change of local coordinates, $(g_{ij}) = (a^{ij})^{-1}$ defines a positive definite covariant tensor. Hence M is regarded as a Riemannian manifold with metric tensor (g_{ij}) and A can be expressed in the following intrinsic form

$$A = \tfrac{1}{2}\Delta + \boldsymbol{e} \tag{4.12}$$

where Δ is the Laplace-Beltrami operator on M and $\boldsymbol{e}$ is a tangent vector field on M, which is regarded as a differential operator

$$\boldsymbol{e} = \sum_i e^i(x)\frac{\partial}{\partial x^i}$$

written in local coordinates.

In § 3 we have mentioned that the diffusion on M with generator A starting at a fixed point is constructed locally as the solution of a certain stochastic differential equation and globally by patching together the local diffusions. To find out a *single* stochastic differential equation to determine the diffusion globally we will introduce the notion of lifted diffusions.

Let $T_x^{(m)}$ be the product (vector) space of m copies of the tangent space $T_x(M)$ of M at $x \in M$ and $T^{(m)}(M)$ the vector bundle generated by $T_x^{(m)}$, $x \in M$. A point r of $T^{(m)}(M)$ is represented as

$$r = (x, [\boldsymbol{e}_1, \boldsymbol{e}_2, \cdots, \boldsymbol{e}_m]) , \quad x \in M , \quad \boldsymbol{e}_i \in T_x(M) .$$

Let $O(M)$ be the orthonormal frame bundle of M whose point r is represented as above with an orthonormal basis $[\boldsymbol{e}_1, \boldsymbol{e}_2, \cdots, \boldsymbol{e}_m]$ of $T_x(M)$.

Let $\{x(t)\}$ be a smooth curve on M and $r = (x, [\boldsymbol{e}_1, \boldsymbol{e}_2, \cdots, \boldsymbol{e}_m])$ a point on $T^{(m)}(M)$ with $x(0) = x$. Applying parallel displacement to $\boldsymbol{e}_i$, $i = 1, 2, \cdots, m$ along $x(t)$, we obtain a curve on $T^{(m)}(M)$:

$$r(t) = (x(t), [\boldsymbol{e}_1(t), \boldsymbol{e}_2(t), \cdots, \boldsymbol{e}_m(t)]) ,$$

which is called the *lift* of $\{x(t)\}$ on $T^{(m)}(M)$ starting at r. It is obvious that if $r \in O(M)$, then $r(t) \in O(M)$ for every t. Similarly Malliavian [33] used stochastic parallel displacement to define the lift $\{R(t)\}$ of a diffusion process $\{X(t)\}$ starting at $r = (x, [\boldsymbol{e}_1, \boldsymbol{e}_2, \cdots, \boldsymbol{e}_m])$ where $x(0) = x$.

Let $r = (x, [\boldsymbol{e}_1, \boldsymbol{e}_2, \cdots, \boldsymbol{e}_m]) \in T^{(m)}(M)$ and $\boldsymbol{f} \in T_x(M)$. Let $\{x(t)\}$ be any curve on M with the initial tangent vector $x'(0) = \boldsymbol{f}$ and $\{r(t)\}$ be its lift on $T^{(m)}(M)$ starting at r. Then $r'(0) \in T_r(T^{(m)}(M))$. We call $r'(0)$ the *horizontal lift* of the vector $\boldsymbol{f} \in T_x(M)$, denoted by $L_{\boldsymbol{f}}$. $L_{\boldsymbol{f}}$ is well-defined independently of the choice of the curve $x(t)$. Associating $L_\alpha = L_{\boldsymbol{e}_\alpha}$ with $r = (x, [\boldsymbol{e}_1, \boldsymbol{e}_2, \cdots, \boldsymbol{e}_m])$, we obtain a vector field $L_\alpha(r)$ on $T^{(m)}(M)$. If $r \in O(M)$, then $r(t) \in O(M)$, so $L_\alpha \in T_r(O(M))$. Hence $L_\alpha(r)$ is also a vector field on $O(M)$, when it is restricted to $O(M)$. If $\boldsymbol{f}(x)$ is a vector field on M, the horizontal lift $L_{\boldsymbol{f}}(r)$ of $\boldsymbol{f}(x)$ is also a vector field on $T^{(m)}(M)$ as well as on $O(M)$.

Set

$$A_{O(M)} = \tfrac{1}{2}\Delta_{O(M)} + L$$

where

$$\Delta_{O(M)} = \sum L_\alpha^2 \quad \textit{(Bochner's horizontal Laplacian)}$$

and L is the horizonal lift of the vector field $\boldsymbol{e}$ of (4.12).

Consider a stochastic differential equation on a process $R(t) = (X(t), [\boldsymbol{e}_1(t), \boldsymbol{e}_2(t), \cdots, \boldsymbol{e}_m(t)])$:

$$(4.13) \quad \begin{cases} dX^i(t) = \sum\limits_\alpha e_\alpha^i(t) \circ dB^\alpha(t) + e^i(X(t))dt \\ de_\alpha^i(t) = -\sum\limits_{k,l=1}^{m} \Gamma_{kl}^i(X(t)) e_\alpha^k(t) \circ dX^l(t) , \end{cases}$$

which is clearly of the form independent of the choice of local coordinates, so the equation has a global meaning. Since the second equation means stochastic parallel displacement along $\{X(t)\}$, the solution lies on $O(M)$ whenever $(X(0), [e_1(0), e_2(0), \cdots, e_m(0)]) \in O(M)$. Hence the equation (4.14) also defines a process $\{R(t)\}$ on $O(M)$ which proves to be a diffusion with generator $A_{O(M)}$. In fact $\{R(t)\}$ is the lift of the diffusion $\{X(t)\}$ with generator A, so $\{X(t)\}$ is the projection of $\{R(t)\}$. Thus $\{X(t)\}$ is globally determined by the equation (4.13).

The above process $\{R(t)\}$ on $O(M)$ is also given by the following equation of the intrinsic form:

$$dR(t) = \sum_\alpha L_\alpha(R(t)) \circ dB(t) + L(R(t))dt \tag{4.14}$$

which means that

$$d\varphi(R(t)) = \sum_\alpha L_\alpha[\varphi](R(t)) \circ dB^\alpha(t) + L[\varphi](R(t))dt\ , \qquad \varphi \in C^\infty(O(M))\ ,$$

where L_α and L are interpreted as differential operators.

(f) *Probabilistic approach of harmonic forms*

A harmonic form on an m-dimensional compact Riemannian manifold M is expressed as the limit ω_∞ of the heat equation

$$\frac{\partial\omega}{\partial t} = \frac{1}{2}\Box\omega\ , \tag{4.15}$$

$$\Box = -(\delta d + d\delta) \text{ (the Kodaira-de Rham operator)}\ .$$

The probabilistic method to solve this was initiated (but not completed) by K. Itô [31]. P. Malliavin [33] introduced the *scalarization* method to discuss this problem elegantly.

For simplicity we will explain Malliavin's idea for 1-forms. Let $\omega = (\omega(x), x \in M)$ be a 1-form on M. This induces an R^m-valued function $F_\omega(r)$ on $O(M)$:

$$F_\omega(r) = (\langle\omega, e_\alpha\rangle\ \alpha = 1, 2, \cdots, m)\ ,$$
$$r = (x, e = [e_1, e_2, \cdots, e_m])\ .$$

Let $R(t)$ be the lift of Brownian motion $X(t)$ with $R(0) = r$ as was explained above. Then there exists a unique 1-parameter family $\{\omega_t(x), 0 \le t < \infty\}$ of 1-forms such that

$$F_{\omega_t}(r) = E[F_\omega(R(t))]\ .$$

Also ω_t satisfies

$$\frac{\partial \omega_t}{\partial t} = \frac{1}{2}\Delta \omega_t\,, \qquad \omega_0 = \omega\,.$$

where Δ is the Laplace-Beltrami operator. Observing the Weitzenböck formula: $\Box = \Delta + R$ and modifying $F_{\omega_t}(r)$ by the Kac formula for an appropriately chosen multiplicative operator, we can similarly construct the solution of the equation (4.14). See Malliavin's paper and Ikeda-Watanabe's paper in these proceedings.

(g) *Transformations by drifts (Girsanov's theorem)*

Let $\mathscr{F} = \{\mathscr{F}_t\}$ be a reference family on a probability space (Ω, P) and let $Y \in \mathscr{M}_{c,loc}(\mathscr{F})$ with $Y(0) = 0$. Under a certain integrability condition (for example, $E(\exp(1/2\langle Y\rangle(t))) < \infty$) we can show that

$$M(t) \equiv \exp\{Y(t) - \tfrac{1}{2}\langle Y\rangle(t)\} \tag{4.16}$$

is a martingale relative to $\mathscr{F}$ and that there exists a probability measure Q on Ω with the same domain as P such that

$$\left.\frac{dQ}{dP}\right|_{\mathscr{F}_t} = M(t)\,. \tag{4.17}$$

The reference family $\mathscr{F} = \{\mathscr{F}_t\}$ on (Ω, P) is regarded as one on (Ω, Q) and every process $X = \{X(t)\}$ on (Ω, P) as one on (Ω, Q). It can be shown that

$$X \in \mathscr{M}^P_{c,loc}(\mathscr{F}) \Rightarrow \tilde{X} \equiv X - \langle Y, X\rangle \in \mathscr{M}^Q_{c,loc}(\mathscr{F})\,, \tag{4.18}$$

the indexes P and Q being probability measures on Ω where the process X is observed. Since $\langle Y, X\rangle$ is locally bounded variation,

$$(dX)^2 = (d\tilde{X})^2 \text{ or more generally } dX_1 dX_2 = d\tilde{X}_1 d\tilde{X}_2\,.$$

Hence $\tilde{X}$ is a Wiener martingale relative to $\mathscr{F}$ on (Ω, Q) if X is so on (Ω, P). This fact was first discovered by Girsanov [34] in the special case where $\{Y(t)\}$ is a Wiener martingale.

We can apply this important fact to solve

$$dX = dB + a(t, X(t))dt\,, \qquad X(0) = x \tag{4.19}$$

where $a(t, \xi)$ is measurable and bounded. Take a Wiener process $b = \{b(t)\}$ on a probability space (Ω, P) and $\mathscr{F} = \{\mathscr{F}_t\}$ the reference family generated by b. Then $X(t) = x + b(t)$ is a Wiener martingale relative to $\mathscr{F}$ on (Ω, P). Hence

$$Y(t) = \int_0^t a(s, X(s))db(s) \in \mathcal{M}^P_{c,loc}(\mathcal{F}) .$$

Defining M and Q from Y as above, we can see that

$$B(t) = X(t) - \langle Y, X \rangle(t) = X(t) - \int_0^t a(s, X(s))ds$$

is also a Wiener martingale relative to $\mathcal{F}$ on (Ω, Q), proving that X solves the equation (4.19).

(h) *Stochastic stability*

As the stability problem to investigate the asymptotic behavior of solutions is important for ordinary differential equations, so is that for stochastic differential equations (*stochastic stability*), which is also applicable to study the solutions of degenerate second order partial differential equations. Here we will only give a brief account of the theory, see A. Friedman [35], Vol. 2, Chap. 12, for more details.

Consider a stochastic differential equation:

$$(4.20)\quad dX_t^i = \sum_{\alpha=1}^n \sigma_\alpha^i(X_t)dB_t^\alpha + b^i(X(t))dt , \qquad i = 1, 2, \cdots, m ,$$

where the coefficients are assumed to be smooth. The Kolmogorov generator of the solution $X_t = (X_t^1, X_t^2, \cdots, X_t^m)$ is denoted by L and the solution X_t starting at $X_0 = x$ by $X^x(t)$. Suppose that

$$(4.21)\qquad \sigma_\alpha^i(0) = b^i(0) = 0 .$$

If $x = 0$, then $X^x(t) = 0$ for every t a.s. If $x \neq 0$, $X^x(t)$ never vanishes a.s. Following after Hasminskii [36] we define 0 to be *asymptotically stable* (*AS*) for (4.19) if

$$\lim_{x \to 0} P\left\{\lim_{t \to \infty} X^x(t) = 0\right\} = 1 .$$

For example the solution of the one-dimensional equation

$$dX_t = \sigma \cdot X_t dB_t + b \cdot X_t dt , \qquad \sigma, b\colon \text{constant} ,$$

has a unique solution

$$X^x(t) = x \exp\{\sigma \cdot B(t) + (b - \tfrac{1}{2}\sigma^2)t\} ,$$

so 0 is *AS* if and only if $b - \sigma^2/2 < 0$.

A function defined on $U - \{0\}$ (U being a neighborhood of 0) is called an *S-function* for L if f satisfies the following conditions:

$$\|\sigma \nabla f\| \leqq \text{const.}\,, \quad Lf \leqq -1 \quad \text{and} \quad \lim_{x\to 0} f(x) = -\infty\,.$$

M. Pinsky [37] proved that 0 is *AS* for the equation (4.19) if there exists an *S* function of the Kolmogorov generator L of the process determined by (4.20). $f(x) = \log|x|$ is an *S* function for the example mentioned above if $b - \sigma^2/2 < 0$. More generally consider the case:

$$\sigma(x) = \sigma\cdot x + \varepsilon_1(x)\,, \qquad b(x) = b\cdot x + \varepsilon_2(x)$$

$$\left(\sigma, b\colon \text{constant}, \ \lim_{|x|\to 0} \varepsilon^i(x)/|x| = 0\right),$$

for which $f(x) = \log|x|$ turns out to be an *S*-function if $b - \sigma^2/2 < 0$, so 0 is *AS*.

(i) *Stochastic control*

Let Γ be a compact subset of $\boldsymbol{R}$. A Borel measurable function $u\colon [0, \infty) \times C[0, \infty) \to \Gamma$ is called a *control* (or an *admissible control*), if for every t

$$\xi(s) = \eta(s),\ s \leqq t \ \Rightarrow \ u(t, \xi) = u(t, \eta)\,.$$

A control is called *Markovian*, if for every t

$$\xi(t) = \eta(t) \ \Rightarrow \ u(t, \xi) = u(t, \eta)\,.$$

Consider a stochastic differential equation

$$\text{(4.22)} \quad dX_t = a(X_t, u(t, X))dB_t + b(X_t, u(t, X))dt\,, \qquad X_0 = x\,.$$

Since the coefficients a and b depend only on the past development of the process $X = \{X_t\}$, the equation is meaningful. The solution X contains the control u as a parameter and is called the *response* to u. The equation means that the coefficients a and b are controlled.

Let X be the response to u, and suppose that we gain $f(X_t, u(t, X))dt$ ($f\colon \boldsymbol{R} \times \Gamma \to \boldsymbol{R}$ Borel measurable) during the infinitesimal time interval dt. If the instantaneous interest rate is a constant λ, then the value of fdt estimated at the initial time point 0 equals $e^{-\lambda t}fdt$. Hence the expected total gain is

$$E\left(\int_0^\infty e^{-\lambda t} f(X_t, u(t, X))dt\right),$$

which depends on the starting point $x = X(0)$ and the control u and is called the *gain function*, $V(x, u)$ in notation. The control problem is to maximize $V(x, u)$ when u moves over all admissible controls. The control that maximizes $V(x, u)$ is called an *optimal control*.

Under some conditions which are fullfilled in practical cases Bellman proved that $W(x) = V(x, u_0)$ satisfies the so-called *Bellman equation*:

$$\max_{\gamma \in \Gamma} \left(-\lambda W(x) + \tfrac{1}{2}a^2(x,\gamma)W''(x) + b(x,\gamma)W'(x) + f(x,\gamma)\right) = 0 \tag{4.23}$$

for every x and that if $\gamma = \gamma_0(x)$ maximizes this function of γ for each x, then $u_0(t, \xi) = \gamma_0(\xi(t))$ gives the optimal control, which is clearly Markovian. Bellman also discussed the case where X_t is m-dimensional and Γ is n-dimensional. A rigorous theory of the control problem was established by N. V. Krylov [38]. We will below give a rough idea of the theory for the one-dimensional case.

Let $W(x)$ be a solution of the Bellman equation (4.23) and define $\gamma_0(x)$ as above. We consider a Markovian control

$$u_0(t, \xi) = \gamma_0(\xi(t))$$

and denote the response to the control u_0 by $X^0 = \{X_t^0\}$. The response to a general contral u is denoted by $X = \{X_t\}$. Noting that $W(x)$ satisfies the Bellman equation and applying the stochastic chain rule to $e^{-\lambda t}W(X_t)$, we obtain

$$\begin{aligned}
-W(x) &= E\int_0^\infty d(e^{-\lambda t}W(X_t)) \\
&= E\int_0^\infty e^{-\lambda t}\Big(-\lambda W(X_t) + \frac{1}{2}a^2(X_t, u(t,X))W''(X_t) \\
&\qquad\qquad + b(X_t, u(t,X))W'(X_t)\Big)dt \\
&\leqq -E_x\int_0^\infty f(X_t, u(t, X_t))dt = -V(x,u)
\end{aligned}$$

and similarly

$$\begin{aligned}
-W(x) &= E_x\int_0^\infty d(e^{-\lambda t}W(X_t^0)) \\
&= E_x\int_0^\infty e^{-\lambda t}\Big(-\lambda W(X_t^0) + \frac{1}{2}a^2(X_t^0, u_0(t,X^0))W''(X_t^0) \\
&\qquad\qquad + b(X_t^0, u_0(t,X^0))W'(X_t^0)\Big)dt \\
&= -E\int_0^\infty e^{-\lambda t}f(X_t^0, u_0(t, X_t^0))dt \quad (\text{Note that } u_0(t, X_t^0) = \gamma_0(X_t^0)) \\
&= -V(x, u_0)\,.
\end{aligned}$$

Hence we have

$$V(x, u_0) = W(x) \geqq V(x, u) ,$$

proving that u_0 is an optimal control and that the maximum gain function $V(x, u_0)$ satisfies the Bellman equation.

Krylov [39] also used control theory to construct a Markov process whose generator is a second order differential operator with measurable coefficients or with continuous degenerate coefficients.

(j) *Stochastic filtering*

Let $X = \{X_t\}$ be a stochastic process, called a *signal process*. When we observe this, it happens to obtain another process $Y = \{Y_t\}$ (an *observed process*) by virtue of random disturbances. We want to estimate X_t from the past development of Y. The optimum estimate $\hat{X}_t$ is characterized by the condition:

$$E[(X_t - \hat{X}_t)^2] = \min$$

under all processes $\hat{X}_t$ determined by the past observation, so

$$\hat{X}_t = E(X_t \mid \mathscr{Y}_t) \quad \text{where} \quad \mathscr{Y}_t = \sigma[Y_s, s \leqq t] .$$

The procedure to obtain $\hat{X}_t$ from X_t is called *filterning*. The error of the estimate is measured by the conditional variance $E((X_t - \hat{X}_t)^2 \mid \mathscr{Y}_t)$. The prediction problem is also regarded as a filtering.

The case where X and Y are determined by stochastic differential equations was first discussed by Kalman-Bucy [40] for equations with deterministic coefficients and the general theory has recently been developed extensively. An important tool is the notion of the innovation of the observed process Y, i.e. a Wiener martingale I_t relative to $\{\mathscr{Y}_t\}$ such that $\mathscr{Y}_t$ coincides with $\sigma[I_s, s \leqq t]$. See Bensoussan [41], Liptzer-Shiryaev [42], Fuzisaki-Kallianpur-Kunita [43].

(k) *Stochastic differential equations related to non-linear heat equations*

Let us consider a non-linear heat equation:

$$\frac{\partial}{\partial t} u_t(x) = \frac{1}{2} \frac{\partial^2}{\partial x^2}[e_t^2[u_t](x) u_t(x)] - \frac{\partial}{\partial x}[f_t[u_t](x) u_t(x)] \tag{4.24}$$

$$(\equiv L u_t(x))$$

where $e_t : C(\boldsymbol{R}) \to C^2(\boldsymbol{R})$ and $f_t : C(\boldsymbol{R}) \to C^1(\boldsymbol{R})$ for every t. This equation happens to have a solution which is a probability density function. To find out such a solution we consider a stochastic differential equation:

$$(4.25) \qquad dX(t) = e_t[p^{X(t)}](X(t))dB(t) + f_t[p^{X(t)}](X(t))dt\,,$$

$p^{X(t)}$ being the probability density of $X(t)$. This equation is essentially different from the stochastic differential equations observed above in that it has no sample-wise (or path-wise) meaning because $p^{X(t)}$ does depend not on the sample value of $X(t)$ but on the behavior over the whole probability space (Ω, P). To solve (4.25) we can use successive approximation. Define $X_n(t)$, $n = 0, 1, 2, \cdots$ as follows:

$$X_0(t) \equiv x$$

$$X_{n+1}(t) = x + \int_0^t e_s[p^{X_n(s)}](X_n(s))dB(s) + \int_0^t f_s[p^{X_n(s)}](X_n(s))ds\,,$$

$$n = 1, 2, \cdots.$$

Under some reasonable conditions we can prove that $\{X_n(t)\}$ tends to a continuous process $X(t)$, which satisfies the equation (4.25).

If $X(t)$ satisfies this equation, we can apply the stochastic chain rule to get

$$\varphi(X(t)) - \varphi(X(0))$$

$$= \text{a martingale} + \int_0^t \left[\frac{1}{2} e_s^2[u_s](X_s)\varphi''(X_s) + f_s[u_s](X_s)\varphi'(X_s)\right] ds$$

where $u_s(x) = p^{X(s)}(x)$. Taking the expectations we obtain

$$\int_R \varphi(x)u_t(x)dx - \int_R \varphi(x)u_0(x)dx$$

$$= \int_0^t ds \int_R \left\{\frac{1}{2} e_s^2[u_s](x)\varphi''(x) + f_s[u_s](x)\varphi'(x)\right\} u_s(x)dx$$

$$= \int_R \varphi(x)\,dx \int_0^t L_x u_s(x)ds \qquad \text{(integration by parts)}\,,$$

which proves that u_t satisfies (4.24). To justify this formal computation rigorously we should impose some reasonable conditions on e_t and f_t.

In connection with a statistical mechanical problem H. P. Mckean [44] discussed the case where

$$e[u](x) = \int_{R^m} e(x, y_1, y_2, \cdots, y_m)u(y_1)u(y_2) \cdots u(y_m)dy_1dy_2 \cdots dy_m$$

$$f[u](x) = \int_{R^m} f(x, y_1, y_2, \cdots, y_m)u(y_1)u(y_2) \cdots u(y_m)dy_1dy_2 \cdots dy_m$$

e and f being rapidly decreasing C^∞ functions. See Tanaka's paper in these proceedings for related topics.

Similar problems were discussed by A. Freidlin [45], H. Tanaka [46] and M. Nisio [47].

References

[1] K. Itô: Differential equations determining Markov processes, Zenkoku Shijo Sugaku Danwakai 1077 (1942), 1352–1400 (in Japanese).

[2] A. Kolmogorov: Uber analytische Methoden in der Wahrscheinlichkeitsrechnung, Math. Ann., **104** (1931), 415–458.

[3] J. L. Doob: Stochastic Processes. John Wiley, 1953.

[4] D. L. Fisk: Quasi-martingales and stochastic integrals, Tech. Reports 1, Dept. Math. Michigan State Univ. (1963).

[5] P. Courrège: Intégrales stochastiques et martingales de carré integrable. Sem. Th. Potentials 7, Inst. Henri Poincaré, Paris (1963).

[6] H. Kunita and S. Watanabe: On square integrable martingales, Nagoya Math. J., **30** (1967), 209–245.

[7] P. A. Meyer: Intégrales stochastiques, I∼IV Semi. de Prob. Springer Lecture Notes in Math., **39** (1967), 72–162.

[8] K. Itô: Stochastic differentials, Appl. Math. and Optim., **1** (1975), 347–381.

[9] R. L. Stratonovich: Conditional Markov processes and their application to the theory of optimal control (English translation). Amer. Elsevier, New York, 1968.

[10] K. Itô: On a formula concerning stochastic differentials, Mem. Coll. Science, Univ. Kyoto, Ser. A Math., 28 (1953), 209–223.

[11] E. Wong: Stochastic processes in information and dynamic systems, McGraw-Hill, 1971.

[12] E. J. McShane: Stochastic calculus and stochastic models, Academic Press, 1974.

[13] H. P. McKean: Stochastic integrals, Academic Press, 1969.

[14] K. Itô: On a stochastic integral equation, Proc. Imp. Acad. Tokyo, 22 (1946), 32–35.

[15] K. Itô: On stochastic differential equations, Memoir Amer. Math. Soc., 4 (1951), 1–51.

[16] T. Yamada and S. Watanabe: On the uniqueness of solutions of stochastic differential equations, J. Math. Kyoto Univ., **11** (1971), 155–167.

[17] I. I. Gihman and A. V. Skorohod: Stochastic differential equations, Springer 1972.

[18] See S. Watanabe [23], Chap. 4, §1, Example 1.1 (pp 89–90).

[19] K. Itô: Stochastic differential equations in a differentiable manifold, Nagoya Math. J., **1** (1950), 35–47.

[20] K. Itô: Stochastic differential equations in a differential manifold (2), Mem. Coll. Science, Univ. Kyoto, Ser. A Math., 28 (1953), 81–85.

[21] E. B. Dynkin: Markov processes (English translation), I. II. Academic Press, 1965.

[22] K. Itô and M. Nisio: On stationary solutions of a stochastic differential equation, J. Math. Kyoto Univ., 4 (1964), 1–75.

[23] S. Watanabe: Stochastic Differential Equations, Sangyo Tosho, 1975 (in Japanese).

[24] D. W. Stroock and S. R. S. Varadhan: Diffusion processes with continuous coefficients I, II, Comm. Pure. Appl. Math., 22 (1969), 354–400 and 476–530.

[25] H. Tanaka: Existence of diffusion processes with continuous coefficients, Mem. Proc. Sci. Kyushu Univ. Ser. A, 18 (1964), 89–103.

[26] D. W. Stroock and S. R. Varadhan: On the support of diffusion processes with application to the strong maximum principle, Proc. of 6-th Berkeley Symp. of Prob. and Math. Stat., Vol III (1972), 333–359.

[27] A. D. Wentzell: On boundary conditions for multidimensional diffusion processes, Th. Prob. Appl., 4 (1959), 164–177.

[28] N. Ikeda: On the construction of two-dimensional diffusion processes satisfying Wentzell's boundary conditions and its application to boundary value problems, Mem. Coll. Sci. Kyoto, 33 (1961), 367–427.

[29] S. Watanabe: On stochastic differential equations for multi-dimensional diffusion processes with boundary conditions, J. Math. Kyoto Univ., 11 (1971), 169–180.

[30] D. W. Stroock and S. R. S. Varadhan: Diffusion with boundary conditions, Comm. Pure Appl. Math., 24 (1971), 147–225.

[31] K. Itô: The Brownian motion and tensor fields on Riemannian manifold. Proc. Intern. Congress Mathematicians, Stockholm 1962, 536–539.

[32] K. Itô: Stochastic parallel displacement, Springer, Lecture Notes Math., 451, Probability Methods in Diff. Equat., (1975), 1–7.

[33] P. Malliavin: Formules de la moyenne, calcul de perturbations et théorèmes d'annulation pour les formes harmoniques, J. Funct. Anal., 17-3 (1974), 274–291.

[34] I. V. Girsanov: On transforming a certain class of stochastic processes by absolutely continuous substitution of measures, Th. Prob. Appl., 5 (1960), 285–301.

[35] A. Friedman: Stochastic differential equations and applications Vol 1 and 2, Academic Press, 1975/76.

[36] R. Z. Hasminskii: Necessary and sufficient conditions for the asymptotic stability of linear stochastic systems, Th. Prob. Appl., 12 (1967), 144–147.

[37] M. Pinsky: Stochastic stability and Dirichlet problems, Comm. Pure Appl. Math., 27 (1974), 311–350.

[38] N. V. Krylov: Control of a solution of a stochastic integral equation, Th. Prob. Appl., 17 (1972), 114–131.

[39] N. V. Krylov: On Ito's stochastic integral equation, Th. Prob. Appl. 14 (1969), 330–336 and On the selection of a Markov process from a system of processes and the construction of quasi-diffusion processes, Math. USSR Izvestija, 7 (1973), 691–709.

[40] R. S. Bucy and R. E. Kalman: New results in linear filtering and

prediction theory, J. Basic Engl. ASME Ser. D, 83 (1961), 95–108.
[41] A. Bensoussan: Filtrage optimal des systèmes linéaires. Dunod, Paris, 1971.
[42] R. S. Liptzer and A. N. Shiryaev: Statistics of stochastic processes, Nauka, 1974 (in Russian).
[43] G. Fuzisaki, M. Kallianpur and H. Kunita: Stochastic differential equations for the non-linear filterning problem, Osaka J. Math., 9 (1972), 19–40.
[44] H. P. McKean: A class of Markov processes associated with non-linear parabolic equations, Proc. Nat. Acad. Sci., 56 (1966), 1907–1911.
[45] M. I. Freidlin: Quasilinear parabolic equations and measures in function spaces, Funct. Analys. and Appl., 1 (1956), 157–214 (Engl. translation).
[46] H. Tanaka: Local solutions of stochastic differential equations associated with certain quasilinear parabolic equations, J. Fac., Sci. Univ. Tokyo, 14 (1967), 313–326.
[47] M. Nisio: On stochastic differential equations associated with certain quasilinear parabolic equations, J. Math. Soc. Japan, 22 (1970), 278–292.
[48] I. V. Girsanov: An example of nonuniqueness of the solution of the stochastic differential equation of K. Ito, Th. Prob. Appl., 7 (1962), 325–331.

K. ITÔ, RESEARCH INSTITUTE
FOR MATHEMATICAL SCIENCES
KYOTO UNIVERSITY
KYOTO 606, JAPAN

S. WATANABE
DEPARTMENT OF MATHEMATICS
KYOTO UNIVERSITY
KYOTO 606, JAPAN

Proc. of Intern. Symp. SDE
Kyoto 1976, pp. 1–19

A White Noise Version of the Girsanov Formula

A. V. BALAKRISHNAN

§ 1. Introduction

Much progress has recently been made in filtering problems in which the observed process is modelled as:

$$(1.0) \qquad Y(t) = \int_0^t f(\sigma)d\sigma + W(t)$$

where $f(t)$ is 'adapted' to $Y(t)$, and $W(t)$ is a Wiener process. See [1]. An essential role is played in this theory by the 'Girsanov formula', applicable to the case where

$$(1.1) \qquad Y(t) = \int_0^t g(\sigma)d\sigma + Z(t) \qquad 0 < t < T$$

where $Z(t)$ is a Wiener process and $g(t)$ is adapted to $Z(t)$ and

$$\int_0^T E(\| g(t) \|^2)dt < \infty .$$

In the filtering application, (1.1) arises from (1.0) by setting:

$$g(t) = E[f(t) \,|\, B_Y(t)] .$$

The measure induced by the process $Y(t)$ on $C(0, T)$ is then absolutely continuous with respect to that induced by the Wiener process and the formula of Girsanov for the derivative is: (see (1) for example):

$$(1.2) \qquad \frac{d\mu_Y}{d\mu_W} = \text{Exp}\Big(-1/2\Big\{\int_0^T \| g(\sigma) \|^2 \, d\sigma - 2\int_0^T [g(\sigma), dZ(\sigma)]\Big\}\Big), \cdots .$$

Now let H denote a real separable Hilbert space and let W denote the Hilbert space

$$L_2((0, T) ; H) .$$

Let μ_G denote Gauss measure on W, with elements in W denoted ω. Then ω is a 'white noise' sample function. Let

$$y(\omega) = p(\omega) + \omega \tag{1.3}$$

where $p(\cdot)$ is a [Volterra] function mapping W into W. Note that (1.1) resembles an "integrated" version of (1.3). The author [2] has argued that the white noise model is more realistic [that in fact the Girsanov formula simply does not have operational value with observed data since the latter cannot be non-differentiable with probability one]. In any event, it is of interest to ask for the corresponding formula for the derivative of the (cylinder) measure induced by $y(\cdot)$ to the Gauss measure. Such a problem has already been considered by Gross [3] who views it as a formula for the Jacobian of the transformation. Unfortunately however Gross, who obtained his formula in 1960, does not seem to be motivated by the needs of stochastic filtering theory. In any case, the conditions he puts on the function $p(\omega)$ would appear to be unsuited for our purposes, quite apart from the difficulty of verification.

Here, taking advantage of the Girsanov formula, we give a version for polynomial functions (and their limits), which is a generalization of the formula for the 'linear' case given in [4].

We begin in section 2 with a theory of white noise and white noise integrals. Basically, given a sequence of independent Gaussian variables (of zero mean and unit variance) one has a choise of possible sample spaces: one is ℓ_2 but the measure is only finitely additive and we have "white noise"; and the other is R^∞ where the measure is countably additive and we have a Wiener process. In going to R^∞, one must perforce add many "non-physical" events and variables; however the notion of a random variable is trivial. On the other hand in the white noise case, this question is a difficult one, and a complete characterization is still lacking. Fortunately we can make some progress with polynomials; which we study in most of section 2.

The white noise version of the Girsanov formula is given in section 3 for Volterra polynomials.

§ 2. White noise: Basic notions

Let H denote a real separable Hilbert Space and let

$$W = L_2[0, T; H], \quad 0 < T < \infty,$$

denote the real Hilbert Space of H-valued weakly measurable functions $u(\cdot)$ such that

$$\int_0^T [u(t), u(t)]dt < \infty$$

with inner-product defined by

$$[u, v] = \int_0^T [u(t), v(t)]dt .$$

Let μ_G denote measure on W (on the cylinder sets with Borel basis) with characteristic function

$$C_G(h) = \text{Exp}(-1/2[h, h]) , \qquad h \in W .$$

Elements of W under this (finitely additive) measure will be 'white noise sample functions', denoted ω. This terminology appears to have the sanction of usage; see Skorokhod [5] for example. It is essential for us that W is an L_2-space over a finite interval.

Any function $f(\cdot)$ defined on W into another Hilbert Space H_r such that the inverse images of Borel sets in H_r are cyclinder sets will be called a 'tame' function. See Gross [6]. As is readily seen, the class of tame functions is a linear class. Since the inverse image of the whole space H_r must be cylindrical, it is clear that any tame function has the form $f(P\omega)$ where P is a finite dimensional projection.

To introduce the notion of a 'random variable' let us first confine ourselves to the case where H_r is finite dimensional: $H_r = R^n$ say. We introduce a metric into the linear space of tame functions by

$$|||f - g||| = \int_W \frac{\|f - g\|}{1 + \|f - g\|} d\mu_G$$

and then complete the space, the completion yielding a Frechet Space. Every element of the completed space is called a 'random variable' and if ζ denotes such an element and $f_n(\omega)$ a corresponding Cauchy sequence in probability, then we define the corresponding 'distribution function' or probability measure on R^n to be that induced by the characteristic function

$$C_\zeta(h) = \underset{n}{\text{limit}}\, E(e^{i[f_n(\omega), h]}) . \tag{2.0}$$

The latter limit exists (uniformly on bounded sets of $R^n = H_r$).

In the case where H_r is no longer finite dimensional, we shall still identify Cauchy sequences in probability of tame functions as "weak random variables". The limit in (2.0) still holds, uniformly on bounded sets in H_r, but the limit may in general only define a "weak distribution" on H_r. We recall in this connection the Sazonov theorem [7] that the limit is the characteristic function of a probability measure if and only if it is continuous in the trace-norm topology ('S-topology' see below). This is automatically the case if the sequence is Cauchy in the mean square sense; and we shall then drop the qualification "weak".

Let $f(\omega)$ be any Borel measurable function mapping W into H_r. Then $f(P\omega)$ is tame for every finite-dimensional projection operator P. Let $\{P_n\}$ denote a sequence of finite dimensional projections converging strongly to the Identity; the sequence may be assumed to be monote. If the sequence $f(P_n\omega)$ is Cauchy in probability, then we may associate a (weak, ingeneral) random variable with $f(\cdot)$. Let us denote it by $f^{\sim}$ (a notation used by Gross). This limit of course can depend on the particular projection sequence chosen. Of primary interest to us are those functions $f(\cdot)$ for which $\{f(P_n\omega)\}$ is Cauchy in probability for *every* such sequence of finite dimensional projections and moreover such that all such Cauchy sequences are equivalent so that the limit random variable $f^{\sim}$ is unique. In that case we say that $f(\omega)$ is a (weak) random variable. We shall use the term "random variable" if the corresponding measure is countably additive; we shall be dealing in the sequel only with mean square convergence, when this will be automatic.

The simplest function one can consider is perhaps the linear function:

$$f(\omega) = L\omega$$

where L is a linear bounded transformation mapping W into H_r, where we now allow H_r to be infinite dimensional. Then it is easy to see that if L is Hilbert-Schmit, then $\{LP_m\omega\}$ is Cauchy in the mean square sense, and $L\omega$ is a random variable. Conversely L must be H. S. if $L\omega$ is to be a random variable.

What is the class of functions which are random variables? To answer this question, at least in part, let us introduce the S-topology on W: this is the (locally convex) topology induced by seminorms of the form:

$$\rho(\omega) = \sqrt{[S\omega, \omega]} \tag{2.1}$$

where here (and hereinafter) S will denote a self-adjoint, non-negative definite trace-class operator on W into W. For the case where $H_r = R^1$, Gross [6] has given a sufficient condition: $f(\cdot)$ is a random variable if it is *uniformly* continuous in the S-topology. Uniform continuity means that given $\varepsilon > 0$, we can find $\rho(\cdot)$ such that

$$\|f(x) - f(y)\| < \varepsilon \qquad \text{for all } x, y \text{ such that } \rho(x - y) < 1 .$$

Unfortunately Gross does not seem to discuss non-trivial examples of functions satisfying this condition. Here we shall give a sufficient condition for class of random variables with finite second moment.

Theorem 2.1. *Let $p_n(\omega)$ denote a homogeneous polynomial of degree n mapping W into H_r. Suppose it is continuous at the origin in the S-topology. Let P denote any finite dimensional projection.*

$$\sup_P E(\|p_n(P\omega)\|^2) < \infty \tag{2.2}$$

where the supremum is taken over the class of finite dimensional projections. Conversely, if (2.2) holds, then $p_n(\cdot)$ is continuous at the origin in the S-topology.

Proof. We begin with a simple but useful Lemma.

Lemma 2.1. *Suppose $p_n(\cdot)$ is continuous in the S-topology at the origin. Then there exists a seminorm in the S-topology:*

$$\rho(\omega) = \sqrt{[S\omega, \omega]} \tag{2.3}$$

such that

$$\|p_n(\omega)\| \leq M\rho(\omega)^n \tag{2.4}$$

where M is a constant. Conversely if (2.4) hold, then $p_n(\omega)$ is continuous in the S-topology at the origin.

Proof. Continuity in the S-topology at zero implies this: given $\varepsilon > 0$ we can find a seminorm of the form (2.3) such that

$$\|p_n(\omega)\| < \varepsilon \quad \text{for all } \omega \text{ such that } \rho(\omega) \leq \delta\,. \tag{2.5}$$

Hence for any ω for which

$$\rho(\omega) \neq 0\,,$$

we have that

$$\left\| p_n\left(\frac{\delta\omega}{\rho(\omega)}\right)\right\| < \varepsilon$$

or by the homogeneity of $p_n(\cdot)$,

$$\|p_n(\omega)\| < \left(\frac{\varepsilon}{\delta^n}\right)\rho(\omega)^n,\ \rho(\omega) \neq 0$$

If $\rho(\omega) = 0$, then for any positive number k,

$$\rho(k\omega) = 0$$

and hence from (2.5)

$$\| p_n(\omega) \| < \frac{\varepsilon}{k^n}$$

for all $k > 0$ and hence

$$p_n(\omega) = 0 \ .$$

Hence 2.4 holds. The converse is obvious.

Proof of Theorem. Corresponding to a finite dimensional projections P, we can find an orthonormal basis $\{\phi_i\}$ such that P is the projection operator corresponding to the space spanned by ϕ_i, $i = 1, 2, \cdots, m$. Let

$$p_n(\omega) = k_n(\omega, \cdots, \omega)$$

$k_n(\cdots)$ being the symmetric n-linear form, corresponding to $p_n(\cdot)$. Then

$$p_n(P\omega) = \sum_{L_1=1}^{m} \cdots \sum_{Ln=1}^{m} a_{i_1,\cdots,i_n} \zeta_{i_1} \cdots \zeta_{i_n} \tag{2.6}$$

where

$$a_{i_1,\cdots,i_n} = k_n(\phi_{i_1}, \cdots, \phi_{i_n})$$
$$\zeta_i = [\phi_i, \omega] \ .$$

The $\{\zeta_i\}$ is a sequence of independent zero-mean unit variance Gaussians and (2.6) defines a tame function. Moreover we can readily calculate (by expressing (2.6) in terms of Hermite polynomial for instance) that

$$E(\| P_n(P\omega) \|^2) = \sum_{\nu=0}^{[n/2]} \left(\frac{n!}{(n-2\nu)!\, 2^\nu \nu!} \right)^2 \sum_{i_{2\nu+1}=1}^{m} \sum_{i_n=1}^{m} \left\| \sum_{i_1=1}^{m} \cdots \sum_{i_\nu=1}^{m} a_{i_1,i_1,\cdots,i_\nu,i_\nu,i_{2\nu+1},\cdots,i_n} \right\|^2 . \tag{2.7}$$

But from Lemma 2.1, we have that

$$\| p_n(P\omega) \|^2 \leq [S_m\omega, \omega]^n \tag{2.8}$$

where

$$S_m = PSP$$

and is of course trace-class and finite dimensional. Hence

(2.9) $$E[\|p_n(P\omega)\|^2] \le E([S_m\omega, \omega]^n) .$$

Let ψ_k, $k = 1, \cdots, \nu$, be the orthonormalized eigen-vectors of S_m with corresponding non-zero eigen-values λ_k. Then

$$[S_m\omega, \omega] = \sum_1^\nu \lambda_i[\psi_i, \omega]^2$$

and we have

$$E([S_m\omega, \omega]^n) = f(\text{Tr. } S_m, \text{Tr. } S_m^2, \cdots, \text{Tr. } S_m^n)$$

where $f(\cdots)$ is a fixed continuous function. Of course

$$\text{Tr. } S_m^j$$

is monotone in m for each j and converge to

$$\text{Tr. } S^j .$$

Hence it follows that

$$E[\|p_n(P\omega)\|^2] < \infty$$

for *all* finite dimensional projections.

To prove the converse, suppose (2.2) holds. The (2.7) holds for every m, and taking $\nu = 0$ therein, we obtain that

(2.10) $$\sum_{i_1}^\infty \cdots \sum_{i_r}^\infty \|k_n(\phi_{i_1}, \cdots, \phi_{i_n})\|^2 < \infty$$

for every orthonormal sequence $\{\phi_i\}$. Hence $p_n(\cdot)$ is Hilbert-Schmidt. Of course

(2.11) $$\|p_n(\omega)\|^2 \le M\|\omega\|^{2n} .$$

Define now S by:

$$[S\omega, \omega] = (\|p_n(\omega)\|^2)^{1/n} .$$

Then S is Hilbert-Schmidt by (2.10).

For any finite dimensional projection P,

$$\begin{aligned} E[SP\omega, P\omega] &= E[PSP\omega, \omega] \\ &= E((\|p_n(P\omega)\|^2)^{1/n}) \end{aligned}$$

and hence

$$\sup_P E[PSP\omega, \omega] < \infty .$$

But taking the orthonormal basis of eigen-vectors of S, it follows that S is trace-class.

It follows from Theorem 2.1 that if a homogeneous polynomial is uniformly continuous in the S-topology, the corresponding random variable has finite second moment.

For a homogeneous polynomial of degree 2 with range in $R^1(H_r = R^1$ or R^m more generally) we can prove that continuity at the origin in the S-topology is sufficient to make it a random variable. For from (2.7) we have

$$E[\| p_2(P\omega) \|^2] = \sum_1^m \sum_1^m |k_2(\phi_i, \phi_j)|^2 + \left| \sum_1^m k_2(\phi_i, \phi_i) \right|^2 < \infty$$

and hence

$$\sum_1^\infty |k_2(\phi_i, \phi_i)| < \infty$$

for any orthonormal system. Hence it follows that

$$E[\| p_2(P_n\omega) - p_2(P_m\omega) \|^2]$$

is Cauchy. However, it is not clear whether it is true generally that for a homogeneous polynomial with range in H_r continuity at the origin is sufficient to ensure uniform continuity (in the S-topology).

Here we shall indicate a sufficient condition:

$$\sum_{\nu=0}^{[n/2]} \sum_{i_{2\nu+1}}^{\infty} \cdots \sum_{i_n}^{\infty} \left(\sum_{i_1}^{\infty} \cdots \sum_{i_\nu}^{\infty} \| k_n(\phi_{i_1}, \phi_{i_1}, \cdots, \phi_{i_\nu}, \phi_{i_\nu}, \phi_{i_{2\nu+1}}, \cdots, \phi_n) \| \right)^2 \tag{2.12}$$

for any orthonormal basis $\{\phi_i\}$. Then we have:

Theorem 2.2. *Suppose* (2.12) *holds for the homogeneous polynomial* $p_n(\omega)$ *with range in* H_r. *Then* $p_n(\cdot)$ *is a random variable with finite second moment.*

Proof. We shall actually prove a little bit more; namely that

$$\{p_n(P_m\omega)\}$$

actually converges in the mean of order 2.

For our proof we have only to note that

$$
\begin{aligned}
&p_n(P_j\omega) - p_n(P_i\omega)\,, \qquad j \geq i \\
&= \sum_{i_1}^{j} \cdots \sum_{i_n}^{j} a_{i_1,\ldots,i_n}\zeta_{i_1}\cdots\zeta_{i_n} - \sum_{i_1}^{i} \cdots \sum_{i_n}^{i} a_{i_1,\ldots,i_n}\zeta_{i_1}\cdots\zeta_{i_n} \\
&= \sum_{i_1}^{j} \cdots \sum_{i_n}^{j} b_{i_1,\ldots,i_n}\zeta_{i_1}\cdots\zeta_{i_n}
\end{aligned}
$$

where the $b_{i_1,\ldots,i_n}$ are zero for $\max_k i_k \leq i$. But we have

$$
\begin{aligned}
&E(\|p_n(P_j\omega)-p_n(P_i\omega)\|^2) \\
&= \sum_{\nu=0}^{[n/2]} \left(\frac{n!}{(n-2\nu)!\,2^\nu\nu!}\right)^2 \sum_{i_2+1}^{j} \cdots \sum_{i_n}^{j} \\
&\left\| \sum_{i_1}^{j} \cdots \sum_{i_\nu}^{j} b_{i_1,i_1,i_2,i_2,\ldots,i_\nu,i_\nu,i_{2\nu+1},\ldots,i_n} \right\|^2
\end{aligned}
\tag{2.13}
$$

from which, exploiting (2.12), it readily follows that

$$p_n(P_i\omega)$$

is Cauchy in the mean of order two.

We remark that (2.12) need only be verified for *some* orthonormal basis, the sum being independent (as may be readily verified by the usual analysis) of the particular orthonormal basis chosen.

Again based on (2.13) we can state a necessary and sufficient condition for $\{p_n(P_m\omega)\}$ to be Cauchy in the mean of order two; namely, given $\varepsilon > 0$ we can find m such that

$$
\sum_{\nu=0}^{[n/2]} \sum_{i_{2\nu+1}>m}^{\infty} \cdots \sum_{i_n>m}^{\infty} \left\| \sum_{i_1>m}^{m+p} \sum_{i_\nu>m}^{m+p} a_{i_1,i_1,i_2,i_2,\ldots,i_\nu,i_\nu,i_{2\nu+1},\ldots,i_n} \right\|^2 < \varepsilon\,,
\tag{2.14}
$$

for all $p > 0$.
But this condition is harder to verify then the sufficient condition (2.12). We are most concerned wiƫh the case:

$$W = L_2[(0, T)\,; R^m]$$

where we can choose a basis of orthonormal functions $\phi_i(\cdot)$ which are continuous and

$$\sup_i \sup_t \|\phi_i(t)\| < \infty\ .$$

For such a basis (2.12) follows, provided, for example

$$(2.15) \qquad \sum_{i_1} \cdots \sum_{i_n} \| k_n(\phi_{i_1}, \cdots, \phi_{i_n}) \| < \infty$$

and thus we can generate a rich enough class of random variables. Taking $R^m = R^1$ (to avoid notational complexity), (2.15) implies that

$$k_n(\omega) = \int_0^T \cdots \int_0^T k(t_1, \cdots, t_n)\omega(t_1)\omega(t_2) \cdots \omega(t_n) dt_1 \cdots dt_n$$

where $k(t_1, \cdots, t_n)$ is continuous and

$$\sum_{i_1} \cdots \sum_{i_\nu} a_{i_1, i_1, \cdots, i_\nu, i_\nu, i_{2\nu+1}, \cdots, i_n}$$
$$= \int_0^T \cdots \int_0^T \left(\int_0^T \cdots \int_0^T k(t_1, t_1, t_2, t_2, \cdots, t_\nu, t_\nu, t_{2\nu+1}, \cdots, t_n) dt_1 \cdots dt_\nu \right)$$
$$\times \phi_{i_{2\nu+1}}(t_{2\nu+1}) \cdots \phi_{i_n}(t_n) dt_{2\nu+1} \cdots dt_n \,.$$

By a "polynomial random variable of degree n with finite second moment" we shall mean the sum of homogeneous polynomial random variables, one each, (at most), of degree zero, one, two,$\cdots$ up to n, each with finite second moment.

Ito multiple integrals

It is convenient to introduce the Ito 'multiple intergrals' [8] in our context. Let $p_n(\cdot)$ denote a Hilbert-Schmidt polynomial (or alternately, let $k_n(\cdots)$ be extendable as a continuous linear operator over the nth order tensor product $W \otimes \cdots \otimes W$). Then let us introduce the special class of tame functions corresponding to the "special elementary functions" of Ito.

Thus consider the class of tame functions of the form:

$$(2.16) \qquad p_n^N(\omega) = \sum_1^N \cdots \sum_1^N a_{i_1, \ldots, i_n} [\phi_{i_1}, \omega] \cdots [\phi_{i_n}, \omega]$$

where the coefficients $a_{i_1, \ldots, i_n}$ are such that they vanish as soon as any two indices are the same; and as a consequence:

$$E[\| p_n^N(\omega) \|^2] = \sum_1^N \cdots \sum_1^N \| (a_{i_1, \ldots, i_n}) \|^2$$

Given a Hilbert-Schmit polynomial, we know that we can express it in the form

$$p_n(\omega) = \sum_1^\infty \cdots \sum_1^\infty k_n(\phi_{i_1}, \cdots, \phi_{i_n}) [\phi_{i_1}, \omega] \cdots [\phi_{i_n}, \omega]$$

corresponding to any orthonormal basis $\{\phi_i\}$. As Ito has shown, we can find 'coefficients' of the form in (2.16) such that

$$\sum_1^\infty \cdots \sum_1^\infty \|k_n(\phi_{i_1}, \cdots, \phi_{i_n}) - a_{i_1,\ldots,i_n}\|^2 \to 0$$

and we define the Ito polynomial

$$I_n(p_n(\omega))$$

as the limit of the approximating (in the mean of order two) tame functions of the form (2.16). For more on this see [9]. Moreover for any polyomial $p_n(\cdot)$ satisfying the sufficient condition (2.12), (2.15) or the necessary and sufficient conditions (2.14), we have the expansion in Ito polynomials:

$$p_n(\omega) = \sum_0^{[n/2]} \frac{n!}{(n-2\nu)!\,2^\nu \nu!} I_{n-2\nu}(p_{n-2\nu}) \tag{2.17}$$

where

$$p_{n-2\nu}(\omega) = \sum_1^\infty \cdots \sum_1^\infty k_n(\phi_{i_1}, \phi_{i_1}, \cdots, \phi_{i_\nu}, \phi_{i_\nu}, \omega, \cdots, \omega)$$

and is Hilbert-Schmidt.

If the sufficient condition (2.15) holds, then $p_{n-2\nu}(\cdot)$ can be expressed in terms of the kernels. To avoid lengthy notation we shall do this only for the case $H^1 = R^1 = H_r$. In that case

$$\begin{aligned} p_{n-2\nu}(\omega) = \int_0^T \cdots \int_0^T \cdots \\ \times \left(\int_0^T \cdots \int_0^T k_n(t_1, t_1, t_2, t_2, \cdots, t_\nu, t_\nu, t_{2\nu+1}, \cdots, t_n)\right. \\ \left. dt_1, \cdots, dt_\nu\right)\omega(t_{2\nu+1}) \cdots \omega(t_n) dt_{2\nu+1} \cdots dt_n\,. \end{aligned}$$

There is some justification in calling $f(\omega)$, when it is a random variable, a 'white noise intergral'. However it is *not* a Stratonovich integral (equivalently, Ito circle differential [10]). For this purpose let $H = H_r = R^1$ and let $T = 1$, and define.

$$p_2(\omega) = \int_0^1 \int_0^1 k(t, s)\omega(s)\omega(t) \tag{2.18}$$

where $k(t, s)$ is a continuous function on $0 \leq s, t \leq 1$ and

$$k(t, s) = g(t - s)$$

where $g(\cdot)$ is defined by the Fourier series

$$g(t) = \sum a_n \cos 2\pi nt \qquad (\text{in } L_2(0, 1) \text{ sense}) .$$

It is possible to find a continuous function $g(\cdot)$ such that

$$\sum |a_n| = \infty .$$

Hence for this choice of $g(\cdot)$, (2.18) does *not* define a white noise integral. On the other hand both the 'circle' differential and the Ito integral are defined.

§ 3. Radon-Nikodym derivatives

With the same notation as in section 2, let μ denote another cylinder measure on W. We wish to examine the question of the derivative of μ with respect to μ_G. First of all we say μ is absolutely continuous with respect to μ_G if given $\varepsilon > 0$ we can find σ such that

$$\mu(C) < \varepsilon \quad \text{whenever} \quad \mu_G(C) < \delta . \tag{3.1}$$

We shall say that the derivative of μ with respect μ_G is $f(\omega)$ if $f(\omega)$ is a random variable and

$$\mu(C) = \operatorname*{limit}_{m \to \infty} \int_C f(P_m \omega) d\mu_G . \tag{3.2}$$

Throughout this section we shall assume that

$$H = R^n ; W = L_2((0, T) ; R^n) ,$$

in order to be able to exploit the well established theory of countably additive measures. Thus let $\{\phi_i\}$ denote an orthonormal basis in W and let

$$\zeta_i(\omega) = [\phi_i, \omega]$$

yielding a sequence of independent zero mean, unit variance Gaussians. We can also map W isomorphically onto ℓ_2 and ℓ_2 then can taken to be sample space for $\{\zeta_i(\omega)\}$. Of course

$$\sum_1^\infty \zeta_i(\omega)^2 = \|\omega\|^2 < \infty$$

in contradistinction to the usual case. Let $\tilde{\omega}$ denote points in R^∞. Then

$\ell_2 \subset R^\infty$; and we can introduce a countably additive measure on (the Borel sets $\mathscr{B}$ of) R^∞ such that the cylinder set measures are identical with those on ℓ_2, with Gauss measure μ_G on ℓ_2. Of course

$$\sum_1^\infty \zeta_i(\tilde{\omega})^2 = \infty \qquad \text{with probability one,}$$

the set ℓ_2 having zero probability as a subset of R^∞. Both measures μ and μ_G can be extended to be countably additive on $\mathscr{B}$. For any real valued random variable $f(\omega)$, let us note that, P_m denoting the projection corresponding to the space spanned by the ϕ_i, $i = 1, \cdots, m$, we have that

$$f(P_m\omega) = \psi(\zeta_1, \cdots, \zeta_m)$$

and can be identified with the corresponding function on R^∞ and hence yields a sequence of random variables converging in probability to a random variable measurable $\mathscr{B}$ (say $f^\sim$) which then has the same distribution as the random variable $f(\omega)$ (and is identifiable one with the other). Now suppose (3.2) holds. Then it is clear that the countably additive extension of μ is absolutely continuous with the countably additive extension of μ_G and that the derivative is $f^\sim$. Conversely if the countably additive extension of μ is absolutely continuous with the countably additive extension μ_G *and* if the derivative is a white noise integral, then (3.2) holds.

Let $p_n(\omega)$ be a homogeneous polynomial of degree n satisfying (2.14) and mapping W into itself. Let

$$y(\omega) = p_n(\omega) + \omega\ . \tag{3.3}$$

Let C be a cylinder set. Define the cylinder measure:

$$\mu_y(C) = \lim_{m\to\infty} \mu_G[\omega \mid p_n(P_m\omega) + \omega \in C]\ . \tag{3.4}$$

For a certain class of such polynomials we shall be interested in showing that $\mu_y(\cdot)$ is absolutely continuous with respect to μ_G and evaluating the derivative in (3.2).

For the case where the polynomial is of degree one, the result is given in [4]. Our technique here will be slightly different and of course we shall be concerned with the non-linear case.

In what follows we shall take

$$H = R^1\,;\ W = L_2((0, T)\,;\ R^1)$$

rather R^n in order to minimise notational complexity since the extension to $n < \infty$ entails nothing essentially new.

Let $\{\phi_i\}$ again denote an orthonormal basis for W. Let

$$\zeta_i(\omega) = [\phi_i, \omega] \ . \tag{3.5}$$

This also provides a 1 : 1 isomorphic map of W into ℓ_2. We shall use ω to denote points in ℓ_2 (under this isomorphism) and $\tilde{\omega}$ to denote $R^\infty \supset \ell_2$; let $\mathscr{B}$ denote the Borel sets in R^∞; with the Guassian sequence (3.5) we can associate the usual countably additive measure on $\mathscr{B}$ with the same cylinder measures as ℓ_2. Let $\tilde{\omega}$ define points in R^∞ and let $\overline{\mathscr{B}}$ denote the completion of $\mathscr{B}$. Then $\zeta_i(\tilde{\omega})$ are also independent Gaussians. Define

$$W(t, \tilde{\omega}) = \sum_1^\infty \zeta_i(\tilde{\omega}) \int_0^t \phi_i(s)ds \qquad 0 < t < T \ .$$

This yields a Wiener process on $C(0, T)$, the usual space of continuous functions with range in $H = R^n$. $W(t, \tilde{\omega})$ for each t is measurable with respect to the σ-algebra generated by the sequence $\{\zeta_i(\tilde{\omega})\}$, which is the same as $\overline{\mathscr{B}}$. The sigma algebra generated by $W(s, \tilde{\omega})$, $0 \leq s \leq t$, will be denoted $\mathscr{B}(t)$. We note that

$$\mathscr{B}(T) = \overline{\mathscr{B}}$$

Volterra Polynomials

We shall only consider 'Volterra' polynomials: namely of the form:

$$p_n(\omega) = h\,;$$

$$h(t) = \int_0^t \cdots \int_0^t k_n(t, s_1, \cdots, s_n)\omega(s_1) \cdots \omega(s_n)ds_1 \cdots ds_n \ . \qquad a.\,e.\ 0 < t < T \ . \tag{3.6}$$

If $p_n(\omega)$ is to be a random variable it is necessary that

$$\int_0^T \int_0^t \cdots \int_0^t k_n(t, s_1, \cdots, s_n)^2 ds_1 \cdots ds_n dt < \infty$$

and further that so is:

$$p_{n-2\nu}(\omega) = \sum_{i_1} \cdots \sum_{i_\nu} k_n(\phi_{i_1}, \phi_{i_1}, \cdots, \phi_{i_\nu}, \phi_{i_\nu}, \omega, \cdots, \omega) \ . \tag{3.7}$$

Here each term is Volterra:

$$k_n(\phi_{i_1}, \phi_{i_1}, \cdots, \phi_{i_\nu}, \phi_{i_\nu}, \omega, \cdots, \omega) = h$$

$$h(t) = \int_0^t \cdots \int_0^t H_{n-2\nu}(t, s_{2\nu+1}, \cdots, s_n)\omega(s_{2\nu+1}) \cdots \omega(s_n)ds_{2\nu+1} \cdots ds_n \qquad a.\,e.\ 0 < t < T,$$

where the knrnel

$$H_{n-2\nu}(t, s_{2\nu+1}, \cdots, s_n)$$
$$= \sum_{i_1} \cdots \sum_{i_\nu} \int_0^t \cdots \int_0^t k_n(t, s_1, s_2, \cdots, s_{2\nu}, s_{2\nu+1}, \cdots, s_n)\phi_{i_1}(s_1)$$
$$\times \phi_{i_1}(s_2) \cdots \phi_{i_\nu}(s_{2\nu})ds_1 ds_2 \cdots ds_{2\nu} \qquad 0 < t < T \ a.e.$$

is Hilbert-Schmidt.

The convergence in (3.7) being in the mean square, it follows $p_{n-2\nu}(\omega)$ is Volterra for each ν also. Let $k_{n-2\nu}(\cdot)$ denote the corresponding multilinear form. Then the Ito integrals:

$$p_{n-2\nu}(t, \tilde{\omega})$$
$$= \int_0^t \cdots \int_0^t k_{n-2\nu}(t, s_{2\nu+1}, \cdots, s_n) dW(s_{2\nu+1}, \tilde{\omega}) \cdots dW(s_n, \tilde{\omega})$$
$$a.e. \ 0 < t < T$$

are well defined and from (2.17) it follows that $p_n(\omega)$ is the same random variable as:

$$(3.8) \quad p_n(t, \tilde{\omega}) = \sum_0^{[n/2]} \frac{n!}{(n-2\nu)!\, 2^\nu \nu!} p_{n-2\nu}(t, \tilde{\omega}) \qquad 0 < t < T \ a.e.$$

and of course

$$\int_0^T E(p_n(t, \tilde{\omega})^2)dt = E[\| p_n(\omega) \|^2] < \infty \ .$$

Let

$$(3.9) \qquad Y(t, \tilde{\omega}) = \int_0^t p_n(s, \tilde{\omega})ds + W(t, \tilde{\omega}) \qquad 0 < t < T \ .$$

Note that for each finite m

$$\int_0^T \phi_i(t) dY(t, \tilde{\omega}) \qquad i = 1, 2, \cdots, m$$

has the same distribution as

$$[\phi_i(\cdot), y(\omega)] \qquad i = 1, 2, \cdots, m \ .$$

Now (3.9) maps into $C(0, T)$ and satisfies the conditions for Girsanov's theorem to apply, and we have that the measure induced by $Y(\cdot, \tilde{\omega})$ is absolutely continuous with respect to that induced by $W(\cdot, \tilde{\omega})$ and the derivative is given by (see [1] for example):

(3.10) $$g(\tilde{\omega}) = \operatorname{Exp}\left(-1/2\left\{\int_0^T p_n(t, \tilde{\omega})^2 dt - 2\int_0^T p_n(t, \tilde{\omega}) pW(t, \tilde{\omega})\right\}\right).$$

Theorem 3.1. *Assume that $p_n(\omega)$ is a homogeneous Volterra polynomial mapping W into W satisfying (2.14) and is such that the linear bounded operator $L(\omega)$ defined by the Frechet derivative:*

$$\frac{d}{d\lambda} p_n(\omega + \lambda h)|_{\lambda=0} = L(\omega)h$$

is nuclear when symmetrised: (that is $(L(\omega) + L(\omega)^*)$ is nuclear): furthermore

$$p_{n-1}(\omega) = \operatorname{Tr.}\ [L(\omega) + L(\omega)^*]\ ,$$

which is a homogeneous polynomial of degree $(n - 1)$, is also a random variable, satisfying (2.14). Then μ_y is absolutely continuous with respect to μ_G and the deriative $g(\tilde{\omega})$ in (3.10) is the same random variable as

$$\operatorname*{limit}_m f(P_m\omega) \qquad \text{(in the mean of order two)}$$

where

(3.11) $$f(\omega) = \operatorname{Exp}(-1/2\{[p(\omega), p(\omega)] - 2[p(\omega), \omega] + \operatorname{Tr.}\ (L(\omega) + L(\omega)^*)\}) .$$

Proof. We have only to prove that

(3.12) $$\operatorname*{Limit}_m \{[p_n(P_m\omega), P_m\omega] - 1/2 \operatorname{Tr}\ (L(P_m\omega) + L(P_m\omega)^*)\} = \int_0^T p_n(t, \tilde{\omega}) dW(t, \tilde{\omega}) .$$

The condition (2.14) implies first of all that

$$p_n(\omega) = \int_0^t \cdots \int_0^t k_n(t, s_1, \cdots, s_n)\omega(s_1) \cdots \omega(s_n) ds_1 \cdots ds_n$$

where the kernel is Hilbert-Schmidt in *all* the variables

$$\int_0^T \int_0^t \cdots \int_0^t |k_n(t, s_1, \cdots, s_n)|^2\, ds_1 \cdots ds_n dt < \infty .$$

First let us examine the implication of the trace-class property of the derivative:

$$L(\omega) + L(\omega)^*$$

being trace class implies that

$$\sum_1^\infty |[L(\omega)\phi_i, \phi_i]| < \infty$$

for any orthornormal basis $\{\phi_i\}$. But

$$[L(\omega)\phi_i, \phi_i] = n[k_n(\omega, \cdots, \omega, \phi_i), \phi_i]$$

(remember that $k_n(\cdots)$ is symmetric). Hence we have that

$$\sum_1^\infty |[k_n(\omega, \cdots, \omega, \phi_i), \phi_i]| < \infty \; .$$

Let

$$q_{n-k}(\omega) = \mathrm{Tr}\,(L(\omega) + L(\omega)^*) \; .$$

Since this is a random variable by assumption we can express it in terms of Ito integrals (*CF* (2.17)):

$$(3.13) \quad \lim_{m\to\infty} q_{n-1}(P_m\omega) = \sum_0^{[(n-1)/2]} \frac{(n-1)!}{(n-1-2j)!\,2^j j!} I_{n-1-2j}(q_{n-1-2j})$$

where (with $k_n(\cdot)$ denoting the n-linear from corresponding to $p_n(\cdot)$):

$$q_{n-1-2j}(\omega) = n \sum_{i_1} \cdots \sum_{i_j} \left(\sum_k [k_n(\phi_{i_1}, \phi_{i_1}, \cdots, \phi_{i_\nu}, \phi_{i_\nu}, \phi_k, \omega, \cdots, \omega), \phi_k] \right) .$$

Next let

$$\bar{p}_{n+1-2\nu}(\omega) = \sum_{i_1} \cdots \sum_{i_\nu} [p(\phi_{i_1}, \phi_{i_1}, \cdots, \phi_{i_\nu}, \phi_{i_\nu}, \omega, \cdots, \omega), \omega] \; .$$

Note that by expanding as in (3.8), we can obtain:

$$(3.14) \quad \int_0^T p_n(t, \tilde{\omega}) dW(t, \tilde{\omega}) = \sum_0^{[n/2]} I_{n+1-2\nu}(\bar{p}_{n+1-2\nu}) \; .$$

Finally define

$$\bar{p}_{n+1}(\omega) = [p_n(\omega), \omega] \; .$$

[Note that $\bar{p}_{n+1}(\omega) = \tilde{p}_{n+1}(\omega)$].

Then

$$(3.15)\quad \lim_m\,[p_n(P_m\omega), P_m\omega] = \sum_0^{[(n+1)/2]} \frac{(n+1)!}{(n+1-2\nu)!\,2^\nu\nu!} I_{n+1-2\nu}(\tilde{p}_{n-1-2\nu})$$

(using (2.17) again), where $\tilde{p}_{n+1-2\nu}(\omega)$ is to be obtained by first symmetrising the multilinear form corresponding to $\tilde{p}_{n+1}$:

$$\tilde{p}_{n+1}(\phi_{i_1}, \cdots, \phi_{i_n}, \phi_{i_{n+1}}) = \frac{1}{(n+1)!}\Big\{n!\,[p_n(\phi_{i_1}, \cdots, \phi_{i_n}, \cdots, \phi_{i_{n+1}}]$$
$$+ \sum_{j=1}^{n} [p_n(\phi_{i_1}, \cdots, \phi_{i_{j-1}}, \phi_{i_{n+1}}, \cdots, \phi_{i_n}, \phi_{i_j}]\Big\}$$

and hence

$$(3.16)\quad \tilde{p}_{n+1-2\nu}(\omega) = \frac{1}{(n+1)!}\{n!\,(n-2\nu)[p_{n-2\nu}(\omega), \omega] + (n-1)!\,(2\nu)q_{n+1-2\nu}(\omega)\}$$

and is Hilbert-Schmidt, proving that $\tilde{p}_{n+1}(\omega)$ is a random variable and justifying (3.15).

To prove (3.12) it is clearly enough to show that (3.15) is the sum of (3.13) and or, equivalently, that

$$(3.17)\quad \frac{(n+1)!}{(n+1-2\nu)!\,2^\nu\nu!}\tilde{p}_{n+1-2\nu}(\omega) = \frac{(n-1)!}{(n+1-2\nu)!\,2^{\nu-1}(\nu-1)!} q_{n+1-2\nu}(\omega) + \frac{n!}{(n-2\nu)!\,2^\nu\nu!}\bar{p}_{n+1-2\nu}(\omega)\,.$$

But this is clear from (3.16) using the fact that, by definition,

$$[p_{n-2\nu}(\omega), \omega] = \bar{p}_{n+1-2\nu}(\omega)\,.$$

Hence (3.12) is proved.

Theorem 3.1 generalizes in an obvious way to polynomials, under the sufficient condition that each constituent is a homogeneous polynomial. This condition is clearly not necessary; indeed since the key condition is that involving the derivative, one would expect the formula to go thru so long as that is verified, whether or not the function is a polynomial. We also remark that in the case $p(\omega)$ is linear, $p(\omega) = L\omega$, we must have that $L + L^*$ is trace-class; which is satisfied in the filtering application (1.1).

References

[1] Lipscer, Y. and Shiryayev, A., Stastistics of random process (in Russian) Nauka, 1974.

[2] Balakrishnan, A. V., White noise vs. Wiener process, Proceeding of IFAC Symposium on Adaptive Control, Budapest, 1975.

[3] Gross, L., Integration and non-linear transformations in Hilbert space, Trans. Amer. Math. Soc., **94** (1960), 404–440.

[4] Balakrishnan, A. V., Applied functional analysis, Springer-Verlag, 1976.

[5] Skorokhod, A. V., Integration of Hilbert space, Springer-Verlag, 1975.

[6] Gross, L., Harmonic analysis on Hilbert space, A.M.S. Memoir No. 46 (1963).

[7] Sazonov, V., Remarks on characteristic functionals, Theor. Probability Appl., **3** (1958), 188–192.

[8] Ito, K., Multiple Wiener integral, J. Math. Soc. Japan, **3** (1951), 157–169.

[9] Balakrishnan, A. V., Stochastic bilinear partial differential equations, Variable Structure Systems, edited by Mohler & Ruberts, Academic Press, 1974.

[10] Ito, K., Stochastic differentials, J. Appl. Math. Optimization, **1** (1975), 374–381.

System Science Department
School of Engineering and Applied Science
University of California
Los Angeles, California 90024, U.S.A.

Proc. of Intern. Symp. SDE
Kyoto 1976, pp. 21-40

Boundary Layer Analysis in Homogeneization of Diffusion Equations with Dirichlet Conditions in the Half Space

A. BENSOUSSAN, J. L. LIONS and G. PAPANICOLAOU

Introduction

Homogeneization of diffusion equation deals with the following type of problems. Let u_ε be the solution of

$$\text{(i)}\qquad \begin{cases} -\dfrac{\partial}{\partial x_i}\left(a_{ij}\left(\dfrac{x}{\varepsilon}\right)\dfrac{\partial u_\varepsilon}{\partial x_j}\right) = f & \text{in } \mathcal{O} \\ u_\varepsilon|_\Gamma = 0 \end{cases}$$

where $\mathcal{O}$ is an open bounded subset of R^n, whose boundary is denoted by Γ. In (i) the symbol of summation is omitted. The coefficients a_{ij} are strongly elliptic and *periodic* in all variables (with period 1 for simplicity). The problem is the following: what is the behaviour of u_ε as ε goes to 0.

It is well known (see Babuska [1], de Giorgi and Spagnolo [6], Bensoussan, Lions and Papanicolaou [2]) that

$$\text{(ii)}\qquad u_\varepsilon \to u \quad \text{in} \quad L^2(\mathcal{O}) \quad \text{strongly and} \quad H_0^1(\mathcal{O}) \quad \text{weakly}$$

where u is solution of a *homogeneized* diffusion equation, namely

$$\text{(iii)}\qquad \begin{cases} -q_{ij}\dfrac{\partial^2 u}{\partial x_i \partial x_j} = f \\ u|_\Gamma = 0 \end{cases}$$

where q_{ij} is a positive definite matrix (independent of x), which can be computed explicity (however the q_{ij} are not the mean values of the a_{ij}, which could be considered as a reasonable conjecture).

Now there is no strong convergence of u_ε towards u in $H_0^1(\mathcal{O})$. In other words the derivatives do not converge in L^2 strongly, although they converge weakly.

To improve the convergence (ii), one needs a corrector. The following result has been proved in Babuska [1], Bensoussan, Lions and Papanicolaou [2], Tartar [7]

(iv) $$u_\varepsilon - u - \varepsilon\chi\left(x, \frac{x}{\varepsilon}\right) \to 0 \quad \text{in} \quad H^1(\mathcal{O}) \quad \text{strongly}$$

where $\chi(x, y)$ is a function which is periodic in y.

The next question concerns obtaining some estimates on the rate of convergence. In the above references, estimates of the following type have been given

$$|u_\varepsilon - u|_{L^2(\mathcal{O})} \leq C\varepsilon$$
$$\left\| u_\varepsilon - u - \varepsilon\chi\left(x, \frac{x}{\varepsilon}\right)\right\|_{H^1(\mathcal{O})} \leq C\varepsilon^{1/2} .$$

The problem which is dealt with in this article concerns the improvement of the second estimate, namely to obtain an estimate of order ε. One needs to introduce a new corrector, which is, this time, a boundary layer term.

Such a result has been given already by Babuska [1], but we use here an entirely different approach, based upon some estimates obtained by probabilistic techniques, which are interesting by themselves.

Furthermore, our method can also be applied to the study of the convergence of non homogeneous problems, with rapidly varying values, in a way similar to what we did in a previous paper on Transport problems (see Bensoussan, Lions and Papanicolaou [3]).

However a drawback in our method, as well as in Babuska [1], is that so far only the case of a half space can be considered.

§ 1. Assumptions—The main result

Let $a_{ij}(x)$ be functions of $x \in R^n$, satisfying the following conditions

(1.1) $$\begin{cases} a_{ij} \text{ are periodic in all variables with period } 1\,;\ a_{ij} = a_{ji} \\ a_{ij} \in C^n(R^n) \\ a_{ij}\xi_i\xi_j \geq \beta\xi_i\xi_i\,, \qquad \beta > 0\,^{1)}\,. \end{cases}$$

Let $\mathcal{O} = \{x \mid x_n > 0\}$, $\Gamma = \partial\mathcal{O}$. We considera function $f(x)$ and a number α such that

(1.2) $$f \in H^3(\mathcal{O}) \cap L^1(\mathcal{O})\,; \quad \alpha > 0\,.$$

We denote by $u_\varepsilon(x)$ the solution of

1) The symbol of summation is omitted, when the indices are repeated.

$$
(1.3) \qquad \begin{cases} -\dfrac{\partial}{\partial x_i}\left(a_{ij}\left(\dfrac{x}{\varepsilon}\right)\dfrac{\partial u_\varepsilon}{\partial x_j}\right) + \alpha u_\varepsilon = f \\ u_\varepsilon|_\Gamma = 0 \end{cases}
$$

which exists and is unique in $H^5(\mathcal{O}) \cap H_0^1(\mathcal{O})$.

We will also need the Sobolev spaces with weights

$$
\begin{aligned}
&L^{p,\mu}(R^{n-1} \times R^+) \\
&\quad = \left\{ z(x, \lambda) \,\middle|\, \int_{R^{n-1}} dx \int_0^{+\infty} d\lambda\, |z|^p \exp(-p\mu|x|) \exp(-p\mu\lambda) < \infty \right\} \\
&W^{m,p,\mu}(R^{n-1} \times R^+) = \{z \,|\, z, Dz, \cdots, D^m z \in L^{p,\mu}\} \\
&H^{m,\mu}(R^{n-1} \times R^+) = W^{m,2,\mu}(R^{n-1} \times R^+) .
\end{aligned}
$$

Let now $\Pi = (0, 1)^n$. For $l = 1, \cdots, n$, we consider $\chi^l(y)$, $y \in R^n$, solution of the following problem

$$
(1.4) \qquad \begin{cases} \dfrac{\partial}{\partial y_i}\left(a_{ij}\dfrac{\partial \chi^l}{\partial y_j}\right) + \dfrac{\partial a_{il}}{\partial y_i} = 0 \\ \chi^l \text{ periodic}, \displaystyle\int_\Pi \chi^l dy = 0, \; \chi^l \in W^{2,p,\mu}(R), \quad p \geq 2 . \end{cases}
$$

The functions $\chi^l(y)$ are uniquely defined by (1.4). This is a consequence (for instance) of the Fredholm alternative in the subspace of $L^2(\Pi)/R$ of periodic Borel functions, or directly can be seen by using variational techniques. The regularity is a consequence of the regularity of the coefficients.

We next define

$$
(1.5) \qquad q_{ij} = \int_\Pi \left(a_{ij} + a_{ik}\frac{\partial \chi^j}{\partial y_k}\right) dy .
$$

It can be easily seen that q_{ij} is a positive definite matrix. It will be convenient to write $y = (y'_n, y_n)$ where y'_n stands for the $n - 1$ first coordinates.

We next introduce the following problem: to find functions $v^l(y)$ and constants d^l, $l = 1, \cdots, n$ such that

$$
(1.6) \qquad \begin{cases} \dfrac{\partial}{\partial y_i}\left(a_{ij}\dfrac{\partial v^l}{\partial y_j}\right) = 0 \\ v^l(y'_n, 0) = \chi^l(y'_n, 0) - d^l \\ v^l \in W^{2,p,\mu}(R^{n-1} \times R^+), \quad \text{periodic in } y'_n \end{cases}
$$

$$\begin{cases} |v^l(y_n', y_n)| \le k_l e^{-\delta y_n} \\ \left|\dfrac{\partial v^l}{\partial y_i}(y_n', y_n)\right| \le K_l' e^{-\delta_1} y_n \ , \ \delta, \delta_1 > 0 \ . \end{cases}$$

The last conditions mean that v^l as well as its derivatives decrease exponentially as y_n increases, but the constants K_l, K_l', δ, δ_1 are not specified a priori.

We will prove that the problem (1.6) has one solution in v^l, d^l (which is necessarily unique). We finally define u as the solution of

$$\text{(1.7)} \qquad \begin{cases} -q_{ij}\dfrac{\partial^2 u}{\partial x_i \partial x_j} + \alpha u = f \\ u|_\Gamma = 0 \ . \end{cases}$$

Our main result is the following.

Theorem 1.1. *Under the assumptions* (1.1) *we have*

$$\text{(1.8)} \qquad \left\| u_\varepsilon(x) - u(x) - \varepsilon \frac{\partial u}{\partial x_l}(x)\chi^l\left(\frac{x}{\varepsilon}\right) - \varepsilon \frac{\partial u}{\partial x_l}(x_n', 0)v^l\left(\frac{x}{\varepsilon}\right)\right\|_{H^1} \le C\varepsilon \ .$$

Remark 1.1. We have also as already said in the introduction

$$\text{(1.9)} \qquad \left\| u_\varepsilon(x) - u(x) - \varepsilon \frac{\partial u}{\partial x_l}(x)\chi^l\left(\frac{x}{\varepsilon}\right)\right\|_{H^1} \le C\varepsilon^{1/2} \ .$$

The term $\varepsilon(\partial u/\partial x_l)(x_n', 0)v^l(x/\varepsilon)$ is a corrector term, which is a boundary layer term, since it decreases exponentially away from the boundary. ■

The proof of Theorem 1.1 will rely on two Theorems which are interesting in themselves, Theorems 1.2 and 1.3 below. For convenience, we will denote the coefficients $a_{ij}(x_n', x_n)$ by $a_{ij}(x, \lambda)$, $x \in R^{n-1}$, $\lambda \in R$.

We consider the following boundary value problems

$$\text{(1.10)} \qquad \begin{cases} \displaystyle\sum_{i,j=1}^{n-1} \frac{\partial}{\partial x_i}\left(a_{ij}(x, \lambda)\frac{\partial z}{\partial x_j}\right) + \sum_{i=1}^{n-1} \frac{\partial}{\partial x_i}\left(a_{in}(x, \lambda)\frac{\partial z}{\partial \lambda}\right) \\ \displaystyle\qquad + \sum_{i=1}^{n-1} \frac{\partial}{\partial \lambda}\left(a_{ni}(x, \lambda)\frac{\partial z}{\partial x_i}\right) + \frac{\partial}{\partial \lambda}\left(a_{nn}(x, \lambda)\frac{\partial z}{\partial \lambda}\right) = 0 \\ z(x, 0) = \chi(x) \end{cases}$$

in the domain $x \in R^{n-1}$, $\lambda > 0$,

$$(1.11)\quad \begin{cases} -\sum_{i,j=1}^{n-1} \dfrac{\partial}{\partial x_i}\left(a_{ij}(x,\lambda)\dfrac{\partial z}{\partial x_j}\right) - \sum_{i=1}^{n-1} \dfrac{\partial}{\partial x_i}\left(a_{in}(x,\lambda)\dfrac{\partial z}{\partial \lambda}\right) \\ \qquad -\sum_{i=1}^{n-1} \dfrac{\partial}{\partial \lambda}\left(a_{ni}(x,\lambda)\dfrac{\partial z}{\partial x_i}\right) - \dfrac{\partial}{\partial \lambda}\left(a_{nn}(x,\lambda)\dfrac{\partial z}{\partial \lambda}\right) = h(x,\lambda) \\ z(x,0) = 0 \end{cases}$$

also for $x \in R^{n-1}$, $\lambda > 0$.

Theorem 1.2. *Under the assumptions* 1.1, *and*

$$(1.12)\qquad \chi(x) \quad \textit{periodic, Borel, bounded.}$$

There exists one and only one solution of (1.10) *such that*

$$(1.13)\quad \begin{cases} z \text{ is periodic in } x, \text{ bounded} \\ z \in C^2(R^{n-1} \times \{\lambda > 0\}) , \end{cases}$$

$$(1.14)\quad \begin{cases} \text{there exist constants } C_\chi \text{ (depending on } \chi), \gamma, \delta > 0 \text{ independent} \\ \text{of } \chi, \text{ such that} \\ \quad |z(x,\lambda) - C_\chi| \le \|\chi\|_{L^\infty}\gamma e^{-\delta\lambda} , \quad \forall x, \lambda > 0 \\ \quad \left|\dfrac{\partial z}{\partial x_i}(x,\lambda)\right|, \ \left|\dfrac{\partial z}{\partial \lambda}(x,\lambda)\right| \le B_{\delta_1} e^{-\delta_1\lambda} , \quad \forall \delta_1 < \delta \\ \text{where } B_{\delta_1} \text{ depends on } \delta_1 \text{ and } \chi, \text{ and } \lambda \ge \lambda_0 > 0, \lambda_0 \text{ arbitrary.} \end{cases}$$

Theorem 1.3. *Under the assumptions* 1.1, *and*

$$(1.15)\quad \begin{cases} h(x,\lambda) \text{ is measurable in both variables, continuous in } x, \text{ periodic} \\ \text{in } x \text{ and } |h(x,\lambda)| \le h_0 e^{-\beta\lambda}, \beta > 0 . \end{cases}$$

There exists one and only one solution of (1.11) *such that*

$$(1.16)\quad z \text{ is periodic in } x, \text{ bounded, } z \in W^{2,p,\mu}(R^{n-1} \times R_+), \ p \ge 2$$

$$(1.17)\quad \begin{cases} |z(x,\lambda) - C_h| \le Re^{-\rho\lambda} \\ \text{where } C_h \text{ is a constant depending only on } h, \text{ and } \rho \text{ is a positive} \\ \text{constant} \\ \left|\dfrac{\partial z}{\partial x_i}(x,\lambda)\right|, \left|\dfrac{\partial z}{\partial \lambda}(x,\lambda)\right| \le R_{\beta_1} e^{-\rho_1\lambda} , \quad \forall \rho_1 < \rho . \end{cases}$$

The proofs of Theorems 1.2 and 1.3 will require probabilistic techniques and will be given in § 3.

§ 2. Proof of Theorem 1.1

Let us denote by A^ε the differential second order operator

(2.1) $$A^\varepsilon = -\frac{\partial}{\partial x_i}\left(a_{ij}\left(\frac{x}{\varepsilon}\right)\frac{\partial}{\partial x_j}\right) + \alpha .$$

We use a method of asymptotic expansions. We will search a function $\varphi_\varepsilon(x)$ such that

(2.2) $$\varphi_\varepsilon(x) = u(x) + \varepsilon u_1\left(x, \frac{x}{\varepsilon}\right) + \varepsilon^2 u_2\left(x, \frac{x}{\varepsilon}\right) + \varepsilon^3 u_3\left(x, \frac{x}{\varepsilon}\right)$$

where $u(x)$, $u_1(x, y)$, $u_2(x, y)$, $u_3(x, y)$ will be identified later on, in order that φ_ε satisfy

(2.3) $$\begin{cases} A^\varepsilon \varphi_\varepsilon = \varepsilon^2 g_\varepsilon + f\,, & x_n > 0 \\ \varphi_\varepsilon(x'_n, 0) = \varepsilon^2 h_\varepsilon \end{cases}$$

where g_ε, h_ε are measurable functions such that

(2.4) $$\begin{cases} |\varphi_\varepsilon|_{L^2}, |g_\varepsilon|_{L^\infty}, |h_\varepsilon|_{L^\infty} \leq C\,; |g_\varepsilon|_{L^2(\mathcal{O})}\,, \\ |h_\varepsilon|_{L^2(R^{n-1})} \leq C\,; |\varphi_\varepsilon(x'_n, 0)|_{L^2(R^{n-1})} \leq C\,. \end{cases}$$

We will show that such an identification is possible, by virtue of Theorems 1.2 and 1.3. Assuming for a while that this has been done, then defining

(2.5) $$z_\varepsilon = \varphi_\varepsilon - u_\varepsilon$$

we have

(2.6) $$\begin{cases} A^\varepsilon z_\varepsilon = \varepsilon^2 g_\varepsilon\,, \\ z_\varepsilon(x'_n, 0) = \varepsilon^2 h_\varepsilon\,. \end{cases}$$

By the Maximum Principle (since $\alpha > 0$), we have

(2.7) $$|z_\varepsilon(x)| \leq C\varepsilon^2\,.$$

Let us set, for $\theta_1, \theta_2 \in H^1(\mathcal{O})$

$$a(\theta_1, \theta_2) = \sum_{i,j=1}^{n-1} \iint a_{ij}\frac{\partial\theta_1}{\partial x_j}\frac{\partial\theta_2}{\partial x_i}dx'_n dx_n$$
$$+\sum_{i=1}^{n-1}\iint a_{in}\frac{\partial\theta_1}{\partial\lambda}\frac{\partial\theta_2}{\partial x_i}dx'_n dx_n + \sum_{i=1}^{n-1}\iint a_{ni}\frac{\partial\theta_1}{\partial x_i}\frac{\partial\theta_2}{\partial\lambda}dx'_n dx_n$$

$$+ \iint a_{nn} \frac{\partial \theta_1}{\partial \lambda} \frac{\partial \theta_2}{\partial \lambda} dx_n' dx_n + \alpha \iint \theta_1 \theta_2 dx_n' dx_n \ .$$

Now, using the fact that $u_\varepsilon|_\Gamma = 0$, it follows from Green's formula that

$$a(u_\varepsilon, u_\varepsilon) = \iint f u_\varepsilon dx_n' dx_n \tag{2.8}$$

$$a(\varphi_\varepsilon, u_\varepsilon) = \iint A^\varepsilon \varphi_\varepsilon u_\varepsilon dx_n' dx_n = \iint u_\varepsilon (f + \varepsilon^2 g_\varepsilon) dx_n' dx_n \tag{2.9}$$

$$\begin{aligned} a(\varphi_\varepsilon, \varphi_\varepsilon) &= \iint A^\varepsilon \varphi_\varepsilon \varphi_\varepsilon dx_n' dx_n \\ &\quad + \int a_{nj}\left(\frac{x_n'}{\varepsilon}, 0\right) \frac{\partial \varphi_\varepsilon}{\partial x_j}(x_n', 0) \varphi_\varepsilon(x_n', 0) dx_n' \\ &= \iint \varphi_\varepsilon (f + \varepsilon^2 g_\varepsilon)\, dx_n' dx_n \\ &\quad + \varepsilon^2 \int a_{nj}\left(\frac{x_n'}{\varepsilon}, 0\right) \frac{\partial \varphi_\varepsilon}{\partial x_j}(x_n', 0) h_\varepsilon dx_n' \ . \end{aligned} \tag{2.10}$$

Now, by the identity

$$a(z_\varepsilon, z_\varepsilon) = a(\varphi_\varepsilon - u_\varepsilon, \varphi_\varepsilon - u_\varepsilon) = a(\varphi_\varepsilon, \varphi_\varepsilon) + a(u_\varepsilon, u_\varepsilon) - 2a(\varphi_\varepsilon, u_\varepsilon)$$

it follows that

$$\begin{aligned} a(z_\varepsilon, z_\varepsilon) &= \iint z_\varepsilon (f + \varepsilon^2 g_\varepsilon) dx_n' dx_n \\ &\quad + \varepsilon^2 \int a_{nj}\left(\frac{x_n'}{\varepsilon}, 0\right) \frac{\partial \varphi_\varepsilon}{\partial x_j}(x_n', 0) h_\varepsilon dx_n' \\ &\quad - \varepsilon^2 \iint u_\varepsilon g_\varepsilon dx_n' dx_n \ . \end{aligned} \tag{2.11}$$

But $|u_\varepsilon|_{L^2(\mathcal{O})} \le C$ (this follows from (2.8)).
Now using (1.2), (2.7), (2.4), it follows from (2.11) that

$$\|z_\varepsilon\|^2_{H^1(\mathcal{O})} \le C\varepsilon^2 \ . \tag{2.12}$$

To pursue the proof, we now need to identify the functions u, u_1, u_2, u_3. One can easily check that u, u_1, u_2, u_3 must satisfy the equations

$$\frac{\partial}{\partial y_i}\left(a_{ij} \frac{\partial u_1}{\partial y_j}\right) + \frac{\partial a_{ij}}{\partial y_i} \frac{\partial u}{\partial x_j} = 0 \ , \quad u_1(x_n', 0, y_n', 0) = 0 \tag{2.13}$$

$$(2.14) \qquad a_{ij}(y)\frac{\partial^2 u}{\partial x_i \partial x_j} + \frac{\partial}{\partial y_i}\left(a_{ij}\frac{\partial u_1}{\partial x_j}\right) + \frac{\partial}{\partial x_i}\left(a_{ij}\frac{\partial u_1}{\partial y_j}\right) - \alpha u + \frac{\partial}{\partial y_i}\left(a_{ij}\frac{\partial u_2}{\partial y_j}\right) = f$$

$$(2.15) \qquad a_{ij}\frac{\partial^2 u_1}{\partial x_i \partial x_j} + \frac{\partial}{\partial y_i}\left(a_{ij}\frac{\partial u_2}{\partial x_j}\right) + \frac{\partial}{\partial x_i}\left(a_{ij}\frac{\partial u_2}{\partial y_j}\right) - \alpha u_1 + \frac{\partial}{\partial y_i}\left(a_{ij}\frac{\partial u_3}{\partial x_j}\right) = 0 .$$

We will skip some details and give only the solution of (2.13), (2.14), (2.15) in order that conditions (2.4) be satisfied. We write

$$(2.16) \qquad u_1(x, y) = \frac{\partial u}{\partial x_l}(\chi^l(y) - d^l) - \frac{\partial u}{\partial x_l}(x_n', o)v^l(y)$$

where χ^l is a solution of (1.4), and v^l is a solution of (1.6). Note that the existence of the solution of (1.6) is a consequence of Theorem 1.2, provided the constant d^l is defined in order that $C_{\chi^l - d^l} = 0$. We will see that this is possible.

We plug the value of u_1 defined by (2.16) in (2.14). After some calculations one obtains

$$(2.17) \qquad u_2(x, y) = \frac{\partial^2 u}{\partial x_i \partial x_l}\chi^{il}(y) + \frac{\partial^2 u}{\partial x_l \partial x_{i_n}}(x_n', 0)v^{l i_n}(y)$$

where i_n means any value of i except n. The functions χ^{il} are solutions of

$$(2.18) \qquad \begin{cases} \dfrac{\partial}{\partial y_i}\left(a_{ij}\dfrac{\partial \chi^{mp}}{\partial y_j}\right) + a_{mp} + a_{mj}\dfrac{\partial \chi^p}{\partial y_j} - q_{mp} \\ \qquad + \dfrac{\partial}{\partial y_k}(a_{km}(\chi^p - d^p)) = 0 , \\ \chi^{mp} \text{ periodic } \in W^{2,q,\mu}(R^n) \end{cases}$$

and the constants q_{mp} are defined by (1.5).

The functions v^{mp} are then defined by

$$(2.19) \qquad \begin{cases} \dfrac{\partial}{\partial y_i}\left(a_{ij}\dfrac{\partial v^{mp}}{\partial y_j}\right) = a_{pj}\dfrac{\partial v^m}{\partial y_j} + \dfrac{\partial}{\partial y_k}(a_{kp}v^m) \\ v^{mp}(y_n', 0) = e^{mp} \end{cases}$$

where e^{mp} are constants chosen in such a way that the following estimates hold true

$$(2.20)\quad \begin{cases} v^{mp} \in W^{2,q,\mu}(R^{n-1} \times R^+) \\ |v^{mp}(y'_n, y_n)|\,, \left|\dfrac{\partial v^{mp}}{\partial y_i}(y'_n, y_n)\right| \leq Re^{-\rho y_n}\,. \end{cases}$$

The possibility of solving (2.19), (2.20) follows from Theorem 1,3, provided the constant e^{mp} is chosen in order that $C_h = 0$ (see (1.17)), where h is the right hand side of (2.19).

Finally the function $u_3(x, y)$ is defined as follows

$$(2.21)\quad \begin{cases} u_3(x, y) = \dfrac{\partial^3 u}{\partial x_i \partial x_k \partial x_l}\chi^{ikl}(y) + \dfrac{\partial u}{\partial x_l}\zeta^l(y) \\ \qquad + \dfrac{\partial^3 u}{\partial x_{i_n} \partial x_{k_n} \partial x_l}(x'_n, 0)v^{i_n k_n l}(y) + \dfrac{\partial u}{\partial x_l}(x'_n, 0)w^l(y)\,. \end{cases}$$

The functions $\chi^{ikl}(y), \zeta^l(y)$ are periodic $W^{2,p,\mu}(R^n)$ functions. The functions $v^{i_n k_n l}(y)$ and $w^l(y)$ are solutions of non homogeneous Dirichlet problems with adequate constants as boundary conditions. The right hand side will depend on functions v^l and v^{mp}. By virtue of estimates (2.20), (1.6) one can check that one is again in the conditions of application of Theorem 1.3 and thus $v^{i_n k_n l}(y)$ and $w^l(y)$ can be determined and satisfy estimates like (2.20). Details are omitted.

Now the function u being solution of (1.7) with $f \in H^2(\mathcal{O})$, belongs to $H^5(\mathcal{O})$. The functions g_ε and h_ε are obtained from formulas

$$(2.22)\quad g_\varepsilon = -a_{ij}\left(\frac{\partial^2 u_2}{\partial x_j \partial x_i} + \frac{\partial^2 u_3}{\partial y_j \partial x_i}\right) - \frac{\partial}{\partial y_i}\left(a_{ij}\frac{\partial u_3}{\partial x_j}\right) - \varepsilon a_{ij}\frac{\partial^2 u_3}{\partial x_i \partial x_j}$$

$$(2.23)\quad h_\varepsilon = u_2(x'_n, 0, y'_n, 0) + \varepsilon u_3(x'_n, 0, y'_n, 0)\,.$$

From the regularity of u and the fact that the functions of y are bounded as well as their derivatives, it follows that $g_\varepsilon, h_\varepsilon$ satisfy conditions (2.4). Now φ_ε also satisfies (2.4) and moreover $\|\varepsilon u_2 + \varepsilon^2 u_3\|_{H^1(\mathcal{O})} \leq C$. From this and (2.12) which is valid, we obtain

$$\|u_\varepsilon - u - \varepsilon u_1\|_{H^1(\mathcal{O})} \leq C\varepsilon\,.$$

Finally, noting that $\|d^l(\partial u/\partial x_l)\|_{H^1(\mathcal{O})} \leq C$, we complete the proof of Theorem 1.1.

§ 3. Proof of Theorem 1.2

We now must prove Theorems 1.2 and 1.3 and in particular make precise the values of constants C_χ and C_h. The proof will be based on probabilistic techniques and in particular, in applying results from ergodic theory.

We begin by the proof of Theorem 1.2. We define on $H^{1,\mu}$ the bilinear form

$$\text{(3.1)}\quad a_\mu(\theta_1,\theta_2) = \sum_{i,j=1}^{n-1}\iint a_{ij}\frac{\partial\theta_1}{\partial x_j}\frac{\partial}{\partial x_i}(\theta_2\exp(-2\mu|x|)\exp(-2\mu\lambda))dxd\lambda$$
$$+\sum_{i=1}^{n-1}\iint a_{in}\frac{\partial\theta_1}{\partial\lambda}\frac{\partial}{\partial x_i}(\theta_2\exp(-2\mu|x|)\exp(-2\mu\lambda))dxd\lambda$$
$$+\sum_{i=1}^{n-1}\iint a_{ni}\frac{\partial\theta_1}{\partial x_i}\frac{\partial}{\partial\lambda}(\theta_2\exp(-2\mu|x|)\exp(-2\mu\lambda))dxd\lambda$$
$$+\iint a_{nn}\frac{\partial\theta_1}{\partial\lambda}\frac{\partial}{\partial\lambda}(\theta_2\exp(-2\mu|x|)\exp(-2\mu\lambda))dxd\lambda\,.$$

Some routine calculations show that

$$\text{(3.2)}\quad a_\mu(\theta,\theta)+\gamma|\theta|^2_{L^{2,\mu}}\ge\alpha\|\theta\|^2_{H^{1,\mu}}\,,\qquad\forall\theta\in H^{1,\mu}\,,\ \alpha>0\,,\ \gamma\ge0\,.$$

It follows that one can solve the variational inequality (V.I)

$$\text{(3.3)}\quad\begin{cases}a_\mu(\theta,w-\theta)+\gamma(\theta,w-\theta)\ge(L,w-\theta)\\ \forall w\in H^{1,\mu}\,|\,w(x,0)=\chi(x)\,,\qquad\theta(x,0)=\chi(x)\end{cases}$$

where we first assume that $\chi(x)\in\mathscr{D}(R^{n-1})$, besides (1.12). The solution of (3.3) is easily checked to be the solution of the non homogeneous Dirichlet problem

$$\text{(3.4)}\qquad A\theta+\gamma\theta=L\,,\qquad\theta(x,0)=\chi(x)$$

where A is the operator

$$\text{(3.5)}\quad A=-\frac{\partial}{\partial x_i}\left(a_{ij}\frac{\partial}{\partial x_j}\right)-\frac{\partial}{\partial x_i}\left(a_{in}\frac{\partial}{\partial\lambda}\right)-\frac{\partial}{\partial\lambda}\left(a_{nj}\frac{\partial}{\partial x_j}\right)-\frac{\partial}{\partial\lambda}\left(a_{nn}\frac{\partial}{\partial\lambda}\right).$$

The first equation holds true in the sense of distributions. We now pass to the solution of problem

$$\text{(3.6)}\qquad Az=0\,,\qquad z(x,0)=\chi(x)\,.$$

One uses an iterative technique, defining a sequence z^n by

$$(3.7)\qquad a_\mu(z^{n+1}, w - z^{n+1}) + \gamma(z^{n+1}, w - z^{n+1}) \geq \gamma(z^n, w - z^{n+1})$$

$$\forall w \in H^{1,\mu} \mid w(x, 0) = \chi(x)\,; \qquad z^{n+1}(x, 0) = \chi(x)\,,$$

with $z^0 = \inf \chi$. By the Maximum Principle, one proves that

$$z^0 \leq z^1 \leq \cdots \leq z^n \leq \cdots \leq \sup \chi$$

and from (3.7) we also have $\|z^n\|_{H^{1,\mu}} \leq C$. Hence $z^n \uparrow z$ solution of (3.6). We thus have proved the existence of (3.7) which satisfies

$$(3.8)\qquad z \in H^{1,\mu}\,, \qquad \inf \chi \leq z \leq \sup \chi\,.$$

However, since χ is smooth, we see that $z - \chi$ satisfies

$$(A + \gamma)(z - \chi) = \gamma(z - \chi)\,; \qquad (z - \chi)(x, 0) = 0\,.$$

From the regularity results of the solution of homogeneous Dirichlet problems, we obtain that $z - \chi \in W^{2,p,\mu}$. From the regularity of the coefficients we also have $z - \chi \in W^{3,p,\mu}$, hence $z \in C^2(R^{n-1} \times R^+)$.

We next introduce the Ito equation

$$(3.8)'\qquad \begin{cases} d\zeta = g_{n-1}(\zeta, \mu)ds + \sigma_{n-1}(\zeta, \mu)dw \\ d\mu = g_n(\zeta, \mu)ds + \sigma_n(\zeta, \mu)dw \\ \zeta(0) = x\,, \qquad \mu(0) = \lambda \end{cases}$$

where we have the decomposition

$$g(x, \lambda) = \begin{pmatrix} g_{n-1} \\ g_n \end{pmatrix}; \ \sigma(x, \lambda) = \begin{pmatrix} \sigma_{n-1} \\ \sigma_n \end{pmatrix}$$

$$a = (a_{ij}) = \frac{\sigma^2}{2}\,; g_j = \frac{\partial a_{ij}}{\partial x_i}\,.$$

Let $T_{x\lambda}$ be the exit time of $\mu_{x\lambda}(s)$ from 0 ($\lambda \geq 0$). It can be proved that $T_{x\lambda} < \infty$ a.s.. Indeed defining firstly $T_{x\lambda N} = \inf\{s \geq 0 \mid \mu_{x\lambda}(s) \notin (0, N)\}$, and applying Ito's formula to the process (3.8) and the function χ^n, solution of (1.4) with $l = n$, one obtains

$$\chi^n(x, \lambda) + \lambda = E\chi^n(\zeta_{x\lambda}(T_{x\lambda N}), \mu_{x\lambda}(T_{x\lambda N})) + NP_N$$

where $P_N = \text{Prob}\,(\mu_{x\lambda}(T_{x\lambda N}) = N)$.

Since χ^n is bounded, it follows that $P_N \to 0$ as $N \to \infty$, from which one easily deduces that $T_{x\lambda} < \infty$, a.s..

Turning to the solution of (3.6), still assuming $\chi \in \mathscr{D}(R^{n-1})$, we obtain from Ito's formula that

$$z(x, \lambda) = E\chi(\zeta_{x\lambda}(T_{x\lambda})) . \tag{3.9}$$

We thus have proved that when $\chi \in \mathscr{D}(R^{n-1})$ and satisfies (1.12), there exists a solution of (1.10) which satisfies (1.13). It is given explicitly by (3.9), and it is easily checked that function (3.9) is indeed the unique solution.

The next step consists in considering the case when χ satisfies (1.12) and is only continuous. One then considers a sequence of functions

$$\chi_n \to \chi \quad \text{in} \quad C^0(R^{n-1}) , \quad \chi_n \quad \text{smooth and periodic} .$$

One defines by z_n the solution of

$$Az_n = 0 , \qquad z_n(x, 0) = \chi_n(x)$$

which is given explicitly by

$$z_n(x, \lambda) = E\chi_n(\zeta_{x\lambda}(T_{x\lambda})) .$$

Clearly, $z_n \to E\chi(\zeta_{x\lambda}(T_{x\lambda}))$ in $C^0(R^{n-1} \times [0, \infty))$.

Since z_n is harmonic, it follows that z_n remains in a bounded subset of $W^{2,p}_{\text{loc}}(R^{n-1} \times R^+)$. Therefore z defined by (3.9) belongs to $W^{2,p}_{\text{loc}}(R^{n-1} \times R^+)$ and satisfies $Az = 0$. By differentiation in x_i and λ, one checks that $z \in W^{3,p}_{\text{loc}}(R^{n-1} \times R^+)$.

We finally turn to the case when χ only satisfies (1.12). We then define $z(x, \lambda)$ by formula (3.9). We first show that

$$z(x, \lambda) \text{ is continuous in } x, \lambda \text{ for } \lambda > 0 . \tag{3.10}$$

This follows from the strong Feller property of the process (ζ, μ). For $u > 0$, we have

$$z(x, \lambda) = E\chi(\zeta_{x\lambda}(T_{x\lambda}))\delta_{T_{x\lambda}>u} + E\chi(\zeta_{x\lambda}(T_{x\lambda}))\delta_{T_{x\lambda}\leq u}$$

where $\delta_{T_{x\lambda}>u} = 1$ if $T_{x\lambda} > u$ and 0 otherwise.

From the strong Markov property we have

$$z(x, \lambda) = Ez(\zeta_{x\lambda}(u), \mu_{x\lambda}(u)) + E\delta_{T_{x\lambda}\leq u}[\chi(\zeta_{x\lambda}(T_{x\lambda})) - z(\zeta_{x\lambda}(u), \mu_{x\lambda}(u))] .$$

Now

$$P(T_{x\lambda\leq u}) \leq P\left(\sup_{0\leq s\leq u} |\mu(s) - \lambda| \geq \lambda\right) \leq \frac{Cu}{\lambda^2} .$$

From this estimate and the fact that $Ez(\zeta_{x\lambda}(u), \mu_{x\lambda}(u))$ is continuous in x, λ, when $u > 0$, by the strong Feller property, one easily proves (3.10).

Taking $\lambda_0 > 0$, one can then solves by a previous argument the problem

$$\begin{aligned} &A\tilde{z} = 0 \qquad \text{in} \quad R^{n-1} \times \{\lambda > \lambda_0\} \\ &\tilde{z}(x, \lambda_0) = z(x, \lambda_0) \, ; \, \tilde{z} \in W^{2,p}_{\text{loc}}(R^{n-1} \times (\lambda_0, \infty)) \\ &\tilde{z} \in C^2(R^{n-1} \times (\lambda_0, \infty)) \, , \qquad \tilde{z} \quad \text{periodic in } x \, . \end{aligned}$$

If

$$T^0_{x\lambda} = \inf \{s \geq 0 \,|\, \mu_{x\lambda}(s) = \lambda_0\}$$

one has by an easy probabilistic argument

$$\tilde{z}(x, \lambda) = E^{x\lambda} z(\zeta(\tau^0), \mu(\tau^0)) = z(x, \lambda) .$$

This proves that the function (3.9) is indeed a solution of (1.10), (1.13). We furthermore have the following property

(3.11) if $\chi \geq 0$, then $z(x, \lambda) > 0 \qquad \forall \lambda > 0$ unless $z(x, \lambda) \equiv 0$.

This is a consequence of the strong Maximum Principle. Assuming that $z \not\equiv 0$, there exists a point (x_0, λ_0) where $z(x_0, \lambda_0) > 0$. If $\inf_x z(x, \lambda_0) > 0$, then for $\lambda \geq \lambda_0$, it easily follows from the probabilistic interpretation that $z(x, \lambda) > 0$. Let us next consider the case when $\inf_x z(x, \lambda_0) = 0$.

Since $M = \sup z(x, \lambda_0) > 0$, we see that $z - M \in C^2(R^{n-1} \times (\lambda_0, \infty))$ $z - M \not\equiv$ constant, $A(z - M) = 0$ and $z - M \geq -M$. By the strong Maximum Principle, it follows that $z - M > -M$ for $\lambda > \lambda_0$.

We thus have $z(x, \lambda) > 0 , \qquad \forall \lambda > \lambda_0$.

If $\lambda_0 = 0$, the result is proved. If not, take $0 < \lambda' < \lambda_0$, then one cannot have $z(x, \lambda') \equiv 0 \; \forall x$, since otherwise considering the Dirichlet problem on $R^{n-1} \times (\lambda', \infty)$, one would have $z(x, \lambda) = 0 \; \forall \lambda \geq \lambda'$, which contradicts the fact that $z(x_0, \lambda_0) \neq 0$. We can then take λ' instead of λ_0, and show that $z(x, \lambda) > 0$, $\forall \lambda > \lambda'$. Since λ' is arbitrary > 0 we obtain (3.11).

We can now prove (1.14). It is a consequence of ergodic results. One defines a family of operators on the set of Borel periodic bounded functions by setting

(3.12) $$P(n)\chi(x) = z(x, n) \qquad n \in N\,.$$

Using the strong Markov property, one can prove that

(3.13) $$P(n) = P(1)^n = P^n\,.$$

This defines a Markov chain on the $n - 1$ dimensional torus, whose transition probability is denoted by

(3.14) $$P(x, n; E) = P^n \chi_E(x)\,,$$

where E is a borel subset of the $n - 1$ dimensional torus. From what has been already proved, one obtains

(3.15) $$\begin{cases} x \to P(x, n; E) \text{ is } \quad C^2(R^{n-1}) \quad \text{periodic and} \\ P(x, n; E) > 0 \quad \forall x \quad \text{unless } P(x, n; E) = 0 \quad \forall x\,. \end{cases}$$

In particular

(3.16) $$P(x, n; U) > 0 \qquad \forall U \quad \text{ball of } \tilde{\Pi} \ (n - 1 \text{ dimensional torus})\,.$$

It is then a standard result for Markov chains with compact phase space for which (3.15), (3.16) are satisfied that the following ergodic results hold true

(3.17) $$\begin{cases} \text{there exists one and only one invariant probability measure } \bar{m} \\ \text{on the } n - 1 \text{ dimensional torus such that} \\ \left| z(x, n) - \int \chi d\bar{m} \right| \le \|\chi\| \gamma e^{-\delta n}\,, \end{cases}$$

where γ, δ are positive constants independent of χ.

For $\lambda > n$, it follows from the Maximum Principle that

$$\left| z(x, \lambda) - \int \chi d\bar{m} \right| \le \sup_x \left| z(x, n) - \int \chi d\bar{m} \right| \le \|\chi\| \gamma e^{-\delta n}\,.$$

In particular, this relation is true for $n = [\lambda]$. Hence changing γ into γe^δ, the first property (1.14) follows, with

(3.18) $$C_\chi = \int \chi d\bar{m}\,.$$

Let $\lambda_0 > 0$ be fixed. Let us set

$$\tilde{z}(x, \lambda) = (z(x, \lambda) - C_\chi) - (z(x, \lambda_0) - C_\chi) \exp(-\delta(\lambda - \lambda_0))\,,$$
$$\lambda \ge \lambda_0\,, \qquad x \in R^{n-1}$$

$$\hat{z}(x, \lambda) = \tilde{z}(x, \lambda) \exp \delta\lambda .$$

Then $\hat{z}(x, \lambda)$ is solution of an homogeneous Dirichlet problem on $R^{n-1} \times (\lambda_0, \infty)$, and one can easily show that $\hat{z}$ has a finite norm in $W^{2,p,\mu}(R^{n-1} \times (\lambda_0, \infty))$. In particular $(\partial\tilde{z}/\partial x_i) \exp(\delta - \mu)\lambda$, $(\partial\tilde{z}/\partial\lambda) \exp(\delta - \mu)\lambda$ are Hölder functions on $R^{n-1} \times (\lambda_0, \infty)$ and $L^p(\Pi \times (\lambda_0, \infty))$, and periodic in x. They are necessarily bounded on $\bar{\Pi} \times [\lambda_0, \infty)$. This implies the second estimate (1.14).

The proof of Theorem 1.2 is complete. ■

Theorem 1.2 implies the existence of the solution of (1.6), taking into account that the solution is globally in $W^{2,p,\mu}(R^{n-1} \times R^+)$, since $\chi^l(y'_n, 0)$ is a regular function. The last estimate (1.6) holds true on $\lambda \geq 0$, $x \in R^{n-1}$ and not only on $\lambda \geq \lambda_0 > 0$, $x \in R^{n-1}$, since on $[0, \lambda_0]$ the derivatives are bounded. Moreover, since

$$C_\chi = \int \chi^l(y'_n, 0) d\bar{m}(y'_n) - d^l$$

we have $C_\chi = 0$, if and only if $d^l = \int \chi^l(y'_n, 0) d\bar{m}(y'_n)$.

§ 4. Proof of Theorem 1.3

The proof uses the following property

$$(4.1) \qquad E \int_0^{T_{xp}} e^{-\beta\mu_{xp}(s)} ds < C , \quad \forall x \in R^{n-1} , \quad p \quad \text{integer} \geq 0 , \quad \beta > 0 .$$

The proof of (4.1) will be given later. It is not a trivial fact, since $ET_{xp} = +\infty$. One first solves

$$(4.2) \qquad \begin{cases} Az_\gamma + \gamma z_\gamma = h , \quad z_\gamma(x, 0) = 0 , \quad z_\gamma \in W^{2,p,\mu}(R^{n-1} \times R^+) \\ z_\gamma \quad \text{periodic in } x . \end{cases}$$

At least for γ large enough, (4.2) has one and only one solution, given by

$$(4.3) \qquad z_\gamma(x, \lambda) = E \int_0^{T_{x\lambda}} e^{-\gamma t} h(\zeta_{x\lambda}(t), \mu_{x\lambda}(t)) dt .$$

Using (1.15) and (4.1), one can check that $|z_\gamma(x, \lambda)| \leq C$, independent of γ, x, λ. Using then the equation, one obtains $\|z_\gamma\|_{W^{2,p,\mu}(R^{n-1}\times R^+)} \leq C$.

Letting γ go to 0, one then shows the existence of a solution of

(4.4) $Az = h\,,\quad z(x,0) = 0\,,\quad z \in W^{2,p,\mu}(R^{n-1} \times R^+)\,, z$ periodic in x.

This solution is unique, since it has the probabilistic interpretation

$$(4.5)\qquad z(x,\lambda) = E\int_0^{T_{x\lambda}} h(\zeta_{x\lambda}(t), \mu_{x\lambda}(t))dt\,.$$

It remains to prove (1.17). Let us first assume that $h \geq 0$, and consider $z(x, n)$ which we rewrite as

$$z(x,n) = E^{xn}\int_0^T h(\zeta(t), \mu(t))dt$$

taking canonical processes and $T = \inf\{s \mid \mu(s) \leq 0\}$. Defining $S = \inf\{s \geq 0 \mid \mu(0) - \mu(1) \geq 1\}$, and

$$g(x,\lambda) = E^{x\lambda}\int_0^S h(\zeta(t), \mu(t))dt$$

then one can show that

$$z(x,n) = \sum_{i=1}^n P(i-1)g_{n-i+1}(x) = \sum_{j=1}^n P(n-j)g_j(x)$$

where

$$g_{n-i+1}(x) = g(x, n-i+1)\,.$$

Using (1.15) and (4.1), one obtains $|g(x,\lambda)| \leq Ce^{-\beta\lambda}$. Setting $G_n(x) = \sum_{j=1}^n g_j(x)$, one has $G_n(x) \uparrow G(x)$ and $|G(x) - G_n(x)| \leq C_1e^{-\beta n}$. Rewriting

$$z(x,n) = \sum_{j=1}^k P(n-j)g_j + \sum_{j=k+1}^n P(n-j)g_j$$

with $k = [n/2]$, we have

$$\left|\sum_{j=k+1}^n P(n-j)g_j(x)\right| \leq C_2e^{-\beta(n/2)}$$

and

$$\sum_{j=1}^k P(n-j)g_j = P^{n-k}G_k + \sum_{j=1}^{k-1}(P^{n-j} - P^{n-j-1})G_j\,.$$

By ergodic properties

$$|P^{n-j}G_j - \overline{G}_j| \leq \|G_j\|\gamma e^{-\delta(n-j)}$$

where $\bar{G}_j$ is the mean of G_j with respect to the invariant measure. Hence

$$\left|\sum_{j=1}^{k-1} (P^{n-j} - P^{n-j-1})G_j\right| \leq \sum_{j=1}^{k-1} \gamma \|G_j\| (e^{-\delta/(n-j)} + e^{-\delta(n-j-1)})$$
$$\leq C_3 e^{-\delta(n/2)} .$$

Similarly $|P^{n-k}G_k - \bar{G}| \leq C_4(e^{-\beta(n/2)} + e^{-\delta(n/2)})$.

Gathering estimates, one obtains

$$|z(x,n) - \bar{G}| \leq Ce^{-(n/2)\delta\wedge\beta} . \tag{4.6}$$

Noticing that $\bar{G}$ is a linear functional of h, it is easy to check (decomposing h into $h^+ - h^-$) that (4.6) holds true for general h. Thus the first property (1.17) is proved for λ integer. Using (4.6) and again (4.1), one can prove (4.6) for λ instead of n. Thus the first part of (1.17) is proved. The second part is similar to the corresponding one of Theorem 1.2.

Proof of (4.1).

It remains to prove (4.1). It would be easy to prove it by direct methods, if β is large enough. However the difficulty arises from the fact that β can be arbitrarily small. It follows from a probabilistic argument using a random walk derived from the process $\zeta(s)$, $\mu(s)$. For $x \in R^{n-1}$ and p integer, one defines

$$\tau_{xp}^1 = \inf\{s \geq 0 \,|\, \mu_{xp}(s) - p = \pm 1\}$$
$$\cdots$$
$$\tau_{xp}^{k+1} = \inf\{s \geq \tau_{xp}^k \,|\, \mu_{xp}(s) - \mu_{xp}(\tau_{xp}^k) = \pm 1\}$$
$$\cdots$$
$$\zeta_{xp}^k = \zeta_{xp}(\tau_{xp}^k) ; \qquad \mu_{xp}^k = \mu_{xp}(\tau_{xp}^k) .$$

The process ζ_{xp}^k, μ_{xp}^k is called the induced random walk. We introduce $\nu_{xp} = \inf\{k \geq 0 \,|\, \mu_{xp}^k \leq 0\}$. It is not difficult to see that (4.1) will be a consequence of the property

$$E^{xp} \sum_{k=0}^{\infty} e^{-\beta\mu^k} \chi_{k<\nu} < C \tag{4.7}$$

where we have used the canonical process, and some obvious changes of notation. To the induced random walk is associated a discrete time semigroup, setting

$$B\varphi(x,m) = E\varphi(\zeta_{xm}^1, \mu_{xm}^1) \tag{4.8}$$

where $\varphi(x,m)$ is a bounded measurable functions on $R^{n-1} \times Z$, which is periodic in x. One can see that

(4.9) $$B^k\varphi(x, m) = E\varphi(\zeta^k_{xm}, \mu^k_{xm}) .$$

Now, if $g(x, m)$ is a positive continuous bounded periodic function in x, ($m \in Z$) then considering the solution u of

(4.10) $$(I - B)u = g \quad x \in R^{n-1}, \quad m > 0; \quad u(x, 0) = 0$$

one has the following representation

(4.11) $$u(x, m) = E^{xm} \sum_{k=0}^{\infty} g(\zeta^k, \mu^k)\chi_{k<\nu} .$$

From this, it is easy to see that (4.7) will be a consequence of the following property

(4.12) $$\begin{cases} \text{there exists a function } u(x, m) \text{ continuous, } \geq 0, \text{ bounded,} \\ \text{periodic in } x \text{ satisfying } (I - B)u \geq e^{-\beta m}. \end{cases}$$

To prove (4.12), one introduces a Markov chain on the $n - 1$ dimensional torus, setting

(4.13) $$Pf(x) = Ef(\zeta_{xp}(\tau^1_{xp})) = Ef(\zeta_{x0}(\tau^1_{x0}))$$

where f is a Borel bounded periodic function on R^{n-1}. We have the decomposition $P = P^+ + P^-$, where

$$P^+f(x) = Ef(\zeta_{x0}(\tau^1_{x0}))\chi_{\mu_{x0}(\tau^1_{x0})=1}$$
$$P^-f(x) = Ef(\zeta_{x0}(\tau^1_{x0}))\chi_{\mu_{x0}(\tau^1_{x0})=-1} .$$

One can show, using arguments similar to those of Theorem 1.2, that P is an ergodic Markov chain, and thus, one has

(4.14) $$\begin{cases} \text{if } \varphi(x) \text{ is Borel bounded periodic on } R^{n-1}, \text{ with } \int \varphi d\bar{P} = 0, \\ \text{where } \bar{P} \text{ is the unique invariant measure of } P, \text{ then the} \\ \text{equation } (I - P)v = \varphi \text{ has a solution which is Borel periodic} \\ \text{bounded, given by} \\ v = \sum_{k=0}^{\infty} P^k\varphi \quad \text{(convergence uniform in } x). \end{cases}$$

Now we note that

(4.15) $$B\varphi(x, m) = P^+\varphi_{m+1}(x) + P^-\varphi_{m-1}(x)$$

where $\varphi_m(x) = \varphi(x, m)$.

We search the function u satisfying (4.12) in the following form

$$u(x, m) = u_0(m) + \varphi(m)v(x) + \psi(m)w(x) \tag{4.16}$$

where

$$u_0(m) = A - Be^{-rm} \qquad (A, B, \gamma \text{ to be defined})$$
$$\varphi(m) = \gamma Be^{-rm}$$
$$\psi(m) = \gamma^2 Be^{-rm} .$$

We choose v to be the solution of

$$(I - P)v = P^{+}1 - P^{-}1 . \tag{4.17}$$

This is possible, by virtue of (4.14), using the property

$$\int P^{+}1 d\bar{P} = \int P^{-}1 d\bar{P} = \frac{1}{2} .$$

Next $w(x)$ is chosen as follows

$$(I - P)w = -(P^{+}v - P^{-}v)(x) + \overline{P^{+}v - P^{-}v} . \tag{4.18}$$

With the above choices, one obtains

$$(I - B)u = Be^{-rm}\left[\frac{\gamma^2}{2} + \gamma^2\overline{P^{+}v - P^{-}v} + o(\gamma^2)\right] .$$

We then notice that

$$\frac{1}{2} + \overline{P^{+}v - P^{-}v} = \frac{1}{2} + \overline{(P^{+} - P^{-})(I - P)^{-1}(P^{+} - P^{-})1} = \sigma^2 > 0 .$$

Hence one can choose γ small enough in order that (4.12) holds true, chosing also the constants A and B in an adequate way. The proof of (4.1) is thus complete.

Remark 4.1. Techniques of the type used in the proof of Theorem 1.2 and 1.3 are similar to those of Bromberg [5] and Williams [8] in a different context.

References

[1] Babuska, I., Homogeneization approach in engineering, Proceedings of the 2nd IRIA Symposium on Dec. 1975, Paris. Springer Verlag Lecture Notes, to appear.

[2] Bensoussan, A., Lions, J. L. and Papanicolaou, G., C. R. Acad. Sci. Paris, 281 (1975), 89–94 and 317–322.

[3] ——, Boundary layers and homogeneization of transport processes, Publ. Res. Inst. Math. Sci., to appear.

[4] ——, Book in preparation.

[5] Bromberg, N., Boundary layer analysis of equations of Brownian motion in force field, Thesis, Courant Institute, 1975.

[6] De Giorgi, E.-Spagnolo, S., Sulla convergenza degli integrali dell' energia per operatori ellitici del secundo ordine, 8. Boll. U.M.I. (1973), 391–411.

[7] Tartar, L., C. R. Acad. Sci. Paris, to appear.

[8] Williams, M. Homogeneization of linear transport problems, Courant Institute, Feb. 1976.

LABORATOIRE DE RECHERCHE EN INFORMATIQUE
ET AUTOMATIQUE
DOMAINE DE VOLUCEAU-ROCQUENCOURT
78150 LE CHESNAY, FRANCE

Proc. of Intern. Symp. SDE
Kyoto 1976, pp. 41–47

On the Principal Eigenvalue of Elliptic Second Order Differential Operators

M. D. DONSKER* and S. R. S. VARADHAN*

§ 1. Introduction

Let L be an elliptic second order partial differential operator of the form

$$L=\frac{1}{2}\sum_{i,j=1}^{n}\frac{\partial}{\partial x_i}a_{ij}(x)\frac{\partial}{\partial x_j}+\sum_{j=1}^{n}b_j(x)\frac{\partial}{\partial x_j}\,. \tag{1.1}$$

We assume that the coefficients $\{a_{ij}(x)\}$ and $\{b_j(x)\}$ are all infinitely differentiable functions of x in R^n. In addition the matrix $\{a_{ij}(x)\}$ is assumed to be symmetric and positive definite for each $x\in R^n$. No assumption is made regarding the growth of the coefficients near ∞.

Although there may not be a unique diffusion process corresponding to L, due to possible explosion, for any bounded region G the process is well defined until the exit time from G.

Let $V(x)$ be a smooth function on R^n and G any bounded open set with a smooth boundary. Let τ_G be the first exit time from the region G. Then

$$u(t,x)=E_x\left[\exp\left[\int_0^t V(x(s))\,ds\right]f(x(t))\,;\,\tau_G>t\right] \tag{1.2}$$

solves the partial differential equation

$$\begin{cases}\dfrac{\partial u}{\partial t}=Lu+Vu & \text{for } t>0 \text{ and } x\in G\\ u(0,x)=f(x) & \text{for } x\in G\\ u(t,x)=0 & \text{for } t>0 \text{ and } x\in\partial G\,.\end{cases} \tag{1.3}$$

The solution $u(t,x)$ defines a semigroup $\{T_t^{V,G}\}$ of bounded linear transformations of $C(\bar{G})$ which is strongly continuous on $C_0(G)$, the space of continuous functions on $\bar{G}$ which vanish on the boundary ∂G of G.

By the submultiplicative property of the norm

* This research was supported by NSF Grant Number MCS74-01921 AO3.

(1.4) $$\lim_{t\to\infty} \frac{1}{t} \log \| T_t^{V,G} \| = \lambda_G(V)$$

exists and is finite. One can show that the spectrum $\sigma_{V,G}$ of the infinitesimal generator $L_{V,G}$ of the semigroup $T_t^{V,G}$, which is a restriction of the operator $L + V$ to a suitable class of functions vanishing on the boundary ∂G, has the following properties. The spectrum $\sigma_{V,G}$ is contained in the half space $[z: \operatorname{Re} z \leq \lambda_G(V)]$. Moreover $\lambda_G(V) \in \sigma_{V,G}$. Therefore, although the spectrum in general is complex, $\lambda_G(V)$ can be thought of as the principal eigenvalue of the operator $L + V$ with Dirichlet boundary conditions in the region G.

If the coefficients $\{b_j(x)\}$ are identically zero then L is formally self-adjoint with respect to the Lebesgue measure on R^n and with Dirichlet boundary conditions in G the infinitesimal generator $L + V$ of the semigroup $T_t^{V,G}$ is actually self adjoint. From the classical variational formula for the principal eigenvalue we can write

(1.5) $$\lambda_G(V) = \sup_{\substack{f \in L_2 \\ \|f\|_2 = 1 \\ f = 0 \text{ off } G}} \left[\int f^2(x) V(x) dx - \frac{1}{2} \int \langle a \nabla f, \nabla f \rangle dx \right].$$

The aim of this article is to indicate a probabilistic interpretation of the formula (1.5) which admits a generalization to the non-self adjoint case. In this process we obtain a generalization of the formula (1.5) to such cases. The details of the proofs will be omitted. See [3] for details.

§ 2. The probabilistic interpretation

The probabilistic interpretation of the variational formula (1.5) explains why such a formula should exist in the non-self adjoint case. The actual derivation of the formula does not use probabilistic methods.

Once the region G is given the coefficients $\{a_{ij}(x)\}$ and $\{b_j(x)\}$ outside the region G, have no bearing on the problem. One can therefore assume without loss of generality that the coefficients are sufficiently well behaved at ∞. Therefore the process corresponding to L can be assumed to exist uniquely without explosion. We shall denote by P_x, the measure corresponding to the process starting from the point x, on the space of continuous trajectories on $[0, \infty]$ with values in R^d. Let us denote by Ω the space of all trajectories and by w any particular trajectory. We denote as usual by $x(t) = x(t, w)$ the position of the trajectory at any time $t \geq 0$.

From the positivity property of the semigroup $T_t^{V,G}$ one can see immediately that

$$\| T_t^{V,G} \| = \sup_{x \in \bar{G}} (T_t^{V,G} 1)(x)$$

$$= \sup_{x \in \bar{G}} E^{P_x}\left[\exp\left[\int_0^t V(x(s))\,ds\right];\ \tau_G > t\right]$$

where

$$\tau_G = \inf\,[t\colon x(t) \notin G]$$

is the first exit time from G.

For each trajectory w and each $t > 0$, we introduce the occupation distribution of the trajectory w up to time t. It is a probability distribution $L(t, w, \cdot)$ defined for any Borel set A by

$$L(t, w, A) = \frac{1}{t}\int_0^t \chi_A(x(s))ds\ .$$

$L(t, w, A)$ is therefore the proportion of time spent by the trajectory in the set A during the interval $[0, t]$ of time. If $\mathcal{M}$ denotes the space of all probability distributions on R^n then for each $t > 0$ $L(t, \cdot, \cdot)$ gives us a map of Ω into $\mathcal{M}$. With a natural (weak) topology on $\mathcal{M}$ and the corresponding Borel-field we get a measure $Q_{t,x}$ on $\mathcal{M}$ induced by $L(t, \cdot, \cdot)$ from the measure P_x on Ω.

Let us first look at the case $V \equiv 0$. Then the set

$$\{\tau_G > t\} = \{w\colon \operatorname{supp} L(t, w, \cdot) \subset G\}\ .$$

Therefore

$$(T_t^{0,G}1)(x) = Q_{t,x}[\mu\colon \operatorname{supp} \mu \subset G]\ .$$

Moreover from the regularity of every boundary point of G we conclude that

$$P_x[\tau_G > t] = P_x[x(s) \in \bar{G} \text{ for } 0 \le s \le t]\ .$$

Hence for each $t > 0$

$$Q_{t,x}[\mu\colon \operatorname{supp} \mu \subset G] = Q_{t,x}[\mu\colon \mu(\bar{G}) = 1]\ .$$

If we denote by $\mathcal{M}_{\bar{G}} = \{\mu\colon \mu(\bar{G}) = 1\}$, then the exponential rate of decay of

$$\sup_x Q_{t,x}[\mathcal{M}_{\bar{G}}]$$

is the principal eigenvalue $\lambda_G(0)$. The supremum over x is rather harmless due to the compactness of the region $\bar{G}$. Fixing x for the moment the mass of $Q_{t,x}$ decays exponentially at a rate $I(\mu)$ locally around the measure $\mu \in \mathcal{M}_{\bar{G}}$. The rate at which $Q_{t,x}[\mathcal{M}_{\bar{G}}]$ decays is of course the slowest local rate. Keeping track of the signs

(2.1) $$\lambda_G(0) = -\inf_{\mu \in \mathcal{M}_{\bar{G}}} I(\mu) .$$

Comparison with formula (1.5) suggests that μ should be identified with the density $f^2(x)$. If we denote $f^2(x)$ by $g(x)$ then

$$\frac{1}{2}\int \langle a\nabla f, \nabla f\rangle dx = \frac{1}{8}\int \frac{\langle a\nabla g, \nabla g\rangle}{g} dx .$$

If we define $I(\mu)$ by

$$I(\mu) = \frac{1}{8}\int \frac{\langle a\nabla g, \nabla g\rangle}{g} dx$$

whenever μ has a density g with the integral above existing and finite, and by $+\infty$ in all other cases then the variational formula (2.1) is the same as (1.5).

Let us now consider the case of an arbitrary function $V(x)$.

We see from the definition that

$$\|T_t^{V,G}\| = \sup_{x \in \bar{G}} E^{Q_{t,x}}\left[\exp\left[t\int V(x)\mu(dx)\right]; \mathcal{M}_{\bar{G}}\right]$$

since the $Q_{t,x}$ measure decays locally at an exponential rate determined by $I(\mu)$, a little reflection indicates that the rate of decay of the integral on the right involves a fight between $\exp\left[t\int V(x)\mu(dx)\right]$ and $\exp[-tI(\mu)]$. It is therefore natural to guess that

(2.2) $$\lim_{t\to\infty} \frac{1}{t} \log \|T_t^{V,G}\| = \lambda_G(V) = \sup_{\mu \in \mathcal{M}_{\bar{G}}} \left[\int V(x)\mu(dx) - I(\mu)\right] .$$

This of course is just the same as formula (1.5).

To see what happens in the general case one must have an independent characterization of $I(\mu)$. If we can derive it by other means then formula (2.2) will hold.

§ 3. The general case

We saw in earlier works [1], [2] that for a fairly general Markov process with transition probabilities $p(t, x, dy)$ and infinitesimal generator L, one should define $I(\mu)$ for probability measures μ on the state space by the formula

$$I(\mu) = -\inf_{\substack{u \in \mathcal{D}(L) \\ u > 0}} \int \left(\frac{Lu}{u}\right)(x)\mu(dx) .$$

In our context we can afford to take μ to have compact support and we can therefore define

$$I(\mu) = - \inf_{\substack{u \in C^\infty(R^n) \\ u>0}} \int \left(\frac{Lu}{u}\right)(x)\mu(dx) .$$

The first theorem is the computation of $I(\mu)$ in case μ has a density $\varphi(x)$ with respect to the Lebesgue measure which is an infinitely differentiable function with compact support. In such a case we shall denote $I(\mu)$ by $I(\varphi)$.

Theorem 1. *Let L be given by (1.1). Let $\varphi(x)$ be an infinitely differentiable nonnegative function having compact support on R^n. Then*

$$I(\varphi) = - \inf_{\substack{u \in C^\infty(R^n) \\ u>0}} \int \left(\frac{Lu}{u}\right)(x)\varphi(x)dx$$

$$(3.1) \qquad = \frac{1}{8}\int \frac{\langle a\nabla\varphi, \nabla\varphi\rangle}{\varphi}dx + \frac{1}{2}\int [\langle ba^{-1}b\rangle + \nabla\cdot b]\varphi dx$$

$$- \frac{1}{2} \inf_{F \in C^\infty(R^n)} \int (b - a\nabla F)a^{-1}(b - a\nabla F)\varphi dx .$$

Remark 1. This theorem is nothing more than a computation. We first replace u by $u(x) = \exp[U(x)]$ where $U(x)$ is an arbitrary C^∞ function. We then formally replace $U(x)$ by $\frac{1}{2}\log\varphi(x) - F(x)$ and carry out the variation over C^∞ functions $F(\cdot)$. Simply algebraic manipulations and an integration by parts on one of the terms leads to the answer. To justify it we must take care of the term $\frac{1}{2}\log\varphi(x)$. We replace it by $\frac{1}{2}\log(\varepsilon + \varphi(x))$ where $\varepsilon > 0$ is arbitrary and then let $\varepsilon \to 0$. The details can be found in [3].

Remark 2. If b were zero identically then the formula for $I(\varphi)$ reduces to

$$I(\varphi) = \frac{1}{8}\int \frac{\langle a\nabla\varphi, \nabla\varphi\rangle}{\varphi}dx$$

which coincides with the formula in § 2.

Remark 3. If $b = a\nabla F$ for some C^∞ function $F(x)$ on R^n, then

$$I(\varphi) = \frac{1}{8}\int \frac{\langle a\nabla\varphi, \nabla\varphi\rangle}{\varphi}dx + \frac{1}{2}\int [\langle ba^{-1}b\rangle + \nabla\cdot b]\varphi dx ,$$

In such a case the operator L is self adjoint with respect to the measure exp $[F(x)]dx$ and $I(\varphi)$ is again the Dirichlet norm of $\sqrt{\varphi}$ with respect to this measure.

Remark 4. The last term therefore is a correction depending on how non-self adjoint L is. We note that the last term is of one sign only and one gets

$$\frac{1}{8}\int \frac{\langle a\nabla\varphi, \nabla\varphi\rangle}{\varphi}dx + \int [\nabla \cdot b]\varphi dx$$
$$\leq I(\varphi) \leq \frac{1}{8}\int \frac{\langle a\nabla\varphi, \nabla\varphi\rangle}{\varphi}dx + \frac{1}{2}\int [\langle ba^{-1}b\rangle + \nabla \cdot b]\varphi x \,.$$

The next theorem is the variational formula:

Theorem 2. *For any smooth region G and any smooth function $V(x)$ on R^n*

$$\lambda_G(V) = \sup_{\substack{\varphi \in C^\infty \\ \varphi = 0 \text{ off } G \\ \int \varphi dx = 1}} \left[\int V(x)\varphi(x)dx - I(\varphi)\right]. \tag{3.2}$$

Remark 5. This formula is clearly a generalization of formula (1.5).

Remark 6. One consequence of Theorem 2 and Remark 4 is that if two operators L_1 and L_2 differ only by a first order term then the corresponding principal eigenvalues $\lambda_G^1(V)$ and $\lambda_G^2(V)$ differ by an amount which can be estimated in terms of the coefficients involved but independent of the V.

Remark 7. The proof of Theorem 2 uses only the definition $I(\varphi)$. It is proved using the definition and a mini-max type theorem. The details can be found in [3].

§ 4. An inverse problem

One can now investigate the conditions on two operators L_1 and L_2 in order that $\lambda_G^1(V) \equiv \lambda_G^2(V)$ for all regions G and all functions $V(\cdot)$.

We first have

Theorem 3. *Let L_1, L_2 be two operators of the type considerd here, $\lambda_G^1(V), \lambda_G^2(V)$ the corresponding principal eigenvalues, and, $I_1(\varphi)$ and $I_2(\varphi)$ the corresponding I-functionals then*

$$\lambda_G^1(V) = \lambda_G^2(V) \qquad \textit{for all } G \textit{ and } V$$

if and only if

$$I_1(\varphi) = I_2(\varphi) \qquad \textit{for all } \varphi \ .$$

[*Here* G, V *and* φ *are restricted to the relevent classes considered before*].

Remark 8. The first part is a trivial consequence of Theorem 3. The second part is proved by explicitly inverting (3.2) to get a formula for $I(\varphi)$ in terms of $\lambda_G(V)$. The relation between $I(\cdot)$ and $\lambda_G(\cdot)$ is one that exists between a pair of conjugate convex functionals and the inversion is again a theorem of the type mini-max equals max-min.

We can therefore ask the equivalent question: when can $I_1(\varphi) \equiv I_2(\varphi)$ for all C^∞-densities with compact support?

Theorem 4. *We have* $I_1(\varphi) \equiv I_2(\varphi)$ *if and only if one of the two alternatives hold: either there exists a positive harmonic function* $U(x)$ *for* L_1 (*i.e.* $L_1 U \equiv 0$) *such that*

$$(L_2 f)(x) = \left(\frac{L_1(Uf)}{U}\right)(x)$$

or there is a positive invariant density $\Phi(x)$ *for* L_1 (*i.e.* $L_1^* \Phi \equiv 0$) *such that*

$$(L_2 f)(x) = \left(\frac{L_1^*(\Phi f)}{\Phi}\right)(x) \ .$$

This theorem is proved by analysing formula (3.1). One proves successively that for each φ each of three terms must separately be the same. This leads to the conclusion that the two operators can only differ by a first order term. The nature of the difference is then investigated. The details can again be found in [3].

References

[1] Donsker, M. D. and Varadhan, S. R. S., On a variational formula for the principal eigenvalue for operators with maximum principle, Proc. Nat. Acad. Sci. U.S.A., March (1975), 780–783.

[2] ——, Asymptotic evaluation of certain Markov process expectations for large time-III, Comm. Pure Appl. Math., **29** (1976), 389–461.

[3] ——, On the principal eigenvalue of second order elliptic differential operators, Comm. Pure Appl. Math., **29** (1976), 595–621.

COURANT INSTITUTE OF MATHEMATICAL SCIENCES
NEW YORK UNIVERSITY, 251 MERCER STREET
NEW YORK, N.Y. 10012, U.S.A.

Proc. of Intern. Symp. SDE
Kyoto 1976, pp. 49–56

Quality Control and Quasi Variational Inequalities

Avner FRIEDMAN

§ 1. Let $w(t)$ be an m-dimensional Brownian motion and let $\mathcal{F}_t = \sigma(w(s), 0 \leq s \leq t)$. Given functions f, g, consider the cost function

$$J_x(\tau) = E^x\left[g(w(\tau)) + \int_0^\tau f(w(t))\,dt\right] \tag{1.1}$$

where τ is any stopping time with respect to $\mathcal{F}_t$. Let

$$V(x) = \inf_\tau J_x(\tau) \ . \tag{1.2}$$

Under suitable assumptions on f, g the following result is true (see [5], [6]).

Theorem 1. *There exists a unique solution $u(x)$ of the variational inequality*

$$\Delta u(x) + f(x) \geq 0 \qquad \text{for a.a. } x \in R^m \ , \tag{1.3}$$

$$u(x) \leq g(x) \qquad \text{for all } x \in R^m \ , \tag{1.4}$$

$$(\Delta u(x) + f(x))(g(x) - u(x)) = 0 \qquad \text{for a.a. } x \in R^m \ , \tag{1.5}$$

$$u \in C(R^m) \cap W^{2,p}_{\text{loc}}(R^m) \qquad \text{for any } 1 < p < \infty \ ,$$

and

(i) $V(x) = u(x)$,
(ii) $V(x) = J_x(\tau^*)$ *where* τ^* = *hitting time of the set*

$$S = \{x \in R^m \,;\, V(x) = g(x)\} \ .$$

The existence of u is established by P. D. E. methods. The crucial step in establishing (i) and (ii) is based on the following relation, which follows from the strong Markov property:

$$V(x) = \inf_\tau E^x\left[\int_0^\tau f(w(t))\,dt + V(w(\tau))\right] . \tag{1.6}$$

If τ is a stopping time with respect to a family $\mathscr{G}_t$ of σ-fields such that $\mathscr{G}_t \subset \mathscr{F}_t$, $\mathscr{G}_t \neq \mathscr{F}_t$, then (1.6), and consequently also (i), (ii), are not true in general. We may refer to this case as the case where the stopping times τ are based on the *partial observations* $\mathscr{G}_t$.

§2. Theorem 1 extends to the case where $w(t)$ is replaced by a general Markov process. In this talk we consider a Markov process $(\theta(t), x(t))$, where $\theta(t)$ is a jump Markov process with n states $1, 2, \cdots, n$, and

$$x(t) = x + w(t) + \int_0^t g(\theta(s))ds \, , \tag{2.1}$$

where $g(i) = \lambda_i$, λ_i a given m-vector, and $\lambda_i \neq \lambda_j$ if $i \neq j$. The stopping times will henceforth be taken with respect to $\mathscr{F}_t = \sigma(w(s), 0 \leq s \leq t)$ instead of with respect to $\mathscr{M}_t = \sigma(w(s), \theta(s), 0 \leq s \leq t)$. Thus we have here a stopping time problem with partial observation.

If we designate by $P_x^{\lambda_i}$ the Brownian motion with drift λ_i, then the process $(\theta(t), x(t))$ is called *random evolution* of the processes $P_x^{\lambda_1}, \cdots, P_x^{\lambda_n}$ as dictated by the Markov process $\theta(t)$. Note that whereas $(\theta(t), x(t))$ is a Markov process, the process $x(t)$ alone is not a Markov process.

In the following sections we shall describe a model problem which leads to the introduction of a certain cost function based on $\mathscr{F}_t$ stopping times, and then study this problem in some detail. The results are taken from recent work by Anderson and Friedman [1], [2], [3].

§3. In this section we describe a model arising in the theory of Quality Control.

Suppose a machine can be in n different states $1, \cdots, n$. When it is in state i, it manufactures a product $P_x^{\lambda_i}$. The machine shifts from position i to position j in accordance with the law of the Markov process $\theta(t)$, and the state n is absorbing in some sense; for instance

$$q_{i,n} > 0 \quad \text{if} \quad 1 \leq i \leq n-1, \; q_{n,n} = 0 \tag{3.1}$$

where $(q_{i,j})$ is the infinitesimal matrix of $\theta(t)$.

The manufacturer observes the product but he is unable to deduce from this observation the state of the machine (i.e., he observes $\mathscr{F}_t$, not $\mathscr{M}_t$).

Now, each product has a different worth and this is measured (after a normalization) by the fact that manufacturing $P_x^{\lambda_i}$ entails a cost c_{i-1} with $c_0 = 0$, $c_j > 0$ if $j > 0$. Suppose for definiteness that $c_{n-1} > c_{j-1}$ if $j < n$. Then we can view the machine as being partially inefficient in states i, $1 < i < n$, and being broken when in state n. Thus the manufacturer would like to make, from time to time, inspections of the state of the machine; if he discovers that the machine has gone haywire (i.e., it is in state n) then he stops production.

Of course, an inspection entails a cost, given by K_i if in the preceding inspection it was determined that the machine was at state i.

After the first inspection τ_1 the total cost is

$$E^{i,x}\left[K_i + \int_0^{\tau_1} f(\theta(s))ds\right] \qquad \text{where } f(j) = c_{j-1} ,$$

and where $i = \theta(0)$, $x = x(0)$. In the event that $\theta(\tau_1) \neq n$ the manufacturer continues to run the machine until the second inspection, etc.

To write down a formula for the cost entailed by a sequence of inspection $\{\tau_j\}$ we first define what we mean by such a sequence.

Let ϕ be the shift operator. An $\mathscr{F}_t$ stopping time σ is said to be *regular* if $\sigma(\omega_t) \to \sigma(\omega_s)$ when $t \downarrow s$, where $\omega_t = \phi_t\omega$. A *sequence of inspections* $\tau = (\tau_1, \tau_2, \cdots, \tau_j, \cdots)$ is given by

$$\tau_{j+1} = \tau_j + \sum_{l=1}^{n-1} I_{\theta(\tau_j)=l}\sigma_{j+1,l}(\phi_{\tau_j}) \qquad (j \geq 1) \tag{3.2}$$

where τ_1 and the $\sigma_{i,l}$ are regular stopping times.

We can now write the total cost per a sequence $\tau = \{\tau_j\}$:

$$\begin{aligned} J_x^i(\tau) &= E^{i,x}\left[K_i + \sum_{j=1}^{n-1} K_j \sum_{l=1}^{\infty} I_{\theta(\tau_l)=j}\right] \\ &\quad + E^{i,x}\left[\int_0^{\tau_1} f(\theta(s))ds + \sum_{j=1}^{n-1}\sum_{l=1}^{\infty} I_{\theta(\tau_l)=j}\int_{\tau_l}^{\tau_{l+1}} f(\theta(s))ds\right] . \end{aligned} \tag{3.3}$$

Problem. Characterize

$$V^i(x) = \inf J_x^i(\tau)$$

and find optimal sequences τ_*^i of inspection, i.e., τ_*^i such that $V^i(x) = J_x^i(\tau_*^i)$.

Remark. It can be shown that the $V^i(x)$ do not depend on x, i.e.,

$$V^i(0) = \inf J_x^i(\tau) .$$

§ 4. In order to solve the above quality control problem, we introduce, for each i, the process

$$y_{i,j}(t) = \frac{P^{i,x}[\theta(t) = j \mid \mathscr{F}_t]}{P^{i,x}[\theta(t) = i \mid \mathscr{F}_t]} \qquad (j = 1, 2, \cdots, n) . \tag{4.1}$$

More generally, we define probabilities

$$Q_i^{x,Y_1,\cdots,Y_n} = \sum_{l=1}^{n} \frac{Y_l}{Y_1 + \cdots + Y_n} P^{l,x} \tag{4.2}$$

where $Y_i \equiv 1$ and $Y_j \geq 0$ $(j \neq i)$

and processes

$$y_{i,j}(Y_1, \cdots, Y_n, t) = \frac{Q_i^{x,Y_1,\cdots,Y_n}[\theta(t) = j \mid \mathscr{F}_t]}{Q_i^{x,Y_1,\cdots,Y_n}[\theta(t) = i \mid \mathscr{F}_t]} . \tag{4.3}$$

Thus $y_{i,j}(Y_1, \cdots, Y_n, t) = y_{i,j}(t)$ if $Y_1 = \cdots = Y_{i-1} = Y_{i+1} = \cdots = Y_n = 0$. The $y_{i,j}$ can be computed explicitly in terms of $(q_{i,j})$ and the λ_i.

Theorem 2. *The process*

$$(y_{i,1}(Y_1, \cdots, Y_n, t), \cdots, y_{i,n}(Y_1, \cdots, Y_n, t)) \tag{4.4}$$

together with the measures

$$Q_i^{0,Y_1,\cdots,Y_n} \tag{4.5}$$

form a strong Markov process. Its generator M_i is given by

$$\begin{aligned} M_i u(y) = {} & \frac{1}{2} \sum_{j,l=1}^{n} (\lambda_j - \lambda_i) \cdot (\lambda_l - \lambda_i) y_j y_l \frac{\partial^2 u}{\partial y_j \partial y_l} \\ & + \sum_{j=1}^{n} \Big[(\lambda_i - \lambda_j) \cdot \lambda_i y_i + \sum_{l=1}^{n} (q_{l,j} - q_{l,i} y_j) y_l \\ & + y_j(\lambda_j - \lambda_i) \cdot \frac{\lambda_1 y_1 + \cdots + \lambda_n y_n}{y_1 + \cdots + y_n} \Big] \frac{\partial u}{\partial y_j} \end{aligned} \tag{4.6}$$

where $y = (y_1, \cdots, y_n)$ and $y_i \equiv 1$.

We can express the terms appearing in $J_x^i(\tau)$ by means of the y-process. We mention, for illustration, some of the formulas used in this connection:

$$E^{i,x}\left[\int_0^\sigma f(\theta(s)) ds\right] = E^{i,x}\left[\int_0^\sigma \frac{\sum_{j\neq i} c_{j-1} y_{i,j}(s)}{1 + \sum_{j \neq i} y_{i,j}(s)} ds\right] , \tag{4.7}$$

$$\begin{aligned} & E^{l,x}[I_{\theta(\tau_1)=i} I_{\theta(\tau_2)=j} h(x(\tau_2))] \\ & \quad = E^{l,x}[I_{\theta(\tau_1)=i} E^{i,x(\tau_1)}[I_{\theta(\sigma_2)=j} h(x(\sigma_2))]] \end{aligned} \tag{4.8}$$

where $\tau_2 = \tau_1 + \sigma_2(\phi_{\tau_1})$, τ_1 is $\mathscr{M}_t$ stopping time, σ_2 is an $\mathscr{F}_t$ stopping time and $h(x)$ is any bounded continuous function,

$$(4.9) \qquad E^{l,x}\left[I_{\theta(\tau_1)=i}\int_{\tau_1}^{\tau_2} f(\theta(s))ds\right] = E^{l,x}\left[I_{\theta(\tau_1)=i}E^{i,x(\tau_1)}\left[\int_0^{\sigma_2} f(\theta(s))ds\right]\right]$$

where τ_1, σ_2, τ_2 are as in (4.8).

§5. The above results are used in proving the following theorem.

Theorem 3. *Let $u^i(y)$ be functions in $W^{2,\infty}_{\text{loc}}(A_i) \cap C(\bar{A}_i)$ where $A_i = \{y \in R^n, y_1>0, \cdots, y_n>0 \text{ and } y_i \equiv 1\}$, satisfying the system of quasi variational inequalities*

$$(5.1) \qquad M_i u^i(y) + \frac{\sum_{j=1}^{n-1} c_j y_{j+1}}{\sum_{j=1}^{n} y_j} \geq 0 \qquad \text{a.e. in } A_i\,,$$

$$(5.2) \qquad u^i(y) \leq K_i + \frac{\sum_{j=1}^{n-1} y_j u^j(e_j)}{\sum_{j=1}^{n} y_j} \qquad \text{in } A_i$$

where $e_j = (0, \cdots, 0, 1, 0, \cdots, 0)$ with 1 in the j-th place, and

$$(5.3) \qquad \left(M_i u^i(y) + \frac{\sum_{j=1}^{n-1} c_j y_{j+1}}{\sum_{j=1}^{n} y_j}\right)\left(K_i + \frac{\sum_{j=1}^{n-1} y_j u^j(e_j)}{\sum_{j=1}^{n} y_j} - u^i(y)\right) = 0 \qquad \text{a.e. in } A_i\,.$$

Let

$$(5.4) \qquad S_i = \left\{y \in A_i\,;\, u^i(y) = K_i + \frac{\sum_{j=1}^{n-1} y_j u^j(e_j)}{\sum_{j=1}^{n} y_j}\right\}.$$

Then

$$(5.5) \qquad V^i(0) = \inf_\tau J^i_x(\tau) = u^i(e_i)$$

and the optimal sequence of inspections is given by (3.2) where $\sigma_{j,l}$ is the hitting time $\hat{\tau}_l$ of the set S_l by the process $(y_{l,1}(t), \cdots, y_{l,n}(t))$, and $\tau_1 = \hat{\tau}_i$ for the cost $J^i_x(\tau)$.

Each M_i is an elliptic operator in A_i with rank equal to the dimension of the space spanned by

$$\lambda_j - \lambda_i\,, \qquad j = 1, 2, \cdots, n\,.$$

Thus, if $m < n-1$ then the M_i degenerate throughout A_i, whereas if $m \geq n-1$ then when the λ_j are in "general position", the M_i do not degenerate in A_i. However the M_i do degenerate on ∂A_i. The degeneracy is of the worst kind, i.e., the normal diffusion vanishes and the Fichera drift points inward. Thus, Dirichlet boundary data should not be prescribed on ∂A_i. The $V^i(0)$ are the values of $u^i(y)$ at the vertex points of the A_i.

If the M_i do not degenerate in A_i, then the assertions of Theorem 3 remain valid assuming that $u^i \in W^{2,2}_{\text{loc}}(A^i)$ (instead of $W^{2,\infty}_{\text{loc}}(A^i)$).

We would like to explain that the concept "quasi variational inequality" used in Theorem 3 means a variational inequality in which the constraint depends on the unknown function (i.e., $u \leq H(u)$ rather than $u \leq g$, for instance).

The known techniques are insufficient at present to prove the existence of a solution u^i of (5.1)–(5.3), in general. Some results in special cases will be mentioned below.

The last remark regarding (5.1)–(5.3) is that when the transition probabilities $p_{ij}(t)$ of $\theta(t)$ are such that $p_{ij}(t) = 0$ if $j < i$, the system (5.1)–(5.3) can be totally decoupled. We then have to solve successively q.v.i. in dimension 1, then 2, etc., up to dimension $n - 1$.

§ 6. We specialize to 2 states with 1-dimensional Brownian motion. The q.v.i. reduces to

$$(6.1)\qquad \begin{gathered} Lu \equiv \frac{1}{2}y^2u'' + \left(\lambda(1+y) + \frac{y^2}{1+y}\right)u' \geq -\frac{\gamma y}{1+y}, \\ u(y) \leq K + \frac{u(0)}{1+y}, \\ \left(Lu + \frac{\gamma y}{1+y}\right)\left(K + \frac{u(0)}{1+y} - u(y)\right) = 0 \end{gathered}$$

for $0 < y < \infty$, where

$$\begin{pmatrix} q_{1,1} & q_{1,2} \\ q_{2,1} & q_{2,2} \end{pmatrix} = \begin{pmatrix} -\alpha & \alpha \\ 0 & 0 \end{pmatrix},$$

$$\lambda = \frac{\alpha}{(\lambda_2 - \lambda_1)^2}, \qquad \gamma = \frac{c}{(\lambda_2 - \lambda_1)^2}.$$

Theorem 4. *There exists a unique bounded solution $u(y)$ in $C^1[0, \infty) \cap C^2[0, b]$ (for some $b > 0$) of (6.1) and $u(y) = K + u(0)/(1 + y)$ if $y \geq b$. The process $y_{1,2}(t)$ is given by*

$$(6.2)\qquad \begin{aligned} y_{1,2}(t) = e^{\alpha t}\int_0^t du\,\alpha e^{-\alpha u} \\ \cdot \exp\left[(\lambda_2 - \lambda_1)(w(t) - w(u)) - \tfrac{1}{2}(\lambda_2^2 - \lambda_1^2)(t - u)\right]. \end{aligned}$$

Thus $V(0) = u(0)$ and the optimal inspections are taken at times $\tau, 2\tau, 3\tau, \cdots$, where τ is the first time t such that $y_{1,2}(t) = b$.

In this special case the y-process was first introduced by Shiryaev [7] and the solution of (6.1) was obtained by Bather [4]. However these papers do not establish rigorously the steps needed for justifying the last assertion of Theorem 4.

§ 7. The approach outlined in the previous sections extends to other types of Markov processes (instead of Brownian motion). The case when the products $P_x^{\lambda_i}$ (of the machine) are Markov chains (and θ is a Markov chain) was studied in [3]. We shall give here an example when the number of states is two and the products are Poisson processes $P_x^{\lambda_1}, P_x^{\lambda_2}$ with parameters λ_1 and λ_2. The space of paths is the space $D[0, \infty)$ of right continuous functions $x(t)$ having discontinuities of the first kind only. Let

$$Lu = \beta(y)u'(y) + \gamma(y)(u(\lambda y) - u(y)) \tag{7.1}$$

where

$$\beta(y) = (\alpha - \lambda_2 + \lambda_1)y + \alpha \,,$$
$$\gamma(y) = \frac{\lambda_1 + \lambda_2 y}{1 + y}\,, \qquad \lambda = \frac{\lambda_2}{\lambda_1}\,.$$

Consider the q.v.i.

$$\begin{aligned} &Lu + \frac{cy}{1+y} \geq 0 \\ &u(y) \leq K + \frac{u(0)}{1+y} \\ &\left(Lu + \frac{cy}{1+y}\right)\left(K + \frac{u(0)}{1+y} - u(y)\right) = 0\,, \end{aligned} \tag{7.2}$$

for $y > 0$.

Theorem 5. *If $\lambda_2 > \lambda_1$, $\alpha - \lambda_2 + \lambda_1 > 0$ then there exists a unique solution u in $C^1[0, \infty)$ of (7.2) and $u(y) = K + u(0)/(1 + y)$ if and only if $y > b$, for some $b > 0$. Further, setting*

$$\begin{aligned} y(t) = e^{\alpha t} &\int_0^t du\alpha e^{-\alpha u} \\ &\cdot \exp\left[ln\frac{\lambda_2}{\lambda_1}(x(t) - x(u)) - (\lambda_2 - \lambda_1)(t - u)\right], \end{aligned} \tag{7.3}$$

we have $V(0) = u(0)$ and the optimal sequence of inspections is $\tau, 2\tau, 3\tau, \cdots$ where τ is the first time such that $y(t) = b$.

The existence of a solution of (7.2) and the second assertion of Theorem 5 can be established also in case $\lambda_2 < \lambda_1$, provided $(\lambda_2/\lambda_1) > 1/2$ and

$$\lambda_1 - \lambda_2 < (2\lambda - 1)\lambda\alpha .$$

References

[1] Anderson, R. F. and Friedman, A., A quality control problem and quasi variational inequalities, Arch. Rational Mech. Anal., 63 (1977), 205–252.

[2] ——, Multi-dimensional quality control problem and quasi variational inequalities, Trans. Amer. Math. Soc. to appear.

[3] ——, Quality control for Markov chains and free boundary problems, to appear.

[4] Bather, J. A., On a quickest detection problem, Ann. Math. Statist., 38 (1967), 711–724.

[5] Bensoussan, A. and Lions, J. L., Problèmes de temps d'arrêt optimal et inéquations varietionelles paraboliques, Applicable Anal., 3 (1973), 267–294.

[6] Friedman, A., Stochastic differential equations and applications, 2, Academic Press, New York, 1976.

[7] Shiryaev, A. N., On optimum methods in quickest detection problems, Theor. Probality Appl., 8 (1963), 22–46.

DEPARTMENT OF MATHMATICS
COLLEGE OF ARTS AND SCIENCES
NORTHWESTERN UNIVERSITY
EVANSTON, ILLINOIS 60201, U.S.A.

Proc. of Intern. Symp. SDE
Kyoto 1976, pp. 57-74

On Stochastic Integrals with Respect to an Infinite Number of Brownian Motions and its Applications

Masuyuki HITSUDA and Hisao WATANABE

§ 1. Introduction

In this paper, we will give a definition of stochastic integrals with respect to an infinite number of Brownian motions. Here, a vector composing of an infinite number of Brownian motions, say, $B(t) = (B^{(1)}(t), \cdots, B^{(n)}(t), \cdots)$ will not necessarily be an element of some Hilbert space and will be considered only as an element of R^∞. In some situations, such a consideration seems to be more convenient for applications. In fact, we will use such stochastic integrals in a forthcoming paper [3]. In § 2, we define stochastic integrals. In § 3, we will describe formulas concerning stochastic differentials based upon $\{B(t), t \in [0, 1]\}$. In § 4, we will state a generalization of a theorem of Girsanov [1] which may be interesting to compare with the result of Ouvrard [9]. In § 5, we discuss orthogonal development of L^2-functionals of $B(t) = (B^{(1)}(t), \cdots, B^{(n)}(t), \cdots)$ in series of Fourier-Hermite functionals and also its representations. In § 6, representation theorems of martingales with respect to the σ-field generated by an infinite number of Brownian motions and also by more general Gaussian processes will be discussed.

This paper includes the complementary results to our joint paper [3]. Therefore, results are described in the convenient form for our use in [3]. So that, we don't pursue the final generalizations.

§ 2. Itô Integrals with respect to an infinite number of Brownian motions

Let $(\Omega, \mathscr{F}, P)$ be a probability space and $\{\mathscr{F}_t, t \in [0, 1]\}$ be a system of increasing sub-σ-fields. In this paper, by a Brownian motion $B(t)$ we mean a vector of infinite independent Gaussian processes with independent increments, that is to say, $B(t) = (B^{(1)}(t), B^{(2)}(t), \cdots, B^{(n)}(t), \cdots)$. So that, $(B^{(1)}(t), \cdots, B^{(n)}(t), \cdots)$ may be considered as an element in R^∞. Let $\mathscr{F}_t(B) = \sigma(B^{(1)}(s), \cdots, B^{(n)}(s), \cdots; s \leq t)$, where $\sigma(\,\cdot\,)$ will denote the σ-field generated by $\cdot$. We assume that $E(dB^{(i)}(t)) = 0$ $(i = 1, 2, \cdots)$ and $E((dB^{(i)}(t))^2) = dm^i(t)$, where $m^1(t)$ is absolutely continuous with respect to the Lebesgue measure and $m^{i+1}(t)$ is absolutely continuous with respect to $m^i(t)$ for $i = 1, 2, \cdots$. We assume that if we put $\mathcal{N}_t(B) = \sigma\{B^{(i)}(u) - B^{(i)}(v),\ u, v \geq t,\ i = 1, 2, \cdots\}$, $\mathcal{N}_t(B)$ and $\mathcal{F}_t$ are independent and $\mathcal{F}_t(B) \subset \mathcal{F}_t$ for any t. Therefore, we may consider that $B(t)$ is a Brownian motion adapted to $\{\mathcal{F}_t\}$.

In this paper, we put $L^2[0, 1]$[1] $= \{f = (f^1, f^2, \cdots);\ f^i(t)$ are measurable and $\sum_{i=1}^{\infty} \int_0^1 (f^i(s))^2 dm^i(s) < \infty\}$.

Definition 1.1. A vector function $f(t, \omega) = (f^1(t, \omega), \cdots, f^n(t, \omega), \cdots)$ on $[0, 1] \times \Omega$ will be called *an element of the class M* if the following conditions are satisfied:

(M.1) $f^i(t, \omega)$'s are measurable in (t, ω) on $([0, 1] \times \Omega, \tau \otimes \mathcal{F})$, where τ is the σ-field generated by intervals in $[0, 1]$,

(M.2) $f^i(t, \omega)$ are $\mathcal{F}_t$-measurable,

(M.3) $f(t, \omega) \in L^2[0, 1]$, with probability 1, namely,

$$P\Big(\sum_{i=1}^{\infty} \int_0^1 (f^i(t, \omega))^2 dm^i(t) < \infty\Big) = 1 .$$

We will define stochastic integrals for elements in the class M. For this purpose, at first, we define stochastic integrals for a rather restricted class.

Namely, we assume that in place of (M.3)

(S.3) $$\sum_{i=1}^{\infty} \int_0^1 E((f^i(s, \omega))^2) dm^i(s) < \infty .$$

Now, we put $f^n = (f^1, \cdots, f^n, 0, 0, \cdots)$ for $f = (f^1, \cdots, f^n, \cdots)$ satisfying (M.1), (M.2) and (M.3). Then, we can define

$$I(t, \omega, f^n) = \sum_{i=1}^{n} \int_0^t (f^i(s, \omega)) dB^{(i)}(\omega) .$$

as in K. Itô [4]. Also, we have for every c with $0 < c < \infty$,

$$\begin{aligned} c^2 P\Big(\sup_{0 \leq t \leq 1} |I(t, \omega; f^n - f^m)| \geq c\Big) &\leq \sum_{i=n+1}^{m} E\Big(\int_0^1 (f^i(s, \omega))^2 dm^i(s)\Big) \\ &\to 0 \quad (n, m \to \infty) . \end{aligned}$$

---

[1] In the following, we assume that $0 \leqq t \leqq 1$, for simplicity. $[0, 1]$ may be replaced by any finite interval $0 \leqq t \leqq T < \infty$.

Therefore, we can easily see that $I(t, \omega, f^{n_k})$ converges uniformly with respect to t for some subsequence $\{n_k\}$ with probability 1, thus, we can define, the stochastic integral for f, $I(t, \omega, f)$ as $\lim_{k\to\infty} I(t, \omega, f^{n_k})$ and we shall write $I(t, \omega, f) = \sum_{i=1}^{\infty} \int_0^t f^i(s, \omega) dB^{(i)}(s)$ for t with $0 \leqq t \leqq 1$.

Secondly, we extend the definition of stochastic integrals for the class M. Take an $f \in M$. Put

$$\tau_N(\omega) = \begin{cases} \inf\left\{t \leq 1\,;\ \sum_{i=1}^{\infty} \int_0^1 (f^i(s, \omega))^2 dm^i(s) \geq N\right\} \\ 1, \qquad \text{if } \sum_{i=1}^{\infty} \int_0^1 (f^i(s, \omega))^2 dm^i(s) < N\,. \end{cases}$$

and

$$f_N(s, \omega) = f(s, \omega) \chi_{\{s \leqq \tau_N(\omega)\}}\,,$$

where χ_A is the indicator function of a set A.

Then, $f_N(s, \omega)$ satisfies (M.1), (M.2) and (S.3). Therefore, we can define the stochastic integral

$$\sum_{i=1}^{\infty} \int_0^t f_N^i(s, \omega) dB^{(i)}(s, \omega)\,.$$

Since $f_N(t, \omega) = f(t, \omega)$, $0 \leqq t \leqq 1$ for all sufficiently large N, with probability 1, by (M.3), we can define the limit in probability

$$\sum_{i=1}^{\infty} \int_0^t f^i(s, \omega) dB^{(i)}(s, \omega) = \lim_{N\to\infty} \sum_{i=1}^{\infty} \int_0^t f_N^i(s, \omega) dB^{(i)}(s, \omega)\,.$$

We will state several properties of stochastic integrals.

(i) $\sum_{i=1}^{\infty} \int_s^t f^i(\tau, \omega) dB^{(i)}(\tau, \omega)$ is continuous in s, t with probability 1 for f satisfying (M.1)–(M.3).

(ii) If $f_1(t, \omega)$ and $f_2(t, \omega)$ satisfy (M.1), (M.2) and (M.3), then $af_1(t, \omega) + bf_2(t, \omega)$ (a, b are arbitrary constants) satisfies (M.1), (M.2) and (M.3), and it holds

$$\sum_{i=1}^{\infty} \int_s^t (af_1^i(\tau, \omega) + bf_2^i(\tau, \omega)) dB^{(i)}(\tau, \omega)$$
$$= a \sum_{i=1}^{\infty} \int_s^t f_1^i(\tau, \omega) dB^{(i)}(\tau, \omega) + b \sum_{i=1}^{\infty} \int_s^t f_2^i(\tau, \omega) dB^{(i)}(\tau, \omega)$$

for every s, t such that $0 \leqq s, t \leqq 1$ with probability 1.

(iii) If each of $f_n(t, \omega)$ satisfies (M.1), (M.2) and (M.3) and f satisfies

(M.1) and (M.2) and if the sequence $\{f_n(t, \omega), n = 1, 2, \cdots\}$ converges in probability to f in L^2, namely, for every $\varepsilon > 0$,

$$\lim_{n\to\infty} P\left(\sum_{i=1}^{\infty}\int_0^1 |f_n^i(\tau, \omega) - f^i(\tau, \omega)|^2\, m^i(d\tau) \geq \varepsilon\right) = 0\,,$$

then

$$\sup_{0\leqq u\leqq v\leqq t}\left|\sum_{i=1}^{\infty}\int_u^v (f_n^i(\tau, \omega) - f^i(\tau, \omega))dB^{(i)}(\tau, \omega)\right|$$

converges to 0 in probability.

The proof of properties (i)–(iii) is done by usual methods, therefore, we omit it.

§ 3. Itô's formula concerning stochastic differentials

Let $a^i(t, \omega)$ $(i = 1, 2, \cdots)$ and $(b_j^i(t, \omega), (i, j = 1, 2, \cdots))$ satisfy the properties (M.1) and (M.2). We assume that

$$P\left(\int_0^1 |a^i(\tau, \omega)|\, dm^i(\tau) < \infty\right) = 1 \qquad (i = 1, 2, \cdots) \tag{3.1}$$

and

$$P\left(\sum_{j=1}^{\infty}\int_0^1 |b_j^i(\tau, \omega)|^2\, dm^j(\tau) < \infty\right) = 1 \quad (i = 1, 2, \cdots) \tag{3.2}$$

Then, the following stochastic processes are defined

$$\xi^i(s, \omega) - \xi^i(t, \omega) = \int_t^s a^i(\tau, \omega)dm^i(\tau) + \sum_{j=1}^{\infty}\int_t^s b_j^i(\tau, \omega)dB^{(j)}(\tau, \omega)$$

for s, t with $0 \leqq t \leqq s \leqq 1$ and for $i = 1, 2, \cdots$.

Furthermore, if

$$P\left(\sum_{i=1}^{\infty}\int_0^1 |a^i(\tau, \omega)|\, dm^i(\tau) < \infty\right) = 1 \tag{3.3}$$

and

$$P\left(\sum_{i=1}^{\infty}\sum_{j=1}^{\infty}\int_0^1 |b_j^i(\tau, \omega)|^2\, dm^j(\tau) < \infty\right) = 1\,, \tag{3.4}$$

then, by the property (iii) of stochastic integrals in § 2, $\sum_{i=1}^{\infty} \xi^i(\tau)$ converges with probability 1 and consequently,

$$P\Big(\sum_{i=1}^{\infty} (\xi^i(\tau))^2 < \infty\Big) = 1 .$$

Proposition 3.1. *Assume that $a^i(t,\omega)$ $(i = 1, 2, \cdots)$ and $b^i_j(t,\omega)$, $(i, j = 1, 2, \cdots)$ satisfy* (M.1), (M.2), (3.1) *and* (3.2). *Then the right hand side of the following equation have the meaning and it holds*

(3.5)
$$\begin{aligned}
&(\xi^i(s) - \xi^i(t))(\xi^j(s) - \xi^j(t)) \\
&= \int_t^s (\xi^i(\tau) - \xi^i(t))a_j(\tau)\,dm^j(\tau) + \int_t^s (\xi^j(\tau) - \xi^j(t))a_i(\tau)dm^i(\tau) \\
&\quad + \sum_{k=1}^{\infty} \int_t^s b^i_k(\tau)b^j_k(\tau)dm^k(\tau) \\
&\quad + \sum_{k=1}^{\infty} \int_t^s (\xi^i(\tau) - \xi^i(t))b^j_k(\tau)dB^{(k)}(\tau) \\
&\quad + \sum_{k=1}^{\infty} \int_t^s (\xi^j(\tau) - \xi^j(t))b^i_k(\tau)dB^{(k)}(\tau) .
\end{aligned}$$

Proof. Let $d\xi^i_N(\tau) = a_i(t)dm^i(t) + \sum_{j=1}^N b^i_j(t)dB^{(j)}(t)$, for $i = 1, 2, \cdots$. Then, we have, by K. Itô [5],

(3.6)
$$\begin{aligned}
&(\xi^i_N(s) - \xi^i_N(t))(\xi^j_N(s) - \xi^j_N(t)) \\
&= \int_t^s (\xi^i_N(\tau) - \xi^i_N(t))a_j(\tau)dm^j(\tau) + \int_t^s (\xi^j_N(\tau) - \xi^j_N(t))a_i(\tau)dm^i(\tau) \\
&\quad + \sum_{k=1}^{N} \int_t^s b^i_k(\tau)b^j_k(\tau)dm^k(\tau) \\
&\quad + \sum_{k=1}^{N} \int_t^s (\xi^i_N(\tau) - \xi^i_N(t))b^j_k(\tau)dB^{(k)}(\tau) \\
&\quad + \sum_{k=1}^{N} \int_t^s (\xi^j_N(\tau) - \xi^j_N(t))b^i_k(\tau)dB^{(k)}(\tau) .
\end{aligned}$$

Now, we note that

$$(\xi^i_N(s) - \xi^i_N(t)) - (\xi^i(s) - \xi^i(t))) = \sum_{j=N+1}^{\infty} \int_t^s b^i_j(\tau)dB^{(j)}(\tau)$$

and

$$\sum_{j=N+1}^{\infty} \int_t^s (b^i_j(\tau))^2 dm^j(\tau) \to 0 \qquad \text{a.s. } (N \to \infty) .$$

Therefore, by means of the property (iii) of stochastic integrals, by taking some subsequence N_k, we can make $\xi^i_{N_k}(\tau) - \xi^i_{N_k}(t)$ tend to $\xi^i(\tau)$

$-\xi^i(t)$ uniformly on $[t, s]$ a.s.. Therefore, if we put $N_k \equiv N$,

$$\sum_{k=1}^{N}\int_t^s [(\xi_N^i(\tau) - \xi_N^i(t)) - (\xi^i(\tau) - \xi^i(t))]^2 (b_k^j(\tau))^2 (b_k^j(\tau))^2 dm^k(\tau)$$
$$+ \sum_{k=N+1}^{\infty}\int_t^s (\xi^i(\tau) - \xi^i(t))^2 b_k^j(\tau)^2 dm^k(\tau)$$
$$\leqq \max_{t\leqq\tau\leqq s} |(\xi_N^i(\tau) - \xi_N^i(t)) - (\xi^i(\tau) - \xi^i(t))|^2 \sum_{k=1}^{\infty}\int_t^s (b_k^j(\tau))^2 dm^k(\tau)$$
$$+ \max_{t\leqq\tau\leqq s} |\xi^i(\tau) - \xi^i(t)|^2 \sum_{k=N+1}^{\infty}\int_t^s |b_k^j(\tau)|^2\, dm^k(\tau) \to 0 \ \text{ a.s. } (N \to \infty)\ .$$

Hence, each term of the right hand side of (3.6) tends to respective term of the right hand side of (3.5).

Also, we have the following theorem.

Theorem 3.2. (Itô's formula) *Let a process* $\{\xi^i(t, \omega)\,;\, i = 1, 2, \cdots, n\}$ *be defined by*

$$d\xi^i(t, \omega) = a^i(t, \omega) dm^i(t) + \sum_{j=1}^{\infty} b_j^i(t, \omega) dB^{(j)}(t) \qquad (i = 1, 2, \cdots, n)\ ,$$

where $\{a^i(t, \omega)\ i = 1, \cdots, n\}$, $\{b_j^i(t, \omega)\ i = 1, \cdots, n,\ j = 1, \cdots\}$ *satisfy* (M.1), (M.2), (3.1) *and* (3.2). *Let* G *be an open subset of the* n-*dimensional space* R^n *which contains all the points* $\{(\xi^i(t, \omega),\ i = 1, 2, \cdots, n)\}$ *for* $0 \leqq t \leqq 1$, $\omega \in \Omega$. *Let* $F(t, x^1, x^2, \cdots, x^n)$ *be a continuous function defined in* $0 \leqq t \leqq 1$ *and* $(x^1, \cdots, x^n) \in G$ *such that partial derivatives* $F'_t(t, x^1, \cdots, x^n) = (\partial F / \partial m^1(t))(t, x^1, \cdots, x^n)$, $F'_{x^i}(t, x^1, \cdots, x^n) = (\partial F / \partial x^i)\,(t, x^1, \cdots, x^n)$, $(i = 1, 2, \cdots, n)$, $F''_{x^i x^j}(t, x^1, \cdots, x^n) = (\partial^2 F)/(\partial x^i \partial x^j)(t, x^1, \cdots, x^n)$ $(i, j = 1, 2, \cdots, n)$ *are all continuous. Then, the differential of* $Y(t, \omega) = F(t, \xi^1(t, \omega), \cdots, \xi^n(t, \omega))$ *is given by*

(3.7)

$$dY(t, \omega) = F'_t(t, \xi^1, \cdots, \xi^n) dm^1(t) + \sum_{k=1}^{n} F'_{x^k}(t, \xi^1, \cdots, \xi^n)) a^k(t, \omega) dm^k(t)$$
$$+ \sum_{j=1}^{\infty}\Big(\sum_{i=1}^{n} F'_{x^i}(t, \xi^1, \cdots, \xi^n)\cdot b_j^i(t, \omega)\Big) dB^{(j)}(t)$$
$$+ \sum_{k=1}^{\infty}\Big(\frac{1}{2}\sum_{i,j=1}^{n} F''_{x^i, x^j}(t, \xi^1, \cdots, \xi^n) b_k^i(t, \omega) b_k^j(t, \omega)\Big) dm^k(t)\ .$$

Proof. Let $d\xi_N^i(t) = a_i(t) dm^i(t) + \sum_{j=1}^{N} b_j^i(t) dB^{(j)}(t)$ for $i = 1, 2, \cdots, n$. Then, if we put $Y_N(t, \omega) = F(t, \xi_N^1(t, \omega), \cdots, \xi_N^n(t, \omega))$, we have, by K. Itô [5],

(3.8)

$$
\begin{aligned}
dY_N(t,\omega) = {} & F'_t(t, \xi^1_N, \cdots, \xi^n_N)dm^1(t) \\
& + \sum_{k=1}^{n} F'_{x^k}(t, \xi^1_N, \cdots, \xi^n_N)a^k(t)dm^k(t) \\
& + \sum_{j=1}^{N} \left(\sum_{i=1}^{n} F'_{x^i}(t, \xi^1_N, \cdots, \xi^n_N) b^i_j(t) \right) dB^{(j)}(t) \\
& + \sum_{k=1}^{N} \left(\frac{1}{2} \sum_{i,j=1}^{n} F''_{x^i x^j}(t, \xi^1_N, \cdots, \xi^n_N) b^i_k(t) b^j_k(t) \right) dm^k(t) .
\end{aligned}
$$

As in Proposition 3.1, we can prove that $F'_{x^k}(t, \xi^1_N, \cdots, \xi^n_N)$ and $F''_{x^i x^j}(t, \xi^1_N, \cdots, \xi^n_N)$ uniformly (with respect to t) converge to $F'_{x^i}(t, \xi^1, \cdots, \xi^n)$ and $F''_{x^i x^j}(t, \xi, \cdots, \xi^n)$ respectively as $N \to \infty$ a.s..

Since

$$
\begin{aligned}
& \sum_{j=1}^{N} \int_0^1 \left| \sum_{i=1}^{n} (F'_{x^i}(t, \xi^1_N, \cdots, \xi^n_N) - (F'_{x^i}(t, \xi^1, \cdots, \xi^n)) b^i_j(t) \right|^2 dm^j(t) \\
& \qquad + \sum_{j=N+1}^{\infty} \int_0^1 \left| \sum_{i=1}^{n} (t, \xi^1, \cdots, \xi^n) b^i_j(t) \right|^2 dm^j(t) \\
& \leqq \max_{0 \le t \le 1} \max_{1 \le i \le n} |F'_{x^i}(t, \xi^1_N, \cdots, \xi^n_N) \\
& \qquad - F'_{x^i}(t, \xi^1, \cdots, \xi^n)| \sum_{j=1}^{\infty} \sum_{i=1}^{n} \int_0^1 |b^i_j(t)|^2 \, dm^j(t) \\
& \qquad + \max_{0 \le t \le 1} \max_{1 \le i \le n} |F'_{x^i}(t, \xi^1, \cdots, \xi^n)| \sum_{j=N+1}^{\infty} \sum_{i=1}^{n} \int_0^1 |b^i_j(t)|^2 \, dm^j(t) \\
& \to 0 \qquad \text{a.s. } (N \to \infty) ,
\end{aligned}
$$

we have that the third term of the right hand side of (3.8) converges to the corresponding term of (3.7). We can easily show that the other terms of (3.8) also converge to the corresponding terms of (3.7).

We can generalize Theorem 3.2. Let $F(t, x^1, x^2, \cdots)$ be a real valued function on defined on $0 \leqq t \leqq 1$ and $(x^1, x^2, \cdots) \in l^2$. F is said to be *twice differentiable* if there exist continuous partial derivatives $F'_t(t, x) = (\partial F/\partial m^1(t))(t, x)$, $F'_{x^i}(t, x)$ and $F''_{x^i x^j}(t, x)$ defined on $0 \leqq t \leqq 1$ and $x = (x^1, \cdots, x^n, \cdots) \in l^2$ such that $\sum_{i=1}^{\infty} (F'_{x^i})^2$ and $\sum_i \sum_j (F''_{x^i x^j}(t, x^1, \cdots, x^n, \cdots))^2$ are convergent and continuous with respect to $(t, x^1, \cdots, x^n, \cdots) \in [0, 1] \times l^2$.

Theorem 3.3. *Let $F(t, x^1, x^2, \cdots, x^n, \cdots)$ be a twice differentiable real valued function defined on* $[0, 1] \times l^2$. *Let $\xi^i(t)$ be a stochastic process with differential $a^i(t, \omega)dm^i(t) + \sum_{j=1}^{\infty} b^i_j(t, \omega)dB^{(j)}(t, \omega)$ with properties* (3.3), (3.4) *and* $\sum_{i=1}^{\infty} \int_0^1 (a^i(t, \omega))^2 dm^i(t) < \infty$ *with probability* 1.

Then $Y(t,\omega) = F(t, \xi^1(t,\omega), \cdots, \xi^n(t,\omega), \cdots)$ *satisfies*

$$\begin{aligned} dY(t,\omega) &= (F'_t(t, \xi^1, \cdots, \xi^n, \cdots)dm^1(t) \\ &\quad + \sum_{k=1}^{\infty} F'_{x^k}(t, \xi^1, \cdots, \xi^n, \cdots))a^k(t,\omega)dm^k(t) \\ &\quad + \sum_{j=1}^{\infty}\left(\sum_{i=1}^{\infty} F'_{x^i}(t, \xi^1, \cdots, \xi^n, \cdots)b^i_j(t,\omega)\right)dB^{(j)}(t) \\ &\quad + \frac{1}{2}\sum_{k=1}^{\infty}\sum_{i,j=1}^{\infty}(F''_{x^ix^j}(t, \xi^1, \cdots, \xi^n, \cdots)b^i_k(t,\omega)b^j_k(t,\omega))dm^k(t)\,. \end{aligned} \tag{3.9}$$

Proof. Let $F_N(x^1, \cdots, x^N) = F(x^1, \cdots, x^N, 0, \cdots)$ and $Y_N = F_N(t, \xi^1, \cdots, \xi^N, 0, \cdots)$. Then, by Theorem 3.2, we have

$$\begin{aligned} dY_N(t,\omega) &= (F_N)'_t(t, \xi^1, \cdots, \xi^N)dm^1(t) \\ &\quad + \sum_{k=1}^{N}(F_N)'_{x^k}(t, \xi^1, \cdots, \xi^N)a^k(t,\omega)dm^k(t) \\ &\quad + \sum_{j=1}^{\infty}\left(\sum_{i=1}^{N}(F_N)'_{x^i}(t, \xi^1, \cdots, \xi^N)b^i_j(t)\right)dB^{(j)}(t) \\ &\quad + \frac{1}{2}\sum_{k=1}^{\infty}\left(\sum_{i,j=1}^{N}(F_N)''_{x^ix^j}(t, \xi^1, \cdots, \xi^N)b^i_k(t)b^j_k(t)\right)dm^k(t)\,. \end{aligned} \tag{3.10}$$

Since we can prove as in Proposition 3.1 that $\sum_{i=N+1}^{\infty}(\xi^i(t))^2$ uniformly (with respect to $0 \leqq t \leqq 1$) converges to zero with probability one as $N \to \infty$, we can see that $\max_{0\le t\le 1}\sum_{k=1}^{N}|(F_N)'_{x^k}(t, \xi^1, \cdots, \xi^N) - F'_{x^k}(t, \xi^1, \cdots, \xi^N, \xi^{N+1}, \cdots)|^2$ and $\max_{0\le t\le 1}\sum_{i,j=1}^{N}|(F_N)''_{x^ix^j}(t, \xi^1, \xi^N) - F''_{x^ix^j}(t, \xi^1, \cdots, \xi^N, \cdots)|^2$ tends to zero with probability one as $N \to \infty$. Therefore, we have

$$\begin{aligned} &\sum_{j=1}^{\infty}\int_0^t\Bigg|\sum_{i=1}^{N}((F_N)'_{x^i}(s, \xi^1, \cdots, \xi^N) - F'_{x^i}(s, \xi^1, \cdots, \xi^N, \xi^{N+1}, \cdots))b^i_j(s) \\ &\qquad + \sum_{i=N+1}^{\infty}F'_{x^i}(s, \xi^1, \cdots, \xi^N, \xi^{N+1}, \cdots)b^i_j(s)\Bigg|^2 dm^j(s) \\ &\leqq 2\sum_{j=1}^{\infty}\int_0^t\sum_{i=1}^{N}((F_N)'_{x^i}(s, \xi^1, \cdots, \xi^N) - F'_{x^i}(s, \xi^1, \cdots, \xi^{N+1}, \cdots))^2 \\ &\qquad\qquad \times \sum_{i=1}^{N}(b^i_j(s))^2dm^j(s) \\ &\quad + 2\sum_{j=1}^{\infty}\int_0^t\sum_{i=N+1}^{\infty}(F'_{x^i}(s, \xi^1, \cdots, \xi^N, \xi^{N+1}, \cdots))^2\sum_{i=N+1}^{\infty}(b^i_j(s))^2dm^j(s) \\ &\leqq 2\max_{0\le s\le 1}\sum_{i=1}^{N}((F_N)'_{x^i}(s, \xi^1, \cdots, \xi^N) - F'_{x^i}(s, \xi^1, \cdots, \xi^N, \xi^{N+1}, \cdots))^2 \\ &\qquad\qquad \times \sum_{j=1}^{\infty}\sum_{i=1}^{\infty}\int_0^t(b^i_j(s))^2dm^j(s) \end{aligned}$$

$$+ 2 \max_{0 \le s \le 1} \sum_{i=N+1}^{\infty} (F'_{x^i}(s, \xi^1, \cdots, \xi^N, \xi^{N+1}, \cdots))^2$$
$$\times \sum_{j=1}^{\infty} \sum_{i=N+1}^{\infty} \int_0^t (b_j^i(s))^2 dm^j(s)$$
$$\to 0 \qquad (N \to \infty) .$$

Consequently, we can prove that the third term of (3.10) tends to the corresponding term of (3.9). Also, we can show easily that other terms of (3.10) tend with probability one to the corresponding terms of (3.9) as $N \to \infty$.

§ 4. A generalization of Girsanov's theorem

Let $\{f^i(t, \omega) ; i = 1, 2, \cdots, 0 \le t \le 1\}$ satisfy (M.1), (M.2) and (M.3). Also, let $\{B^{(i)}(t) ;\ i = 1, 2, \cdots, 0 \le t \le 1\}$ be independent Brownian motions defined in § 1. Let M_t $(0 \leqq t \leqq 1)$ be an $\{\mathscr{F}_t\}$-adapted process of the form

$$M_t = \exp\left\{\left(\sum_{i=1}^{\infty} \int_0^t f^i(s, \omega) dB^{(i)}(s, \omega)\right.\right.$$
$$\left.\left. - \frac{1}{2} \sum_{i=1}^{\infty} \int_0^t (f^i(s, \omega))^2 dm^i(s)\right)\right\} .$$

Theorem 4.1. *Assume that $\{dm^i(t) ; i = 1, 2, \cdots\}$ is a system of measures satisfying conditions in § 2 and $E(M_1) = 1$. If we put*

$$\tilde{B}^{(i)}(t, \omega) = B^{(i)}(t, \omega) - \int_0^t f^i(s, \omega) dm^i(s) \qquad (i = 1, 2, \cdots) .$$

then the vector valued stochastic process $\tilde{\boldsymbol{B}}(t) = \{\tilde{B}^{(i)}(t, \omega) ; i = 1, 2, \cdots, 0 \le t \le 1\}$ with respect to the measure $\tilde{P}$ have the same law as $\{B^{(i)}(t, \omega) ; i = 1, 2, \cdots, 0 \le t \le 1\}$ with respect to the measure P, where $d\tilde{P} = M_1(\omega) dP(\omega)$. In other words, $\tilde{\boldsymbol{B}}(t)$ is a Brownian motion adapted to $\{\mathscr{F}_t\}$ on the probability space $(\Omega, \mathscr{F}, \tilde{P})$.

Proof. By Itô's formula and the assumption (M.3), we have

$$M_t = 1 + \sum_{i=1}^{\infty} \int_0^t M_s f^i(s, \omega) dB^{(i)}(s, \omega) .$$

It is enough to show that

$$\tilde{E}\left(\exp\left(i \sum_{i=1}^{N} z_{\nu_i} (\tilde{B}^{(i)}(t) - \tilde{B}^{(i)}(s)) \,|\, \mathscr{F}_s\right)\right)$$

$$= \exp\left(-\frac{1}{2}\sum_{i=1}^{N} z_{\nu_i}^2 (m^{\nu_i}(t) - m^{\nu_i}(s))\right)$$

for every system of integers $\nu_1, \nu_2, \cdots, \nu_N$, any real numbers $z_{\nu_1}, \cdots, z_{\nu_N}$ and s, t with $0 \leqq s \leqq t \leqq 1$, under the condition

$$P\left(0 < c_1 < \inf_{0 \leq t \leq 1} M_t \leq \sup_{0 \leq t \leq 1} M_t \leq c_2 < \infty\right) = 1\,, \tag{4.1}$$

where c_1 and c_2 are constants (see [7]).

If we put

$$\eta(t, s) = \exp\left(i \sum_{i=1}^{N} z_{\nu_i} (\tilde{B}^{(\nu_i)}(t) - \tilde{B}^{(\nu_i)}(s))\right),$$

then we have, by virtue of Theorem 3.2,

$$\begin{aligned} \eta(t, s) M_t = M_s &+ \sum_{i=1}^{N} \int_s^t \eta(u, s) M_u f^{\nu_i}(u, \omega) dB^{(\nu_i)}(u) \\ &+ i \sum_{i=1}^{N} z_{\nu_i} \int_s^t \eta(u, s) M_u dB^{(\nu_i)}(u) \\ &- \frac{1}{2} \sum_{i=1}^{N} \int_s^t z_{\nu_i}^2 \eta M_u dm^{\nu_i}(u) \end{aligned}$$

and we have, with $\tilde{P}$-measure 1, by the same method as in Lemma 6.6 of Liptzer and Shiryaev [7],

$$M_s^{-1} E(\eta(t, s) M_t \,|\, \mathscr{F}_s) = \tilde{E}(\eta(t, s) \,|\, \mathscr{F}_s)\,.$$

Therefore, with P-measure 1,

$$M_s^{-1} E(\eta(t, s) M_t \,|\, \mathscr{F}_s) = 1 - \frac{1}{2} \sum_{i=1}^{N} \int_s^t z_{\nu_i}^2 M_s^{-1} E(\eta(u, s) M_u \,|\, \mathscr{F}_s) dm^{\nu_i}(u)$$

If we put $f(t, s) = M_s^{-1} E(\eta(t, s) M_t \,|\, \mathscr{F}_s)$, $f(t, s)$ satisfies the following integral equation

$$f(t, s) = 1 - \sum_{i=1}^{N} \frac{z_{\nu_i}^2}{2} \int_s^t f(u, s) dm^{\nu_i}(u)\,.$$

Hence, $f(t, s) = \exp\{-\sum_{i=1}^{N} (z_{\nu_i}^2/2)(m^{\nu_i}(t) - m^{\nu_i}(s))\}$.

Reduction to general cases without the condition (4.1) is easily done as in [7, p 264–266].

§ 5. Representation theorems of L^2-functionals of $B(t)$

Let $\mathscr{B} = \mathscr{F}_1(B)$ and let $L^2(\mathscr{B}, P) = \{F; F = F(\omega)$ is $\mathscr{B}$-measurable and $E(F^2) < \infty\}$. The main purpose in this section is to show the following result.

Theorem 5.1. *Every element in $L^2(\mathscr{B}, P)$ with $E(F) = 0$ can be expressed in the following manner:*

$$F(\omega) = \sum_{i=1}^{\infty} \int_0^1 f^i(t, \omega) dB^{(i)}(t, \omega) , \tag{5.1}$$

where $f(t, \omega) = (f^1(t, \omega), f^2(t, \omega), \cdots)$ is a sequence of $\mathscr{F}_t(B)$-measurable process satisfying (M.1), (S.3) *of* § 2 *and*

$$E(F^2(\omega)) = \sum_{i=1}^{\infty} \int_0^1 E(f^i(s, \omega)^2) dm^i(s) < + \infty . \tag{5.2}$$

We prepare several lemmas before proving the above theorem. Define the Hermite polynomials:

$$H_n(t, x) = \frac{1}{n!} (-t)^n \exp(x^2/2t) \frac{\partial^n}{\partial x^n} \exp(-x^2/2t)$$

and

$$H_n(0, x) = \lim_{t \downarrow 0} H_n(t, x) ,$$

where $n = 0, 1, 2, \cdots$ and $t > 0$.
We note that $H_0(0, x) = 1$ and $H_n(0, x) = 0$ $(n > 0)$. As is well known, the following lemma holds.

Lemma 5.2. *For fixed $t_i > 0$ $(i = 1, 2, \cdots, N)$, the system $\{H_{n_1}(t_1, x_1) H_{n_2}(t_2, x_2) \cdots H_{n_N}(t_N, x_N); n_1, n_2, \cdots, n_N = 0, 1, 2, \cdots\}$ is a complete orthonormal one with respect to the Gaussian measure $N(0, V)$, where the covariance matrix V is given by*

$$\begin{bmatrix} t_1, & & & 0 \\ & t_2, & & \\ & & \ddots & \\ 0 & & & t_N \end{bmatrix}$$

Lemma 5.3. *Let $B(t)$ be the Brownian motion defined as in* § 2. *Then, we have the following formulas:*

$$\text{(I)}\quad H_n(m^i(t)-m^i(s), B^{(i)}(t)-B^{(i)}(s)) = \int_s^t H_{n-1}(m^i(u)-m^i(s), B^{(i)}(u)-B^{(i)}(s))dB^{(i)}(u)$$

and

$$\text{(II)}\quad \prod_{l=1}^{L} H_{n_l}(m^{i_l}(t)-m^{i_l}(s), B^{(i_l)}(t)-B^{(i_l)}(s))$$
$$= \sum_{k=1}^{L}\int_s^t \Big\{\prod_{l\neq k} H_{n_l}(m^{i_l}(u)-m^{i_l}(s), B^{(i_l)}(u)-B^{(i_l)}(s))$$
$$\times H_{n_k-1}(m^{i_k}(u)-m^{i_k}(s), B^{(i_k)}(u)-B^{(i_k)}(s))\Big\} dB^{(i_k)}(u)$$

for $1 \leqq i_1 < i_2 < \cdots < i_L$.

For the proof it is enough to note the formulas

$$\frac{\partial}{\partial t}H_n(t,x) = -\frac{1}{2}H_{n-2}(t,x)$$

and

$$\frac{\partial}{\partial x}H_n(t,x) = H_{n-1}(t,x)\ ,$$

and to apply Itô formula for $B(t)$ (§ 3) (see McKean [8] p. 37).

Lemma 5.4. *Let* $Q = \{\tau_0 = 0, \tau_1, \tau_2, \cdots\}$ *be a countable dense subset of* $[0, 1]$ *and let* $\mathscr{B}_Q = \sigma(B^{(n)}(\tau_i) - B^{(n)}(\tau_j)\,;\, \tau_i, \tau_j \in Q, n = 1, 2, \cdots)$. *Then* $\mathscr{B} = \mathscr{B}_Q$

Proof. It follows from the fact that each sample path of $B^{(n)}(t)$, $n = 1, 2, \cdots$ is continuous.

Proof of Theorem 5.1. Put $\mathscr{B}_{Q_l} = \sigma(B^{(n)}(\tau_i) - B^{(n)}(\tau_j)\,;\, i, j = 0, 1, \cdots, l,\ n = 1, 2, \cdots)$. Then the sub σ-algebra $\mathscr{B}_{Q_l}$ increases to $\mathscr{B}_Q = \mathscr{B}$ as $l \to \infty$ (i.e. $\bigvee_l \mathscr{B}_{Q_l} = \mathscr{B}$). Therefore, an element $F(\omega)$ of $L^2(\mathscr{B}, P)$ with $E(F) = 0$ can be approximated by a sequence $\{F_l(\omega)\,;\, l = 1, 2, \cdots\}$, where $F_l(\omega)$ is in $L^2(\mathscr{B}_{Q_l}, P) = \{F(\omega) \in L^2(\mathscr{B}, P)\,;\, F(\omega)$ is $\mathscr{B}_{Q_l}$-measurable$\}$ with $E(F_l) = 0$. Now, if the set $\{0, s_1, s_2, \cdots, s_l\}$ is the rearrangement of the set $Q_l = \{0, \tau_1, \cdots, \tau_l\}$ in the increasing order, namely, $s_0 = 0 < s_1 < s_2 < \cdots < s_l$, then we have $\mathscr{B}_{Q_l} = \sigma(B^{(n)}(s_i) - B^{(n)}(s_{i-1})\,;\, i = 1, \cdots, l,\ n = 1, \cdots, l)$. The Gaussian random variables $B_{i,n} = B^{(n)}(s_i) - B^{(n)}(s_{i-1})$ $(i = 1, 2, \cdots, l,\ n = 1, 2, \cdots, l)$ are mutually independent.

If we apply Lemma 5.2 with $N = l^2$, then we can see that

$$\left\{\prod_{i=1}^{l}\prod_{n=1}^{l} H_{k_{i,n}}(m^n(s_i) - m^n(s_{i-1}),\, B^{(n)}(s_i) - B^{(n)}(s_{i-1}))\,;\quad k_{i,n} = 0, 1, 2, \cdots\right\}$$

forms a complete orthogonal system in $L^2(\mathscr{B}_{Q_l}, P)$.

Therefore, each element $F_l(\omega)$ of $L^2(\mathscr{B}_{Q_l}, P)$ is expanded in the form

(5.3)

$$F_l(\omega) = \sum_K a_K \prod_{j=1}^{l}\prod_{n=1}^{l} H_{k_{j,n}}(m^n(s_j) - m^n(s_{j-1}),\, B^{(n)}(s_j) - B^{(n)}(s_{j-1}))\,,$$

where $K = (k_{j,n}) = (k_{1,1}, \cdots, k_{1,l}\,; k_{2,1}, \cdots, k_{2,l}\,; \cdots; k_{l,1}, \cdots, k_{l,l})$ takes over all possible l^2-dimensional vectors with components of non-negative integers. If we put $j_0 = j_0(K) = \max\{j\,; k_{j,n} \neq 0 \text{ for some } n\}$, then by the Lemma 5.2, we can rewrite each term of (5.3) as follows:

$$\begin{aligned}
&\prod_{j=1}^{l}\prod_{n=1}^{l} H_{k_{j,n}}(m^n(s_j) - m^n(s_{j-1}),\, B^{(n)}(s_j) - B^{(n)}(s_{j-1}))\\
&\quad = \sum_{n=1}^{l}\int_{s_{j_0-1}}^{s_{j_0}} \Bigg[\left\{\prod_{j=1}^{j_0-1} H_{k_{j,n}}(m^n(s_j) - m^n(s_{j-1}),\, B^{(n)}(s_j) - B^{(n)}(s_{j-1}))\right\}\\
&\quad\quad \times \left\{\prod_{d\neq n} H_{k_{j_0,d}}(m^d(u) - m^d(s_{j_0-1}),\, B^{(d)}(u) - B^{(d)}(s_{j_0-1}))\right\}\\
&\quad\quad \times H_{k_{j_0,n}-1}(m^n(u) - m^n(s_{j_0-1}),\, B^{(n)}(u) - B^{(n)}(s_{j_0-1}))\Bigg]dB^{(n)}(u)\\
&\quad = \sum_{n=1}^{l}\int_{s_{j_0-1}}^{s_{j_0}} \phi_K^n(u, \omega)dB^{(n)}(u)\,.
\end{aligned}$$

Therefore, we have

$$\begin{aligned}
(5.4)\qquad F_l(\omega) &= \sum_K a_K \sum_{n=1}^{l}\int_{s_{j_0-1}}^{s_{j_0}} \phi_K^n(u, \omega)dB^{(n)}(u)\\
&= \sum_{n=1}^{l}\sum_K \int_{s_{j_0-1}}^{s_{j_0}} a_K\phi_K^n(u, \omega)dB^{(n)}(u)\\
&= \sum_{n=1}^{l}\sum_{j=1}^{l}\sum_{K;\, j_0(K)=j} \int_{s_{j-1}}^{s_j} a_K\phi_K^n(u, \omega)dB^{(n)}(u)\,.
\end{aligned}$$

The summation $\sum$ in the right hand side of (5.4) has to be considered in the $L^2(\mathscr{B}, P)$ sense. Therefore, we have

$$(5.5)\qquad F_l(\omega) = \sum_{n=1}^{l}\int_0^1 f_l^n(u, \omega)dB^{(n)}(u)\,,$$

where we put $f_l^n(u, \omega) = \sum_{K;\, j_0(K)=j} a_K \phi_K^n(u, \omega)$, for $u \in [s_{j-1}, s_j)$.

Clearly (5.5) is the desired expression (5.1) for $F_l(\omega)$ of $L^2(\mathscr{B}_Q, P)$. If we put $f_l^n(u, \omega) = 0$ for $n \geq l+1$, then (5.5) can be written in the following form:

$$F_l(\omega) = \sum_{n=1}^{\infty} \int_0^1 f_l^n(u, \omega) dB^{(n)}(u) = \int_0^1 \boldsymbol{f}_l(u, \omega) \cdot d\boldsymbol{B}(u) .$$

Since $F_l(\omega)$ converges to $F(\omega)$ in $L^2(\mathscr{B}, P)$, there exists $\boldsymbol{f}(u, \omega) \in \boldsymbol{L}^2[0, 1]$ such that $|||f - f_l||| \to 0$ (as $l \to \infty$), where $|||f|||^2 = \sum_{n=1}^{\infty} \int_0^1 E|f^n(u, \omega)|^2 dm^n(u)$ and $\boldsymbol{f}(t, \omega) = (f^1(t, \omega), f^2(t, \omega), \cdots, f^n(t, \omega), \cdots)$. Consequently, we can see that the limit $\boldsymbol{f}(t, \omega)$ satisfies (M.1), (M.2) and (S.3), so that we have

$$F(\omega) = \int_0^1 \boldsymbol{f}(u, \omega) \cdot d\boldsymbol{B}(u) = \sum_{n=1}^{\infty} \int_0^1 f^n(u, \omega) dB^{(n)}(u) .$$

and also (5.2). Thus, the proof of Theorem 5.1 has been completed.

As a corollary of Theorem 5.1, we can deduce a Wiener type expansion for $L^2(\mathscr{B}, P)$.

Corollary 5.5. *Every element $F(\omega)$ in $L^2(\mathscr{B}, P)$ can be expressed in the following manner:*

$$F(\omega) = \sum_{i=1}^{\infty} \sum_{\alpha} \int_0^1 \int_0^{t_1} \cdots \int_0^{t_{i-1}} k_\alpha(t_1, \cdots, t_i) dB^{(\alpha_i)}(t_i) dB^{(\alpha_{i-1})}(t_{i-1}) \cdots dB^{(\alpha_1)}(t_1) ,$$

where $\sum_\alpha$ is taken over all i-tuples $(\alpha_1, \cdots, \alpha_i)$ of positive integers and $k_\alpha(t_1, \cdots, t_i)$'s are Volterra kernels (namely, $k_\alpha(t_1, \cdots, t_i) = 0$ unless $t_1 \geq t_2 \geq \cdots \geq t_i$) such that

$$\sum_{i=0}^{\infty} \sum_{\alpha} \int_0^1 \int_0^{t_1} \cdots \int_0^{t_{i-1}} k_\alpha(t_1, \cdots, t_i)^2 dm^{\alpha_i}(t_i) \cdots dm^{\alpha_1}(t_1) < \infty .$$

Proof. The case $F = F_l \in L^2(\mathscr{B}_{Q_l}, P)$ for some l is an immediate consequence of Lemma 5.3, and for the general case the usual limit procedure can be applied.

Let us consider more general situations. Let Λ be a countable subset of [0, 1], say $\Lambda = \{t_j,\ j = 1, 2, \cdots\}$. Suppose that a system of independent Gaussian random variables $\boldsymbol{B}_1 = \{B_{t_j}^l,\ t_j \in \Lambda,\ l = 1, 2, \cdots, L_j\}$ $(L_j \leqq \infty)$ are given on the probability space and suppose that $E(B_{t_j}^l) = 0$

and $E\{(B_{t_j}^l)^2\} = 1$ and that the system $\boldsymbol{B}_1$ and the Brownian motion $\boldsymbol{B}(t)$ defined in § 2 are mutually independent. Now, we put

$$B_{t_j}^1(t) = \begin{cases} B_{t_j}^1 & \text{if } t \geqq t_j \\ 0 & \text{if } t < t_j \end{cases}$$

and for $l \geq 2$

$$B_{t_j}^l(t) = \begin{cases} B_{t_j}^l & \text{if } t > t_j \\ 0 & \text{if } t \leqq t_j \, . \end{cases}$$

We consider a Gaussian process

$$\begin{aligned} \boldsymbol{C}(t) &= (B^{(1)}(t), \cdots, B^{(n)}(t), \cdots ; B_{t_1}^1(t), \cdots, B_{t_1}^{L_1}(t) ; B_{t_2}^1(t), \\ &\qquad\qquad \cdots, B_{t_2}^{L_2}(t) ; \cdots) \\ &= (\boldsymbol{B}(t) ; \boldsymbol{C}_{t_1}(t) ; \boldsymbol{C}_{t_2}(t) ; \cdots) , \end{aligned}$$

where

$$\boldsymbol{C}_{t_j}(t) = (B_{t_j}^1(t), \cdots, B_{t_j}^l(t), \cdots) .$$

Such a Gaussian process is important as an innovation process for general Gaussian process (see Hida [2] and Hitsuda-Watanabe [3]). The following theorem is an extension of the Theorem 5.1 which gives the representation of $L^2(\mathscr{C}, P)$ for $\mathscr{C} = \mathscr{F}_1(\boldsymbol{C})$.

Theorem 5.6. *Every element $F(\omega)$ in $L^2(\mathscr{C}, P)$ with $E(F) = 0$ can be expressed in the following manner:*

$$F(\omega) = \sum_{n=1}^{\infty} \int_0^1 f^n(t, \omega) dB^{(n)}(t) + \sum_{t_j \in \Lambda} \sum_{l=1}^{L_j} \sum_{n=1}^{\infty} g_n^{j,l}(\omega) \sqrt{n!}\, H_n(1, B_{t_j}^l) , \tag{5.6}$$

where $\{f^n(t, \omega)\}$ is a sequence of $\mathscr{F}_t(\boldsymbol{C})$-measurable process satisfying (M.1), (S.3) *of* § 2, *the random variables $g_n^{j,l}(\omega)$ are $\mathscr{F}_{t-}(\boldsymbol{C}) \vee \sigma(B_{t_j}^m ; m < l)$ measurable and $H_n(1, x)$'s are Hermite polynomials, and moreover*

$$E(F^2(\omega)) = \sum_{n=1}^{\infty} \int_0^1 E(f^{(n)}(s, \omega)^2) dm^n(s) + \sum_{t_j \in \Lambda} \sum_{l=1}^{L_j} \sum_{n=1}^{\infty} E(g_n^{j,l}(\omega)^2) < \infty .$$

Remark. $\{\sqrt{n!}\,H_n(1, x)\,;\, n = 0, 1, 2, \cdots\}$ forms the complete orthonormal system in $L^2((-\infty, \infty), \mu)$, where

$$\mu(dx) = \frac{1}{\sqrt{2\pi}} e^{-x^2/2} dx\,.$$

Proof. The proof of Theorem 5.6 is analogous to that of Theorem 5.1. Here, we only note that we need the following lemma in place of Lemma 5.4.

Lemma 5.4′. *Let* $P = Q \cup \Lambda$, *where* Q *is the same set as in Lemma* 5.4. *Also let* $\mathscr{B}_P = \{B^{(n)}(\tau_i) - B^{(n)}(\tau_j),\ \tau_i, \tau_j \in Q\ \ n = 1, 2, \cdots\} \vee \sigma\{B^l_{t_j}\,;\, t_j \in \Lambda\ \ l = 1, 2, \cdots\}$. *Then we have* $\mathscr{C} = \mathscr{B}_P$.

§ 6. Representation theorems of martingales

The results in this section can be regarded as an application of the discussions in § 5.

Kunita-Watanabe [6] obtained the corresponding representation theorems of martingales with respect to diffusion processes (especially the Brownian motion). It seems that their method is not applicable for our processes $\boldsymbol{B}(t)$ and $\boldsymbol{C}(t)$, because they are Markovian but inhomogeneous in time.

Theorem 6.1. *Every martingale* $X_t(\omega)$ *with respect to* $(\mathscr{F}_t(\boldsymbol{B}), P)$ *such that* $E(X_1^2(\omega)) < \infty$ *can be represented in the following form*:

$$X_t = X_0 + \sum_{i=1}^{\infty} \int_0^t f^i(s, \omega) dB^{(i)}(s, \omega)\,,$$

where $\boldsymbol{f}(s, \omega) = (f^1(s, \omega), f^2(s, \omega), \cdots)$ $s \in [0, 1]$ *satisfies the conditions* (M.1), (S.3) *and* (M.2′) *as in Theorem* 5.1.

Proof. If we put $F(\omega) = X_1 - X_0$, then we can apply Theorem 5.1 and we can obtain

$$F(\omega) = \sum_{i=1}^{\infty} \int_0^1 f^i(s, \omega) dB^{(i)}(s)$$

for some $\boldsymbol{f}(s, \omega) = (f^1(s, \omega), f^2(s, \omega), \cdots)$. From the martingale property of X_t, it follows that

$$E(F(\omega) \,|\, \mathscr{F}_t(\boldsymbol{B})) = X_t - X_0\,.$$

On the other hand, by the uniform integrability of

$$\left\{F_n(\omega) = \sum_{i=1}^{n} \int_0^1 f^i(s, \omega) dB^{(i)}(s),\ n = 1, 2, \cdots \right\},$$

we have

$$\begin{aligned} E(F(\omega) \mid \mathscr{F}_t(\boldsymbol{B})) &= \lim_{n\to\infty} \mathrm{E}(F_n(\omega) \mid \mathscr{F}_t(\boldsymbol{B})) \\ &= \sum_{i=1}^{\infty} \int_0^t f^i(s, \omega) dB^{(i)}(s) . \end{aligned}$$

Thus, the proof of Theorem 6.1 is finished.

Theorem 6.2. *Every martingale $X_t(\omega)$ with respect to $(\mathscr{F}_t(\boldsymbol{B}), P)$ can be represented in the form*

$$X_t = X_0 + \sum_{i=0}^{\infty} \int_0^t f^i(s, \omega) dB^{(i)}(s, \omega) ,$$

where $\boldsymbol{f}(s, \omega) = (f^1(s, \omega),\ f^2(s, \omega), \cdots),\ s \in [0, 1]$ satisfies the conditions (M.1), (M.2) *and* (M.3).

Proof. The proof is established by the usual method (for example, see Liptzer-Shiryaev [7], section 5). Hence, we omit it.

By using Theorem 5.6, we can obtain the analogous theorem for the process $\boldsymbol{C}(t)$ introduced in § 5.

Theorem 6.3. *Every martingale $X_t(\omega)$ with respect to $(\mathscr{F}_t(\boldsymbol{C}), P)$ such that $E(X_t(\omega)^2) < \infty$ can be represented in the following form:*

$$X_t = X_0 + \sum_{i=1}^{\infty} \int_0^t f^i(s, \omega) dB^{(i)}(s, \omega) + \sum_{t_j \le t} (\phi_{t_j}^{+}(\omega))(t) + \sum_{t_j \le t} (\phi_{t_j}^{-}(\omega))(t) ,$$

where $\boldsymbol{f}(s, \omega) = (f^1(s, \omega),\ f^2(s, \omega), \cdots)$ is a sequence of $\mathscr{F}_t(\boldsymbol{C})$-measurable process satisfying (M.1), (S.3) *of* § 2, *and where*

$$(\phi_{t_j}^{-}(\omega))(t) = \begin{cases} \phi_{t_j}^{-}(\omega) , & t \geqq t_j , \\ 0 , & t < t_j , \end{cases}$$

$$(\phi_{t_j}^{+}(\omega))(t) = \begin{cases} \phi_{t_j}^{+}(\omega) , & t > t_j , \\ 0 , & t \leqq t_j , \end{cases}$$

and $\phi_{t_j}^{\pm}(\omega)$ satisfy:

$$\phi^+_{t_j}(\omega) \text{ is } \mathscr{F}_{t_j+}(C)\text{-measurable with } E(\phi^+_{t_j}(\omega)\,|\,\mathscr{F}_{t_j}(C)) = 0\,,$$
$$\phi^-_{t_j}(\omega) \text{ is } \mathscr{F}_{t_j}(C)\text{-measureable with } E(\phi^-_{t_j}(\omega)\,|\,\mathscr{F}_{t_j-}(C)) = 0\,,$$

and

$$\sum_{t_j \in \Lambda} E(\phi^+_{t_j}(\omega))^2 + E(\phi^-_{t_j}(\omega))^2 < +\infty \text{ (a.e.)}\,.$$

References

[1] Girsanov, I. V., On the transforming a certain class of stochastic processes by absolutely continuous substitution of measures, Theor. Probability Appl. (English translation), **5** (1960), 285–301.

[2] Hida, T., Canonical representations of Gaussian processes and their applications, Mem. Coll. Sci. Univ. Kyoto, **33** (1960), 109–155.

[3] Hitsuda, M. and Watanabe, H., On a causal and causally invertible representation of equivalent Gaussian processes, in Multivariate Analysis-IV, North-Holland Publishing Company, 1977, Amsterdam.

[4] Itô, K., Stochastic differential equations in a differentiable manifold, Nagoya Math. J., **1** (1950), 35–47.

[5] ——, On a formula concerning stochastic differentials, Nagoya Math. J., **3** (1951), 55–65.

[6] Kunita, H. and Watanabe, S., On square integrable martingales, Nagoya Math. J., **30** (1967), 209–245.

[7] Liptzer, P. C. and Shiryaev, A. N., Statistics of stochastic processes, Nauka, 1974, Moscow. (in Russian)

[8] McKean, H. P. Jr., Stochastic integrals, Academic Press, 1969, New York and London.

[9] Ouvrard, Jean-Yves, Martingales locales et théorème de Girsanov dans les espaces de Hilbert réels séparables, Ann. Inst. H. Poincaré Sect. **B9** (1973), 351–368.

Masuyuki HITSUDA
DEPARTMENT OF MATHEMATICS
NAGOYA INSTITUTE OF TECHNOLOGY
NAGOYA 466, JAPAN

Hisao WATANABE
DEPARTMENT OF APPLIED SCIENCE
FACULTY OF ENGINEERING
KYUSHU UNIVERSITY
FUKUOKA 812, JAPAN

Proc. of Intern. Symp. SDE
Kyoto 1976, pp. 75-94

Heat Equation and Diffusion on Riemannian Manifold with Boundary

Nobuyuki IKEDA and Shinzo WATANABE

§ 1. Introduction

A stochastic solution of the heat equation for differential forms on a manifold was constructed by K. Itô [6] by introducing the notion of *stochastic parallel displacement*. P. Malliavin [10] introduced an equivalent notion of *lifted diffusion* on the bundle of orthonormal frames and then the stochastic solution is given by a matrix analog of Feynman-Kac formula.

In the case of a manifold with boundary, there is a classical work of P. E. Conner [5] showing that a semigroup corresponds to the heat equation on forms satisfying the absolute boundary conditions. Recently, H. Airault ([1], [2], [3]) constructed a stochastic solution by a method of singular perturbation. She found a stochastic differential equation for a multiplicative operator functional of the lifted diffusion which is not a usual stochastic differential equation. In this paper, we will give a somewhat different approach to Airault's stochastic differential equation using results in [14].

Finally, we will summarize the contents of this paper. In section 2, we collect, for the future use, a number of notions and known facts about Riemannian manifold with boundary. In section 3, we construct the paths of the lifted diffusion of the reflecting Brownian motion on the bundle of orthonormal frames. We will call the lifted diffusion itself as the *reflecting Brownian motion on the bundle of orthonormal frames*. In section 4, we construct a multiplicative operator functional of the reflecting Brownian motion on the bundle of orthonormal frames by solving Airault's stochastic differential equation. In section 5, we construct a semigroup corresponding to the heat equation using the multiplicative operator functional.

§ 2. Definitions and notations

The setting of this section will be essentially that of Malliavin [10] and Airault [3]. Let M be a compact oriented n-dimensional Riemannian manifold with boundary ∂M and let $O(M)$ be the set of $(n+1)$-tuples $(x, e_1, e_2, \cdots, e_n)$ where $x \in M$ and $(e_1, e_2, \cdots, e_n)$ is an orthonormal basis of T_xM. Then $a = (a_{ij}) \in O(n)$ acts on $O(M)$ as $R_a(x, e_1, e_2, \cdots, e_n) = (x, \sum_{i=1}^n a_{i1}e_i, \sum_{i=1}^n a_{i2}e_i, \cdots, \sum_{i=1}^n a_{in}e_i)$. Thus we have a bundle of orthonormal frames $(O(M), O(n), M)$, (cf. [11]). We will often denote the bundle by its bundle space $O(M)$ alone. It is sometimes convenient to view $O(M)$ as the set of orthogonal linear transformations of R^n into the tangent space of M, that is, we identify $r = (x, e_1, e_2, \cdots, e_n)$ with the map $r: (\xi_1, \xi_2, \cdots, \xi_n) \in R^n \mapsto \sum_{i=1}^n \xi_i e_i \in T_xM$, ([4]).

Definition 2.1. Let F be an R^n-valued function on $O(M)$. F is said to be *equivariant* if

$$F(R_a r) = a^{-1}F(r), \quad \text{for any} \quad a \in O(n). \tag{2.1}$$

Let $\Lambda^1(M)$ be the linear space of differential 1-forms on M and let $(e_1^0, e_1^0, \cdots, e_n^0)$ be the canonical basis of R^n.

Lemma 2.1 (Malliavin [10]). *There is a one-to-one correspondence $\theta \leftrightarrow F_\theta$ between the set of differential forms on M and the set of R^n-valued functions on $O(M)$ which are equivariant, given by*

$$F_\theta(r) = (\theta_{\pi(r)}(r(e_j^0)))_{j=1,2,\cdots,n}, \quad \text{for } \theta \in \Lambda^1(M), \tag{2.2}$$

where $\pi: O(M) \to M$ is the projection.

Before proceeding to the heat equation, we will prepare some notations. For a differential p-form θ, let θ_{norm} be the normal component of θ. Let η be the inwardpointing unit normal in the cotangent space at a boundary point. Then, we can write

$$\theta_{\text{norm}} = \varphi \wedge \eta \quad \text{and} \quad *\varphi = (*\theta) \wedge \eta,$$

where $*$ denotes the usual duality operator carrying a differential form of degree p into its adjoint of degree $n-p$.

Definition 2.2 (Ray and Singer [12]). A differential form θ is said to satisfy *absolute boundary conditions* if $\theta_{\text{norm}} = (d\theta)_{\text{norm}} = 0$.

Set $\Delta = -(d\delta + \delta d)$. Consider the following initial value problem on differential 1-forms

$$(2.3)\qquad \begin{cases} \dfrac{\partial u}{\partial t} = \dfrac{1}{2}\Delta u\,, & \lim_{t\downarrow 0} u(t,\cdot) = \theta\,, \\ u_{\text{norm}} = 0,\ (du)_{\text{norm}} = 0 & \text{on}\quad \partial M\,, \end{cases}$$

where θ is a given smooth differential 1-form on M. If we take a local coordinate $(x^1, x^2, \cdots, x^n)$ in a coordinate neighborhood U of M, every orthonormal frame $r \in \pi^{-1}(U)$ may be expressed in the form

$$r = (x, e_1, e_2, \cdots, e_n)\,, \quad e_i = \sum_{k=1}^{n} e_i^k \frac{\partial}{\partial x^k}\,, \quad i = 1, 2, \cdots, n\,,$$

where $e = (e_i^k)$ is such that

$$\sum_{k,l=1}^{n} e_i^k e_j^l g_{kl}(x) = \delta_{ij}\,, \qquad i, j = 1, 2, \cdots, n\,,$$

or equivalently,

$$\sum_{i=1}^{n} e_i^k e_i^l = g^{kl}(x)\,, \qquad k, l = 1, 2, \cdots, n\,.$$

Here $(g_{ij}(x))$ is the metric tensor in this local coordinate and $(g^{ij}(x)) = (g_{ij}(x))^{-1}$. Conversely, every $e = (e_i^k)$ with this property defines a orthonormal frame r as above. For the sake of brevity, we introduce the following notation:

$$(2.4)\qquad r = (x, e_1, e_2, \cdots, e_n) = [x, e]$$

where $x = (x^1, x^2, \cdots, x^n)$ and $e = (e_i^k)$.[1] Then we have

Lemma 2.2. *Let* $\theta = \sum_{i=1}^{n} \theta_i(x)dx^i \in \Lambda^1(M)$ *and let* F_θ *be the* R^n-*valued function on* $O(M)$ *given by* (2.2). *Then*

$$(2.5)\qquad F_\theta(r) = e\begin{pmatrix} \theta_1(x) \\ \vdots \\ \theta_n(x) \end{pmatrix}, \qquad r = [x, e] \in O(M)\,.$$

Near the boundary ∂M, we can take a local coordinate $(x^1, x^2, \cdots, x^n)$ satisfying the following conditions, (cf. Ray and Singer [12]):

(i) $(x^1, x^2, \cdots, x^{n-1})$ is a local coordinate in ∂M

(ii) $\partial M = \{(x^1, x^2, \cdots, x^n);\ x^n = 0\}$ in the coordinate neighborhood,

---

[1] As a matrix, we regard $e = (e_{ik})$ with $e_{ik} = e_i^k$. In general, when we regard (a_i^k) as a matrix $a = (a_{ik})$, we set $a_{ik} = a_i^k$.

(iii) $g_{in} = 0$, for $i < n$.

Lemma 2.3 (Conner [5]). *Let $\theta = \sum_{i=1}^n \theta_i(x)dx^i \in \Lambda^1(M)$. Then θ satisfies the absolute boundary conditions if and only if*

$$\theta_n(x) = 0\,, \quad \frac{\partial}{\partial x^n}\theta_i(x) = 0\,, \quad i = 1, 2, \cdots, n-1\,, \quad \text{for } x \in \partial M\,. \tag{2.6}$$

Proof. By (i), (ii) and (iii), we have

$$\theta_{\text{norm}} = \theta_n(x)dx^n\,, \qquad (d\theta)_{\text{norm}} = \sum_{i<n}\left(\frac{\partial\theta_i}{\partial x^n} - \frac{\partial\theta_n}{\partial x^i}\right)dx^n \wedge dx^i$$

and the assertion follows at once from this.

Set $P = (p_{ij})$ where $p_{ij} = \delta_{in}\delta_{jn}$, $1 \le i,\ j \le n$ and $Q = I - P$. Then we have at once the following:

Lemma 2.4. *Let $\theta \in \Lambda^1(M)$ and let F_θ be the R^n-valued function on $O(M)$ given by (2.2). Then (2.6) holds if and only if*

$$Pe^{-1}F_\theta(r) = 0\,, \qquad Qe^{-1}\frac{\partial}{\partial x^n}F_\theta(r) = 0\,, \quad \text{for } r = [x, e] \in \partial(O(M))\,, \tag{2.7}$$

where $\partial(O(M)) = \{r = [x, e]\,;\, x \in \partial M\}$.

Let $\Delta_{O(M)}$ be the horizontal Laplacian of Bochner[2)] and let $R_k^h(x) = \sum_{i=1}^n R^{h\cdot i\cdot}_{\cdot i\cdot k}(x)$ be the Ricci tensor. Consider the matrix $R(x) = (R_k^h(x))$ and define J by $J(r) = eR(x)e^{-1}$ for $r = [x, e] \in O(M)$. Then, by the Weitzenböck formula, we have, for a differential form $\theta \in \Lambda^1(M)$,

$$F_{\Delta\theta}(r) = \boldsymbol{\Delta}_{O(M)}F_\theta(r)\,, \qquad r \in O(M)\,, \tag{2.8}$$

where $\boldsymbol{\Delta}_{O(M)} = \Delta_{O(M)} + J$, (cf. Malliavin [10]). Using Lemmas 2.3, 2.4 and (2.8), the initial value problem (2.3) can be rewritten in the following initial value problem for R^n-valued functions on $O(M)$ which are equivariant:

$$\begin{cases} \dfrac{\partial u}{\partial t} = \dfrac{1}{2}\boldsymbol{\Delta}_{O(M)}u\,, & \lim_{t\downarrow 0} u(t, \cdot) = F\,, \\ Pe^{-1}u + Qe^{-1}\dfrac{\partial u}{\partial x^n} = 0\,, & \text{for } r = [x, e] \in \partial(O(M))\,, \end{cases} \tag{2.9}$$

2) For the definition, see §3, (3.4).

where F is a given R^n-valued function on $O(M)$ which is equivariant.

§ 3. The reflecting Brownian motion on the bundle of orthonormal frames

In this section, we will construct and derive some properties of path functions of the reflecting Brownian motion on the bundle of orthonormal frames $O(M)$. Following Malliavin [10], we obtain it by lifting on $O(M)$ the reflecting Brownian motion on M through the Riemannian connection and this is one way of realizing the parallel displacement along the Brownian paths which was first introduced by Itô ([6], [8]). For this purpose, we use a stochastic differential equation and, as usual, we can restrict our construction in a coordinate neighborhood by a standard localization argument. Hence, in the remainder of the paper, we will consider under the following circumstances:

Assumption (A). M is the upper half space of R^n:

$$M = \{x\,;\, x = (x^1, x^2, \cdots, x^n) \in R^n,\ x^n \geq 0\}\,,$$

and

$$\partial M = \{x\,;\, x = (x^1, x^2, \cdots, x^n) \in M,\ x^n = 0\}\,.$$

Assumption (B). The Riemannian metric tensor $(g_{ij}(x))$ under the coordinate $x = (x^1, x^2, \cdots, x^n)$ consists of C^∞-functions which are bounded together with all of their partial derivatives. Furthermore $(g_{ij}(x))$ is uniformly positive definite and satisfies the following condition: $g_{in} = 0$, for $i = 1, 2, \cdots, n-1$.

We consider the following stochastic differential equation for the process $(X(t), e(t))$ on $R^n_+ \times R^{n^2}$:

$$(3.1)\qquad \begin{aligned} dX^i(t) &= \sum_{k=1}^n e^i_k(t) \circ dB^k(t) + \delta_{in} d\psi(t)\,, \\ de^i_\alpha(t) &= -\sum_{k,l=1}^n \Gamma^i_{lk}(X(t)) e^k_\alpha(t) \circ dX^l(t)\,. \end{aligned} \qquad i, \alpha = 1, 2, \cdots, n\,.$$

Here the symbol $\circ$ denotes the symmetric Q-multiplication in the sense of Itô [7] and $\Gamma^i_{lk}(x)$ are the coefficients of the connection on M in the coordinate $(x^1, x^2, \cdots, x^n)$. Also $B(t) = (B^i(t))$ is an n-dimensional standard Brownian motion and $\psi(t)$ is a continuous non-decreasing process which increases only when $X(t) \in \partial M$. A precise formulation is as follows:

Definition 3.1. By a solution $(X(t), e(t)) = (X^i(t), e^i_\alpha(t))_{i,\alpha=1,2,\cdots,n}$ of the equation (3.1), we mean a family of stochastic processes $(X(t), e(t))$

defined on a probability space $(\Omega, \mathscr{F}, P)$ with a right-continuous family $(\mathscr{F}_t)_{t\geq 0}$ of increasing sub σ-fields of $\mathscr{F}$ such that

(i) $X(t) = (X^i(t))_{i=1,2,\cdots,n}$ is an n-dimensional continuous process adapted to $(\mathscr{F}_t)$ such that $X^n(t) \geq 0$ for all t, a.s.,

(ii) $e(t) = (e^i_\alpha(t))_{i,\alpha=1,2,\cdots,n}$ is an n^2-dimensional continuous process adapted to $(\mathscr{F}_t)$,

(iii) there exists an n-dimensional standard $(\mathscr{F}_t)$-Brownian motion $B(t) = (B^k(t))_{k=1,2,\cdots,n}$ and a non-decreasing continuous process $\psi(t)$ adapted to $(\mathscr{F}_t)$ such that $\psi(0) = 0$ and

$$\int_0^t I_{\partial M}(X(s))d\psi(s) = \psi(t) \qquad \text{for all } t, \text{ a.s.,}$$

such that, with probability one, the following holds:

$$(3.2)\qquad \begin{aligned} X^i(t) - X^i(0) &= \sum_{k=1}^n \int_0^t e^i_k(u) \circ dB^k(u) + \delta_{in}\psi(t)\,, \\ e^i_\alpha(t) - e^i_\alpha(0) &= -\sum_{k,l=1}^n \int_0^t \Gamma^i_{lk}(X(u))e^k_\alpha(u) \circ dX^l(u)\,, \\ & \qquad\qquad i, \alpha = 1, 2, \cdots, n\,. \end{aligned}$$

Here, the integrals by $\circ dB^k$ and $\circ dX^l$ are understood in the sense of symmetric Q-multiplication of Itô [7] by the quasimartingales $B^k(t)$ and $X^l(t)$ respectively.

By [13], we have

Lemma 3.1. *For any Borel probability measure μ on $R^n_+ \times R^{n^2}$, there exists a unique solution $(X(t), e(t))$ of the equation* (3.1) *such that the law of $(X(0), e(0))$ coincides with μ.*

In exactly the same way as in [9], we can show that if

$$\sum_{k,l=1}^n e^k_i(0)e^l_j(0)g_{kl}(X(0)) = \delta_{ij}\,,$$

then, for all $t \geq 0$,

$$\sum_{k,l=1}^n e^k_i(t)e^l_j(t)g_{kl}(X(t)) = \delta_{ij}$$

holds almost surely. Thus we have

Lemma 3.2. *Let $(X(t), e(t))$ be a solution of* (3.1) *such that* $[X(0), e(0)] \in O(M)$. *Then* $[X(t), e(t)] \in O(M)$ *for all* $t \geq 0$.

Using Lemmas 3.1 and 3.2, we can show by a standard argument that $[X(t), e(t)]$ defines a diffusion process on $O(M)$.

Definition 3.2. The diffussion process $r(t) = [X(t), e(t)]$ on $O(M)$ constructed above is called the *reflecting Brownian motion on the bundle of orthonormal frames* $O(M)$.

Now, we will compute the infinitesimal generator of the reflecting Brownian motion on $O(M)$. For this, we introduce the following notations. Take $r = (x, e_1, e_2, \cdots, e_n) \in O(M)$ and let ψ_j be the geodesic curve starting at x with the tangent vector e_j. Let ψ_j^* be the horizontal lift in $O(M)$ of the curve ψ_j in M through the Riemannian connection such that $\psi_j^*(0) = r$ and let A_r^j be the tangent vector of ψ_j^* at r. Then the map $A_j : r \mapsto A_r^j$ defines a vector field on $O(M)$ and we denote by L_{A_j} the differential operator defined by A_j. If F is a function defined on $O(M)$, then, for any $r = (x, e_1, e_2, \cdots, e_n) \in O(M)$,

$$(3.3) \quad (L_{A_m}F)(r) = \left(\sum_{i=1}^{n} e_m^i \frac{\partial}{\partial x^i} - \sum_{i,j,k,l=1}^{n} \Gamma_{kl}^i(x) e_m^l e_j^k \frac{\partial}{\partial e_j^i}\right) F(r) , \qquad m = 1, 2, \cdots, n .$$ [3)]

The horizontal Laplacian is, by definition,

$$(3.4) \qquad \Delta_{O(M)} = \sum_{j=1}^{n} L_{A_j}^2 ,$$

(cf. [10]). Set

$$(3.5) \quad \alpha^{nn}(r) = g^{nn}(x) , \qquad \alpha^{n,(m,i)}(r) = -\sum_{k=1}^{n} e_m^k \Gamma_{nk}^i(x) g^{nn}(x) , \qquad r = [x, e] \in O(M) .$$

Theorem 3.1. *Let* $r(t) = [X(t), e(t)]$ *be the reflecting Brownian motion on* $O(M)$ *constructed above by the solution of* (3.1). *Then,*

(i) *for any smooth function* $F(t, r)$ *on* $[0, \infty) \times O(M)$,

$$(3.6) \quad \begin{aligned} dF(t, r(t)) = {} & \sum_{m=1}^{n} (L_{A_m}F)(t, r(t)) dB^m(t) \\ & + \left\{\frac{1}{2}(\Delta_{O(M)}F)(t, r(t)) + \frac{\partial F}{\partial t}(t, r(t))\right\} dt \\ & + (X_n^* F)(t, r(t)) d\psi(t) , \end{aligned}$$

3) To be precise, $O(M)$ is smoothly imbedded in $R_+^n \times R^{n^2}$ and the differential operator given by (3.3) on $R_+^n \times R^{n^2}$, if restricted on $O(M)$, may be regarded as a differential operator on $O(M)$.

where X_n^* is the horizontal lift of the vector field $X_n = \partial/\partial x^n$ which is given explicitly as

$$(3.7)\qquad (X_n^*F)(t,r) = \frac{\partial F}{\partial x^n}(t,r) + \sum_{i,m=1}^{n} \frac{\alpha^{n,(m,i)}(r)}{\alpha^{nn}(r)} \frac{\partial F}{\partial e_m^i}(t,r)\,,$$

(ii)

$$(3.8)\qquad \begin{cases} dX^n(t)\cdot dX^n(t) = \alpha^{nn}(r(t))dt\,, \\ dX^n(t)\cdot de_m^i(t) = \alpha^{n,(m,i)}(r(t))dt\,, \qquad m, i = 1, 2, \cdots, n\,. \end{cases}$$

(As for the definition of $dX\cdot dY$ for quasi-martingales X and Y, we refer to [7].*)*

Proof. Using Itô's formula in (3.1), we have

$$\begin{aligned} dF(t,r(t)) = & \sum_{m,i=1}^{n} e_m^i(t)\frac{\partial F}{\partial x^i}(t,r(t))\circ dB^m \\ & - \sum_{i,j,k,l,m=1}^{n} \frac{\partial F}{\partial e_j^i}(t,r(t))\Gamma_{ik}^i(x(t))e_j^k(t)e_m^l(t)\circ dB^m(t) \\ & + \frac{\partial F}{\partial t}(t,r(t))dt + \frac{\partial F}{\partial x^n}(t,r(t))d\psi(t) \\ & - \sum_{i,k,m=1}^{n} \frac{\partial F}{\partial e_m^i}(t,r(t))\Gamma_{nk}^i(r(t))e_m^k(t)d\psi(t) \\ = & \sum_{m=1}^{n}(L_{A_m}F)(t,r(t))\circ dB^m(t) + \frac{\partial F}{\partial t}(t,r(t))dt \\ & + (X_n^*F)(t,r(t))d\psi(t)\,. \end{aligned}$$

From this formula, we have

$$\begin{aligned} d\Big\{\sum_{m=1}^{n}(L_{A_m}F)(t,r(t))\Big\} & \\ = & \sum_{m,\alpha=1}^{n} L_{A_\alpha}(L_{A_m}F)(t,r(t))\circ dB^\alpha(t) \\ & + \sum_{m=1}^{n}\frac{\partial(L_{A_m}F)}{\partial t}(t,r(t))dt + \sum_{m=1}^{n} X_n^*(L_{A_m}F)(t,r(t))d\psi(t)\,. \end{aligned}$$

Hence,

$$\sum_{m=1}^{n}(L_{A_m}F)(t,r(t))\circ dB^m(t)$$

$$= \sum_{m=1}^{n} (L_{A_m}F)(t, r(t))dB^m(t) + \frac{1}{2}d\Big\{\sum_{m=1}^{n} (L_{A_m}F)(t, r(t))\Big\}\cdot dB^m(t)$$
$$= \sum_{m=1}^{n} (L_{A_m}F)(t, r(t))dB^m(t) + \frac{1}{2}\sum_{m=1}^{n} (L^2_{A_m}F)(t, r(t))dt\ ,$$

and the proof of (3.6) is complete.

(ii) is proved easily if we notice

$$\sum_{\alpha=1}^{n} e^i_\alpha(t)e^j_\alpha(t) = g^{ij}(X(t))\ , \qquad i, j = 1, 2, \cdots, n\ .$$

Theorem 3.1 implies that our reflecting Brownian motion $r(t)$ on $O(M)$ is the diffusion process on $O(M)$ with the infinitesimal generator $\Delta_{O(M)}/2$ whose domain is characterized by the boundary condition

$$X^*_n F = 0 \quad \text{on} \quad \partial(O(M))\ .$$

Note that, from (3.7) and (3.8), this diffusion $r(t)$ is a *normally reflecting diffusion process* in the sense of [14], § 5.1. This fact plays an important role in the discussion of the section 5. Also we have the following:

Lemma 3.3. *The reflecting Brownian motion on $O(M)$ is invariant under the action of R_a, $a \in O(n)$. To be precise, let $r \in O(M)$ and $r(t)$ be the solution of* (3.1) *such that $r(0) = r$. Denoting by P_r and $R_a[P_r]$ the probability laws of the processes $r(t)$ and $R_a(r(t))$ respectively, we have*

(3.9) $$P_{R_a r} = R_a[P_r]\ , \qquad a \in O(n)\ , \quad r \in O(M)\ .$$

Proof. Let $r(t)$ be a solution of (3.1) with $B(t)$ and $\psi(t)$ such that $r(0) = r$. Then, for $a \in O(n)$, $\tilde{r}(t) = R_a r(t)$ is a solution of (3.1) with $\tilde{B}(t) = a^{-1}B(t)$ and $\tilde{\psi}(t) = \psi(t)$ such that $\tilde{r}(0) = R_a r(0)$. $\tilde{B}(t)$ is another n-dimensional $(\mathscr{F}_t)$-Brownian motion and hence, by the uniqueness of the solution, (3.9) holds.[4]

§ 4. Multiplicative operator functionals

In this section, we will construct a multiplicative operator functional (MOF) of the reflecting Brownian motion on the bundle of orthonormal frames $O(M)$ by solving a stochastic differential equation introduced by Airault [1].

---

[4] By this lemma, we see that $X(t) = \pi[r(t)]$ is a diffusion process on M and it is not difficult to verify that $X(t)$ is the Brownian motion on M with reflecting boundary, i.e. it has as the generator the half Laplace-Beltrami operator with Neumann's boundary condition.

Let $r(t) = [X(t), e(t)]$ be the reflecting Brownian motion on $O(M)$. Thus, $r(t)$ is a solution of (3.1) defined on a probability space $(\Omega, \mathscr{F}, P)$ with an increasing family $(\mathscr{F}_t)_{t\geq 0}$. Note that the non-decreasing continuous process $\psi(t)$ appearing in (3.1) is just the *local time* of $X(t)$ on ∂M:

$$\psi(t) = \lim_{\varepsilon \downarrow 0} \frac{1}{2\varepsilon} \int_0^t g^{nn}(X(s)) I_{[0,\varepsilon)}(X^n(s)) ds . \tag{4.1}$$

Let $A(t)$ be the right continuous inverse of $t \mapsto \psi(t)$ and set

$$D = \{s \geq 0 ; A(s-) < A(s)\} . \tag{4.2}$$

Let $R^n \otimes R^n$ be the algebra of all $n \times n$ real matrices $a = (a_{ij})$ endowed with a norm $\|a\| = \sqrt{\sum_{i,j=1}^n |a_{ij}|^2}$. As above, $P = (p_{ij})$ with $p_{i,j} = \delta_{in}\delta_{jn}$ and $Q = I - P$. Let $K(t)$ be an $R^n \otimes R^n$-valued process adapted to $(\mathscr{F}_t)$. We set $K^1(t) = K(t)P$ and $K^2(t) = K(t)Q$: hence, $K(t) = K^1(t) + K^2(t)$. Consider a stochastic differential equation of the following form for the process $K(t)$:

(4.3) (i) for any $t \geq 0$ such that $X(t) \in M^\circ = M \backslash \partial M$,

$$dK^1(t) = K(t)\{e(t)^{-1}de(t) + \tfrac{1}{2}R(X(t))dt\}P , \tag{4.3$_a$}$$

(ii) for any $t \geq 0$

$$dK^2(t) = K(t)\{e(t)^{-1}de(t) + \tfrac{1}{2}R(X(t))dt\}I_{M^\circ}(X(t))Q , \tag{4.3$_b$}$$

(iii) with probability one, $t \mapsto K^1(t)$ is right continuous with left-hand limits; furthermore,

$$K^1(t) = 0 \quad \text{if} \quad X(t) \in \partial M . \tag{4.3$_c$}$$

Remark 4.1. (a) (4.3), (ii) implies automatically that, with probability one, $t \mapsto K^2(t)$ is continuous.

(b) A precise formulation of (4.3), (i) is as follows: let a continuous process $Y(t)$ be defined as

$$Y(t) = \int_0^t K(s)\left\{e(s)^{-1}de(s) + \frac{1}{2}R(X(s))ds\right\}P$$

by a quasi-martingale stochastic integral. Then, with probability one, if $u \in D$ then $K^1(t) - K^1(s) = Y(t) - Y(s)$ holds for all $s \leq t$ such that $[s, t] \subset (A(u-), A(u))$.

The above stochastic differential equation is given also in an equivalent form of a stochastic integral equation. Namely, we have

Lemma 4.1. *Let $r = [x, e] \in O(M)$ be given. Then an $R^n \otimes R^n$-valued process $K(t)$ adapted to (F_t) is a solution of (4.3) with the initial condition*

$$K^1(0) = I_{M^\circ}(x)eP\,, \qquad K^2(0) = eQ \tag{4.4}$$

if and only if $K(t)$ satisfies the following stochastic integral equation

$$\begin{aligned} K^1(t) &= I_{\{t<\sigma\}}\Big(e + \int_0^t K(u)\Big[e(u)^{-1}de(u) + \frac{1}{2}R(X(u))du\Big]\Big)P \\ &\quad + I_{\{t\geq\sigma\}}\int_{\tau(t)}^t K(u)\Big[e(u)^{-1}de(u) + \frac{1}{2}R(X(u))du\Big]P\,, \\ K^2(t) &= eQ + \int_0^t K(u)\Big[e(u)^{-1}de(u) + \frac{1}{2}R(X(u))du\Big]I_{M^\circ}(X(u))Q\,, \end{aligned} \tag{4.5}$$

where σ is the first hitting time of $X(t)$ to ∂M:

$$\sigma = \begin{cases} \inf\{s\,;\, X(s) \in \partial M\}\,, \\ \infty\,, \quad \text{if } \{\ \} = \varnothing\,, \end{cases} \tag{4.6}$$

and $\tau(t)$ is the last exit time before t from ∂M:

$$\tau(t) = \begin{cases} \sup\{s\,;\, s \leq t,\ X(s) \in \partial M\}\,, \\ 0\,, \quad \text{if } \{\ \} = \varnothing\,. \end{cases} \tag{4.7}$$

The proof is easy and omitted.

Remark 4.2. $\int_{\tau(t)}^t \cdot$ is understood, of course, as $Y(t) - Y(\tau(t))$ where $Y(t)$ is the continuous process defined by $Y(t) = \int_0^t \cdot$.

Let Ξ be the totality of $R^n \otimes R^n$-valued processes $\xi(t)$ defined on $(\Omega, \mathscr{F}, P)$ and adapted to $(\mathscr{F}_t)$ such that $t \mapsto \xi(t)$ is right continuous with left-hand limits, a.s. and satisfies

$$\sup_{t\in[0,T]} E[\|\xi(t)\|^2] < \infty\,, \qquad \text{for all } \ T > 0\,. \tag{4.8}$$

Let $r = [x, e] \in O(M)$ be given. Define a map $\Phi: \Xi \to \Xi$ by

(a) $$\begin{aligned} &\Phi^1(\xi)(t)\ (= \Phi(\xi)(t)P) \\ &\quad = I_{\{\sigma>t\}}\Big(e + \int_0^t \xi(u)\Big[e(u)^{-1}de(u) + \frac{1}{2}R(X(u))du\Big]\Big)P \\ &\qquad + I_{\{\sigma\leq t\}}\int_{\tau(t)}^t \xi(u)\Big[e(u)^{-1}de(u) + \frac{1}{2}R(X(u))du\Big]P\,, \end{aligned}$$

(b) $\Phi^2(\xi)(t) \,(= \Phi(\xi)(t)Q)$

$$= eQ + \int_0^t \xi(u)\left[e(u)^{-1}de(u) + \frac{1}{2}R(X(u))du\right]I_{M^\circ}(X(u))Q \,.$$

If $t > 0$ is fixed, then $\tau(t) = A(\psi(t)-)$ a.s.. By Theorem 1′ of [14], if $g(t)$ is an $(\mathscr{F}_t)$-adapted measurable process such that $t \mapsto E[g(t)^2]$ is locally bounded, then

$$\text{(4.9)}\qquad \begin{aligned} E\left[\left\{\int_{\tau(t)}^t g(s)dB^k(s)\right\}^2\right] &\le E\left[\sum_{u \in D}\left\{\int_{A(u-)}^{A(u)\wedge t} g(s)dB^k(s)\right\}^2\right] \\ &= E\left[\int_0^t g(s)^2 ds\right], \qquad k = 1, 2, \cdots, n \,, \end{aligned}$$

and hence, it is easy to see that, for any $T > 0$, there exists a constant $K = K(T)$ such that

$$\text{(4.10)}\qquad E[\|\Phi(\xi)(t)\|^2] \le K\left(1 + \int_0^t E[\|\xi(u)\|^2]du\right), \qquad t \in [0, T] \,.$$

This proves $\Phi(\xi) \in \Xi$ if $\xi \in \Xi$. Again, using (4.9), we can show that, for any $T > 0$, there exists a constant $K = K(T)$ such that, for $\xi, \eta \in \Xi$,

$$\text{(4.11)}\qquad \begin{aligned} &E[\|\Phi(\xi)(t) - \Phi(\eta)(t)\|^2] \\ &\qquad \le K\int_0^t E[\|\xi(s) - \eta(s)\|^2]ds \,, \qquad t \in [0, T] \,. \end{aligned}$$

Theorem 4.1. *For any $r = [x, e] \in O(M)$, stochastic differential equation* (4.3) *has one and only one solution $K(t) \in \Xi$ satisfying the initial condition* (4.4).

Proof. Let $\xi_n \in \Xi$, $n = 0, 1, 2, \cdots$, be defined by $\xi_0 \equiv 0$ and $\xi_n = \Phi(\xi_{n-1})$, $n = 1, 2, \cdots$. Using (4.11), we can show that $\xi \in \Xi$ exists such that $E[\sup_{0 \le t \le T}\|\xi_n(t) - \xi(t)\|^2] \to 0$. Then cleary ξ is a solution of (4.5). The uniqueness follows also from (4.11). Arguments are standard and so we omit the details.

Now we turn to the construction of MOF. For this, it is convenient to formulate our reflecting Browian motion on $O(M)$ in a canonical form. Let $W = C([0, \infty) \to O(M))$ be the set of all continuous functions $w : t \in [0, \infty) \mapsto w(t) \in O(M)$, $\mathscr{B}(W)$ be the σ-field on W generated by Borel cylinder sets and $\mathscr{B}_t(W)$ be that generated by Borel cylinder sets up to time t. Let $r = [x, e] \in O(M)$ and $r(t) = [X(t), e(t)]$ be a solution of (3.1) such that $r(0) = r$ and let P_r be the probability law on $(W, \mathscr{B}(W))$ of the process $r(t)$. Let $\mathscr{F}$ and $\mathscr{F}_t$ be the usual completions of $\mathscr{B}(W)$ and $\mathscr{B}_t(W)$ by the system of probability measures

$$P_\mu(\cdot) = \int_{O(M)} P_r(\cdot)\mu(dr) .$$

Let $\theta_t: W \to W$ be defined by $(\theta_t w)(s) = w(t + s)$. In this way, we have a canonical realization $(W, \mathscr{F}, P_r; \mathscr{F}_t, \theta_t)$ of the reflecting Brownian motion on $O(M)$. We write $w(t) = r(t, w) = [X(t, w), e(t, w)]$. For every Borel probability measure μ on $O(M)$, we can consider the equation (4.3) on the space $(W, \mathscr{F}, P_\mu; \mathscr{F}_t)$ with the initial condition

$$K(0) = I_{M^\circ}(X(0, w))e(0, w)P + e(0, w)Q .$$

Denote the solution as $K(t, w)$ and set

$$M(t, w) = K(t, w)e(t, w)^{-1} , \qquad t \geq 0 . \tag{4.12}$$

Theorem 4.2. *$M = \{M(t, w)\}_{t\geq 0}$ is an $R^n \otimes R^n$-valued MOF of the reflecting Brownian motion on $O(M)$; i.e.,*

(i) *$M(t, w)$ is $(\mathscr{F}_t)$-adapted,*

(ii) *for $t, s \geq 0$,*

$$M(t + s, w) = M(s, w)M(t, \theta_s w) , \quad a.s. . \tag{4.13}$$

Proof. (i) is clear from the way of our construction. To prove (ii), we fix s and set $\tilde{K}(t) = K(t + s, w)$, $\tilde{X}(t) = X(t + s, w)$ and $\tilde{e}(t) = e(t + s, w)$. Then it is clear that $\tilde{K}(t)$ satisfies the equation (4.3) with respect to $[\tilde{X}(t), \tilde{e}(t)]$. On the otherhand, by applying the shift operator θ_s to the equation (4.3), we see that $\bar{K}(t) = K(t, \theta_s w)$ satisfies the equation (4.3) with respect to $[X(t, \theta_s w), e(t, \theta_s w)] = [\tilde{X}(t), \tilde{e}(t)]$. If we set

$$\tilde{K}' = K(s, w)e(s, w)^{-1}\bar{K}(t) ,$$

then

$$\begin{aligned} \tilde{K}'(0) &= K(s, w)e(s, w)^{-1}(I_{M^\circ}(X(s, w))e(s, w)P + e(s, w)Q) \\ &= K(s, w)(I_{M^\circ}(X(s, w))P + Q) \\ &= K(s, w) \end{aligned}$$

by $(4.3)_c$. Therefore, $\tilde{K}(t)$ and $\tilde{K}'(t)$ satisfy the same equation and the same initial condition. Hence $\tilde{K}(t) \equiv \tilde{K}'(t)$ by the uniqueness of solutions. That is

$$K(t + s, w) = K(s, w)e(s, w)^{-1}K(t, \theta_s w) .$$

By multiplying $e(t + s, w)^{-1} = e(t, \theta_s w)^{-1}$ from the right, we have (4.13).

We will list some properties of $M = \{M(t, w)\}$ in the following lemmas.

Lemma 4.2. *If $X(0) \in \partial M$, then*

$$Pe(0)^{-1}M(t) = 0\,, \qquad \textit{for all} \quad t \geq 0\,.$$

Proof. It is enough to prove that $Pe(0)^{-1}K(t) = 0$ for all $t \geq 0$. If $X(0) \in \partial M$, then

$$Pe(0)^{-1}K(0) = Pe(0)^{-1}(I_{M^\circ}(X(0))e(0)P + e(0)Q) = PQ = 0\,.$$

Since $\tilde{K}(t) = Pe(0)^{-1}K(t)$ satisfies (4.3), $\tilde{K}(t) = 0$ by the uniqueness of solutions.

Let $a \in O(n)$ and let $R_a : W \to W$ be defined by

$$(R_a w)(t) = R_a[w(t)]\,, \qquad t \geq 0\,.$$

Lemma 4.3. $M(t, R_a w) = a^{-1}M(t, w)a\,, \qquad t \geq 0,\ a \in O(n).$

Proof. Since $X(t, R_a w) = X(t, w)$ and $e(t, R_a w) = a^{-1}e(t, w)$ we see at once $K(t, R_a w) = a^{-1}K(t, w)$ by the uniqueness of solutions of (4.3). Then $M(t, R_a w) = a^{-1}K(t, w)[a^{-1}e(t, w)]^{-1} = a^{-1}K(t, w)e(t, w)^{-1}a$ $= a^{-1}M(t, w)a$ which completes the proof.

§ 5. A stochastic solution of the heat equation on forms

In this section, we will discuss on the following initial value problem for heat equation on R^n-valued functions on $O(M)$ which are equivariant:

$$(2.9)\quad \begin{cases} \dfrac{\partial u}{\partial t} = \dfrac{1}{2}\Delta_{O(M)}u\,, & \lim\limits_{t\downarrow 0} u(t, \cdot) = F\,, \\[2ex] Pe^{-1}u + Qe^{-1}\dfrac{\partial u}{\partial x^n} = 0 & \text{on} \quad \partial(O(M))\,. \end{cases}$$

As we saw in section 2, this initial value problem is equivalent to the initial value problem (2.3) for 1-forms. More generally, we consider the following initial value problem for heat equation on R^n-valued functions on $O(M)$:

$$(5.1)\quad \begin{cases} \dfrac{\partial u}{\partial t} = \dfrac{1}{2}\Delta_{O(M)}u\,, \qquad \lim\limits_{t\downarrow 0} u(t, \cdot) = F \\[2ex] Pe^{-1}u + Qe^{-1}\Bigg[\dfrac{\partial u}{\partial x^n} - \sum\limits_{i,k,m=1}^{n} \dfrac{\partial u}{\partial e^i_m}\Gamma^i_{nk}(x)e^k_m + e\Gamma_n(x)e^{-1}u\Bigg] = 0\,, \\[2ex] \qquad\qquad\qquad\qquad\qquad\qquad \text{on} \quad r = [x, e] \in \partial(O(M))\,, \end{cases}$$

where $\Gamma_n(x) = (\Gamma^j_{ni}(x))$. If $u(t, r)$ is equivariant, then $u(t, r) = e\bar{u}(x)$, $r = [e, x]$, where $\bar{u}(x)$ is a smooth R^n-valued function on M and hence

$$\sum_{i,k,m=1}^{n} \frac{\partial u}{\partial e^i_m} \Gamma^i_{nk}(x) e^k_m = e\Gamma_n(x)e^{-1}u .$$

Thus, the initial value problem (5.1) is reduced to the initial value problem (2.9) in the case of equivariant functions. In the following, we will construct a semigroup corresponding to the initial value problem (5.1) by a probabilistic method.

Let $(r(t, w) = [X(t, w), e(t, w)], W, \mathscr{F}, P_r ; \mathscr{F}_t, \theta_t)$ be the canonical realization of the reflecting Brownian motion on the bundle of orthonormal frames $O(M)$ and let $M = \{M(t, w)\}_{t \geq 0}$ be the MOF of the process constructed in section 4. Let $C_0(M \to R^n)$ be the set of all bounded continuous functions $F(r)$ on $O(M)$ taking values in R^n such that

$$(5.2) \qquad Pe^{-1}F(r) = 0 \quad \text{if} \quad r = [x, e] \in \partial(O(M)) , \quad (\text{i.e. } x \in \partial M) .$$

For $F \in C_0(M \to R^n)$ and $t \geq 0$, set

$$(5.3) \qquad (H_t F)(r) = E_r[M(t, w)F(r(t, w))] .$$

Theorem 5.1. (i) *H_t defines a one parameter semigroup of operators on $C_0(M \to R^n)$.*

(ii) *If F is equivariant, then so is $H_t F$ for all $t \geq 0$.*

Proof. If $r = [x, e] \in \partial(O(M))$, then by Lemma 4.2, $Pe^{-1}(H_t F)(r) = 0$. The continuity in r of functions $H_t F$, $t \geq 0$ is proved in a usual way as follows. $\{X(t), e(t), M(t)\}$ are determined by the equations (3.1) and (4.3) (or equivalently (4.5)) and as we know, the uniqueness of solutions holds. Then the probability law of the above system is weakly continuous in the initial value r of $r(t)$ by a standard argument and this implies the continuity of $r \mapsto H_t F(r)$. The semigroup property of H_t is obvious since M is MOF.

The assertion (ii) follows from Lemmas 3.3 and 4.3.

Theorem 5.2. *Let $F(t, r)$ be a smooth function on $[0, \infty) \times M$ taking values in R^n such that for each $t \geq 0$, $r \mapsto F(t, r)$ is a function in $C_0(M \to R^n)$. Then, with probability one,*

$$M(t)F(t, r(t)) - M(0)F(0, r(0))$$
$$= \sum_{i=1}^{n} \int_0^t M(u)(L_{A_i}F)(u, r(u))dB^i(u)$$

$$
\begin{aligned}
&+\int_0^t M(u)\left(\left[\frac{\partial}{\partial t}+\frac{1}{2}\Delta_{O(M)}\right]F\right)(u,r(u))du \\
(5.4)\qquad &+\int_0^t M(u)e(u)Qe(u)^{-1}\left[\frac{\partial F}{\partial x^n}(u,r(u))\right. \\
&-\sum_{i,k,m=1}^{n}\frac{\partial F}{\partial e_m^i}(u,r(u))\Gamma_{nk}^i(X(u))e_m^k(u) \\
&\left.+e(u)\Gamma_n(X(u))e(u)^{-1}F(u,r(u))\right]d\psi(u)\,.
\end{aligned}
$$

Remark 5.1. $\psi(t)$ and $B(t)=(B^i(t))$ are as in the equation (3.1):

$$\psi(t)=\lim_{\varepsilon\downarrow 0}\frac{1}{2\varepsilon}\int_0^t I_{[0,\varepsilon)}(X(s))g^{nn}(X(s))ds$$

and

$$dB^i(t)=\sum_{k=1}^{n}[{}^te(u)]_{ik}^{-1}\circ[dX^k(u)-\delta_{kn}d\psi(u)]\,.$$

$B(t)$ is an n-dimensional $(\mathscr{F}_t)$-Brownian motion and hence the first term in the right-hand side of (5.4) is a martingale.

Proof. As we remarked in section 3, the diffusion $r(t)$ is a normally reflecting diffusion on $O(M)$ in the sense of [14], § 5.1. A characteristic feature of such a diffusion is that, if $f(t,r)$ is a smooth function on $[0,\infty)\times O(M)$ and if $g(t)$ is an $(\mathscr{F}_t)$-adapted process such that $s\mapsto g(s)$ is right continuous with left-hand limits and $s\mapsto E[g(s)^2]$ is locally bounded, then the following identity holds:

$$(5.5)\qquad \sum_{\substack{s\le\psi(t)\\ s\in D}}{}^{*}\int_{A(s-)}^{A(s)\wedge t}g(s)df(s,r(s))=\int_0^t g(s)df(s,r(s))\,,$$

where the integral is understood in the sense of stochastic integral by a quasi-martingale $s\mapsto f(s,r(s))$ and the sum $\sum^*_{\substack{s\le\psi(t)\\ s\in D}}$ is understood as the limit in probability of a finite sum $\sum_{\substack{s\le\phi(t)\\ A(s)-A(s-)>\varepsilon}}$ as $\varepsilon\downarrow 0$.

First we prove the following:

Lemma 5.1. *For any t such that $X_t\in M^\circ$,*

$$(5.6)\qquad dM(t)=\tfrac{1}{2}M(t)J(r(t))dt\,,$$

i.e., if $u\in D$, then for every $s\le t$ such that $[s,t]\subset(A(u-),A(u))$, we have

$$(5.6)'\qquad M(t)-M(s)=\frac{1}{2}\int_s^t M(\eta)J(r(\eta))d\eta\,.$$

Proof. By (4.3) and Itô's formula, if $X_t \in M^\circ$,

$$\begin{aligned} dM(t) &= K(t)e(t)^{-1}\{de(t)\cdot e(t)^{-1} + e(t)\cdot d[e(t)^{-1}] + de(t)\cdot d[e(t)^{-1}]\} \\ &\quad + \tfrac{1}{2}K(t)e(t)^{-1}J(r(t))dt \\ &= K(t)e(t)^{-1}d(e(t)e(t)^{-1}) + \tfrac{1}{2}K(t)e(t)^{-1}J(r(t))dt \\ &= \tfrac{1}{2}M(t)J(r(t))dt\ . \end{aligned}$$

Now we return to the proof of (5.4). We have, by Lemma 5.1 and Itô's formula,

$$\begin{aligned} &\sum_{\substack{s\le\psi(t)\\ s\in D}}{}^{*} \{M(t\wedge A(s))F(t\wedge A(s), r(t\wedge A(s))) \\ &\qquad\qquad - M(A(s-))F(A(s-), r(A(s-)))\} \\ &= \sum_{\substack{s\le\psi(t)\\ s\in D}}{}^{*} \int_{A(s-)}^{A(s)\wedge t} d[M(u)F(u, r(u))] \\ &= \sum_{\substack{s\le\psi(t)\\ s\in D}}{}^{*} \int_{A(s-)}^{A(s)\wedge t} M(u)dF(u, r(u)) \\ &\quad + \frac{1}{2}\sum_{\substack{s\le\psi(t)\\ s\in D}}{}^{*} \int_{A(s-)}^{A(s)\wedge t} M(u)J(r(u))F(u, r(u))du\ . \end{aligned} \tag{5.7}$$

Using (5.5) and (3.6), (5.7) is equal to

$$\begin{aligned} &\int_0^t M(u)\,dF(u, r(u)) + \frac{1}{2}\int_0^t M(u)J(r(u))F(u, r(u))du \\ &= \sum_{m=1}^{n}\int_0^t M(u)(L_{A_m}F)(u, r(u))dB^m(u) \\ &\quad + \int_0^t M(u)\Big[\frac{\partial F}{\partial t}(u, r(u)) + \frac{1}{2}(\Delta_{O(M)}F)(u, r(u)) \\ &\qquad\qquad + J(r(u))F(u, r(u))\Big]du \\ &\quad + \int_0^t M(u)\Big[\frac{\partial F}{\partial x^n}(u, r(u)) \\ &\qquad\qquad - \sum_{i,k,m=1}^{n}\frac{\partial F}{\partial e_m^i}(u, r(u))\Gamma_{nk}^i(X(u))e_m^k(u)\Big]d\psi(u)\ . \end{aligned} \tag{5.8}$$

On the otherhand, for every $s < t$,

$$\begin{aligned} &M(t)F(t, r(t)) - M(s)F(s, r(s)) \\ &\quad = K^1(t)Pe(t)^{-1}F(t, r(t)) - K^1(s)Pe(s)^{-1}F(s, r(s)) \\ &\qquad + K^2(t)Qe(t)^{-1}F(t, r(t)) - K^2(s)Qe(s)^{-1}F(s, r(s))\ . \end{aligned}$$

Then, noting that $Pe^{-1}F(t,r) = 0$ if $r \in \partial(O(M))$ and that $r(A(u)) \in \partial(O(M))$ if $u \in D$ and $r(A(u-)) \in \partial(O(M))$ if $u \in D$ and $u > 0$, we see that the first line of (5.7) is equal to

$$\begin{aligned}
&[K^1(t)Pe(t)^{-1}F(t, r(t)) - K^1(0)Pe(0)^{-1}F(0, r(0))] \\
(5.9)\qquad &+ \sum_{\substack{s \le \psi(t) \\ s \in D}}^{*} \{K^2(t \wedge A(s))Qe(t \wedge A(s))^{-1}F(t \wedge A(s), r(t \wedge A(s))) \\
&\qquad - K^2(A(s-))Qe(A(s-))^{-1}F(A(s-), r(A(s-)))\} .
\end{aligned}$$

Since, by (4.3),

$$\begin{aligned}
dK^2(u) = &K(u)\{e(u)^{-1}de(u) + \tfrac{1}{2}R(X(u))du\}Q \\
&- K(u)e(u)^{-1}de(u)I_{\partial M}(X(u))Q ,
\end{aligned}$$

it is clear that $d\{K^2(u)Qe(u)^{-1}F(u, r(u))\}$ has the form

$$\begin{aligned}
&g_1(u)de(u) + g_2(u)du + g_3(u)d(e(u)^{-1}) + g_4(u)dF(u, r(u)) \\
&\quad - I_{\partial M}(X(u))K(u)e(u)^{-1}de(u)Qe(u)^{-1}F(u, r(u)) ,
\end{aligned}$$

where $u \mapsto g_i(u)$, $i = 1, 2, 3, 4$, are all right continuous $(\mathscr{F}_t)$-adapted processes with left-hand limits. Hence, using again (5.5), the second term of (5.9) is equal to

$$\begin{aligned}
&\sum_{\substack{s \le \psi(t) \\ s \in D}}^{*} \int_{A(s-)}^{A(s)\wedge t} d\{K^2(u)Qe(u)^{-1}F(u, r(u))\} \\
&\quad = \int_0^t g_1(u)de(u) + \int_0^t g_2(u)du + \int_0^t g_3(u)d(e(u)^{-1}) \\
&\qquad + \int_0^t g_4(u)dF(u, r(u)) \\
&\quad = K^2(t)Qe(t)^{-1}F(t, r(t)) - K^2(0)Qe(0)^{-1}F(0, r(0)) \\
&\qquad + \int_0^t I_{\partial M}(X(u))K(u)e(u)^{-1}de(u)Qe(u)^{-1}F(u, r(u)) .
\end{aligned}$$

Because of (3.1) and $I_{\partial M}(X(u))Pe(u)^{-1}F(u, r(u)) = 0$,

$$\begin{aligned}
&\int_0^t I_{\partial M}(X(u))K(u)e(u)^{-1}de(u)Qe(u)^{-1}F(u, r(u)) \\
&\quad = -\int_0^t K(u)e(u)^{-1}e(u)\Gamma_n(X(u))Qe(u)^{-1}F(u, r(u))d\psi(u) \\
&\quad = -\int_0^t K(u)\Gamma_n(X(u))e(u)^{-1}F(u, r(u))d\psi(u) .
\end{aligned}$$

Thus, we have

$$
\begin{aligned}
&M(t)F(t, r(t)) - M(0)F(0, r(0)) \\
&\quad - \int_0^t M(u)e(u)\Gamma_n(X(u))e(u)^{-1}F(u, r(u))d\psi(u) \\
&= \sum_{m=1}^{n} \int_0^t M(u)(L_{A_m}F)(u, r(u))dB^m(u) \\
&\quad + \int_0^t M(u)\left[\frac{\partial F}{\partial t}(u, r(u)) + \frac{1}{2}(\Delta_{O(M)}F)(u, r(u))\right]du \qquad (5.10) \\
&\quad + \int_0^t M(u)\left[\frac{\partial F}{\partial x^n}(u, r(u))\right. \\
&\quad \left. - \sum_{i,k,m=1}^{n} \frac{\partial F}{\partial e_m^i}(u, r(u))\Gamma_{nk}^i(X(u))e_m^k(u)\right]d\psi(u) .
\end{aligned}
$$

Finally we remark that, if $g(u)$ is a measurable $(\mathscr{F}_t)$-adapted process, then

$$
\begin{aligned}
&\int_0^t M(u)g(u)d\psi(u) \\
&\qquad = \int_0^t K^1(u)e(u)^{-1}g(u)d\psi(u) + \int_0^t K^2(u)e(u)^{-1}g(u)d\psi(u) \\
&\qquad = \int_0^t K^2(u)e(u)^{-1}g(u)d\psi(u) \\
&\qquad = \int_0^t M(u)e(u)Qe(u)^{-1}g(u)d\psi(u) ,
\end{aligned}
$$

because of $I_{\partial M}(X(u))K^1(u) = 0$. Now (5.4) follows from (5.10).

Theorem 5.2 may be regarded as a martingale version of the statement that $u(t, r) = H_tF(r)$ solves (5.1). In fact, by (5.4), we can conclude that any smooth solution of (5.1), if it exists, must coincide with $u(t, r) = H_tF(r)$.

References

[1] Airault, H., Résolution stochastique d'un problème de Dirichlet-Neumann, C. R. Acad. Sci. Paris, **280** (1975), 781–784.

[2] ——, Problèmes de Dirichlet-Neumann étalés et fonctionnelles multiplicatives associeés, Séminaire sur les équations aux dérivées partielles III, Collége de France, 1974–1975.

[3] ——, Perturbations singulières et solutions stochastiques de problèmes de D. Neumann-Spencer, J. Math. Pures Appl., **54** (1976).

[4] Bishop, R. L. and Crittenden, R. J., Geometry of manifolds, Academic Press, 1964.

[5] Conner, P. E., The Neumann's problem for differential forms on Riemannian manifolds, Mem. Amer. Math. Soc., **20** (1956).
[6] Itô, K., The Brownian motion and tensor fields on Riemannian manifold, Proc. Internat. Congress of Math. Stockholm (1962), 536–539.
[7] ——, Stochastic differentials, Appl. Math. Opt., **1** (1975), 374–381.
[8] ——, Stochastic parallel displacement, Probabilistic methods in differential equations, Lecture Notes in Math., **451** (1975), Springer, 1–7.
[9] ——, The Brownian motion and harmonic tensor fields on Riemannian manifold, Sugaku, Math. Soc. Japan, **28** (1976), 137–146 (in Japanese).
[10] Malliavin, P., Formules de la moyenne, calcul de perturbations et théorèmes d'annulations pour les formes harmoniques, J. Functional Analysis, **17** (1974), 274–291.
[11] ——, Elliptic estimates and diffusions in Riemannian geometry and complex analysis, Probabilistic methods in differential equations, Lecture Notes in Math., **451** (1975), Springer, 26–33.
[12] Ray, D. B. and Singer, I. M., R-torsion and the Laplacian on Riemannian manifolds, Advances in Math., **7** (1971), 145–210.
[13] Watanabe, S., On stochastic differential equations for multi-dimensional diffusion processes with boundary conditions, J. Math. Kyoto Univ., **11** (1971), 169–180.
[14] ——, Excursion point process of diffusion and stochastic integral, These Proceedings, 437–461.

NOBUYUKI IKEDA
DEPARTMENT OF MATHEMATICS
OSAKA UNIVERSITY
TOYONAKA 560, JAPAN

SHINZO WATANABE
DEPARTMENT OF MATHEMATICS
KYOTO UNIVERSITY
KYOTO 606, JAPAN

Proc. of Intern. Symp. SDE
Kyoto 1976, pp. 95–109

Extension of Stochastic Integrals

Kiyosi Itô

§ 1. Introduction

Let $B = \{B_t, 0 \leqq t < \infty\}$ be a Brownian motion. From the intuitive meaning of integrals we want to have

$$\int_s^t B_1 dB_u = B_1(B_t - B_s) . \tag{1.1}$$

Unfortunately the left hand side has no meaning in the sense of Brownian stochastic integrals [1] because $Y_t \equiv B_1$ depends on the future development of B for $t < 1$. However, we can interpret the integral changing our view from Brownian stochastic integrals into quasimartingale integrals. Let $\mathscr{F} = \{\mathscr{F}_t, 0 \leqq t < \infty\}$ be the reference family (i.e. the filtration) generated by two processes B and $Y_t \equiv B_1$. Then $Y = \{Y_t, 0 \leqq t < \infty\}$ is $\mathscr{F}$-adapted and B is an $\mathscr{F}$-quasimartingale with the canonical $\mathscr{F}$-decomposition:

$$B_t = M_t + A_t\,, \quad A_t = \int_0^{t\wedge 1} \frac{B_1 - B_u}{1-u} du\,, \quad M_t = B_t - A_t\,. \tag{1.2}$$

Hence the left hand side of (1) is meaningful in the sense of quasimartingale stochastic integrals and

$$\begin{aligned}\int_s^t B_1 dB_u &= \int_s^t B_1 dM_u + \int_s^t B_1 dA_u \\ &= B_1(M_t - M_s) + B_1(A_t - A_s) \\ &= B_1(B_t - B_s)\,,\end{aligned}$$

as we desire. The purpose of this paper is to develop this idea and present some of its applications.

In Section 2 we will review the theory of quasimartingale stochastic integrals from our view-point. This theory was first initiated by D. L. Fisk [2] and later developed extensively by P. Courrège [3], H. Kunita and S. Watanabe [4] and P. A. Meyer [5]. In Section 3 we will observe different aspects of Stratonovich's symmetric stochastic integrals [6]. The forward symmetric stochastic integrals discussed in this section was introduced first by Fisk [2]; see also K. Itô [7]. In the last section we will discuss stochastic paralled displacement on a Riemannian manifold introduced by K. Itô [8], [9]. E. B. Dynkin [10] extended it to the case of a manifold with connection. This case can be discussed in the same way as we do here.

§ 2. Stochastic integrals

Let $\Omega = (\Omega, \mathscr{F}_\Omega, P)$ be a complete probability space. A stochastic process $X = \{X_t(\omega), 0 \leqq t < \infty\}$ on Ω is called *continuous* (resp. *locally bounded, locally finite variation*) if its sample function is continuous a.s. (resp. bounded on every finite interval a.s., of finite absolute variation on every finite interval a.s.).

Let $\mathscr{F} = \{\mathscr{F}_t, 0 \leqq t < \infty\}$ be a right-continuous increasing family of sub-σ-algebras of $\mathscr{F}_\Omega$ such that $\mathscr{F}_t$ contains all P-null sets. Such a family is simply called a *reference family* in this paper. A process adapted to $\mathscr{F}$ is called an *$\mathscr{F}$-adapted process* or an *$\mathscr{F}$-process*. Similarly we use the terms *$\mathscr{F}$-martingales, $\mathscr{F}$-stopping times, $\mathscr{F}$-well-measurable processes*, etc.. We may omit the prefix $\mathscr{F}$ in case we refer to a fixed reference family. Let X be a stochastic process. The σ-algebra generated by the sets

$$\{\omega: X_t(\omega) < a\}, \quad t \in [0, \infty), \quad a \in (-\infty, \infty)$$

and the P-null sets is denoted by $\bar{\sigma}_t[X]$. The reference family

$$\bar{\sigma}_{t+}[X] \equiv \bigcap_n \bar{\sigma}_{t+1/n}[X]$$

is called the *reference family generated by X*, $\mathscr{F}[X] = \{\mathscr{F}_t[X]\}$ in notation. $\mathscr{F}[X]$ is the least of all reference families to which X is adapted. Similarly for the reference family $\mathscr{F}[X, Y, \cdots]$ generated by a family of processes $X, Y, \cdots$.

A processes $X = \{X_t\}$ is called a *continuous local $\mathscr{F}$-martingale* if X is continuous and if there exists a sequence of $\mathscr{F}$-stopping times $\theta_1 \leqq \theta_2 \leqq \cdots \to \infty$ a.s. such that the stopped process $\{X_{t\wedge\theta_n}\}$ is an $\mathscr{F}$-martingale for every n. X is called a *continuous local $\mathscr{F}$-quasimartingale* if X is expressible as

$$X = M + A, \quad A(0) = 0 \tag{2.1}$$

where X is a continuous local $\mathscr{F}$-martingale and A is a continuous, locally bounded variation $\mathscr{F}$-process. Since the expression (2.1) is uniquely determined by X and $\mathscr{F}$ if exists, it it is called the *canonical $\mathscr{F}$-decomposi-*

tion of X. The family of all continuous local $\mathscr{F}$-quasimartingales is denoted by $\mathscr{Q}_{\mathscr{F}}$.

Let $\mathscr{W}_{\mathscr{F}}$ denote the $\mathscr{F}$-well-measurable processes and $\mathscr{B}_{\mathscr{F}}$ the locally bounded $\mathscr{F}$-well-measurable processes. For every $X \in \mathscr{Q}_{\mathscr{F}}$ with the canonical $\mathscr{F}$-decomposition $X = M + A$ we denote by $\mathscr{L}_{\mathscr{F}}(dX)$ the family of all $Y \in \mathscr{W}_{\mathscr{F}}$ such that

$$\int_0^t Y^2(dM)^2 < \infty \quad \text{and} \quad \int_0^t |Y||dA| \qquad \text{for every } t \tag{2.2}$$

a.s., where $(dM)^2$ is the Lebesgue-Stieltjes measure induced by the quadratic variation $\{\langle M\rangle_t\}$ of $\{M_t\}$ and $|dA|$ is that induced by the absolute variation $\{|A|_t\}$ of $\{A_t\}$.

Suppose that $X \in \mathscr{Q}_{\mathscr{F}}$. For every $Y \in \mathscr{L}_{\mathscr{F}}(dx)$ we define

$$\mathscr{F}\text{-}\int_0^t YdX = \mathscr{F}\text{-}\int_0^t YdM + \int_0^t YdA\ , \quad 0 < t < \infty\ , \tag{2.3}$$

where $\mathscr{F}\text{-}\int YdM$ is the stochastic (martingale) $\mathscr{F}$-integral and $\int YdA$ is the Lebesgue-Stieltjes integral (for each sample). The process

$$\mathscr{F}\text{-}\int_0^t YdX\ , \qquad 0 \leq t < \infty\ ,$$

is called the *stochastic* (*quasimartingale*) $\mathscr{F}$-*integral* of Y based on dX. It is obviously a continuous local $\mathscr{F}$-quasimartingale whose canonical decomposition is given by (2.1). The properties of quasi-martingale integrals will follow immediately from those of martingale integrals and Lebesgue-Stieltjes integrals. For example,

$$\begin{aligned} &\int_0^t (Y_n - Y)^2(dM)^2 + \int_0^t |Y_n - Y||dA| \to 0,\ \forall t, \text{ a.s. }. \\ &\qquad \Longrightarrow \mathscr{F}\text{-}\int_0^t Y_n dX \to \mathscr{F}\text{-}\int_0^t YdX,\ \forall t, \text{ a.s.} \end{aligned} \tag{2.4}$$

Since $\mathscr{B}_{\mathscr{F}} \subset \mathscr{L}_{\mathscr{F}}(dX)$ for every $X \in \mathscr{Q}_{\mathscr{F}}$, the integral $\mathscr{F}\text{-}\int YdX$ is defined for every $Y \in \mathscr{B}_{\mathscr{F}}$.

Thus far we have fixed the reference family $\mathscr{F}$. Now we will discuss what happens on when we replace $\mathscr{F}$ by another reference family $\tilde{\mathscr{F}}$ finer than $\mathscr{F}$; $\tilde{\mathscr{F}}$ is called *finer* than $\mathscr{F}$, $\tilde{\mathscr{F}} \succ \mathscr{F}$ in notation, if $\tilde{\mathscr{F}}_t \supset \mathscr{F}_t$ for every t. Suppose that $\tilde{\mathscr{F}} \succ \mathscr{F}$. Then $\mathscr{W}_{\tilde{\mathscr{F}}} \supset \mathscr{W}_{\mathscr{F}}$ and $\mathscr{B}_{\tilde{\mathscr{F}}} \supset \mathscr{B}_{\mathscr{F}}$, but there is no inclusion relation between $\mathscr{Q}_{\tilde{\mathscr{F}}}$ and $\mathscr{Q}_{\mathscr{F}}$. Even if $X \in \mathscr{Q}_{\mathscr{F}} \cap \mathscr{Q}_{\tilde{\mathscr{F}}}$, there is no inclusion relation between $\mathscr{L}_{\mathscr{F}}(dX)$ and $\mathscr{L}_{\tilde{\mathscr{F}}}(dX)$, but we have

$$\mathscr{B}_{\mathscr{F}} \subset \mathscr{L}_{\mathscr{F}}(dX) \cap \mathscr{L}_{\tilde{\mathscr{F}}}(dX) .$$

Theorem 2.1. *Suppose that* $\tilde{\mathscr{F}} \succ \mathscr{F}$. *If* $X \in \mathscr{Q}_{\mathscr{F}} \cap \mathscr{Q}_{\tilde{\mathscr{F}}}$, then

$$\mathscr{F}\text{-}\int_0^t YdX = \tilde{\mathscr{F}}\text{-}\int_0^t YdX , \qquad 0 \leqq t < \infty, \text{ a.s.} \tag{2.5}$$

for every $Y \in \mathscr{L}_{\mathscr{F}}(dX) \cap \mathscr{L}_{\tilde{\mathscr{F}}}(dX)$ *(and hence for every* $Y \in \mathscr{B}_{\mathscr{F}}$*).*

Proof. Let $\mathscr{L}$ denote the family of all Y's in $\mathscr{L}_{\mathscr{F}}(dX) \cap \mathscr{L}_{\tilde{\mathscr{F}}}(dX)$ for which the equation (2.5) holds. It is obvious that $\mathscr{L}$ is closed under linear combinations. Using the property (2.4) we can easily prove that if (i) $Y^{(n)} \in \mathscr{L}$, $n = 1, 2, \cdots$, (ii) for each ω, $Y_t^{(n)}(\omega)$ is bounded on every finite t-interval and (iii) $Y_t^{(n)}(\omega) \to Y_t(\omega)$ for every (t, ω), then $Y \in \mathscr{L}$. By the definition $\mathscr{L}$ contains all right-continuous step $\mathscr{F}$-processes with possible jumps on a discrete subsets of $[0, \infty)$ independent of ω. Hence $\mathscr{L} \supset \mathscr{B}_{\mathscr{F}}$. If $Y \in \mathscr{L}_{\mathscr{F}}(dX) \cap \mathscr{L}_{\tilde{\mathscr{F}}}(dX)$, then the sequence of the truncated processes

$$Y_t^{(n)}(\omega) = Y_t(\omega)1_{[-n,n]}(Y_t(\omega)) , \qquad n = 1, 2, \cdots$$

satisfies the assumption of (2.4) and

$$Y^{(n)} \in \mathscr{B}_{\mathscr{F}} \subset \mathscr{L} , \qquad n = 1, 2, \cdots$$

as was proved above. Hence we have $Y \in \mathscr{L}$, completing the proof.

Example 2.1. (The example mentioned in the introduction). Let $B = \{B_t, 0 \leqq t < \infty\}$ be a Brownian motion. We consider two reference families

$$\mathscr{F} = \mathscr{F}[B] \quad \text{and} \quad \tilde{\mathscr{F}} = \mathscr{F}[B_1, B]$$

where B_1 denotes the process equal to $B_1(\omega)$ for every t. Since $B = \{B_t\}$ is a continuous $\mathscr{F}$-martingale (so $B \in \mathscr{Q}_{\mathscr{F}}$), we can define

$$\mathscr{F}\text{-}\int_0^t YdB$$

for every Y in the family

$$\mathscr{L}_{\mathscr{F}}(dB) = \left\{ Y \in \mathscr{W}_{\mathscr{F}} : \int_0^t Y_s^2 ds < \infty, \forall t, \text{ a.s.} \right\} ,$$

noting that $(dB)^2 = dt$. In fact this $\mathscr{F}$-integral is essentially the same as the classical Brownian stochastic integral. Since $B = \{B_t\}$ is a continuous $\tilde{\mathscr{F}}$-quasi-martingale with the canonical decomposition

$$B_t = M_t + A_t\,, \quad A_t = \int_0^{t\wedge 1} \frac{B_1 - B_s}{1-s} ds\,, \quad M_t = B_t - A_t\,,$$

we can define

$$\tilde{\mathscr{F}}\text{-}\int_0^t Y dB$$

for every Y in the family

$$\mathscr{L}_{\tilde{\mathscr{F}}}(dB) = \left\{ Y \in \mathscr{W}_{\tilde{\mathscr{F}}} : \int_0^t Y_s^2 ds + \int_0^{t\wedge 1} |Y_s| \frac{|B_1 - B_s|}{1-s} ds < \infty,\ \forall t, \text{ a.s.} \right\}.$$

Noting that

$$\langle M \rangle_t = \langle B \rangle_t = t$$

and

$$E(|A|_1) = E\left(\int_0^1 \left| \frac{B_1 - B_s}{1-s} \right| ds \right) = \sqrt{\frac{2}{\pi}} \int_0^1 \frac{\sqrt{1-s}}{1-s} ds < \infty\,,$$

so

$$|A|_t \leqq |A|_1 < \infty \qquad \text{a.s.}\,.$$

It is obvious that

$$\text{(2.6)} \qquad Y = \{Y_t \equiv B_1,\ 0 \leqq t < \infty\} \in \mathscr{L}_{\tilde{\mathscr{F}}}(dB) \backslash \mathscr{L}_{\mathscr{F}}(dB)\,.$$

If we could prove that $\mathscr{L}_{\tilde{\mathscr{F}}}(dB) \supset \mathscr{L}_{\mathscr{F}}(dB)$, the $\tilde{\mathscr{F}}$-integral would be a proper extension of the $\mathscr{F}$-integral by virtue of Theorem 2.1, but we can neither prove nor disprove this inclusion relation at present.

Example 2.2. This is another example which is slightly more complicated than the previous one. Let $B = \{B_t\}$ be a Brownian motion as above. We want to define the stochastic integral

$$\text{(2.7)} \qquad \int_0^t f_s \left(\int_{s\wedge 1}^1 B_u du \right) dB_s$$

where $f_s(x)$ is a locally bounded Borel function of $(s, x) \in [0, \infty) \times (-\infty, \infty)$. It is obvious that this integral is meaningless in the sense of the classical Brownian stochastic integral. Let $\tilde{\mathscr{F}} = \{\tilde{\mathscr{F}}_t, 0 \leqq t < \infty\}$ be the reference family generated by the processes

$$B = \{B_t, 0 \leqq t < \infty\} \quad \text{and} \quad I = \left\{ I_t \equiv \int_{t\wedge 1}^{1} B_u du,\ 0 \leqq t < \infty \right\}.$$

Observing that for every $t \in (0, 1)$

$$\begin{aligned} \sigma[B_s, I_s, s \leqq t] &= \sigma\left[B_s, \int_s^1 B_u du,\ s \leqq t\right] \\ &= \sigma\left[B_s,\ s \leqq t, \int_t^1 B_u du\right] \\ &= \sigma\left[B_s,\ s \leqq t, \int_t^1 (B_u - B_t) du\right] \end{aligned}$$

and recalling that B is a Gaussian process with $E(B_t) = 0$ and $E(B_t B_s) = s \wedge t$, we can prove that B is a continuous local $\tilde{\mathscr{F}}$-quasimartingale with the canonical $\tilde{\mathscr{F}}$-decomposition

$$B_t = M_t + A_t$$

where

$$A_t = \int_0^{t\wedge 1} \frac{3}{(1-s)^2} \int_s^1 |B_u - B_s|\, duds \quad \text{and} \quad M_t = B_t - A_t\,.$$

It should be noted that

$$\begin{aligned} E(|A|_1) &\leqq E\left(\int_0^1 \frac{3}{(1-s)^2} \int_s^1 |B_u - B_s|\, duds\right) \\ &= \int_0^1 \frac{3}{(1-s)^2} \int_s^1 \sqrt{u-s}\, duds \\ &= \int_0^1 \frac{2}{\sqrt{1-s}} ds < \infty\,, \end{aligned}$$

so $|A|_t \leqq |A|_1 < \infty$ for every t a.s.. It is obvious that

$$\langle M \rangle_t = \langle B \rangle_t = t\,.$$

Hence

$$\mathscr{L}_{\mathscr{F}}(dB) = \left\{ Y \in \mathscr{W}_{\mathscr{F}} : \int_0^t Y^2 dt + \int_0^{t\wedge 1} |Y| \frac{3}{(1-s)^2} \left| \int_s^1 (B_u - B_s) du \right| ds < \infty,\ \forall t,\ \text{a.s.} \right\}.$$

Since the integrand process of (2.7) belongs to $\mathscr{B}_{\mathscr{F}} \subset \mathscr{L}_{\mathscr{F}}(dX)$, the stochastic integral (2.7) is well-defined.

Example 2.3. The above examples show that we may define the stochastic integral of an integrand depending on the future development of the base process by taking a finer reference family to which the integrand and the base process are adapted. But this is not always possible. Let $B = \{B_t\}$ be a Brownian motion as above. Since the space C of all continuous functions on $[0, \infty)$ is a Polish space with non-denumerable points, there exists a Borel isomorphism $f: C \to [0, 1]$. Let

$$Y_t(\omega) = f(B_u(\omega), 0 \leqq u < \infty) \qquad \text{for every } t .$$

In order to define the stochastic integral $\int YdB$ we must choose a reference family $\mathscr{F}$ finer than $\mathscr{F}[B, Y]$, but B cannot be a continuous local $\mathscr{F}$-quasimartingale for such an $\mathscr{F}$. Suppose that it had a canonical $\mathscr{F}$-decomposition

$$B = M + A .$$

Since $\mathscr{F}_t \supset \sigma[Y_t] = \sigma[B_u, 0 \leqq u < \infty]$ for every t, we have

$$(B_{t+\Delta} - B_t) - E(B_{t+\Delta} - B_t | \mathscr{F}_t) = (B_{t+\Delta} - B_t) - (B_{t+\Delta} - B_t) = 0 ,$$

which implies $M_t \equiv 0$, so

$$B_t \equiv A_t .$$

Hence it follows that $dt = (dB_t)^2 = (dA_t)^2 = 0$, which is a contradiction.

Let $\mathscr{F} = \{\mathscr{F}_t, 0 \leqq t < \infty\}$ be a reference family. Then so is the family $\mathscr{F}^{(s)} = \{\mathscr{F}_{s+t}, 0 \leqq t < \infty\}$. If $X \in \mathscr{Q}_{\mathscr{F}}$ and $Y \in \mathscr{L}_{\mathscr{F}}(dX)$, then

$$X^{(s)} = \{X_{s+t}, 0 \leqq t < \infty\} \in \mathscr{Q}_{\mathscr{F}^{(s)}}$$

and

$$Y^{(s)} = \{Y_{s+t}, 0 \leqq t < \infty\} \in \mathscr{L}_{\mathscr{F}^{(s)}}(dX^{(s)}) ,$$

so the stochastic integral

$$\mathscr{F}^{(s)}\text{-}\int_0^t Y^{(s)} dX^{(s)}$$

is well-defined and is denoted by

$$\mathscr{F}\text{-}\int_t^{s+t} YdX .$$

If $Z \in \mathscr{Q}_{\mathscr{F}}$ and if

$$Z_t - Z_s = \mathscr{F}\text{-}\int_s^t YdX\,, \qquad 0 \leqq s \leqq t < \infty\,,$$

then we denote this relation by

$$dZ = \mathscr{F}\text{-}YdX\;;$$

the prefix will often be omitted if there is no possibility of confusion.

If $dZ_i = Y_i dX_i$, $i = 1, 2, \cdots, n$ and if $F = f(Z)$ where

$$Z = (Z_1, Z_2, \cdots, Z_n) \quad \text{and} \quad f \in C^2(R^n)\,,$$

then the following stochastic chain rule holds:

$$dF = \sum_i \partial_i f(Z) dZ_i + \frac{1}{2} \sum_{i,j} \partial_i \partial_j f(Z) dZ_i dZ_j\,, \tag{2.8}$$

where ∂_i denotes the partial derivative with respect to the i-th variable and

$$dZ_i dZ_j = d\langle Z_i, Z_j\rangle$$
$$(\langle Z_i, Z_j\rangle = \tfrac{1}{4}(\langle Z_i + Z_j\rangle - \langle Z_i - Z_j\rangle))\,.$$

Let us mention one word about the case where the time interval is the whole real line $(-\infty, \infty)$. Let $\mathscr{F} = \{\mathscr{F}_t, -\infty < t < \infty\}$ be a right-continuous increasing family of sub-σ-algebras of $\mathscr{F}_\Omega$ such that $\mathscr{F}_t \supset \mathbf{2}$ for every t. Such a family is called a reference family on the time interval $(-\infty, \infty)$. A stochastic process $X = \{X_t, -\infty < t < \infty\}$ is called a *continuous local $\mathscr{F}$-quasimartingale*, $X \in \mathscr{Q}_{\mathscr{F}}$ in notation, if

$$X^{(s)} = \{X_{s+t}, 0 \leqq t < \infty\}$$

is a continuous local quasimartingale relative to the reference family

$$\mathscr{F}^{(s)} = \{\mathscr{F}_{s+t}, 0 \leqq t < \infty\}\,.$$

A process $Y = \{Y_t, -\infty < t < \infty\}$ is said to belong to $\mathscr{L}_{\mathscr{F}}(dX)$, if

$$Y^{(s)} = \{Y_{s+t}, 0 \leqq t < \infty\} \in \mathscr{L}_{\mathscr{F}^{(s)}}(dX^{(s)})$$

for every $s \in (-\infty, \infty)$. For $X \in \mathscr{Q}_{\mathscr{F}}$ and $Y \in \mathscr{L}_{\mathscr{F}}(dX)$ we define

$$\mathscr{F}\text{-}\int_s^{s+t} YdX = \mathscr{F}^{(s)}\text{-}\int_0^t Y^{(s)} dX^{(s)}\,.$$

This stochastic integral on $(-\infty, \infty)$ has the same properties as the above stochastic integral on $[0, \infty)$. It is obvious that

$$\mathscr{F}\text{-}\int_s^t YdX + \mathscr{F}\text{-}\int_t^u YdX = \mathscr{F}\text{-}\int_s^u YdX\,, \qquad s \leqq t \leqq u\,.$$

§ 3. Symmetric stochastic integrals

In the last section we have defined the stochastic integral

$$\mathscr{F}\text{-}\int_s^t YdX\,, \quad -\infty < s \leqq t < \infty\,, \quad X \in \mathscr{Q}_{\mathscr{F}}\,, \quad Y \in \mathscr{L}_{\mathscr{F}}(dX)\,.$$

By reversing the time order we can define *time reversed stochastic integrals*. The family

$$\mathscr{G} = \{\mathscr{G}_t,\ \infty > t > -\infty\}$$

is called a *time-reversed reference family*, if

$$\mathscr{G}^- = \{\mathscr{G}_t^- \equiv \mathscr{G}_{-t},\ -\infty < t < \infty\}$$

is a reference family in the sense of the last section. For a time-reversed reference family $\mathscr{G} = \{\mathscr{G}_t,\ \infty > t > -\infty\}$ we define a process $X = \{X_t,\ \infty > t > -\infty\}$ to be a *time reversed continuous local $\mathscr{G}$-quasi-martingale*, $X \in \mathscr{Q}_{\mathscr{G}}$ in notation, if

$$X^- = \{X_t^- \equiv X_{-t},\ -\infty < t < \infty\}$$

is a continuous local $\mathscr{G}^-$-quasimartingale. Similarly we define

$$Y = \{Y_t,\ \infty > t > -\infty\} \in \mathscr{L}_{\mathscr{G}}(dX)$$

if $Y^- \in \mathscr{L}_{\mathscr{G}^-}(dX^-)$. The (*time-reversed*) *stochastic $\mathscr{G}$-integral* is defined as follows:

$$\mathscr{G}\text{-}\int_t^s YdX = \mathscr{G}^-\text{-}\int_{-s}^{-t} Y^-dX^-\,, \qquad \infty > t \geqq s > -\infty$$

for $X \in \mathscr{Q}_{\mathscr{G}}$ and $Y \in \mathscr{L}_{\mathscr{G}}(dX)$.

Let $(\mathscr{F}, \mathscr{G})$ be a pair consisting of a reference family $\mathscr{F}$ and a time reversed reference family $\mathscr{G}$. We define

$$\mathscr{Q}_{(\mathscr{F},\mathscr{G})} = \mathscr{Q}_{\mathscr{F}} \cap \mathscr{Q}_{\mathscr{G}}$$

and

$$\mathscr{L}_{(\mathscr{F},\mathscr{G})}(dX) = \mathscr{L}_{\mathscr{F}}(dX) \cap \mathscr{L}_{\mathscr{G}}(dX) \qquad \text{for } X \in \mathscr{Q}_{(\mathscr{F},\mathscr{G})}\,.$$

For $X \in \mathscr{Q}_{(\mathscr{F},\mathscr{G})}$ and $Y \in \mathscr{L}_{(\mathscr{F},\mathscr{G})}(dX)$ we define

$$(\mathscr{F},\mathscr{G})\text{-}\int_s^t Y\circ dX = \frac{1}{2}\left[\mathscr{F}\text{-}\int_s^t YdX - \mathscr{G}\text{-}\int_t^s YdX\right]$$

or

$$= \frac{1}{2}\left[\mathscr{G}\text{-}\int_s^t YdX - \mathscr{F}\text{-}\int_t^s YdX\right]$$

according to whether $s \leqq t$ or $s \geqq t$ and call this integral the *symmetric stochastic integral* of Y based on dX relative to $(\mathscr{F},\mathscr{G})$. We will omit the prefix $(\mathscr{F},\mathscr{G})$ if there is no possibility of confusion.

For symmetric integrals we have

$$\int_s^t Y\circ dX + \int_t^u Y\circ dX = \int_s^t Y\circ dX$$

irrespectively of the order of s, t and u, so

$$\int_s^t Y\circ dX = -\int_t^s Y\circ dX\,.$$

Theorem 3.1. *Let $X\in\mathscr{Q}_{(\mathscr{F},\mathscr{G})}$ and $Y\in\mathscr{L}_{(\mathscr{F},\mathscr{G})}(dX)$. If Y is a continuous process, then*

$$(3.1)\qquad (\mathscr{F},\mathscr{G})\text{-}\int_s^t Y\circ dX = \underset{|\Delta|\to 0}{\text{l.i.p.}}\sum_{i=1}^n \frac{1}{2}(Y_{t_{i-1}} + Y_{t_i})(X_{t_i} - X_{t_{i-1}})\,,\qquad s\leqq t$$

where $\Delta = \{s = t_0 < t_1 < \cdots < t_n = t\}$ is an arbitrary partition of the interval $[s, t]$ and

$$|\Delta| = \max_{1\leqq i\leqq n}|t_i - t_{i-1}|\,.$$

Proof. Since Y is continuous,

$$Y_u(\omega) = \lim_{|\Delta|\to 0}\sum_i Y_{t_{i-1}}(\omega)1_{[t_{i-1},t_i]}(u)\,,\qquad s\leqq u < t\quad \text{a.s.},$$

we obtain

$$\mathscr{F}\text{-}\int_s^t YdX = \underset{|\Delta|\to 0}{\text{l.i.p.}}\sum_{i=1}^n Y_{t_{i-1}}(X_{t_i} - X_{t_{i-1}})\,.$$

Similarly

$$\mathscr{G}\text{-}\int_t^s YdX = \underset{|\Delta|\to 0}{\text{l.i.p.}}\sum_{i=1}^n Y_{t_i}(X_{t_{i-1}} - X_{t_i})\,.$$

Substracting and dividing by 2, we obtain (3.1).

We denote by $dZ = Y \circ dX$ the relation

$$Z_t - Z_s = \int_s^t Y \circ dX, \quad -\infty < s, t < \infty \quad (Z \in \mathcal{Q}_{(\mathcal{F},\mathcal{G})}).$$

Using Theorem 2.1 we obtain

Theorem 3.2 (*The chain rule for symmetric stochastic integrals*). *If* $dZ_i = Y_i \circ dX_i$, $i = 1, 2, \cdots, n$ *and if* $F = f(Z)$ *where* $Z = (Z_1, Z_2, \cdots, Z_n)$ *and* $f \in C^3(\boldsymbol{R}^n)$, *then*

$$dF = \sum_i \partial_i f(Z) \circ dZ_i. \tag{3.2}$$

Proof. Observing that

$$\begin{aligned} f(b) - f(a) &\qquad (b = (b_i),\ a = (a_i)) \\ &= \sum_i \frac{1}{2}(\partial_i f(a) + \partial_i f(b))(b_i - a_i) + (|b - a|^3) \end{aligned}$$

we can prove (3.2) by the argument used in proving (2.8).

Example 3.1. Let $B = \{B_t, 0 \leqq t < \infty\}$ be a Brownian motion with $B_0 = 0$. By defining $B_t = 0$ for $t < 0$ we have $B = \{B_t, -\infty < t < \infty\}$. Let

$$\mathcal{F}_t = \bigcap_n \sigma\left[B_s, s \leqq t + \frac{1}{n}\right] \vee \mathbf{2} \qquad (\mathbf{2} = \{A \in \mathcal{F}_\Omega : P(A) = 0 \text{ or } 1\})$$

and

$$\mathcal{G}_t = \bigcap_n \sigma\left[B_s, s \geqq t - \frac{1}{n}\right] \vee \mathbf{2}.$$

Then $\mathcal{F} = \{\mathcal{F}_t, -\infty < t < \infty\}$ is a reference family and $\mathcal{G} = \{\mathcal{G}_t, \infty > t > -\infty\}$ is a time-reversed reference family. $B = \{B_t\}$ is a continuous $\mathcal{F}$-martingale. Observing that

$$E(B_s - B_t | \mathcal{G}_t) = \frac{s-t}{t} B_t, \qquad 0 < s < t < \infty,$$

we can prove that $B = \{B_t\}$ is a continuous time reversed $\mathcal{G}$-quasimartingale with the canonical decomposition

$$B = M + A,$$

where

$$A_t = \int_{\infty}^{t\vee 0} \frac{1}{s} B_s ds\,, \qquad M = B - A\,.$$

Hence B belongs to $\mathscr{Q}_{(\mathscr{F},\mathscr{G})}$. Let h be any Borel function on $\boldsymbol{R}^1$. Then Y belongs to $\mathscr{L}_{(\mathscr{F},\mathscr{G})}(dB)$ (so the symmetric stochastic integral

$$\int_s^t h(B_u) \circ dB_u\,, \qquad -\infty < s, t < \infty$$

is well-defined), if and only if

$$\int_0^n h(B_t)^2 dt + \int_0^n |h(B_t)| \frac{1}{t} |B_t|\, dt < \infty\,,\ \forall n\,. \tag{3.3}$$

This condition is satisfied if h is a locally bounded Borel function. By Theorem 3.2 we have

$$f(B_t) - f(B_s) = \int_s^t f'(B_u) \circ dB_u \tag{3.4}$$

for $f \in C^3(\boldsymbol{R}^1)$, but we can prove this (identity) for every absolutely continuous function f whose Lebesgue derivative f' satisfies (3.3), observing that the symmetric stochastic integral commutes with limits by virtue of the definition.

Example 3.2. Let $X = (X_t, 0 \leqq t < \infty)$ be a diffusion process with generator

$$A = \frac{1}{2} a\partial^2 + b\partial\,, \quad a > 0, \quad \partial = \frac{d}{dx}\,. \tag{3.5}$$

Then X is determined by the stochastic differential equation:

$$dX = \sqrt{a(X)} dB + b(X) dt\,, \qquad t \geqq 0\,. \tag{3.6}$$

By defining $X_t = X_0$ for $t \leqq 0$ we have a diffusion process on the time interval $(-\infty, \infty)$. Let

$$\mathscr{F}_t = \bigcap_n \sigma\left[X_s,\ s \leqq t + \frac{1}{n}\right] \vee \mathbf{2}$$

and

$$\mathscr{G}_t = \bigcap_n \sigma\left[X_s,\ s \geqq t - \frac{1}{n}\right] \vee \mathbf{2}\,.$$

Using (3.6) we can easily prove that $X \in \mathcal{Q}_{\mathcal{F}}$. X is a diffusion process by reversing the time order. Under some reasonable conditions we can prove that the generator of this backward diffusion is also a second order differential operator. Hence $X \in \mathcal{Q}_{\mathcal{G}}$, so $X \in \mathcal{Q}_{(\mathcal{F},\mathcal{G})}$. As in Example 3.1 we can define a symmetric stochastic integral of the type

$$\int_s^t h(X_u) \circ dX_u$$

and prove the chain rule

$$f(x_t) - f(x_s) = \int_s^t f'(X_u) \circ dX_u$$

as in Example 3.1.

In the above discussions we considered symmetric stochastic integrals relative to a pair $(\mathcal{F}, \mathcal{G})$ of a reference family $\mathcal{F}$ and a time-reversed reference family $\mathcal{G}$. In view of Theorem 3.1 we may define *symmetric stochastic integral* as follows:

$$(3.7) \qquad \int_s^t Y \circ dX = \operatorname*{l.i.p.}_{|\Delta| \to 0} \sum_{i=1}^n \frac{1}{2}(Y_{t_i} + Y_{t_{i-1}})(X_t - X_{t_{i-1}})$$

if this limit exists for every $(s, t) \in \boldsymbol{R}^2$ and has a version continuous in (t, s), where X is a continuous process with

$$(3.8) \qquad \lim_{|\Delta| \to 0} \sum_{i=1}^n |X_{t_i} - X_{t_{i-1}}|^3 \to 0$$

and Y is a continuous process. This definition has nothing to do with reference families. The proof of Theorem 3.2 actually proves that the chain rule holds for such a symmetric stochastic integral. Writing the integral of (3.7) as $Z_t - Z_s$, we can easily show that $Z = \{Z_t\}$ satisfies (3.8).

As a special case we can define the *forward symmetric stochastic integral* as follows. If $\mathcal{F}$ is a reference family and if $X, Y \in \mathcal{Q}_{\mathcal{F}}$, the stochastic integral (3.6) is well-defined and

$$(3.9) \qquad \int_s^t Y \circ dX = \mathcal{F}\text{-}\int_s^t Y dX + \frac{1}{2} \int_s^t dY dX .$$

Similarly for the *backward symmetric stochastic integral* relative to a time-reversed reference family $\mathcal{F}$.

To distinguish the symmetric stochastic $(\mathcal{F}, \mathcal{G})$-integral from others we call it a *two-sided symmetric stochastic integral.*

§ 4. Stochastic parallel displacement

Let E be a d-dimensional Riemannian manifold with metric tensor (g_{ij}) and $X = \{X_t, -\infty < t < \infty\}$ be a Brownian motion on E, i.e., a diffusion process whose generator is $\Delta/2$ where Δ is the Laplace-Beltrami operator on E. The sample function of X can be constructed from the Brownian motion B on $\boldsymbol{R}^d$ by a stochastic differential equation:

$$dX^i = \sum_\alpha \sigma^i_\alpha dB^\alpha + b^i dt \tag{4.1}$$

where

$$\sum_\alpha \sigma^i_\alpha \sigma^j_\alpha = g^{ij}, \qquad (g^{ij}) = (g_{ij})^{-1}$$

and

$$b^i = \frac{1}{2} \sum_{\alpha,\beta} g^{\alpha\beta} \Gamma^i_{\alpha\beta} .$$

Hence X^i is a continuous quasimartingale relative to the reference family $\mathscr{F}$ generated by X (if it is stopped at the exit time from the neighborhood of the local coordinates). Since the tensor (g^{ij}) is strictly positive definite and belongs to the class C^∞, X is a continuous, time-reversed quasimartingale relative to the time reversed reference family $\mathscr{G}$ generated by X in the same way as in Example 3.2. Hence the symmetric stochastic integral of the type

$$(\mathscr{F}, \mathscr{G})\text{-}\int_s^t h_i(X_t) \circ dX^i_t, \qquad i = 1, 2, \cdots, n$$

is well-defined. Since symmetric stochastic differentials are subject to the chain rule of Theorem 3.2 that takes the same form as in the ordinary calculus, we can easily prove that the above integral along the sample curve of X is independent of the choice of local coordinates whenever $h = (h_1, h_2, \cdots, h_d)$ is a covariant vector field.

The stochastic parallel displacement along the Brownian curve X is determined by the symmetric stochastic differential equation:

$$dU^i_t = -\sum_{j,k} \Gamma^i_{jk}(X_t) U^j \circ dX^k_t, \qquad i = 1, 2, \cdots, d . \tag{4.2}$$

For simplicity we will discuss such a parallel displacement along the conditional Brownian motion starting at $X_0 = a$ and ending at $X_1 = b$. Such a conditional Brownian motion is also a diffusion whose generator is an elliptic differential operator. Hence $X^i \in \mathscr{Q}_\mathscr{F} \cap \mathscr{Q}_\mathscr{G}$ where $\mathscr{F}$ and $\mathscr{G}$ are defined relative to the conditional Brownian motion X. Interpreting the above differential equation in the forward (resp. backward) sense we obtain U_1 from U_0 (resp. U_0 from U_1). Noticing that the chain rule for symmetric stochastic differentials takes the same form as in ordinary calculus we can prove that both the map $\Phi: U_0 \to U_1$ and the map $\Psi: U_1 \to U_0$ is orthogonal and depends on the curve X (forward for Φ and backward for Ψ). Since the equation (4.2) is invariant under the time reversal, we can also prove that Φ and Ψ are inverse to each other.

References

[1] McKean, H. P., Jr., Stochastic integrals, Academic Press, 1969.

[2] Fisk, D. L., Quasi-martingales and stochastic integrals, Technical Report No. 1, Dept. Math. Michigan State Univ., 1963.

[3] Courrège, P., Intégrales stochastiques et martingales de carré integrable, Sem. Th. Potent. 7 Inst. Henri Poincare, Paris, 1963.

[4] Kunita, H. and Watanabe, S., On square integrable martingales, Nagoya Math. J., **30** (1967), 209–245.

[5] Meyer, P. A., Intégrales stochastiques, I–IV Sém. de Prob., Lecture Notes in Math., **39** (1967), 72–162, Springer.

[6] Stratonovich, R. L., Conditional Markov processes and their application to optimal control, English Translation, Elsevier, New York, 1968.

[7] Itô, K., Stochastic differentials, Appl. Math. and Optimization, **1** (1975), 347–381.

[8] ——, The Brownian motion and tensor fields on Riemannian manifold, Proc. Internat. Congress of Math. Stockholm, 1962.

[9] ——, Stochastic parallel displacement, Lecture Notes in Math., **451** (1975), 1–7, Springer.

[10] Dynkin, E. B., Diffusion of tensors, Dokl. Akad. Nauk SSSR, **179**-6 (1968), 532–535.

Research Institute
for Mathematical Sciences
Kyoto University
Kyoto 606, Japan

Proc. of Intern. Symp. SDE
Kyoto 1976, pp. 111-126

Necessary and Sufficient Conditions for Absolute Continuity of Measures Corresponding to Point (Counting) Processes

Yu. KABANOV, R. LIPTSER, and A. SHIRYAEV

§ 1. Introduction

Let $(X, \mathscr{B})$ be a measurable space where X is the set of piece-wise constant functions $x = (x_t)$, $t \geq 0$, such that $x_0 = 0$, $x_t = x_{t-} + 0$ or 1, $\mathscr{B}$ be the σ-algebra $\sigma\{x: x_s, s \geq 0\}$. Denote $\mathscr{B}_t = \sigma\{x: x_s, s \leq t\}$,

$$\tau_i(x) = \inf\{s \geq 0: x_s = i\}$$

putting $\tau_i(x) = \infty$ if $\lim_{t\to\infty} x_t < i$, and let $\tau_\infty(x) = \lim_{i\to\infty} \tau_i(x)$.

Note, that, for each function $x = (x_t)$, $t \geq 0$,

$$x_t = \sum_{i\geq 1} I(\tau_i(x) \leq t) \ . \tag{1}$$

Let μ be a probability measure on $(X, \mathscr{B})$. Since $\mu\{x_{t\wedge\tau_n} \leq n\} = 1$, the process $X^{(n)} = (x_{t\wedge\tau_n}, \mathscr{B}_t, \mu)$, $t \geq 0$, as any bounded and non-decreasing process, is a submartingale of class D, and hence it has a unique (μ — a.s.) Doob-Meyer decomposition[1)]

$$x_{t\wedge\tau_n} = m_t^{(n)} + A_t^{(n)} \ , \tag{2}$$

where $(m_t^{(n)}, \mathscr{B}_t)$ is a uniformly intergable martingale and $(A_t^{(n)}, \mathscr{B}_t)$ is an increasing previsible process.

From the fact that decomposition (2) holds for each $n = 1, 2, \cdots$ it follows that there exist such an increasing previsible process A_t and a τ_∞-local martingale m_t such that the point process $X = (x_t, \mathscr{B}_t, \mu)$ has the representation

$$x_t = m_t + A_t \qquad (\{t < \tau_\infty\}\,;\, \mu - \text{a.s.}) \ . \tag{3}$$

Besides

$$m_{t\wedge\tau_n} = m_t^{(n)} \ , \qquad A_{t\wedge\tau_n} = A_t^{(n)} \ .$$

1) Note that the family $(\mathscr{B}_t)$, $t>0$, is right-continuous.

The process $A = (A_t, \mathscr{B}_t, \mu)$, in the sequel called the *compensator* of the *point* (counting) *process* $X = (x_t, \mathscr{B}_t, \mu)$, satisfies the conditions (cf. [1], [2]): $A_0 = 0$, $\Delta A_t \leq 1$ ($\Delta A_t = A_t - A_{t-}$). It is also known ([1], [2]) that any previsible process with such properties uniquely determines a measure μ with respect to which this process is the compensator of a process $X = (x_t, \mathscr{B}_t, \mu)$, $t \geq 0$. Moreover the compensator uniquely determines this measure in the sense that if $X = (x_t, \mathscr{B}_t, \mu)$ and $\tilde{X} = (x_t, \mathscr{B}_t, \tilde{\mu})$ are two processes with the compensators $A = (A_t, \mathscr{B}_t, \mu)$ and $\tilde{A} = (\tilde{A}_t, \mathscr{B}_t, \mu)$ such that $A_{t\wedge\theta} = \tilde{A}_{t\wedge\theta}$ (μ or $\tilde{\mu}$-a.s.) where θ is a Markov time with respect to the family $(\mathscr{B}_t)$, $t \geq 0$, then the restrictions $\mu|\mathscr{B}_\theta$ and $\tilde{\mu}|\mathscr{B}_\theta$ of the measures μ and $\tilde{\mu}$ onto the σ-algebra $\mathscr{B}_\theta$ coincide:

$$\mu|\mathscr{B}_\theta = \tilde{\mu}|\mathscr{B}_\theta$$

(cf. [1], [2]).

§ 2. Main results

Now let $X = (x_t, \mathscr{B}_t, \mu)$ and $\tilde{X} = (x_t, \mathscr{B}_t, \tilde{\mu})$ be two point processes. As noted above, the measures μ and $\tilde{\mu}$ are completely determined by the compensators $A_t = A_t(x)$ and $\tilde{A}_t = \tilde{A}_t(x)$ of these processes. It is therefore natural to express conditions for the absolute continuity of $\tilde{\mu}$ with respect to μ ($\tilde{\mu} \ll \mu$) in terms of the compensators A_t and $\tilde{A}_t$.

The main result of this paper is the following

Theorem 1. *For $\tilde{\mu} \ll \mu$, it is necessary and sufficient that,* $\tilde{\mu}$ — a.s.,

I. $$\tilde{A}_t(x) = \int_0^t \lambda_s(x) dA_s(x), \qquad t < \tau_\infty(x);$$

II. $$\Delta A_t(x) = 1 \Rightarrow \Delta\tilde{A}_t(x) = 1, \qquad t < \tau_\infty(x);$$

III. $$\int_0^{\tau_\infty(x)} I(\Delta A_t(x) = 0) \frac{|1 - \lambda_t(x)|^2}{1 + |1 - \lambda_t(x)|} dA_t(x) < \infty;$$

IV. $$\sum_{t<\tau_\infty(x)} I(\Delta\tilde{A}_t(x) = 1) \, |\ln \Delta A_t(x)| < \infty;$$

V. $$\sum_{t\in T(x)\cap\Gamma(x)} \left[\Delta\tilde{A}_t(x) \ln \frac{\Delta\tilde{A}_t(x)}{\Delta A_t(x)} + (1 - \tilde{A}_t(x)) \ln \frac{1 - \Delta\tilde{A}_t(x)}{1 - \Delta A_t(x)}\right] < \infty;$$

VI. $$\sum_{t\in T(x)\cap([0,\tau_\infty(x))\setminus\Gamma(x))} \left[\Delta\tilde{A}_t(x)\left(\left|\ln \frac{\Delta\tilde{A}_t(x)}{\Delta A_t(x)}\right| \wedge 1\right) + (1 - \Delta\tilde{A}_t(x))\left(\left|\ln \frac{1 - \Delta\tilde{A}_t(x)}{1 - \Delta A_t(x)}\right| \wedge 1\right)\right] < \infty,$$

where $\lambda = (\lambda_t(x), \mathscr{B}_t)$ *is a non-negative previsible process and*

$$T(x) = \{t < \tau_\infty(x)\,;\, 0 < \Delta A_t(x) < 1,\ \Delta\tilde{A}_t(x) < 1\}\ ,$$

$$\Gamma(x) = \left\{t < \tau_\infty(x) : \left|\frac{\Delta\tilde{A}_t(x)}{\Delta A_t(x)} - 1\right| \le \frac{1}{2},\ \left|\ln\frac{1-\Delta\tilde{A}_t(x)}{1-\Delta A_t(x)}\right| \le 2\right\}.$$

Before going into the proof, let us formulate some corollaries of the theorem.

Theorem 2. *Let the compensator* $A_t = A_t(x)$ *of a point process* $X = (x_t, \mathscr{B}_t, \mu)$ *be continuous (in t). Then* $\tilde{\mu} \ll \mu$ *iff,* $\tilde{\mu}$*-a.s., the following conditions are satisfied:*

A. $\tilde{A}_t(x) = \int_0^t \lambda_s(x) dA_s(x)\ , \qquad t < \tau_\infty(x)\ ;$

B. $\int_0^{\tau_\infty(x)} \frac{|1-\lambda_s(x)|^2}{1+|1-\lambda_s(x)|} dA_s(x) < \infty\ ,$

where λ *is a non-negative previsible process.*

The density $(d\tilde{\mu}/d\mu)(t, x)$ *of the measure* $\tilde{\mu}$ *with respect to the measure* μ *on the σ-algebra* $\mathscr{B}_t$ *is given by the formula:*

$$(4)\qquad \frac{d\tilde{\mu}}{d\mu}(t, x) = \exp\left\{\int_0^t \ln\frac{d\tilde{A}_s(x)}{dA_s(x)} dx_s - [\tilde{A}_t(x) - A_t(x)]\right\}.$$

Theorem 3. *Let the compensators* $A_t(x)$ *and* $\tilde{A}_t(x)$ *be such that*

$$\mu\{A_{\tau_\infty}(x) < \infty\} = 1\ , \qquad \tilde{\mu}\{\tilde{A}_{\tau_\infty}(x) < \infty\} = 1\ .$$

Then $\tilde{\mu} \ll \mu$ *iff,* $\tilde{\mu}$*-a.s.,*

A. $\tilde{A}_t(x) = \int_0^t \lambda_s(x) dA_s(x)\ , \qquad t < \tau_\infty(x)\ ;$

B. $\Delta A_t(x) = 1 \Rightarrow \Delta\tilde{A}_t(x) = 1\ ;$

C. $\tilde{\mu}\{A_{\tau_\infty}(x) < \infty\} = 1\ ,$

where λ *is a non-negative previsible process.*

Theorem 2 enables us to describe in terms of the compensators, all the point processes with the corresponding measures absolutely continuous with respect to the Poisson measure (compare with Theorem 7.11 in [3]). Namely, we have the following

Theorem 4. *Let* $X = (x_t, \mathscr{B}_t, \mu_\pi)$ *be a Poisson process with the unit*

parameter and $\tilde{X} = (x_t, \mathscr{B}_t, \tilde{\mu})$ *be a point process with the measure* $\tilde{\mu} \ll \mu_\pi$*. Then the compensator of the process* $\tilde{X}$ *is of the form*

$$\tilde{A}_t(x) = \int_0^t \lambda_s(x) ds \ ,$$

where $\lambda = (\lambda_t(x), \mathscr{B}_t)$ *is a non-negative previsible process such that,* $\tilde{\mu}$ *-a.s.,*

$$\int_0^\infty \frac{|1 - \lambda_t(x)|^2}{1 + |1 - \lambda_t(x)|} dt < \infty \ . \tag{5}$$

The density is given by

$$\frac{d\tilde{\mu}}{d\mu}(t, x) = \exp\left\{\int_0^t \ln \lambda_s(x) dx_s - \int_0^t [\lambda_s(x) - 1] ds\right\} . \tag{6}$$

§ 3. An outline of the proof of Theorem 1

Necessity of conditions I and II follows from an analogue of the well-known Girsanov theorem [2] asserting, in particular, that if $\tilde{\mu} \ll \mu$ then $\tilde{\mu}$ -a.s.

$$\tilde{A}_t(x) = \int_0^t \lambda_s(x) dA_s(x) \ , \qquad t < \tau_\infty(x) \ ; \tag{7}$$

$$\Delta A_t(x) = 1 \Rightarrow \lambda_t(x) = 1 \ , \ t < \tau_\infty(x) \ , \tag{8}$$

where

$$\lambda_t(x) = \hat{Z}_t (Z_{t-})^+ \ , \quad Z_t = E_\mu\left(\frac{d\tilde{\mu}}{d\mu}\bigg| \mathscr{B}_t\right) , \quad a^+ = \begin{cases} a^{-1} \ , & a \neq 0 \\ 0 \ , & a = 0 \end{cases}$$

and $\hat{Z}_t$ is a previsible process such that, for any non-negative previsible process f_t (see [2])

$$E_\mu \int_0^{\tau_\infty} f_t Z_t dx_t = E_\mu \int_0^{\tau_\infty} f_t \hat{Z}_t dA_t \ .$$

Therefore the statement of theorem 1 can be reformulated as follows:

Under assumptions I, II, $\tilde{\mu} \ll \mu$ *iff condition* III–VI *are satisfied*:

$$\tilde{\mu} \ll \mu \Leftrightarrow \text{III–VI} \ . \tag{9}$$

Let $\lambda = (\lambda_t(x), \mathscr{B}_t)$ be a previsible process in (7). Without loss of generality one can always assume that, for each $x \in X$

$$\lambda_t(x) \leq (\Delta A_t(x))^{-1} .$$

Denote[2)]

(10) $$\mathscr{Z}_t(\lambda) = \prod_{\{n:\, \tau_n \leq t\}} \lambda_{\tau_n} \cdot \prod_{\substack{s \leq t \\ s \neq \tau_n}} [1 + (1 - \Delta A_s)^+ (1 - \lambda_s)\Delta A_s] \times \exp\left[\int_0^t (1 - \lambda_s) dA_s^c\right]$$

where

$$\Delta A_s = A_s - A_{s-} , \qquad A_s^c = A_s - \sum_{u \leq s} \Delta A_u .$$

The stochastic process $\mathscr{Z} = (\mathscr{Z}_t(\lambda), \mathscr{B}_t, \mu)$, $t < \tau_\infty$, satisfies the equation

(11) $$\mathscr{Z}_t(\lambda) = 1 + \int_0^t \mathscr{Z}_{s-}(\lambda)(\lambda_s - 1)(1 - \Delta A_s)^+ d(x_s - A_s)$$

is a non-negative supermartingale and τ_∞-local martingale (i.e. there exists such a sequence of Markov times (σ_n), $n = 1, 2, \cdots$, that $\sigma_n \uparrow \tau_\infty$ and the processes $\mathscr{Z}^{(n)}(\lambda) = (\mathscr{Z}_{t\wedge\sigma_n}(\lambda), \mathscr{B}_t, \mu)$, $t \geq 0$, for each $n = 1, 2, \cdots$, are uniformly interable martingales with $E_\mu \mathscr{Z}_{\sigma_n}(\lambda) = 1$). (For details, see [1]–[3]).

Define, on $(X, \mathscr{B})$, probability measures $\tilde{\mu}(n)$, $n = 1, 2, \cdots$, with $d\tilde{\mu}^{(n)} = \mathscr{Z}_{\sigma_n}(\lambda) d\mu$. Then, in accordance with (7), the compensator of the process $\tilde{X}^{(n)} = (x_t, \mathscr{B}_t, \tilde{\mu}^{(n)})$ equals

$$\tilde{A}_t^{(n)} = \int_0^{t\wedge\sigma_n} \lambda_s(x) dA_s(x) .$$

Hence

$$\tilde{A}_{t\wedge\sigma_n}^{(n)} = \tilde{A}_{t\wedge\sigma_n} , \quad t \geq 0 , \quad \tilde{\mu}\text{-a.s.},$$

and (since the compensator determines the measure uniquely) the restrictions of the measures $\tilde{\mu}^{(n)}$ and $\tilde{\mu}$ onto the σ-algebra $\mathscr{B}_{\sigma_n}$ coincide.

Denote by

$$\mu_n = \mu | \mathscr{B}_{\sigma_n} , \quad \tilde{\mu}_n = \tilde{\mu} | \mathscr{B}_{\sigma_n} , \quad \tilde{\mu}_n^{(n)} = \tilde{\mu}^{(n)} | \mathscr{B}_{\sigma_n}$$

the restrictions of the measures μ, $\tilde{\mu}$ and $\tilde{\mu}^{(n)}$ onto $\mathscr{B}_{\sigma_n}$. Since $\tilde{\mu}_n = \tilde{\mu}_n^{(n)}$ and $\tilde{\mu}_n^{(n)} \ll \mu_n$, we have $\tilde{\mu}_n \ll \mu_n$ and

---

2) In what follows, all stochastic integrals of the form $\int_0^t f_s dx_s$ and $\int_0^t f_s dA_s$ are understood as Stieltjes integrals defined for any elementary event (see [3], ch. 18).

$$\frac{d\tilde{\mu}_n}{d\mu_n} = E_\mu(\mathscr{Z}_{\sigma_n}(\lambda)|\mathscr{B}_{\sigma_n}) = \mathscr{Z}_{\sigma_n}(\lambda)\,, \qquad \mu\text{-a.s.}\,.$$

The sequence $(\mathscr{Z}_{\sigma_n}(\lambda),\ \mathscr{B}_{\sigma_n},\ \mu)$, $n = 1, 2, \cdots$ is a non-negative martingale with $E_\mu \mathscr{Z}_{\sigma_n}(\lambda) = 1$. Hence μ-$\lim_n \mathscr{Z}_{\sigma_n}(\lambda)$ exists $(= \mathscr{Z}(\lambda))$.

It is well-known that $\tilde{\mu} \ll \mu$ iff $\tilde{\mu}_n \ll \mu_n$, $n = 1, 2, \cdots$ and $E_\mu \mathscr{Z}(\lambda) = 1$. In the case under consideration, it is rather complicated to check that conditions III–VI are equivalent to the condition $E_\mu \mathscr{Z}(\lambda) = 1$. Therefore we make use of another criterion (Lemma 19.13 in [3]): $\tilde{\mu} \ll \mu$ iff $\tilde{\mu}_n \ll \mu_n$, $n = 1, 2, \cdots$, and $\lim_{n\to\infty} (d\tilde{\mu}_n)/(d\mu_n)$ exists and is finite $\tilde{\mu}$-a.s.

Introducing the condition Λ: $\lim_{n\to\infty} \mathscr{Z}_{\sigma_n}(\lambda)$ $(= \tilde{\mathscr{Z}}(\lambda))$ exists and is finite $\tilde{\mu}$-a.s., one can say that (under assumptions I, II)

$$\tilde{\mu} \ll \mu \Leftrightarrow \Lambda\,. \tag{12}$$

We will show later that condition Λ is equivalent (again under assumptions I, II) to the following eight conditions:

$$\Lambda_j : \lim_{n\to\infty} \mathscr{Z}_{\sigma_n}(\lambda_j) \qquad (= \tilde{\mathscr{Z}}(\lambda_j))\,,$$

$j = 1$–8 exists and in finite $\tilde{\mu}$-a.s., where λ_j are defined by (16) and $\mathscr{Z}_{\sigma_n}(\lambda_j)$ are obtained from $\mathscr{Z}_{\sigma_n}(\lambda)$ by substituting λ_j for λ.

In other words,

$$\Lambda \Leftrightarrow \Lambda_1\text{–}\Lambda_8 \tag{13}$$

and, in view of (12) and (13), to prove (9) it suffices only to show that (under assumptions I, II)

$$\text{III–VI} \Leftrightarrow \Lambda_1\text{–}\Lambda_8\,. \tag{14}$$

Let us now establish (13). To this end let us introduce previsible processes $\alpha_j = (\alpha_j(t), \mathscr{B}_t)$, $j = 1$–8, with

$$\begin{aligned}
&\alpha_1(t) = I\left(\Delta A_t = 0,\ |\lambda_t - 1| \le \frac{1}{2}\right),\\
&\alpha_2(t) = I\left(\Delta A_t = 0,\ |\lambda_t - 1| > \frac{1}{2}\right),\\
&\alpha_3(t) = I(\lambda_t \Delta A_t = 1)\,, \qquad\qquad (15)\\
&\alpha_4(t) = I\Bigg(0 < \Delta A_t < 1,\ \lambda_t \Delta A_t < 1,\ |\lambda_t - 1|\\
&\qquad \le \frac{1}{2}, \left|\ln \frac{1 - \lambda_t \Delta A_t}{1 - \Delta A_t}\right| \le 2\Bigg),
\end{aligned}$$

$$\alpha_5(t) = I\Big(0 < \Delta A_t < 1,\ \lambda_t \Delta A_t < 1,\ 1 \le \lambda_t \le \frac{3}{2},\ \Big|\ln \frac{1-\lambda_t \Delta A_t}{1-\Delta A_t}\Big| > 2\Big),$$

$$\alpha_6(t) = I\Big(0 < \Delta A_t < 1,\ \lambda_t \Delta A_t < 1,\ \frac{1}{2} \le \lambda_t < 1,\ \Big|\ln \frac{1-\lambda_t \Delta A_t}{1-\Delta A_t}\Big| > 2\Big),$$

$$\alpha_7(t) = I\Big(0 < \Delta A_t < 1,\ \lambda_t \Delta A_t < 1,\ \lambda_t < \frac{1}{2}\Big),$$

$$\alpha_8(t) = I\Big(0 < \Delta A_t < 1,\ \lambda_t \Delta A_t < 1,\ \lambda_t > \frac{3}{2}\Big).$$

Obviously, $\alpha_j^2(t) = \alpha_j(t)$, $j = 1$–8, and $\sum_{j=1}^{8} \alpha_j(t) = 1$. Put

(16) $$\lambda_j(t) = (\lambda_t(x))^{\alpha_j(t)},\ j = 1\text{–}8\ ,$$

and consider the processes $\mathscr{Z}(\lambda_j) = (\mathscr{Z}_t(\lambda_j), \mathscr{B}_t, \mu)$, $j = 1$–8, $t \ge 0$. They are non-negative supermartingales and τ_∞-local martingales. Without loss of generality we will assume that the above sequence of Markov times (σ_n), $n = 1, 2, \cdots$, is such that all the processes $(\mathscr{Z}_{t\wedge\sigma_n}(\lambda_j), \mathscr{B}_t, \mu)$, $j = 1 \sim 8$, are uniformly integrable martingales for each $n = 1, 2, \cdots$.

To prove (13) note that, for each $j = 1$–8, the processes $(\mathscr{Z}_{\sigma_n}(\lambda_j), \mathscr{B}_{\sigma_n}, \mu)$, $n = 1, 2, \cdots$, are non-negative martingales with $E_\mu \mathscr{Z}_{\sigma_n}(\lambda_j) = 1$. Therefore $\tilde{\mu}$-$\lim_{n\to\infty} \mathscr{Z}_{\sigma_n}(\lambda_j)$ $(= \mathscr{Z}(\lambda_j))$ exists and is finite. If condition Λ is satisfied, then, in view of (12), $\tilde{\mu} \ll \mu$ and hence all conditions Λ_j, $j = 1$–8, are satisfied.

Conversely, if Λ_1–Λ_8 are satisfied, then, in view of the equality $\mathscr{Z}_{\sigma_n}(\lambda) = \prod_{j=1\sim 8} \mathscr{Z}_{\sigma_n}(\lambda_j)$, condition Λ is also, obviously, satisfied.

Note one more property of the above random variables $\tilde{\mathscr{Z}}(\lambda_j)$:

(17) $$\Lambda_1\text{–}\Lambda_8 \Rightarrow \tilde{\mu}\{\tilde{\mathscr{Z}}(\lambda_j) > 0\} = 1\ , \qquad j = 1 \sim 8\ .$$

In fact, if Λ_1–Λ_8 hold, then $\tilde{\mu} \ll \mu$ and hence $\mu\{\tilde{\mathscr{Z}}(\lambda) = \mathscr{Z}(\lambda)\} = 1$. Therefore

$$\tilde{\mu}\{\tilde{\mathscr{Z}}(\lambda) = 0\} = \tilde{\mu}\{\mathscr{Z}(\lambda) = 0\} = \int_{\{\mathscr{Z}(\lambda)=0\}} \mathscr{Z}(\lambda)\,d\mu = 0$$

and hence

$$\tilde{\mu}\{\tilde{\mathscr{Z}}(\lambda_j) > 0\} = 1\ , \qquad j = 1 \sim 8\ .$$

§ 4. Proof of Theorem 1

To prove (under assumptions I and II) the necessity and sufficiency of conditions III–VI, it suffices to check the following implications:

$$\Lambda_1\text{–}\Lambda_8 \Rightarrow \begin{cases} \text{III} \Rightarrow (\Lambda_1, \Lambda_2) \\ \text{IV} \Rightarrow \Lambda_3 \\ \text{V} \Rightarrow \Lambda_4 \\ \text{VI} \Rightarrow (\Lambda_5\text{–}\Lambda_8)\,. \end{cases} \tag{18}$$

a) Condition III. (Λ_1–$\Lambda_8 \Rightarrow \text{III} \Rightarrow \Lambda_1, \Lambda_2$). Consider the random variables $\mathscr{Z}_{\sigma_n}(\lambda_1)$, $n = 1, 2, \cdots$. According to (11)

$$\begin{aligned} \ln \mathscr{Z}_{\sigma_n}(\lambda_1) &= \int_0^{\sigma_n} \ln \lambda_1(t) dx_t + \int_0^{\sigma_n} (1 - \lambda_1(t)) dA_t^c \\ &= \int_0^{\sigma_n} \ln \lambda_1(t) d[x_t - \tilde{A}_t] + \int_0^{\sigma_n} [\lambda_1(t) \ln \lambda_1(t) + 1 - \lambda_1(t)] dA_t\,, \end{aligned}$$

where we made use of the equality

$$\ln \lambda_1(t) d\tilde{A}_t = \lambda_t \ln \lambda_1(t) dA_t = \lambda_1(t) \ln \lambda_1(t) dA_t\,.$$

Let us introduce the function $g(y) = y \ln y + 1 - y$ ($y \geq 0$). Since $g(0) = 1$, $g(1) = 0$ and $g'(y) = \ln y$, this function is non-negative. It is also easy to see that, if $|y - 1| \leq 1/2$,

$$0 < c_1 \leq g(y) \Big/ \frac{(y-1)^2}{1 + |y - 1|} \leq c_2 < \infty\,, \tag{19}$$

$$0 < c_1 \leq g(y)/y \ln^2 y \leq c_2 < \infty \tag{20}$$

where c_1 and c_2 are some constants.

By condition III

$$\int_0^{\tau_\infty} \frac{|\lambda_1(t) - 1|^2}{1 + |\lambda_1(t) - 1|} dA_t < \infty \qquad (\tilde{\mu}\text{-a.s.})\,. \tag{21}$$

Therefore it follows from (19) that ($\tilde{\mu}$-a.s.)

$$\lim_{n\to\infty} \int_0^{\sigma_n} (\lambda_1(t) \ln \lambda_1(t) + 1 - \lambda_1(t)) dA_t$$

exists and is finite.

Consider now the random variables

$$\mathscr{M}_t = \int_0^t \ln \lambda_1(s) d[x_s - \tilde{A}_s] , \qquad t < \tau_\infty .$$

By (20) and (21) $\mathscr{M} = (\mathscr{M}_{t\wedge\tau_\infty}, \mathscr{B}_t, \tilde{\mu})$ is a τ_∞-locally square integrable martingale ([1], [3]) for which, as can be easily shown, there exists a finite $\tilde{\mu}$-$\lim_{t\to\infty} \mathscr{M}_{t\wedge\tau_\infty}$.

Thus III $\Rightarrow \Lambda_1$.

Let us now consider the random variables

$$\ln \mathscr{Z}_{\sigma_n}(\lambda_2) = \int_0^{\sigma_n} \ln \lambda_2(t) dx_t + \int_0^{\sigma_n} (1 - \lambda_2(t)) dA_t .$$

It is not difficult to see that, if $|y - 1| > 1/2$,

(22) $$0 < c_1 \leq |y - 1| \Big/ \frac{|y - 1|^2}{1 + |y - 1|} \leq c_2 < \infty ,$$

(23) $$0 \leq y(|\ln y| \wedge 1) \Big/ \frac{|y - 1|^2}{1 + |y - 1|} \leq c_2 < \infty .$$

By condition III

(24) $$\int_0^{\tau_\infty} \frac{|\lambda_2(t) - 1|^2}{1 + |\lambda_2(t) - 1|} dA_t < \infty \qquad (\tilde{\mu}\text{-a.s.})$$

which together with (22), shows that

$$\tilde{\mu}\text{-}\lim_{n\to\infty} \int_0^{\sigma_n} (1 - \lambda_2(t)) dA_t$$

exists and is finite.

Now by (23) and (24)

(25) $$\int_0^{\tau_\infty} \lambda_t(|\ln \lambda_2(t)| \wedge 1) dA_t < \infty \qquad (\tilde{\mu}\text{-a.s.}) .$$

Let us now make use of the following property of Stieltjes stochastic integrals (Theorem 18.6 in [3]): if f_t is a non-negative previsible process, then, for any $0 < c < \infty$, up to some set of null measure,

(26) $$\left\{\int_0^{\tau_\infty} f_t dx_t < \infty\right\} = \left\{\int_0^{\tau_\infty} (f_t \wedge c) dA_t < \infty\right\} .$$

Therefore it follows from (25) and (26) that $\tilde{\mu}$-$\lim_{n\to\infty} \int_0^{\sigma_n} \ln \lambda_2(t) dx_t$ exists and is finite. Hence III $\Rightarrow \Lambda_2$.

Let us now prove the implication Λ_1–$\Lambda_8 \Rightarrow$ III.

Denote

$$\tilde{\mathscr{B}}_t = \int_0^t [\lambda_1(s) \ln \lambda_1(s) + 1 - \lambda_1(s)] dA_s\,,$$

$$\tilde{\mathscr{M}}_t = \int_0^t \ln \lambda_1(s) d[x_s - \tilde{A}_s]\,.$$

It is clear that $\ln \mathscr{Z}_t(\lambda) = \tilde{\mathscr{B}}_t + \tilde{\mathscr{M}}_t$. The process $(\tilde{\mathscr{M}}_{t\wedge\tau_\infty}, \mathscr{B}_t, \tilde{\mu})$, $t \geq 0$, is a τ_∞-local martingale with right-continuous sample paths with uniformly bounded jumps, and $\tilde{B}_t$ is a continuous non-decreasing process with $\tilde{\mathscr{B}}_0 = 0$.

It follows from condition Λ_1 that $\tilde{\mu}(\sup_{t\geq 0} \mathscr{Z}_{t\wedge\tau_\infty}(\lambda_1) < \infty) = 1$. Hence, by Lemma 18.13 in [3], $\tilde{\mu}(\tilde{\mathscr{B}}_{\tau_\infty} < \infty) = 1$, i.e.

$$\int_0^{\tau_\infty} [\lambda_1(s) \ln \lambda_1(s) + 1 - \lambda_1(s)] dA_s < \infty \qquad (\tilde{\mu}\text{-a.s.})$$

what, together with (19), leads to (21).

Further, it follows from condition Λ_2 that $\tilde{\mu}\text{-}\lim_{n\to\infty} \ln \mathscr{Z}_{\sigma_n} < \infty$. And, from conditions Λ_1–Λ_8, it follows (cf. (17)) that $\tilde{\mu}\text{-}\lim_{n\to\infty} \ln \mathscr{Z}_{\sigma_n}(\lambda_2) > -\infty$; thus, conditions Λ_1–Λ_8 imply that $\tilde{\mu}\text{-}\lim_{n\to\infty} \ln \mathscr{Z}_{\sigma_n}(\lambda_2)$ exists and is finite.

Consider now more carefully the random variables

$$(27) \quad \ln \mathscr{Z}_t(\lambda_2) = \int_0^t \ln \lambda_2(s) dx_s + \int_0^t (1 - \lambda_2(s)) dA_s\,, \qquad t < \tau_\infty\,.$$

Since $\mathscr{Z} = (\mathscr{Z}_t(\lambda_2), \mathscr{B}_t, \mu)$ is a non-negative supermartingale, $\mu\text{-}\lim_{t\to\tau_\infty} \mathscr{Z}_t(\lambda_2)$ exists, and hence conditions Λ_1–Λ_8, ensuring the absolute continuity $\tilde{\mu} \ll \mu$, imply that $\tilde{\mu}\text{-}\lim_{t\to\tau_\infty} \mathscr{Z}_t(\lambda_2)$ also exists. Together with the fact that $\tilde{\mu}\text{-}\lim_n \ln \mathscr{Z}_{\sigma_n}(\lambda_2)$ is finite, this implies that so is $\tilde{\mu}\text{-}\lim_{t\to\tau_\infty} \ln \mathscr{Z}_t(\lambda_2)$.

Note now that the second integral in (27) is a continuous process in t, and the first one has jumps (at moments τ_n) the absolute values of which are bounded by some constant. Taking into account that $\tilde{\mu}\text{-}\lim_{n\to\infty} \ln \mathscr{Z}_{\tau_n}(\lambda_2)$ is finite and $\lambda_2(t)$ is such that either $|\lambda_2(t) - 1| > 1/2$ or $\lambda_2(t) = 1$, we come to the conclusion that $\lambda_2(\tau_n) \neq 1$ only for a finite number ($\tilde{\mu}$-a.s.) of values n. Hence

$$(28) \qquad \int_0^{\tau_\infty} |\ln \lambda_2(t)|\, dx_t < \infty \qquad (\tilde{\mu}\text{-a.s.})$$

and consequently

$$\int_0^{\tau_\infty} |1 - \lambda_2(t)|\, dA_t < \infty \qquad (\tilde{\mu}\text{-a.s.})\ . \tag{29}$$

The later inequality and (22) obviously imply inequality (24) which, together with (21), guarantee that condition III is satisfied.

b) Condition IV (Λ_1–$\Lambda_8 \Rightarrow$ IV $\Rightarrow \Lambda_3$). By (11)

$$\begin{aligned} \mathscr{L}_t(\lambda_3) &= \prod_{\{n:\ \tau_n \le t\}} \lambda_3(\tau_n) = \exp\left(\int_0^t \ln \lambda_3(s) dx_s\right) \\ &= \exp\left(\int_0^t \ln \lambda_3(s) d[x_s - \tilde{A}_s] + \int_0^t \ln \lambda_3(s) d\tilde{A}_s\right) . \end{aligned} \tag{30}$$

It is not difficult to show (cf. Lemma 19.5 in [3]) that

$$\int_0^t I(\Delta\tilde{A}_s = 1) d[x_s - \tilde{A}_s] = 0\ , \qquad (\{t < \tau_\infty\},\ \tilde{\mu}\text{-a.s.})\ .$$

Hence by the definition of $\lambda_3(t)$ and assumption I ($\tilde{\mu}$-a.s.)

$$0 = \int_0^t I(\Delta\tilde{A}_s = 1) \ln \lambda_3(s) d[x_s - \tilde{A}_s] = \int_0^t \ln \lambda_3(s) d[x_s - \tilde{A}_s]\ . \tag{31}$$

From (30) and (31), we obtain that $\tilde{\mu}$-a.s.

$$\begin{aligned} \ln \mathscr{L}_{\sigma_n}(\lambda_3) &= \int_0^{\sigma_n} \ln \lambda_3(t) d\tilde{A}_t = \int_0^{\sigma_n} I(\Delta\tilde{A}_t = 1) \ln \lambda(t) d\tilde{A}_t \\ &= \int_0^{\sigma_n} I(\Delta\tilde{A}_t = 1) \lambda_t \ln \lambda_t dA_t \\ &= \sum_{t \le \sigma_n} I(\Delta\tilde{A}_t = 1) \frac{\Delta\tilde{A}_t}{\Delta A_t}\left(\ln \frac{\Delta\tilde{A}_t}{\Delta A_t}\right) \Delta A_t \\ &= \sum_{t \le \sigma_n} I(\Delta\tilde{A}_t = 1)\, |\ln \Delta A_t|\ . \end{aligned}$$

It follows obviously that even a stronger property than the stated one holds too, namely: $\Lambda_3 \Leftrightarrow$ III.

c) Condition V (Λ_1–$\Lambda_8 \Rightarrow V \Rightarrow \Lambda_4$). According to (II)

$$\ln \mathscr{L}_t(\lambda_4) = \sum_{s \le t} \ln \lambda_4(s) \Delta x_s + \sum_{s \le t} \ln \frac{1 - \lambda_4(s) \Delta A_s}{1 - \Delta A_s} (1 - \Delta x_s)\ .$$

Let us introduce the functions

$$\Phi(a, y) = \begin{cases} ay \ln y + (1 - ay) \ln \dfrac{1 - ay}{1 - a}\,, & \\ & 0 \le a < 1\,, \quad 0 \le y \le \dfrac{1}{a}\,, \\ y \ln y\,, & a = 1\,, \quad 0 \le y \le \dfrac{1}{a} \end{cases}$$

and

$$\varphi(a, y) = \begin{cases} \ln y - \ln \dfrac{1 - ay}{1 - a}\,, & 0 \le a < 1\,, \quad 0 \le y \le \dfrac{1}{a}\,, \\ \ln y\,, & a = 1\,, \quad 0 \le y \le \dfrac{1}{a}\,, \end{cases}$$

We have

$$\text{(32)} \qquad \ln \mathscr{Z}_t(\lambda_4) = \int_0^t \varphi(\Delta A_s, \lambda_4(s)) d[x_s - \tilde{A}_s] + \sum_{\substack{s \le t \\ \Delta A_s < 1}} \Phi(\Delta A_s, \lambda_4(s))$$

and condition V can be rewritten in the following form

$$\text{(33)} \qquad \sum_{\substack{s < \tau_\infty \\ \Delta A_s < 1}} \Phi(\Delta A_s, \lambda_4(s)) < \infty \qquad (\tilde{\mu}\text{-a.s.})\,.$$

Let us now show that (33) implies the inequality

$$\text{(34)} \qquad \int_0^{\tau_\infty} \varphi^2(\Delta A_s, \lambda_4(s))(1 - \Delta \tilde{A}_s) d\tilde{A}_s < \infty \qquad (\tilde{\mu}\text{-a.s.})\,.$$

To this end note that, by assumptions I and II and by the definition of $\lambda_4(t)$,

$$\int_0^{\tau_\infty} \varphi^2(\Delta A_s, \lambda_4(s))(1 - \Delta \tilde{A}_s) d\tilde{A}_s = \sum_{\substack{s < \tau_n \\ \Delta A_s < 1}} \varphi^2(\Delta A_s, \lambda_4(s))(1 - \lambda_4(s)\Delta A_s)\lambda_4(s)\Delta A_s\,.$$

Therefore, to prove (34), it suffices to show that, if $0 < a < 1$, $1/2 \le y \le 3/2$, $|\ln (1 - ay)/(1 - a)| \le 2$ and $ay < 1$, then

$$\text{(35)} \qquad \varphi^2(a, y)(1 - ay)ay \le K \cdot \Phi(a, y)\,,$$

where K is a constant.

It is not difficult to check that

$$\text{(36)} \qquad \frac{\partial y}{\partial y}[\varphi^2(a, y)(1 - ay)ay]$$

$$= \frac{\partial \Phi(a, y)}{\partial y}\left[\left(1 - \frac{ay}{2}\right)\ln\frac{y(1-a)}{1-ay} + 2a\right].$$

If $y \geq 1$, then $(\partial\Phi(a, y))/\partial y \geq 0$, and under the above conditions on a and y,

$$0 \leq \left[\left(1 - \frac{ay}{2}\right)\ln\frac{y(1-a)}{1-ay} + 2a\right] \leq K,$$

where K is a constant.

Hence, if $y \geq 1$, inequality (35) is fulfilled.

Let now $y < 1$. Then $(\partial\Phi(a, y))/\partial y \leq 0$ and hence

$$\frac{\partial}{\partial y}[\varphi^2(a, y)(1 - ay)ay] \geq \frac{\partial\Phi(a, y)}{\partial y}\left|\left(1 - \frac{ay}{2}\right)\ln\frac{y(1-a)}{1-ay} + 2a\right|$$

$$\geq K\frac{\partial\Phi(a, y)}{\partial y},$$

where K is a constant such that

$$\left|\left(1 - \frac{ay}{2}\right)\ln\frac{y(1-a)}{1-ay} + 2a\right| \leq K.$$

Note that $\Phi(a, 1) = 0$, $\varphi(a, 1) = 0$. Therefore, if $y < 1$

$$\varphi^2(a, y)(1 - ay)ay = -\int_y^1 \frac{\partial}{\partial z}[\varphi^2(a, z)(1 - az)az]dz$$

$$\leq -K\int_y^1 \frac{\partial\Phi(a, z)}{\partial z}dz = K\Phi(a, y).$$

Thus, if condition (33) (i.e. condition V) is fulfilled, then so is inequality (34). The latter implies (see Theorem 18.8 in [3]) that the process

$$\left(\int_0^{t\wedge\tau_\infty} \varphi(\Delta A_s, \lambda_4(s))d[x_s - \tilde{A}_s], \mathscr{B}_t, \tilde{\mu}\right), \qquad t \geq 0,$$

in a τ_∞-locally square integrable martingale. Hence, by Lemma 18.14 in [3],

$$\tilde{\mu}\text{-}\lim_{t\to\infty}\int_0^{t\wedge\tau_\infty} \varphi(\Delta A_s, \lambda_4(s))d[x_s - \tilde{A}_s]$$

exists, and, by assertion c) of Theorem 18.8 in [3], this limit is $\tilde{\mu}$-a.s. finite.

Thus $V \Rightarrow \Lambda_4$.

Let now conditions Λ_1–Λ_8 be satisfied; Then $\tilde{\mu}$-$\lim_{t\to\infty} \ln \mathscr{Z}_{t\wedge\tau_n}(\lambda_4)$ exists and is finite (compare with the corresponding fact proved in a)). Hence, by the condition $|\lambda_4(t) - 1| \leq 1/2$, we have $\tilde{\mu}\{\sup_t \mathscr{Z}_{t\wedge\tau_\infty}(\lambda_4) < \infty\} = 1$.

Denote

$$\tilde{\mathscr{M}}_t = \int_0^t \varphi(\Delta A_t, \lambda_4(s))d[x_s - \tilde{A}_s] , \qquad \tilde{\mathscr{B}}_t = \sum_{\substack{s\leq t\\ \Delta A_s<1}} \Phi(\Delta A_s, \lambda_4(s)) .$$

Then $\ln \mathscr{Z}_t(\lambda_4) = \tilde{\mathscr{M}}_t + \tilde{\mathscr{B}}_t$ where $(\tilde{\mathscr{M}}_{t\wedge\tau_\infty}, \mathscr{B}_t, \mu)$, $t \geq 0$, is a τ_∞-local martingale and $(\tilde{\mathscr{B}}_t, \mathscr{B}_t)$ is a previsible process with non-decreasing right continuous sample paths. Therefore, by Lemma 18.13 in [3], $\tilde{\mu}\{\tilde{\mathscr{B}}_{\tau_\infty} < \infty\} = 1$ which is equivalent to condition V.

d) Condition VI (Λ_1–$\Lambda_8 \Rightarrow$ VI $\Rightarrow \Lambda_5$–Λ_8). By (II) for $j=5, 6, 7, 8$,

$$\ln \mathscr{Z}_t(\lambda_j) = \sum_{s\leq t} \ln \lambda_j(s)\Delta x_s + \sum_{\substack{s\leq t\\ \Delta A_s<1}} \left(\ln \frac{1-\lambda_j(s)\Delta A_s}{1-\Delta A_s}\right)(1-\Delta x_s) . \tag{37}$$

Let condition VI be satisfied. Then $\tilde{\mu}$-a.s., for $j = 5$–8,

$$\sum_{s<\tau_\infty} \lambda_j(s)\Delta A_s(|\ln \lambda_j(s)| \wedge 1) < \infty , \tag{38}$$

$$\sum_{\substack{s<\tau_\infty\\ \Delta A_s<1}} (1-\lambda_j(s)\Delta A_s)\left(\left|\ln \frac{1-\lambda_j(s)\Delta A_s}{1-\Delta}\right| \wedge 1\right) < \infty . \tag{39}$$

It follows from (38) and (26) that $\tilde{\mu}$-a.s.

$$\sum_{s<\tau_\infty} |\ln \lambda_j(s)|\, \Delta x_s = \int_0^{\tau_\infty} |\ln \lambda_j(s)|\, dx_s < \infty . \tag{40}$$

Condition (39) implies that $\tilde{\mu}$-a.s.

$$\sum_{\substack{s<\tau_\infty\\ \Delta A_s<1}} \left|\ln \frac{1-\lambda_j(s)\Delta A_s}{1-\Delta A_s}\right| (1-\Delta x_s) < \infty . \tag{41}$$

In fact,

$$\sum_{\substack{s<\tau_\infty\\ \Delta A_s<1}} \left|\ln \frac{1-\lambda_j(s)\Delta A_s}{1-\Delta A_s}\right| (1-\Delta x_s)$$
$$= \sum_{s<\tau_\infty} I(\Delta A_s < 1) \left|\ln \frac{1-\lambda_j(s)\Delta A_s}{1-\Delta A_s}\right| (1-\Delta x_s)$$

$$= \int_0^{\tau_\infty} I(\Delta A_s < 1) \left| \ln \frac{1 - \lambda_j(s)\Delta A_s}{1 - \Delta A_s} \right| d\tilde{x}_s^j ,$$

where

$$\tilde{x}_t^j = \sum_{s \le t} I\left(\left| \ln \frac{1 - \lambda_j(s)\Delta A_s}{1 - \Delta A_s} \right| > 0,\ \Delta A_s < 1 \right)(1 - \Delta x_s) .$$

For each $j = 5 \sim 8$, $(\tilde{x}_t^j, \mathscr{B}_t, \tilde{\mu})$ is a point process with the compensator

$$\tilde{A}_t^j = \sum_{s \le t} I\left(\left| \ln \frac{1 - \lambda_j(s)\Delta A_s}{1 - \Delta A_s} \right| > 0,\ \Delta A_s < 1 \right)(1 - \Delta \tilde{A}_s) .$$

Therefore (41) follows from (39) and (26).

Thus, if condition VI is fulfilled, then, for $j = 5$–8, inequalities (40) and (41) hold which, in turn, imply obviously Λ_5–Λ_8.

Let now Λ_1–Λ_8 be satisfied. Then, in particular, $\tilde{\mu}$-lim $\ln \mathscr{L}_{t \wedge \tau_\infty}(\lambda_j)$ ($j = 5$–8) exists and is finite. Note that, for $j = 5$ and 6

$$\left| \ln \frac{1 - \lambda_j(s)\Delta A_s}{1 - \Delta A_s} \right| > 2 \quad \text{or} \quad \ln \frac{1 - \lambda_j(s)\Delta A_s}{1 - \Delta A_s} = 0$$

and $\lambda_7(s) < 1/2$ if $\lambda_7(s) \neq 1$ and $\lambda_8(s) > 3/2$ if $\lambda_8(s) \neq 1$. Hence it follows, from the fact that the limits $\tilde{\mu}$-$\lim_{t \to \infty} \ln \mathscr{L}_t(\lambda_j)$ exist and from (37) that $\tilde{\mu}$-a.s. ($j = 5$–8)

$$\sum_{s < \tau_\infty} |\ln \lambda_j(s)|\, \Delta x_s = \int_0^{\tau_\infty} |\ln \lambda_j(s)|\, dx_s < \infty ,$$

$$\sum_{\substack{s < \tau_\infty \\ \Delta A_s < 1}} \left| \ln \frac{1 - \lambda_j(s)\Delta A_s}{1 - \Delta A_s} \right| (1 - \Delta x_s)$$
$$= \int_0^{\tau_\infty} I(\Delta A_s < 1) \left| \ln \frac{1 - \lambda_j(s)\Delta A_s}{1 - \Delta A_s} \right| d\tilde{x}_s^j < \infty .$$

To complete the proof it remains only to note that the required condition VI directly follows from these two inequalities and from (26).

Thus, Λ_1–$\Lambda_8 \Rightarrow$ VI.

The proof is completed.

Corollary. *Under the conditions of Theorem* I,

$$E_\mu\left(\frac{d\tilde{\mu}}{d\mu} \middle| \mathscr{B}_t \right) = \mathscr{L}_t(\lambda) \qquad \{(t < \tau_\infty)\,;\ \mu\text{-a.s}\}$$

where $\mathscr{L}_t(\lambda)$ *is defined by* (II), *and*

$$E_\mu \mathscr{Z}(\lambda) = 1$$

where $\mathscr{Z}(\lambda) = \mu\text{-}\lim_{n\to\infty} \mathscr{Z}_{\sigma_n}(\lambda)$.

§ 5. Proof of Theorems 2–4

Theorem 2 follows directly from Theorem 1.

To prove Theorem 3 it suffices only to show that, in the case under consideration, condition $C: \tilde{\mu}\{A_{\tau_\infty}\{x) < \infty\} = 1$ is equivalent to assumptions III–VI of Theorem 1. But condition C, together with the assumption $\tilde{\mu}\{A_{\tau_\infty}(x) < \infty\} = 1$ means that the number of jumps of the compensators A_t and $\tilde{A}_t$ is $\tilde{\mu}$-a.s. finite. This obviously implies conditions III–VI of Theorem 1, and hence $\tilde{\mu} \ll \mu$. Conversely, let $\tilde{\mu} \ll \mu$. Then, by the assumption $\mu\{A_{\tau_\infty} < \infty\} = 1$, condition is also satisfied. As to conditions A and B, they follow directly from Theorem 1.

To prove Theorem 4 one should only note that, for a Poisson process with the unit parameter, $A_t(x) \equiv t$ and $\tau_\infty(x) \equiv \infty$.

References

[1] Kabanov Yu, Liptser R. and Shiryaev A., Martingale methods in the theory of point processes, Proc. of School-Seminar on the Theory of Random Processes, Druskininkay (25–30 November, 1974), Vilnius, 2 (1975), 296–353.

[2] Jacod J., Multivariate point processes: Predictable projection, Radon-Nikodym derivatives, Representation of martingales, Z. Wahrscheinlichkeitstheorie und Verw. Gebiete, **31** (1975), 235–253.

[3] Liptser R. and Shiryaev A., Statistics of Random Processes, Springer-Verlag, 1977, Vol. I, II.

STEKLOV MATHEMATICAL INSTITUTE
VAVILOVA 42, MOSCOW
V-333, U.S.S.R.

Proc. of Intern. Symp. SDE
Kyoto 1976, pp. 127–140

A Linear Stochastic System with Discontinuous Control

G. KALLIANPUR

§ 1. Introduction

This paper offers some remarks on the problem of control of a linear stochastic system in white the available state information is perturbed by additive Gaussian white noise. Despite the attention devoted to the general theory of stochastic control, this problem has been studied extensively by many authors (e.g. Wonham [11] and Balakrishnan [1, 2]) because of its usefulness in practice and the applicability of filtering theory towards its solution.

Our interest lies in extending the model of [11] to control laws which may be discontinuous. That such models do arise is indicated in [9] (See also [5] and [6] for the corresponding extension of the stochastic equation satisfied by the optimal non-linear filter.) We show that the basic results on conditional normality and the equations governing the conditional mean and variance of the controlled signal carry over for the model (1a′, b) of Section 4. The main tool used is the Bayes formula established in [7] for the case of a signal process independent of the observation (or channel) noise. Recently, Beneš has used the formula to extend Clark's result on the equality of the observation and innovation sigma fields. ([3], [4]). Since Beneš's extension is needed in Proposition 1 we include a proof which is a modification of the original proof given by Clark. For Proposition 1 and the proof of our main conclusions (Theorem 1) we also need an extension of the Bayes formula given in Liptser and Shiryaev's book [8]. In a future paper we hope to apply Theorem 1 to specific problems in which discontinuous control laws occur.

In considering the system (1a′, b) the approach adopted here is similar to Balakrishnan's in that we separate from this study the question of the existence of a (strong) solution to the equations defining the system. The latter is a difficult question in general. At the end of the paper we give an example of an admissible control law (i.e. one which depends non-anticipatively on the observation process) for which the equations (1a′, b) have no strong solution.

§ 2. Proof of $\mathscr{F}_t^\nu = \mathscr{F}_t^z$

Let $\mathscr{F}_t^z$ be the σ-field generated by (Z_s) $(s \leq t)$ where

$$(1) \qquad Z_t = \int_0^t X_s(u)ds + W_t(v) \qquad 0 \leq t \leq T ,$$

$$X \perp\!\!\!\perp W , \qquad \int_0^T E|X_s|\, ds < \infty ,$$

$$\omega = (u, v) \quad \text{and} \quad \int_0^T X_s^2(u)ds < \infty \text{ a.s.}$$

Let $m > 0$ and $S(t, m) = [u\colon \|X(u)\|^2 \vee |X_t(u)| \leq m]$ where

$$\|X(u)\|^2 = \int_0^T X_s^2(u)ds .$$

Let $X_s^m = X_s \cdot I_{S(t,m)}$, $\hat{X}_s = E[X_s | \mathscr{F}_s^z]$ and define

$$(2) \qquad Z_t = \int_0^t \hat{X}_s ds + \nu_t$$

ν being the innovation process associated with Z. (See [4], [5]).

Clark's proof assumes the boundedness of $X_t(u)$. Beneš has recently removed this restriction. The proof here is a modification of Clark's. We follow Meyer's notation. (See [3], [4]).

Let $H = (H_t)$ satisfy the following conditions:

(i) H is progressively measurable with respect to $(\mathscr{F}_t^z)$.

(ii) $|H_t(\omega)| \leq M < \infty$, ($M$ depending on H) for a.e. (t, ω).

Writing

$$R_t^H(u, \omega) = \int_0^t X_s(u)\, d\nu_s(\omega) - \frac{1}{2}\int_0^t X_s^2(u)ds + \int_0^t X_s(u)H_s(\omega)ds ,$$

set

$$(3) \qquad T_t^m(H, \omega) = \frac{\displaystyle\int_{S(t,m)} X_t(u) \exp R_t^H(u, \omega) P^X(du)}{\displaystyle\int_{S(t,m)} \exp R_t^H(u, \omega) P^X(du)} .$$

We choose a progressively measurable version of $\hat{X}^m$ of $T_t^m(\hat{X})$. We have $|\hat{X}_t^m| \leq m$, so that the process $\hat{X}^m$ satisfies (i) and (ii). Furthermore, the Bayes formula gives

$$T_t^m(\hat{X}, \omega) \to \hat{X}_t(\omega) \qquad \text{as } m \to \infty .$$

Before proceeding further we show that the denominator in (3) is positive and finite a.s..

From

$$R_t^H(u,\omega) - R_t^{\hat{X}}(u,\omega) = \int_0^t X_s(u)[H_s(\omega) - \hat{X}_s(\omega)]ds$$

we have

$$|R_t^H(u,\omega) - R_t^{\hat{X}}(u,\omega)| \leq \|X(u)\|\cdot\|H(\omega) - \hat{X}(\omega)\| \leq m\cdot\|H(\omega) - \hat{X}(\omega)\| \quad \text{if } u \in S(t,m)\,.$$

Here, we use the notation $R^{\hat{X}}$ and $T_t^m(\hat{X})$ even though $\hat{X}$ is not bounded and

$$\|H(\omega) - \hat{X}(\omega)\| = \left(\int_0^T [H_s(\omega) - \hat{X}_s(\omega)]^2 ds\right)^{1/2}.$$

Hence

$$\exp[-m\|H(\omega) - \hat{X}(\omega)\|]\cdot\int_{S(t,m)} \exp\{R_t^{\hat{X}}(u,\omega)\}P^X(du)$$
$$\leq \int_{S(t,m)} \exp\{R_t^H(u,\omega)\}P^X(du)$$
$$\leq \exp[m\|H(\omega) - \hat{X}(\omega)\|]\int_{S(t,m)} \exp\{R_t^{\hat{X}}(u,\omega)\}P^X(du)\,.$$

The assertion now follows from the fact that

$$0 < \int_{S(t,m)} \exp\{R_t^{\hat{X}}(u,\omega)\}P^X(du) < \infty, \text{ a.s.}$$

For $0 \leq \alpha \leq 1$ define

$$R_t^{H,\alpha,m}(u,\omega) = R_t^{\alpha H + (1-\alpha)\hat{X}^m}(u,\omega)$$

and

$$T_t^{\alpha,m} = T_t^m[\alpha H + (1-\alpha)\hat{X}^m]\,.$$

$T_t^{\alpha,m}(H,\omega)$ is given by formula (3). Now

$$\exp\{R_t^{H,\alpha,m}(u,\omega)\} \leq \alpha\exp R_t^H(u,\omega) + (1-\alpha)\exp R_t^{\hat{X}^m}(u,\omega)$$
$$\leq \exp\{R_t^H\} + \exp\{R_t^{\hat{X}^m}\},$$

and

$$\left|\frac{d}{d\alpha}\exp R_t^{H,\alpha,m}(u,\omega)\right| \le [\exp R_t^H(u,\omega)+\exp R_t^{\hat{X}^m}(u,\omega)]$$
$$\times\int_0^t |X_s(u)|\,|H_s(\omega)-\hat{X}_s(\omega)|\,ds$$
$$\le \|X(u)\|\cdot\|H(\omega)-\hat{X}^m(\omega)\|\cdot[\exp R_t^H(u,\omega)+\exp R_t^{\hat{X}^m}(u,\omega)]$$
$$\le m\cdot\|H(\omega)-\hat{X}^m(\omega)\|\cdot[\exp R_t^H(u,\omega)+\exp R_t^{\hat{X}^m}(u,\omega)]$$
$$\text{for } u\in S(t,m)\ .$$

For fixed ω, the R. H. Side of the above inequality is P^X-integrable over $S(t,m)$. Hence it is permissible to interchange integration and differentiation with respect to α in the numerator and denominator of formula (3) for $T_t^{\alpha,m}(H,\omega)$. We then obtain

$$\frac{d}{d\alpha}T_t^{\alpha,m}(H,\omega)$$
$$=\frac{\int_{S(t,m)} X_t(u)\exp[R_t^{H,\alpha,m}(u,\omega)]\left\{\int_0^t X_s(u)[H_s(\omega)-\hat{X}_s^m(\omega)]ds\right\}P^X(du)}{\int_{S(t,m)}\exp[R_t^{H,\alpha,m}(u,\omega)]P^X(du)}$$
$$-\,T_t^{\alpha,m}(H,\omega)$$
$$\times\frac{\int_{S(t,m)}\exp[R_t^{H,\alpha,m}(u,\omega)]\left\{\int_0^t X_s(u)[H_s(\omega)-\hat{X}_s^m(\omega)]ds\right\}P^X(du)}{\int_{S(t,m)}\exp[R_t^{H,\alpha,m}(u,\omega)]P^X(du)}\ .$$

$$|\text{1st term on } R\cdot H\cdot S\cdot| \le m^2\left\{\int_0^t [H_s(\omega)-\hat{X}_s^m(\omega)]^2 ds\right\}^{1/2}\ .$$

Since $|T_t^{\alpha,m}(H,\omega)|\le m$,

$$|\text{2nd term on } R\cdot H\cdot S\cdot| \le m^2\left\{\int_0^t [H_s(\omega)-\hat{X}_s^m(\omega)]^2 ds\right\}^{1/2}$$

and we obtain

$$\left|\frac{d}{d\alpha}T_t^{\alpha,m}(H,\omega)\right| \le 2m^2\left\{\int_0^t [H_s(\omega)-\hat{X}_s^m(\omega)]^2 ds\right\}^{1/2}\ .$$

Hence from

$$|T_t^{1,m}(H,\omega)-T_t^{0,m}(H,\omega)| \le K\left\{\int_0^t [H_s(\omega)-\hat{X}_s^m(\omega)]^2 ds\right\}^{1/2}\ ,$$

(where $K=2m^2$), we have

$$(4)\qquad |T_t^m(H,\omega) - \hat{X}_t^m(\omega)| \leq K\left\{\int_0^t [H_s(\omega) - \hat{X}_s^m(\omega)]^2 ds\right\}^{1/2} .$$

Set $H_s^0(\omega) \equiv 0$ for all s. Define $H_t^1(\omega) = T_t^m(H^0, \omega)$ and so recursively, $H_t^{n+1}(\omega) = T_t^m(H^n, \omega)$. From the definition of $T_t^m(H, \omega)$ it is clear that all the processes $H^n = (H_t^n)$ are measurable and adapted to $\mathscr{F}_t^\nu$. (Note that H^n also depends on m which is fixed throughout the argument). Replacing H by H^n in (4) we get

$$(5)\qquad |H_t^{n+1}(\omega) - \hat{X}_t^m(\omega)| \leq K\left\{\int_0^t [H_s^n(\omega) - \hat{X}_s^m(\omega)]^2 ds\right\}^{1/2}$$

for each n. Now (5) holds for arbitrary t in $[0, T]$. Hence for $s \in [0, t]$ (5) gives

$$(6)\qquad |H_s^{n+1}(\omega) - \hat{X}_s^m(\omega)|^2 \leq K^2 \int_0^s [H_{s'}^n(\omega) - \hat{X}_{s'}^m(\omega)]^2 ds' .$$

Substituting (6) (with n replaced by $n - 1$) in (5),

$$\begin{aligned}
|H_t^{n+1}(\omega) - \hat{X}_t^m(\omega)| &\leq K^2\left\{\int_0^t \left(\int_0^s [H_{s'}^{n-1}(\omega) - \hat{X}_{s'}^m(\omega)]^2 ds'\right) ds\right\}^{1/2} \\
&= K^2\left[t \cdot \int_0^t |H_{s'}^{n-1}(\omega) - \hat{X}_{s'}^m(\omega)|^2 ds' - \int_0^t s\,|H_s^{n-1}(\omega) - \hat{X}_s^m(\omega)|^2 ds\right]^{1/2} \\
&= K^2\left[\int_0^t (t - s)\,|H_s^{n-1}(\omega) - \hat{X}_s^m(\omega)|^2 ds\right]^{1/2} \\
&\leq K^3\left[\int_0^t (t - s)\left\{\int_0^s |H_{s'}^{n-2}(\omega) - \hat{X}_{s'}^m(\omega)|^2 ds'\right\} ds\right]^{1/2} \\
&= K^3\left[\int_0^t \frac{(t - s)^2}{2}\,|H_s^{n-2}(\omega) - \hat{X}_s^m(\omega)|^2 ds\right]^{1/2} \\
&\leq \cdots \leq C \cdot K^{n+1}\left[\int_0^t \frac{(t - s)^n}{n!} ds\right]^{1/2} = \frac{C_1 K^{n+1}}{\sqrt{(n + 1)!}} \to 0
\end{aligned}$$

as $n \to \infty$

(C_1 is a constant depending on m and T). It follows that

$$\sup_{t \in [0,T]} |H_t^{n+1}(\omega) - \hat{X}_t^m(\omega)| \to 0$$

and hence $\hat{X}_t^m$ is $\mathscr{F}_t^\nu$ measurable. Since $\hat{X}_t^m(\omega) \to \hat{X}_t(\omega)$ as $m \to \infty$, it follows that $\hat{X}_t$ is $\mathscr{F}_t^\nu$-measurable, and we have equality of $\mathscr{F}_t^\nu$ and $\mathscr{F}_t^z$. It is assumed here that both sigma fields contain all sets of zero probability.

§ 3. The problem of separation and control

Let $(\Omega, \mathscr{A}, P)$ be a probability space on which are given (i) a Wiener martingale $(W_t, \mathscr{G}_t)$ and (ii) a $(\mathscr{G}_t)$ adapted process (α_t) taking values in a measurable space $(X, \mathscr{B}(X))$. Here $(\mathscr{G}_t)$ is an increasing family of sub σ-fields of $\mathscr{A}$ and it is assumed that $(\alpha_t) \perp\!\!\!\perp (W_t)$, $(0 \le t \le T)$. We take the model of the observation process to be

$$(1) \qquad dZ_t = A_t(\alpha, Z)dt + dW_t \,, \qquad Z_0 = 0 \,.$$

The assumptions concerning (A_t) are the following:

(2a) For each $t, A_t(a, g)$ $(a \in X, g \in C,$ the space of real continuous functions on $[0, T])$ is a measurable, non-anticipative functional on $X \times C$.

$$(2b) \qquad A_t(\alpha, Z) = A_t^*(Z) + \tilde{A}_t(\alpha) \,,$$

where

$$A_t^*(Z) \text{ is } \mathscr{F}_t^z\text{-measurable as a function of } \omega \,.$$

$$(3) \qquad \int_0^T E(A_t^2)dt < \infty \,, \qquad \int_0^T E(A_t^{*2})dt < \infty \,.$$

For our purposes we assume that (Z_t) is *given* by (1) in such a way that Z_t is $\sigma[\alpha_s, W_s, s \le t]$-measurable. In other words, (Z_t) is a strong solution of the stochastic differential equation (1). Define the process $(\tilde{Z}_t)$ by

$$(4) \qquad \tilde{Z}_t = \int_0^t \tilde{A}_s(\alpha)ds + W_t \qquad (0 \le t \le T) \,,$$
$$\tilde{Z}_0 = 0 \,.$$

The processes ν *and* $\tilde{\nu}$: Let us define the innovation processes associated with Z and $\tilde{Z}$:

$$(5) \qquad \nu_t = Z_t - \int_0^t \hat{A}_s ds \,, \qquad \text{where } \hat{A}_s = E[A_s | \mathscr{F}_s^Z] \,.$$

Letting $\mathscr{F}_t^{\tilde{z}} = \sigma[\tilde{Z}_s, 0 \le s \le t]$ define

$$(6) \qquad \tilde{\nu}_t = \tilde{Z}_t - \int_0^t E[\tilde{A}_s | \mathscr{F}_s^{\tilde{z}}]ds \,.$$

Note also the following important relation

$$Z_t = \int_0^t A_s^*(Z)ds + \tilde{Z}_t \,. \tag{7}$$

It is well known that $(\nu_t, \mathscr{F}_t^Z, P)$ and $(\tilde{\nu}_t, \mathscr{F}_t^Z, P)$ are Wiener martingales. Let $g(\alpha)$ be an integrable random variable. By the Bayes formula

$$E[g \mid \mathscr{F}_t^Z] = \int_{\Omega_x} g(\alpha)\rho(\alpha, \omega)P^x(d\alpha) \tag{8}$$

where

$$\rho(\alpha, \omega) = \frac{\exp\left\{\int_0^t A_s(\alpha, \omega)dZ_s(\omega) - \frac{1}{2}\int_0^t A_s^2(\alpha, \omega)ds\right\}}{\int_{\Omega_x} \exp\{-\}P^x(d\alpha)} \,. \tag{9}$$

Let us now compute the right hand side of (9) using assumption (2b). The numerator in (9) may be written as

$$\exp\left[\int_0^t A_s d\nu_s - \frac{1}{2}\int_0^t A_s^2 ds + \int_0^t A_s \hat{A}_s ds\right] .$$

Writing $\hat{A}_s$ for $E[\tilde{A}_s \mid \mathscr{F}_s^Z]$, the exponent becomes

$$\begin{aligned}
&\int_0^t (\tilde{A}_s + A_s^*)d\nu_s - \frac{1}{2}\int_0^t (\tilde{A}_s^2 + 2\tilde{A}_s A_s^* + A_s^{*2})ds \\
&\qquad + \int_0^t (\tilde{A}_s + A_s^*)(\hat{A}_s + A_s^*)ds \\
&\quad = \left[\int_0^t A_s^* d\nu_s + \frac{1}{2}\int_0^t A_s^{*2}ds + \int_0^t A_s^* \hat{A}_s ds\right] \\
&\qquad + \left[\int_0^t \tilde{A}_s d\nu_s - \frac{1}{2}\int_0^t \tilde{A}_s^2 ds + \int_0^t \tilde{A}_s \hat{A}_s ds\right] .
\end{aligned}$$

The quantity in the first bracket is $\mathscr{F}_t^Z$ measurable and does not involve α. The factor corresponding to it (in the denominator of (9)) comes out of the integral sign. Hence after a cancellation of the common factor in the numerator and denominator of (9) we obtain

$$\rho(\alpha, \omega) = \frac{\exp\left\{\int_0^t \tilde{A}_s d\nu_s - \frac{1}{2}\int_0^t \tilde{A}_s^2 ds + \int_0^t \tilde{A}_s \hat{A}_s ds\right\}}{\int_{\Omega_x} \exp\{-\}P^x(d\alpha)} \,. \tag{10}$$

Consider now the relation between ν and $\tilde{\nu}$ given by (5), (6) and (7). From these it is easy to see that

$$
\begin{aligned}
d\nu_t &= dZ_t - \hat{A}_t dt = A_t^* dt + d\tilde{Z}_t - \hat{A}_t dt \\
&= d\tilde{Z}_t - \hat{\tilde{A}}_t dt \qquad (11)\\
&= d\tilde{\nu}_t + \delta_t dt
\end{aligned}
$$

where

$$\delta_t = E(\tilde{A} \mid \mathscr{F}_t^Z) - \hat{\tilde{A}}_t .$$

Using (11), the expression

$$
\begin{aligned}
&\int_0^t \tilde{A}_s d\nu_s - \frac{1}{2}\int_0^t \tilde{A}_s^2 ds + \int_0^t \tilde{A}_s \hat{\tilde{A}}_s ds \\
&= \int_0^t \tilde{A}_s d\tilde{\nu}_s + \int_0^t \tilde{A}_s \delta_s ds - \frac{1}{2}\int_0^t \tilde{A}_s^2 ds + \int_0^t \tilde{A}_s [E(\tilde{A}_s \mid \mathscr{F}_s^Z) - \delta_s] ds \\
&= \int_0^t \tilde{A}_s d\tilde{\nu}_s - \frac{1}{2}\int_0^t \tilde{A}_s^2 ds + \int_0^t \tilde{A}_s E(\tilde{A}_s \mid \mathscr{F}_s^Z) ds .
\end{aligned}
$$

Hence from (10),

$$\rho(\alpha, \omega) = \frac{\exp\left\{\int_0^t \tilde{A}_s d\tilde{\nu}_s - \frac{1}{2}\int_0^t \tilde{A}_s^2 ds + \int_0^t \tilde{A}_s E(\tilde{A}_s \mid \mathscr{F}_s^Z) ds\right\}}{\int_{\Omega_x} \exp\{-\} P^x(d\alpha)} . \tag{12}$$

From (9) and (12) it follows that

$$E[g \mid \mathscr{F}_t^Z] = E[g \mid \mathscr{F}_t^{\tilde{Z}}] . \tag{13}$$

Taking $g(\alpha) = \tilde{A}_t(\alpha)$ we get

$$E[\tilde{A}_t \mid \mathscr{F}_t^Z] = E[\tilde{A}_t \mid \mathscr{F}_t^{\tilde{Z}}] . \tag{14}$$

From (11) and (14) we have $\delta_t = 0$ and

$$\nu_t = \tilde{\nu}_t \qquad \text{for all } t . \tag{15}$$

Since $(\tilde{A}_t) \perp\!\!\!\perp (W_t)$ the result of Section 2 applies and combining it with (15) we obtain $\mathscr{F}_t^{\tilde{Z}} = \mathscr{F}_t^{\tilde{\nu}} = \mathscr{F}_t^{\nu}$.

Proposition 1. *Let the process* (Z_t) *and* $(\tilde{Z}_t)$ *be given by* (1) *and* (4) *where* (A_t) *satisfies assumptions* (2) *and* (3). *Let* (ν_t) *be the innovation Wiener process associated with* (Z_t). *The following conclusions then hold*:

(a) $$\mathscr{F}_t^Z = \mathscr{F}_t^{\nu} \qquad \text{for each } t ;$$

(b) *If $g(\alpha)$ is any integrable random variable (so that $g(\alpha) \perp\!\!\!\perp (W_t)$) then,*

$$E[g \mid \mathcal{F}_t^Z] = E[g \mid \mathcal{F}_t^\nu] .$$

Let us now consider the question of the equality of the σ-fields $\mathcal{F}_t^Z$ and $\mathcal{F}_t^\nu$. It is assumed that (Z_t) is a strong solution of (1). Since $\int_0^T \tilde{A}_t^2 dt < \infty$ a.s. and $(\tilde{A}_t) \perp\!\!\!\perp (W_t)$ it follows that Q given by

$$dQ = \exp\left[-\int_0^T \tilde{A}_t dW_t - \frac{1}{2}\int_0^T \tilde{A}_t^2 dt\right] dP$$

is a probability measure. From (4) and Girsanov's theorem $(\tilde{Z}_t, \mathcal{G}_t, Q)$ is a Wiener martingale and moreover P and Q are mutually absolutely continuous. With Q as the probability measure consider the stochastic differential equation

$$d\xi_t = A_t^*(\xi)dt + d\tilde{Z}_t , \qquad \xi_0 = 0 . \tag{16}$$

Proposition 2. *Let (Z_t) be the unique strong solution of* (1). *Then $\mathcal{F}_t^\nu = \mathcal{F}_t^Z$ for each t if and only if (Z_t) is a strong solution of* (16).

Proof. If (16) has a strong solution (ξ_t) then by the definition of $\tilde{Z}$ and (2b) we obtain

$$d\xi_t = A_t(\alpha, \xi)dt + dW_t , \qquad \xi_0 = 0 .$$

By uniqueness, $\xi_t = Z_t$. Thus (Z_t) is a strong solution of (16) and hence $\mathcal{F}_t^Z \subseteq \mathcal{F}_t^{\tilde{Z}}$. The equality $\mathcal{F}_t^Z = \mathcal{F}_t^\nu$ follows from (a) of Proposition 1. On the other hand, suppose $\mathcal{F}_t^Z = \mathcal{F}_t^\nu$ for each t. Now Z_t which is the strong solution of (1) satisfies equation (7). Hence $\mathcal{F}_t^Z = \mathcal{F}_t^{\tilde{Z}}$ by Proposition 1 and it follows that (Z_t) is a strong solution of (16).

Remark. Using Proposition 2 it is easy to show that the following Lipschitz condition is sufficient for $\mathcal{F}_t^\nu = \mathcal{F}_t^Z$: There is a positive constant M such that

$$|A_t(\alpha, f) - A_t(\alpha, g)| \leq M \sup_{0 \leq s \leq t} |f(s) - g(s)| , \qquad (f, g \in C[0, T]) , \tag{17}$$

for each t and α. This condition goes back to Wonham [11].

§ 4. Application to the separation and control problem

Consider the stochastic differential equation

$$dX_t = a_t X_t dt + b_t u_t dt + f_t d\eta_t \tag{1a}$$

(1b) $$dY_t = c_t X_t dt + g_t d\varepsilon_t .$$
$$X_0 = x_0, Y_0 = 0 \qquad (0 \le t \le T)$$

where a_t, b_t, c_t, f_t and g_t are non-random continuous functions of t, (η_t) and (ε_t) are independent Wiener processes, $x_0 \perp\!\!\!\perp (\eta_t), (\varepsilon_t)$ and $g_t \neq 0\ \forall t$. The process $u = (u_t)$ is assumed to be of the form $u_t(\omega) = \Psi[t; \pi_t y(\omega)]$ where $\Psi[t, g]$ is a real valued, Borel measurable functional on $[0, T] \times C$, C being the space of all continuous mappings from $[-T, 0]$ to $\boldsymbol{R}$. (See [5]). It is further assumed that

(2) $$\int_0^T E[u_t]^2 dt < \infty .$$

The process $u_t(\omega)$ is a "control" function which according to our assumption is measurable with respect to $\mathscr{F}_t^Y \equiv \sigma[Y_s, 0 \le s \le t]$. The interdependence between u any y is expressed by the assumption that the system (1) has a unique solution. All control laws we consider are "admissible" in this sense. By the existence of a unique solution of (1) we mean in particular, that for each t, X_t is measurable with respect to the σ-field $\sigma[x_0, \eta_s, Y_s, 0 \le s \le t]$. If we write $\alpha = (x_0, \eta)$ where $\eta = (\eta_s, 0 \le s \le t)$ then $\alpha \perp\!\!\!\perp (\varepsilon_t)$. Before proceeding further let us reduce the second equation in (1) to the standard form ([5] shows how this can be done).

(3) $$dZ_t = K_t X_t dt + dW_t , \qquad (Z_0 = 0) .$$

Here K_t is a continuous non-random function, (W_t) is a Wiener process. It is easy to verify that $\mathscr{F}_t^Z = \mathscr{F}_t^Y$ and that $\alpha \perp\!\!\!\perp (W_t)$. Hence, $K_t X_t$ is of the form $A_t(\alpha, \omega)$ of the last section. Define X_t^* and $\tilde{X}_t$ by the following relations:

(4) $$d\tilde{X}_t = a_t \tilde{X}_t dt + f_t d\eta_t$$
$$\tilde{X}_0 = x_0 ,$$
$$\frac{dX_t^*}{dt} = a_t X_t^* + b_t u_t , \qquad X_0^* = 0 .$$

Then

(5) $$X_t = X_t^* + \tilde{X}_t ,$$

It is clear that $K_t X_t^*$ and $K_t \tilde{X}_t$ can be identified with A_t^* and $\tilde{A}_t$ of Section 4. Furthermore, the observation process (Z_t) in (3) is given by the model (1) of Section 4. A stochastic system which takes into account the possibility of *discontinuous* control is obtained by replacing equation (1a) by

(1a') $$dX_t = a_t X_t dt + b_t dU_t + f_t d\eta ,$$

where $U_t(\omega)$ satisfies the same measurability conditions as $u_t(\omega)$. For a.e. ω, the sample paths $t \to U_t(\omega)$ will be assumed to be of bounded variation in $[0, T]$. Instead of condition (2) we shall assume that

(2a) $$E[\text{Var}_T\, U]^2 < \infty ,$$

where $(\text{Var}_T\, U)(\omega)$ is the total variation over $[0, T]$ of the function $t \to U_t(\omega)$. Suppose we are given the system (1a, b) and condition (2) is satisfied. Defining U_t by $dU_t = u_t dt$ we see that (1a) can be written in the form (1a′) and (2) implies (2a). Therefore we shall consider only the more general system (1a′), (1b) from now on. The assumption now is that (1a′, b) has a unique solution. All the remarks about the measurability of X_t made above now remain in force. As before we have $X_t = X_t^* + \tilde{X}_t$ where $\tilde{X}_t$ is given by (4) and X_t^* is given by

(5′) $$dX_t^* = a_t X_t^* dt + b_t dU_t , \qquad X_0^* = 0 .$$

Note that X_t^* is given explicitly by the expression in (4) and is therefore $(\mathscr{F}_t^Z)$-adapted.

The following result is an easy consequence of Proposition 1 of Section 4.

Theorem 1. *Let (X_t) be the process defined by the system* (1a′, b). *Then*

(a) *The conditional distribution of X_t given $\mathscr{F}_t^Y$ is Gaussian with mean $\hat{X}_t$ and covariance P_t;*

(b) *$(\hat{X}_t)$ satisfies the stochastic equation*

(6) $$d\hat{X}_t = a_t \hat{X}_t dt + b_t dU_t + K_t P_t d\nu_t$$
$$\hat{X}_0 = E(x_0) ,$$

where (ν_t) is the innovation process associated with (Z_t);

(c) *The covariance P_t satisfies the Riccati equation*

(7) $$\frac{dP_t}{dt} = 2a_t P_t + f_t^2 - K_t^2 P_t^2$$
$$P_0 = E(x_0^2) - (Ex_0)^2 .$$

Proof. (a) If λ is any real number, letting $g(\alpha) = e^{i\lambda \tilde{X}_t(\alpha)}$ in Proposition 1 of the previous section we have

(8) $$E[e^{i\lambda X_t} | \mathscr{F}_t^Y] = E[e^{i\lambda X_t} | \mathscr{F}_t^Z] = e^{i\lambda X_t^*} E[e^{i\lambda \tilde{X}_t} | \mathscr{F}_t^Z]$$
$$= e^{i\lambda X_t^*} E[e^{i\lambda \tilde{X}_t} | \mathscr{F}_t^Z] .$$

Recall that $(\tilde{X}_t, \tilde{Z}_t)$ satisfy the equations of the usual Kalman-Bucy theory:

$$d\tilde{X}_t = a_t\tilde{X}_t dt + f_t d\eta_t\,, \qquad \tilde{X}_0 = x_0$$
$$d\tilde{Z}_t = K_t\tilde{X}_t dt + dW_t\,, \qquad \tilde{Z}_0 = 0\,.$$

It is well known from this theory that the conditional distribution of $\tilde{X}_t$ given $\mathcal{F}_t^{\tilde{Z}}$ is Gaussian with mean $E(\tilde{X}_t|\mathcal{F}_t^{\tilde{Z}})$ and covariance $\tilde{P}_t$ which satisfy the following equations

$$\text{(9)} \qquad \begin{aligned} dE(\tilde{X}_t|\mathcal{F}_t^{\tilde{Z}}) &= a_t E(\tilde{X}_t|\mathcal{F}_t^{\tilde{Z}})dt + K_t\tilde{P}_t d\tilde{\nu}_t \\ E(\tilde{X}_0|\mathcal{F}_0^{\tilde{Z}}) &= E(x_0) \end{aligned}$$

and

$$\frac{d\tilde{P}_t}{dt} = 2a_t\tilde{P}_t + f_t^2 - K_t^2\tilde{P}_t^2$$

with

$$\tilde{P}_0 = E(x_0^2) - (E(x_0))^2\,.$$

From Proposition 1, $E(\tilde{X}_t|\mathcal{F}_t^{\tilde{Z}}) = E(\tilde{X}_t|\mathcal{F}_t^{Z}) = \hat{\tilde{X}}_t$, say. Hence it follows from (8) that

$$E[e^{i\lambda X_t}|\mathcal{F}_t^Y] = e^{i\lambda X_t^*}\exp\left(i\lambda\hat{\tilde{X}}_t - \frac{1}{2}\lambda^2\tilde{P}_t\right)$$
$$= \exp\left(i\lambda\hat{X}_t - \frac{1}{2}\lambda^2\tilde{P}_t\right)$$

where we define $\hat{X}_t = E(X_t|\mathcal{F}_t^Z) = X_t^* + \hat{\tilde{X}}_t$. This proves (a). The assertions (b) and (c) follow immediately if we note that

$$P_t = E[(X_t - \hat{X}_t)^2|\mathcal{F}_t^Z] = E[\{\tilde{X}_t - E(\tilde{X}_t|\mathcal{F}_t^{\tilde{Z}})\}^2|\mathcal{F}_t^{\tilde{Z}}] = \tilde{P}_t\,,$$

observe that $\nu_t = \tilde{\nu}_t$ from (15) of Section 4, and use equations (5′) and (9).

The general problem of the equality of the innovation and observation σ-fields has attracted a great deal of attention in recent years. In this connection, Tsirel'son has recently given an example of a stochastic differential equation which admits no strong solution [10]. We shall use his construction to give (a) an example where the state equations (1a′, b) do not have a strong (or causal) solution, (b) an example in which assuming the existence of a strong solution for (1a′, b) we show that $\mathcal{F}_t^\nu \neq \mathcal{F}_t^Z$.

Let $\{t_j\} \subset [0, 1]$, $(j = 0, -1, -2, \cdots)$ with $t_0 = 1$, $t_{j-1} < t_j$ and $t_j \to 0$ as $j \to -\infty$. Let $\{k_j\}$ be a sequence of positive numbers such that $k_j < k_{j+1}$, $\sum_j k_j^{-1} < \infty$ and $\sum_j k_j^4\,(t_j - t_{j-1}) < \infty$. Writing $\{a\}$ for the fractional part of the real number a, define the non-anticipating functional

$$\beta(t, f) = \frac{f_{t_j} - f_{t_{j-1}}}{t_j - t_{j-1}} \tag{10}$$

$$\text{for } t \in [t_j, t_{j+1}),\ j = 0, -1, \cdots \text{ and } f \in C[0, 1] .$$

Let $U_t(f)$ be defined as a right continuous step function with discontinuities at $\{t_j\}$ given by

$$U_{t_j}(f) - U_{t_j - 0}(f) = \frac{k_j^{-1}\beta_j(f) - k_{j-1}^{-1}\beta_{j-1}(f)}{\alpha_{t_j}} \tag{11}$$

$$\text{where } \beta_j(f) = \beta(t_j, f)$$

and $\alpha_t = \exp\left(-\int_0^t a_s ds\right)$. Consider the system

$$\begin{aligned} dX_t &= a_t X_t dt + dU_t + f_t d\eta_t , \qquad X_0 = x_0 , \\ dZ_t &= K_t X_t dt + dW_t , \qquad Z_0 = 0 \end{aligned} \tag{12}$$

where (U_t) is as chosen above, a_t, f_t are continuous non-random functions and $K_t = k_j \alpha_t$ for $t \in [t_j, t_{j+1})$. This choice of K_t ensures that $\int_0^T K_t^4 dt < \infty$ so that in the equation

$$d\tilde{Z}_t = K_t \tilde{X}_t dt + dW_t$$

we have $\int_0^T (K_t \tilde{X}_t)^2 dt < \infty$ a.s. Thus $(\tilde{Z}_t, \mathscr{G}_t)$ is a Wiener martingale under the probability Q introduced in Section 3. It is easy to verify that for each f in $C[0, 1]$, Var $U(f)$ (i.e. the total variation of $U_t(f)$) is finite and has an upper bound independent of f. Condition (2a) is thus satisfied. We also have $\alpha_t X_t^* = k_m^{-1}\beta_m$ for $t_m \leq t < t_{m+1}$. Equation (16) of Section 3 then takes the form

$$dZ_t = \beta(t, Z)dt + d\tilde{Z}_t \tag{16'}$$

where β is Tsirel'son's function (10). Hence (16′) does not have a strong solution.

Example. The stochastic system (12) with $X_0 = x_0$ non-random and $f_t \equiv 0$ does not have a strong solution. In this case, $\tilde{X}_t = (\alpha_t^{-1})x_0$ and the second member of (12) becomes

$$dZ_t = [K_t \alpha_t^{-1} x_0 + \beta(t, Z)]dt + dW_t .$$

Since

$$d\tilde{Z}_t = K_t \alpha_t^{-1} x_0 + dW_t , \qquad \tilde{Z}_0 = 0$$

we have $\mathscr{F}_t^Z = \mathscr{F}_t^W$. If (12) has a strong solution, clearly Z_t is $\mathscr{F}_t^W$-measurable and hence $\mathscr{F}_t^Z$-measurable. But this is impossible because it has been shown that (16′) does not have a strong solution.

If the initial value x_0 is any random variable $\perp\!\!\!\perp (W_t)$ and the coefficient function f_t is not identically zero we do not know whether a strong solution to (12) exists. In any case, if a strong solution (X_t, Z_t) does exist it is easy to see that the equality of the σ-fields $\mathscr{F}_t^\nu$ and $\mathscr{F}_t^Z$ does not hold. This is an immediate consequence of Proposition 2 and the fact that equation (16′) does not have a strong solution.

References

[1] Balakrishnan, A. V., Stochastic control: A function space approach, SIAM J. Control., **10** (1972), 285–297.

[2] ——, A note on the structure of optimal stochastic controls, Applied Mathematics and Optimization, **1** (1974), 87–94.

[3] Beneš V. E., Extension of Clark's innovations equivalence theorem to the case of signal Z independent of noise, with $\int_0^t Z_s^2 ds < \infty$ a.s., to appear.

[4] Clark, J. M. C., Conditions for one-to-one correspondence between an observation process and its innovation, Technical Report, Centre for Computing and Automation, Imperial College, 1969.

[5] Fujisaki, M., Kallianpur, G. and Kunita, H., Stochastic differential equations for the non-linear filtering problem, Osaka J. Math., **9** (1972), 19–40.

[6] Kallianpur, G., A stochastic equation for the optimal non-linear filter, 4th International Symposium on Multivariate Analysis, Dayton, Ohio, 1975.

[7] Kallianpur, G. and Striebel, C., Estimation of stochastic processes: Arbitrary system process with additive white noise observation errors, Ann. Math. Statist., **39** (1968), 785–801.

[8] Liptser, R. S. and Shiryaev, A. N., Statistics of stochastic processes (Russian), Moscow, 1974.

[9] Striebel, C., Martingale conditions for the optimal control of continuous time stochastic systems, (Unpublished).

[10] Tsirel'son, B. S., An example of a stochastic differential equation not having a strong solution (Russian), Theor. Probability Appl., **20** (1975), 427–430.

[11] Wonham, W. M., Random differential equations in control theory, Probabilistic Methods in Applied Mathematics, **2**, Academic Press, 1970.

SCHOOL OF MATHEMATICS
127 VINCENT HALL
UNIVERSITY OF MINNESOTA
MINNEAPOLIS, MINNESOTA 55455
U.S.A.

Proc. of Intern. Symp. SDE
Kyoto 1976, pp. 141–152

The Equivalence of Two Conditions on Weighted Norm Inequalities for Martingales

Norihiko KAZAMAKI

§ 1. Introduction

Let $(\Omega, \mathscr{F}, P)$ be a probability space with a non-decreasing right continuous family $(\mathscr{F}_t)$ of sub σ-fields of $\mathscr{F}$ such that $\mathscr{F} = \bigvee_{t \geq 0} \mathscr{F}_t$. Let M be an L^2-bounded continuous martingale such that $M_0 = 0$, and set

$$Z_t = \exp(M_t - \tfrac{1}{2}\langle M \rangle_t) .$$

Here $\langle M \rangle$ denotes the unique continuous increasing process over $(\mathscr{F}_t)$ such that $M^2 - \langle M \rangle$ is a martingale. As is well-known, the process Z_t is a positive local martingale which is the solution of the stochastic integral equation

$$Z_t = 1 + \int_0^t Z_s dM_s .$$

I. V. Girsanov [1] raised the problem of finding sufficient conditions for the process Z_t to be a martingale. It plays an important role in certain aspects of the theory of stochastic integral equations. Recently, it is proved by A. A. Novikov [5] that if $\exp(\frac{1}{2}\langle M \rangle_t) \in L^1$ for each $t > 0$, then Z_t is a martingale. In section 2 we show that if $\exp(\frac{1}{2}M_t) \in L^1$ for each $t > 0$, then Z_t is a martingale. This result is an improvement of the Novikov theorem. In section 3 we give a necessary and sufficient condition for M_t to be a BMO-martingale. Now we suppose, in section 4, that Z_t is a uniformly integrable martingale; that is, $Z_t = E[Z \mid \mathscr{F}_t]$ for some integrable random variable Z. Then it is shown that there are stopping times $T_n \uparrow \infty$ a.s., such that the weighted normed inequalities

$$E\Big[\sup_{0 \leq t \leq T_n} |X_t|^p Z\Big] \leq C_{p,n} E[|X_{T_n}|^p Z] , \qquad 1 < p < \infty , \qquad n = 1, 2, 3, \cdots$$

are valid for all uniformly integrable martingales X_t. Here $C_{p,n}$ is a positive constant independent of X. From these results, we can deduce a new type martingale inequality.

The reader is assumed to be familiar with the martingale theory as expound in [3]. In what follows we denote by C a positive constant and by C_x a positive constant depending only on the indexed parameter x, both letters are not necessarily the same in each occurence.

§ 2. On the problem of Girsanov

The aim of this section is to give a new sufficient condition for the process Z_t to be a martingale. Now, set $Z'_t = \exp(-M_t - \frac{1}{2}\langle M\rangle_t)$, which is also a local martingale. Of course, under the Novikov condition, Z_t and Z'_t are martingales. But, Z'_t is not necessarily a martingale even if Z_t is a bounded martingale. We start with such an example. Let (B_t) be a one-dimensional Brownian motion such that $B_0 = 0$. We set

$$\xi = \inf\{t > 0;\, B_t \geqslant 1\}$$

which is a stopping time such that $0 < \xi < \infty$ a.s.. Now let $\alpha\colon [0, 1[\to [0, \infty[$ be an increasing homeomorphic function, and set

$$\theta_t = \begin{cases} \alpha(t) \wedge \xi & \text{if } 0 \leqslant t < 1 \\ \xi & \text{if } 1 \leqslant t < \infty\,. \end{cases}$$

Then each θ_t is a stopping time such that $\theta_0 = 0$ and $\theta_1 = \xi$. For a.e. $\omega \in \Omega$ the sample functions $\theta_\cdot(\omega)$ are non-decreasing and continuous, so that $M_t = B_{\theta_t}$ is a continuous local martingale. As $\theta_t \leqslant \xi$, we have $M_t \leqslant 1$ and $Z_t \leqslant e$. Therefore, Z_t is a bounded martingale. On the other hand, as $M_1 = B_\xi = 1$, $E[Z'_1] \leqslant E[\exp(-M_1)] = e^{-1} < 1$. This implies that the process Z'_t is not a martingale.

Theorem 1. *If* $\exp(\frac{1}{2}M_t) \in L^1$ *for each* $t > 0$, *then the process* Z_t *is a martingale.*

Proof. Generally, Z_t is a positive local martingale, so that $E[Z_t] \leqslant 1$ for every t. Therefore, it is a martingale if and only if $E[Z_t] = 1$ for every t.

Now let $\mu_t = \inf\{s > 0;\, \langle M\rangle_s > t\}$. Each μ_t is an $\mathscr{F}_t$-stopping time and we denote by $(\mathscr{G}_t)$ the right continuous family $(\mathscr{F}_{\mu_t})$. Let $(\Omega', \mathscr{F}', P')$ be another probability space which carries a one-dimensional Brownian motion $(B'_t, \mathscr{F}'_t)$ with $B'_0 = 0$. We denote by $(\hat{\Omega}, \hat{\mathscr{F}}, \hat{P})$ the product of $(\Omega, \mathscr{F}, P)$ and $(\Omega', \mathscr{F}', P')$ with π, π' the projections of $\hat{\Omega} = \Omega \times \Omega'$ onto Ω and Ω' respectively. Set $\hat{\mathscr{G}}_t = \mathscr{G}_t \times \mathscr{F}'_t$. Then $\langle M\rangle_t \circ \pi$ is a $\hat{\mathscr{G}}_t$-stopping time. Let $\hat{\mathscr{F}}_t = \hat{\mathscr{G}}_{\langle M\rangle_t \circ \pi}$. The system $(\hat{\Omega}, \hat{\mathscr{F}}, \hat{\mathscr{F}}_t, \hat{P})$ is a lifting of $(\Omega,$

$\mathscr{F}, \mathscr{F}_t, P)$ under π. It is easy to see that $M_{\mu_t} \circ \pi$ and $B'_t \circ \pi'$ are $\hat{\mathscr{G}}_t$-continuous local martingales. As is well-known, from a classical result of P. Lévy,

$$B_t = M_{\mu_j} \circ \pi + B'_t \circ \pi' - B'_{t \wedge (\langle M \rangle_\infty \circ \pi)} \circ \pi'$$

is a Brownian motion over $(\hat{\mathscr{G}}_t)$.

It is clear that $M_t \circ \pi = B_{\langle M \rangle_t \circ \pi}$. Now, set

$$\tau_a = \inf \{t \geqslant 0 ; B_t \leqslant t - a\} , \qquad 0 < a < \infty .$$

As the distribution density of this $\hat{\mathscr{G}}_t$-stopping time τ_a equals

$$\frac{a}{\sqrt{2\pi t^3}} \exp \left(-\frac{(t-a)^2}{2t} \right) ,$$

we have $\hat{E}[\exp(\frac{1}{2}\tau_a)] = \exp(a)$ (see [2]). Here $\hat{E}[\ ]$ denotes expectation with respect to $d\hat{P}$. Since $B_{\tau_a} - \frac{1}{2}\tau_a = \frac{1}{2}\tau_a - a$, it follows that

$$\hat{E}[\exp(B_{\tau_a} - \tfrac{1}{2}\tau_a)] = 1 .$$

This implies that the process $X_t = \exp(B_{t \wedge \tau_a} - \frac{1}{2}(t \wedge \tau_a))$ is a $\hat{P}$-uniformly integrable martingale over $(\hat{\mathscr{G}}_t)$. Then, as $\langle M \rangle_t \circ \pi$ is a $\hat{\mathscr{G}}_t$-stopping time,

$$\hat{E}[X_{\langle M \rangle_t \circ \pi}] = 1 .$$

On the other hand,

$$\begin{aligned} \hat{E}[X_{\langle M \rangle_t \circ \pi}] = \hat{E}[X_{\langle M \rangle_t \circ \pi} &; \tau_a > \langle M \rangle_t \circ \pi] \\ &+ \hat{E}[X_{\langle M \rangle_t \circ \pi} ; \tau_a \leqslant \langle M \rangle_t \circ \pi] . \end{aligned}$$

As $X_{\langle M \rangle_t \circ \pi} = Z_t \circ \pi$ on $\{\tau_a > \langle M \rangle_t \circ \pi\}$, the first term on the right side is smaller than $\hat{E}[Z_t \circ \pi] = E[Z_t]$. And the second term is smaller than

$$\begin{aligned} &\hat{E}[\exp(\tfrac{1}{4}B_{(\langle M \rangle_t \circ \pi) \wedge \tau_a}) \exp(\tfrac{3}{4}B_{\tau_a} - \tfrac{1}{2}\tau_a)] \\ &\quad \leqslant (\hat{E}[\exp(\tfrac{1}{2}B_{(\langle M \rangle_t \circ \pi) \wedge \tau_a})])^{1/2} (\hat{E}[\exp(\tfrac{3}{2}B_{\tau_a} - \tau_a)])^{1/2} . \end{aligned}$$

As $B_{\tau_a} = \tau_a - a$, the second term on the right side is

$$(\hat{E}[\exp(\tfrac{1}{2}\tau_a)] \exp(-\tfrac{3}{2}a))^{1/2} = \exp(-\tfrac{1}{4}a)$$

which converges to 0 as $a \to \infty$. To estimate the first term, set

$$T = \inf \{s \geqslant 0 ; \langle M \rangle_s \circ \pi \geqslant \tau_a\} .$$

For each $t, \{T \leqslant t\} = \{\tau_a \leqslant \langle M \rangle_t \circ \pi\} \in \hat{\mathscr{G}}_{\langle M \rangle_t \circ \pi} = \hat{\mathscr{F}}_t$, so that T is an $\hat{\mathscr{F}}_t$-stopping time. It follows from the definition of T that

$$\langle M \rangle_{t \wedge T} \circ \pi = (\langle M \rangle_t \circ \pi) \wedge \tau_a .$$

As $M_t \circ \pi = B_{\langle M \rangle_t \circ \pi}$ is a martingale over $(\hat{\mathscr{F}}_t)$, by the Doob optional sampling theorem

$$\hat{E}[B_{\langle M \rangle_t \circ \pi} \mid \hat{\mathscr{F}}_{t \wedge T}] = B_{\langle M \rangle_{t \wedge T} \circ \pi} = B_{(\langle M \rangle_t \circ \pi) \wedge \tau_a} .$$

Thus, from the Jensen inequality,

$$\begin{aligned} \hat{E}[\exp(\tfrac{1}{2} B_{(\langle M \rangle_t \circ \pi) \wedge \tau_a})] &\leqslant \hat{E}[\exp(\tfrac{1}{2} B_{\langle M \rangle_t \circ \pi})] \\ &= \hat{E}[\exp(\tfrac{1}{2} M_t \circ \pi)] \\ &= E[\exp(\tfrac{1}{2} M_t)] . \end{aligned}$$

Consequently, we have

$$1 \leqslant E[Z_t] + (E[\exp(\tfrac{1}{2} M_t)])^{1/2} \exp(-a/4) .$$

The right side converges to $E[Z_t]$ as $a \to \infty$. Thus, $E[Z_t] = 1$ for every $t > 0$. This completes the proof.

Remark. We have

$$E[\exp(\alpha M_t)] \leqslant (E[\exp(2\alpha^2 \langle M \rangle_t)])^{1/2}$$

for every $\alpha > 0$. Indeed, applying the Schwarz inequality we get

$$\begin{aligned} E[\exp(\alpha M_t)] &= E[\exp(\alpha M_t - \alpha^2 \langle M \rangle_t) \exp(\alpha^2 \langle M \rangle_t)] \\ &\leqslant (E[\exp(2\alpha M_t - 2\alpha^2 \langle M \rangle_t)])^{1/2} (E[\exp(2\alpha^2 \langle M \rangle_t)])^{1/2} \\ &\leqslant (E[\exp(2\alpha^2 \langle M \rangle_t)])^{1/2} . \end{aligned}$$

Therefore, if $\exp(\frac{1}{2}\langle M \rangle_t) \in L^1$, then $\exp(\frac{1}{2} M_t) \in L^1$. Namely, our result is an improvement of the Novikov theorem. Novikov observed that for every $\delta > 0$ there is a continuous martingale M_t such that $E[\exp((\frac{1}{2} - \delta)\langle M \rangle_t)]$ is finite and the process Z_t is not a martingale. Therefore, our condition may not be essentially weakened.

§ 3. A characterization of BMO-martingales

A martingale M belongs to the class BMO if

$$\|M\|_{\mathrm{BMO}}^2 = \sup_t \operatorname{ess}\cdot\sup_\omega E[\langle M \rangle_\infty - \langle M \rangle_t \mid \mathscr{F}_t] < \infty .$$

The following result is the analog of a real variable result of John and Nirenberg.

Lemma 1. *Let X be a continuous martingale with $\|X\|_{\mathrm{BMO}} < 1$. Then for any stopping time T*

$$E[\exp(\langle X\rangle_\infty - \langle X\rangle_T) \mid F_T] \leqslant \frac{1}{1 - \|X\|_{\mathrm{BMO}}^2} \quad \text{a.s.} .$$

Proof. To prove this lemma, it suffices to show that for every $A \in \mathscr{F}_T$

$$E[\exp(\langle X\rangle_\infty - \langle X\rangle_T); A] \leqslant \frac{1}{1 - \|X\|_{\mathrm{BMO}}^2} P(A) .$$

We may assume that $P(A) > 0$. Let $dP' = I_A(dP/P(A))$ and $\mathscr{F}'_t = \mathscr{F}_{T+t}$. Then $X'_t = X_{T+t} - X_T$ is a P'-continuous local martingale over $(\mathscr{F}'_t)$ such that $\langle X'\rangle = \langle X\rangle_{T+t} - \langle X\rangle_T$. Clearly, we have

$$E[\exp(\langle X\rangle_\infty - \langle X\rangle_T); A] = E'[\exp(\langle X'\rangle_\infty)]P(A)$$

where $E'[\ ]$ denotes expectation over Ω with respect to dP'. Now, let S be any $\mathscr{F}'_t$-stopping time. Then for any $A' \in \mathscr{F}'_S = \mathscr{F}_{T+S}$ we get

$$\begin{aligned}\int_s E'[\langle X'\rangle_\infty - \langle X'\rangle_S \mid F'_S] dP &= \frac{1}{P(A)} \int_{A' \cap A} (\langle X\rangle_\infty - \langle X\rangle_{T+S}) dP \\ &= \frac{1}{P(A)} \int_{A'} E[\langle X\rangle_\infty - \langle X\rangle_{T+S} \mid F_{T+S}] I_A dP \\ &\leqslant \int_{A'} \|X\|_{\mathrm{BMO}}^2 \, dP' ,\end{aligned}$$

from which $E'[\langle X'\rangle_\infty - \langle X'\rangle_S \mid \mathscr{F}'_S] \leqslant \|X\|_{\mathrm{BMO}}^2$. By the energy inequalities,

$$E'[\langle X'\rangle_\infty^n] \leqslant n! \, \|X\|_{\mathrm{BMO}}^{2n} , \qquad n = 0, 1, 2, \cdots .$$

As $\|X\|_{\mathrm{BMO}} < 1$, we get

$$\begin{aligned}E'[\exp(\langle X'\rangle_\infty)] &= \sum_{n=0}^\infty \frac{1}{n!} E'[\langle X'\rangle_\infty^n] \\ &\leqslant \frac{1}{1 - \|X\|_{\mathrm{BMO}}^2}\end{aligned}$$

completing the proof.

The aim of this section is to prove

Theorem 2. (1) *The following conditions are equivalent.*
(a) Z_t *satisfies the condition*

$$(A_p) \qquad \sup_t \text{ess}\cdot\sup_\omega Z_t E[(1/Z_\infty)^{1/(p-1)} \mid \mathscr{F}_t]^{p-1} < \infty$$

for some $p > 1$.
(b) Z_t *satisfies the condition*

$$(B_\lambda) \qquad \sup_t \text{ess}\cdot\sup_\omega E[\exp\{\lambda(\langle M\rangle_\infty - \langle M\rangle_t)\} \mid \mathscr{F}_t] < \infty$$

for some $\lambda > 0$.
(c) M_t *is a BMO-martingale.*
(2) *Set* $Z'_t = \exp(-M_t - \frac{1}{2}\langle M\rangle_t)$. *Then* Z_t *and* Z'_t *satisfy* (A_p) *for all* $p > 1$ *if and only if* Z_t *satisfies* (B_λ) *for all* $\lambda > 0$.

The (A_p) condition has already appeared many times in the literature in connection with the weighted mean convergence of Fourier series. By the Hölder inequality, (A_p) implies (A_r) for $r > p$. Theorem 2 follows at once from the next lemma.

Lemma 2. *Let* $1 < p < \infty$. *Then*
(3) (A_p) *implies* $(B_{(1/2(\sqrt{p}+1)^2)})$,
(4) *If* Z_t *and* Z'_t *satisfy* (A_p), Z_t *satisfies* $(B_{1/2(p-1)})$,
(5) *Conversely, if* Z_t *satisfies* $(B_{(1/2(\sqrt{p}-1)^2)})$, *then* Z_t *and* Z'_t *satisfy* (A_p).

We begin with the proof of Theorem 2. From (3), (a) implies (b). And, by the Jensen inequality, (c) follows at once from (b). Let now M_t be a BMO-martingale. Take $p > 1$ such that

$$\frac{1}{2(\sqrt{p}-1)^2}\|M\|^2_{\text{BMO}} < 1\,.$$

Then, by Lemma 1, we get

$$E\left[\exp\left(\frac{1}{2(\sqrt{p}-1)^2}(\langle M\rangle_\infty - \langle M\rangle_t)\right)\Big| \mathscr{F}_t\right] \leqslant \frac{1}{1 - \dfrac{1}{2(\sqrt{p}-1)^2}\|M\|^2_{\text{BMO}}}\,.$$

That is, by (5), (c) implies (a). The proof of the part (2) is easy. The necessity is clear by (4) and the sufficiency is also clear by (5).

Proof of Lemma 2. To show (3), set $p_0 = \sqrt{p}/(\sqrt{p}-1) > 1$.

Then the exponent conjugate q_0 to p_0 is $\sqrt{p}$, and by a simple computation

$$\frac{1}{2p_0(p-1)} - \frac{q_0}{2p_0^2(p-1)^2} = \frac{1}{2(\sqrt{p}+1)^2}\,.$$

So we get

$$\begin{aligned}
&E\left[\exp\left(\frac{1}{2(\sqrt{p}+1)^2}(\langle M\rangle_\infty - \langle M\rangle_t)\right)\Big|\mathscr{F}_t\right] \\
&\quad = E\left[\exp\left(-\frac{1}{p_0(p-1)}(M_\infty - M_t)\right.\right. \\
&\qquad\qquad\qquad \left. + \frac{1}{2p_0(p-1)}(\langle M\rangle_\infty - \langle M\rangle_t)\right) \\
&\qquad \times \exp\left(\frac{1}{p_0(p-1)}(M_\infty - M_t)\right. \\
&\qquad\qquad\qquad \left.\left. - \frac{q_0}{2p_0^2(p-1)^2}(\langle M\rangle_\infty - \langle M\rangle_t)\right)\Big|\mathscr{F}_t\right]
\end{aligned}$$

which is smaller than

$$\begin{aligned}
&E\left[\exp\left(-\frac{1}{p-1}(M_\infty - M_t) + \frac{1}{2(p-1)}(\langle M\rangle_\infty - \langle M\rangle_t)\right)\Big|\mathscr{F}_t\right]^{1/p_0} \\
&\quad \times E\left[\exp\left(\frac{q_0}{p_0(p-1)}(M_\infty - M_t)\right.\right. \\
&\qquad\qquad\qquad \left.\left. - \frac{q_0^2}{2p_0^2(p-1)^2}(\langle M\rangle_\infty - \langle M\rangle_t)\right)\Big|\mathscr{F}_t\right]^{1/q_0}.
\end{aligned}$$

by the Hölder inequality. By the supermartingale inequality the second term on the right side is dominated by 1. The first term is

$$E[(Z_t/Z_\infty)^{1/(p-1)}\,|\,\mathscr{F}_t]^{1/p_0}\,,$$

which is dominated by some constant from the condition (A_p). Therefore, (3) is proved.

Next we show (4). By the Schwarz inequality

$$\begin{aligned}
&E\left[\exp\left(\frac{1}{2(p-1)}(\langle M\rangle_\infty - \langle M\rangle_t)\right)\Big|\mathscr{F}_t\right] \\
&\leqslant E\left[\exp\left(-\frac{1}{p-1}(M_\infty - M_t)\right.\right. \\
&\qquad\qquad\qquad \left.\left. + \frac{1}{2(p-1)}(\langle M\rangle_\infty - \langle M\rangle_t)\right)\Big|\mathscr{F}_t\right]^{1/2}
\end{aligned}$$

$$\times E\left[\exp\left(\frac{1}{p-1}(M_\infty - M_t) + \frac{1}{2(p-1)}(\langle M\rangle_\infty - \langle M\rangle_t)\right)\Big|\mathscr{F}_t\right]^{1/2}.$$

The first term is $E[(Z_t/Z_\infty)^{1/(p-1)}|\mathscr{F}_t]^{1/2}$ and the second term is $E[(Z'_t/Z'_\infty)^{1/(p-1)}|\mathscr{F}_t]^{1/2}$. Therefore (4) is proved.

Finally we prove (5). For that, set $p_0 = \sqrt{p} + 1$. Then the exponent conjugate q_0 is $(\sqrt{p}+1)/\sqrt{p}$, and by an easy calculation

$$\frac{1}{q_0(\sqrt{p}-1)^2} - \frac{p_0}{(p-1)^2} = \frac{1}{p-1},$$

so that

$$E[(Z_t/Z_\infty)^{1/(p-1)}|\mathscr{F}_t] = E\left[\exp\left(-\frac{1}{p-1}(M_\infty - M_t) - \frac{p_0}{2(p-1)^2}(\langle M\rangle_\infty - \langle M\rangle_t)\right) \times \exp\left(\frac{1}{2q_0(\sqrt{p}-1)^2}(\langle M\rangle_\infty - \langle M\rangle_t)\right)\Big|\mathscr{F}_t\right].$$

By the Hölder inequality, this is smaller than

$$E\left[\exp\left(-\frac{p_0}{p-1}(M_\infty - M_t) - \frac{p_0^2}{2(p-1)^2}(\langle M\rangle_\infty - \langle M\rangle_t)\right)\Big|\mathscr{F}_t\right]^{1/p_0} \times E\left[\exp\left(\frac{1}{2(\sqrt{p}-1)^2}(\langle M\rangle_\infty - \langle M\rangle_t)\right)\Big|\mathscr{F}_t\right]^{1/q_0}.$$

The first term is dominated by 1, and also from the assumption the second term is dominated by some constant. Namely, Z_t satisfies the (A_p) condition. In the same way we can show that Z'_t satisfies (A_p). Therefore the lemma is established.

Obviously, if $\langle M\rangle_\infty$ is bounded, the process Z_t satisfies (B_λ) for all $\lambda > 0$.

§4. A weighted norm inequality for martingales

We suppose here that Z_t is a uniformly integrable martingale; that is, $Z_t = E[Z|F_t]$ for some integrable random variable Z. Now set $d\hat{P} = ZdP$, which is also a probability measure on Ω. As $d\hat{P} = Z_t dP$ on $\mathscr{F}_t$ for each t, we get for any $\hat{P}$-integrable random variable Y

$$(6) \qquad \hat{E}[Y|F_t] = \frac{1}{Z_t} E[ZY|F_t] \qquad \text{a.s., under } dP \text{ and } d\hat{P}$$

where $\hat{E}[\ ]$ denotes expectation with respect to $d\hat{P}$. We use many times this relation. The (A_p) condition is in connection with the weighted norm inequality as follows.

Theorem 3. *If Z_t satisfies (A_p), then the weighted norm inequality*

$$\hat{E}[(X^*)^r] \leqslant C_{p,r}\hat{E}[|X_\infty|^r], \quad r > p, \quad X^* = \sup_t |X_t|$$

is valid for all P-uniformly integrables martingales X_t.

Proof. Let X_t be a P-uniformly integrable martingale. We may assume that the right side is finite. As $X_t = Z_t\hat{E}[X_\infty/Z|F_t]$, by the Hölder inequality

$$|X_t|^p \leqslant (Z_t^p\hat{E}[1/Z^q|\mathscr{F}_t]^{p-1})\hat{E}[|X_\infty|^p|\mathscr{F}_t], \qquad q = p/(p-1).$$

The first term on the right side is $Z_tE[(1/Z)^{1/(p-1)}|\mathscr{F}_t]^{p-1}$ which is dominated by some constant. That is, $|X_t|^p \leqslant C_p\hat{E}[|X_\infty|^p|\mathscr{F}_t]$. By applying the Doob inequality to the $\hat{P}$-martingale $\hat{E}[|X_\infty|^p|\mathscr{F}_t]$ we get

$$\hat{E}[(X^*)^r] \leqslant C_{p,r}\hat{E}\left[\sup_t \hat{E}[|X_\infty|^p|\mathscr{F}_t]^{r/p}\right] \leqslant C_{p,r}\hat{E}[|X_\infty|^r], \qquad r>p.$$

So the theorem is established.

Theorem 4. *Let $1 < p < \infty$, and assume that $E[(1/Z)^{1/(p-1)}]$ is finite. If the weighted norm inequality*

$$\hat{E}[(X^*)^p] \leqslant C_p\hat{E}[|X_\infty|^p]$$

is valid for all P-uniformly integrable martingales X_t, then Z_t satisfies the (A_p) condition.

Proof. Let T be any stopping time, and A be an element of $\mathscr{F}_T$. We set

$$X_t = E[(1/Z)^{1/(p-1)}I_A|\mathscr{F}_t] = Z_t\hat{E}[(1/Z^q)I_A|\mathscr{F}_t], \qquad q = p/(p-1)$$

which is a P-uniformly integrable martingale. So, for this martingale the weighted norm inequality

$$\hat{E}[X_T^p] \leqslant C_p\hat{E}[X_\infty^p] = C_p\hat{E}[\hat{E}[1/Z^q|\mathscr{F}_T]I_A]$$

holds. The left side is $\hat{E}[Z_T^p\hat{E}[1/Z^q|\mathscr{F}_T]^pI_A]$. This implies that

$$Z_T^p \hat{E}[1/Z^q \mid \mathscr{F}_T]^p \leqslant C_p \hat{E}[1/Z^q \mid \mathscr{F}_T] .$$

As $[1/Z^q \mid \mathscr{F}_T]$ is finite a.s., we get $Z_T^p \hat{E}[1/Z^q \mid \mathscr{F}_T]^{p-1} \leqslant C_p$. This completes the proof.

As remarked in the previous section, if $\langle M\rangle_\infty$ is bounded, the martingale Z_t satisfies the (B_λ) condition for all $\lambda > 0$. So we get the following.

Theorem 5. *There are stopping times T_n, increasing to ∞ a.s., such that the weighted norm inequalities*

$$\hat{E}[(X_{T_n}^*)^p] \leqslant C_{p,n}\hat{E}[|X_{T_n}|^p] , \quad 1 < p < \infty , \qquad n = 1, 2, \cdots$$

are valid for all P-uniformly integrable martingales X_t. The stopping times T_n depend only on the martingale Z_t. Here $X_{T_n}^* = \sup_{0 \leqslant t \leqslant T_n} |X_t|$.

Proof. Define the stopping times T_n by

$$T_n = \inf\{t \geqslant 0 ; \langle M\rangle_t > n\} , \qquad n = 1, 2, \cdots .$$

Clearly T_n increases to infinity a.s., and $\langle M\rangle_{T_n} \leqslant n$. So for each n

$$Z_t^{T_n} = \exp(M_{t\wedge T_n} - \tfrac{1}{2}\langle M\rangle_{t\wedge T_n})$$

satisfies (B_λ) for all $\lambda > 0$. Then,as proved in Section 3, each Z^{T_n} satisfies (A_p) for all $p > 1$. Therefore by Theorem 3 we have

$$\begin{aligned} \hat{E}[(X_{T_n}^*)^p] &= E[Z(X_{T_n}^*)^p] \\ &= E[Z_{T_n}(X_{T_n}^*)^p] \\ &\leqslant C_{p,n}E[Z_{T_n}|X_{T_n}|^p] \\ &= C_{p,n}E[Z|X_{T_n}|^p] \\ &= C_{p,n}\hat{E}[|X_{T_n}|^p] . \end{aligned}$$

This completes the proof.

Finally, as an application, we give a new martingale inequality.

Theorem 6. *Suppose that M is a BMO-martingale. Then for every $p > 1$ there is a positive constant $\gamma_p > 0$ depending only on p such that the martingale inequality*

$$E\left[\sup_t |X_t - \gamma_p\langle X, M\rangle_t|^p\right] \leqslant C_p E[|X_\infty - \gamma_p\langle X, M\rangle_\infty|^p]$$

is valid for all L^2-bounded continuous martingales X_t.

Proof. As M is a BMO-martingale, the process Z_t satisfies (B_{λ_0}) for some $\lambda_0 > 0$ by Theorem 2. Then, let's take $\lambda > 0$ such that

$$\lambda > \mathrm{Max}\left\{1 + \frac{1}{(\sqrt{p} - 1)^2}, 8\lambda_0 \|M\|_{\mathrm{BMO}}^2\right\}$$

and set $N_t = \gamma_p M_t$ where $\gamma_p = \sqrt{\lambda_0/\lambda}$. The process N is a BMO-martingale and $\langle N\rangle = \gamma_p^2\langle M\rangle = (\lambda_0/\lambda)\langle M\rangle$. Now we consider

$$W_t = \exp(N_t - \tfrac{1}{2}\langle N\rangle_t)$$

instead of Z_t. As $8\|N\|_{\mathrm{BMO}}^2 = (8\lambda_0/\lambda)\|M\|_{\mathrm{BMO}}^2$ is smaller than 1, by Lemma 1

$$E[\exp(8(\langle N\rangle_\infty - \langle N\rangle_t))\,|\,\mathscr{F}_t] \leqslant \frac{1}{1 - 8\|N\|_{\mathrm{BMO}}^2}.$$

That is, W satisfies the condition (B_8), so that this is an L^2-bounded martingale with respect to dP. Then, define $d\tilde{P} = W_\infty dP$ as before. It is a probability measure on Ω. As is well-known, for any P-continuous local martingale X_t, $\tilde{X}_t = X_t - \langle X, N\rangle_t = X_t - \gamma_p\langle X, M\rangle_t$ is a $\tilde{P}$-continuous local martingale and $\langle\tilde{X}\rangle = \langle X\rangle$. By using the Davis inequality we get

$$\begin{aligned}\frac{1}{4\sqrt{2}}\tilde{E}[\tilde{X}^*] &\leqslant \tilde{E}[\langle\tilde{X}\rangle_\infty^{1/2}]\\ &= E[W_\infty\langle X\rangle_\infty^{1/2}]\\ &\leqslant E[W_\infty^2]^{1/2}E[\langle X\rangle_\infty]^{1/2}.\end{aligned}$$

Therefore, if X_t is an L^2-bounded continuous martingale with respect to dP, then $\tilde{X}_t$ is a $\tilde{P}$-uniformly integrable martingale. Now we set

$$\tilde{W}_t = \exp(-\tilde{N}_t - \tfrac{1}{2}\langle\tilde{N}\rangle_t).$$

Then $\tilde{W}_t = 1/W_t$, so that $\tilde{W}_t$ is a $\tilde{P}$-uniformly integrable martingale and $\tilde{W}_\infty d\tilde{P} = dP$. Therefore, if $\tilde{W}_t$ satisfies the $(A_{p-\varepsilon})$ condition for some $\varepsilon > 0$ with respect to $d\tilde{P}$, then by Theorem 3 we get

$$E[(\tilde{X}^*)^p] \leqslant C_p E[|\tilde{X}_\infty|^p]$$

which is the desired inequality. So it remains to prove that $\tilde{W}_t$ satisfies this condition. As $\alpha = (\lambda - 1)/2 > 1/(2(\sqrt{p} - 1)^2)$, there is some ε, $0 < \varepsilon < p - 1$ such that $1/(2(\sqrt{p - \varepsilon} - 1)^2) < \alpha$. Thus we have

$$\begin{aligned}&\tilde{E}\left[\exp\left(\frac{1}{2(\sqrt{p - \varepsilon} - 1)^2}(\langle\tilde{N}\rangle_\infty - \langle\tilde{N}\rangle_t)\right)\Big|\,\mathscr{F}_t\right]\\ &\qquad\leqslant \tilde{E}[\exp(\alpha(\langle\tilde{N}\rangle_\infty - \langle\tilde{N}\rangle_t))\,|\,\mathscr{F}_t].\end{aligned}$$

The right side is

$$E\left[\frac{W_\infty}{W_t}\exp\left(\alpha(\langle N\rangle_\infty-\langle N\rangle_t)\right)\Big|\mathscr{F}_t\right]$$
$$=E\left[\exp\left((N_\infty-N_t)-(\langle N\rangle_\infty-\langle N\rangle_t)\right)\right.$$
$$\left.\times\exp\left(\frac{2\alpha+1}{2}(\langle N\rangle_\infty-\langle N\rangle_t)\right)\Big|\mathscr{F}_t\right],$$

which is smaller than

$$E[\exp(2(N_\infty-N_t)-2(\langle N\rangle_\infty-\langle N\rangle_t))|\mathscr{F}_t]^{1/2}$$
$$\times E[\exp((2\alpha+1)(\langle N\rangle_\infty-\langle N\rangle_t))|\mathscr{F}_t]^{1/2}$$

by the Schwarz inequality. By the supermartingale inequality the first term is dominated by 1, and since $\lambda=2\alpha+1$, the second term is $E[\exp(\lambda_0(\langle M\rangle_\infty-\langle M\rangle_t))|\mathscr{F}_t]^{1/2}$ which is dominated by some constant from the (B_{λ_0}) condition. Therefore, $\tilde{W}_t$ satisfies the condition $(B_{1/2(\sqrt{p-\varepsilon}-1)^2})$ with respect to $d\tilde{P}$, so that by Lemma 2 it satisfies the condition $(A_{p-\varepsilon})$ with respect to $d\tilde{P}$. This completes the proof.

In particular, if Z_t satisfies (B_λ) for all $\lambda>0$, then the constant γ_p may be taken to be 1.

References

[1] Girsanov, I. V., On transforming a certain class of stochastic processes by absolutely continuous substitution of measures, Theor. Probability Appl., 5 (1960), 285–301.

[2] Izumisawa, M. and Kazamaki, N., Weighted norm inequalities for martingales, Tôhoku Math. J., 29 (1977), 115–124.

[3] Meyer, P. A., Probabilités et potentiel, Hermann, Blaisdell, 1966.

[4] Muckenhoupt, B., Weighted norm inequalities for the Hardy maximal function, Trans. Amer. Math. Soc., **165** (1972), 207–226.

[5] Novikov, A. A., On an identity for stochastic integrals, Theor. Probability Appl., **17** (1972), 717–720.

DEPARTMENT OF MATHEMATICS
COLLEGE OF GENERAL EDUCATION
TÔHOKU UNIVERSITY
SENDAI 980, JAPAN

Proc. of Intern. Symp. SDE
Kyoto 1976, pp. 153–161

On a Growth of Solutions of Second Order Linear Differential Equations with Random Coefficients

Shin-ichi KOTANI

§ 0. Introduction

We are concerned with the study of a stochastic differential equation

$$(0.1)\qquad \begin{cases} d\xi(t) = \eta(t)dt \\ d\eta(t) = -\lambda\xi(t)dt + \xi(t)dq(t)\,, \end{cases}$$

where λ is a real constant and $\{q(t)\}$ is a Lévy process of bounded variation in each finite interval. Thus its characteristic function $\psi(\xi)$ defined by the equation

$$e^{\psi(\xi)} = E(e^{i\xi q(1)})$$

can be represented as

$$\psi(\xi) = \int_{-\infty}^{\infty} (e^{i\xi u} - 1)n(du)$$

where $n(du)$ is a Lévy measure satisfying $\int_{-\infty}^{\infty} (1 \wedge |u|)n(du) < \infty$.

One aim of this note is to prove the exponential growth of the norm of $(\xi(t), \eta(t))$ under some condition on the Lévy measure $n(du)$. This will be done by using the methods in the stability theory. The author has got many ideas of the proof from Frisch-Lloyd [1] who first proved the same result in case that q is a Poisson process. It should be noted that Matsuda-Ishii [2] obtained for a difference equation a similar result by making use of the multiplicative ergodic theorem of H. Furstenberg.

Our differential equation (0.1) is closely related to an operator L defined by

$$(0.2)\qquad L\phi(t) = \frac{d\phi^+(t) - \phi(t)dq(t)}{dt}\,,$$

because for any solution $\phi(t)$ of the equation $L\phi = -\lambda\phi$ the functions

$\xi(t)=\phi(t)$ and $\eta(t)=\phi^+(t)$ satisfy (0.1). Systematic study of such operators with random potentials q is required in connection with solid state physics. This is one motivation of the present study. In fact K. Ishii [3] deduced from the above property of the solutions the lack of the absolutely continuous spectrum with probability one in case of difference operators. We will see in Theorem 2 that the same conclusion can be derived for the differential operator (0.2).

§ 1. Asymptotic behavior of the solutions

Define a new process $\{z(t)\}$ by

$$z(t)=-\frac{\eta(t)}{\xi(t)}.$$

Then $\{z(t)\}$ satisfies a stochastic differential equation

$$dz(t)=(z(t)^2+\lambda)dt-dq(t)$$

up to the explosion time τ, i.e. $\tau=\inf\{t>0\,;z(t)=\infty\}$. In the previous paper [4] the author proved the following

Lemma 1.1. *Suppose that λ is positive. Then the process $\{z(t)\}$ is ergodic and its invariant measure $T(z)dz$ satisfies an equation*

$$(1.1)\qquad (z^2+\lambda)T(z)-\int_{-\infty}^{\infty}n(du)\int_z^{z+u}T(x)dx=N(\lambda)\,,$$

where $N(\lambda)$ is a positive number determined by a condition

$$\int_{-\infty}^{\infty}T(z)dz=1\,.$$

This fact gives us some information about the asymptotic behaviour of the solutions $(\xi(t),\eta(t))$. Here we need a few lemmas for our theorem.

Lemma 1.2. *Let $M(t)$ be a separable square integrable martingale satisfying for every sufficiently large t a condition*

$$EM(t)^2\leqq Kt\,.$$

Then with probability one we have

$$(1.2)\qquad \lim_{t\to\infty}\frac{M(t)}{t}=0\,.$$

Proof. We employ the method in Gihman-Skorohod [5]. For every sufficiently large A and A_1, from the martingale inequality follow inequalities

$$P\left\{\sup_{A<t<A_1} \frac{|M(t)|}{t} > \varepsilon\right\} \leqq P\left\{\sup_{A<t<A_1} |M(t)| > A\varepsilon\right\}$$

$$\leqq \frac{EM(A_1)^2}{A^2\varepsilon^2} \leqq \frac{KA_1}{A^2\varepsilon^2} .$$

Hence we see

$$P\left\{\sup_{t>A} \frac{|M(t)|}{t} > \varepsilon\right\} \leqq \sum_{k=0}^{\infty} P\left\{\sup_{2^kA<t<2^{k+1}A} \frac{|M(t)|}{t} > \varepsilon\right\}$$

$$\leqq \sum_{k=0}^{\infty} \frac{K2^{k+1}A}{2^{2k}A^2\varepsilon^2} = \frac{4K}{A\varepsilon^2} .$$

First letting A to ∞ and then ε to 0, we can prove the lemma.

Denote by $\phi(s)$ the Fourier transform of $T(z)$, i.e.,

$$\phi(s) = \int_{-\infty}^{\infty} e^{-isz} T(z) dz .$$

Lemma 1.3. *Suppose that the measure $n(du)$ satisfies*

$$\int_{|u|>1} \log|u|\, n(du) < \infty .$$

Then we have

(1) $\{(\lambda + z^2)T(z) - N(\lambda)\} \log(1 + z^2) \to 0$ *as* $|z| \to \infty$.
(2) $\phi(s), \phi'(s) \to 0$ *as* $s \to +\infty$.

Proof. First we note that under the assumption on $n(du)$ an estimate

$$(1.3) \qquad \int_{-\infty}^{\infty} n(du) \int_z^{z+u} T(x)dx = O\left(|z|^{-2} + \int_{|u|>|z|} n(du)\right)$$

as $|z| \to \infty$ holds. Hence from (1.1), (1) follows immediately.

On the other hand, we have from (1.1) again

$$\phi(s) = N(\lambda) \int_{-\infty}^{\infty} e^{-isz} \frac{dz}{z^2 + \lambda} + \int_{-\infty}^{\infty} \frac{e^{-isz}}{z^2 + \lambda} dz \int_{-\infty}^{\infty} n(du) \int_z^{z+u} T(x)dx$$

$$= N(\lambda) \frac{\pi}{\sqrt{\lambda}} e^{-\sqrt{\lambda} s} + I(s) ,$$

for $s > 0$. Since (1.3) implies that

$$h(z) = \frac{-iz}{z^2 + \lambda} \int_{-\infty}^{\infty} n(du) \int_{z}^{z+u} T(x)dx$$

is integrable in R^1, we see that

$$I'(s) = \int_{-\infty}^{\infty} e^{-isz} h(z)dz$$

tends to zero as $s \to +\infty$.

It is obvious that $\phi(s) \to 0$ as $s \to +\infty$. This completes the proof.

Now we can prove our theorem.

Theorem 1. *Suppose $\lambda > 0$ and*

$$\int_{|u|>1} (\log|u|)^2 n(du) < \infty \ . \tag{1.4}$$

Then

$$\lim_{t\to\infty} \frac{1}{t} \log(\xi(t)^3 + \eta(t)^2) = 2\int_0^{\infty} |\phi(s)|^2 \frac{ds}{s} \int_{-\infty}^{\infty} (1 - \cos us)n(du) > 0 \ .$$

Proof. Since $q(t)$ can be represented by a stationary Poisson point process $N_p(dt, du)$ with characteristic measure $dtn(du)$ as

$$q(t) = \int_0^{t+} \int_{-\infty}^{\infty} uN_p(ds, du) \ ,$$

we can rewrite the equation (0.1) as

$$\begin{cases} d\xi(t) = \eta(t)dt \\ d\eta(t) = -\lambda\xi(t)dt + \int_{-\infty}^{\infty} u\xi(t)N_p(dt, du) \ . \end{cases}$$

Set $f(\xi, \eta) = \log(\xi^2 + \eta^2)$. Then the generalized Itō's formula gives us an equality

$$\begin{aligned} f(\xi(t), \eta(t)) &- f(\xi(0), \eta(0)) \\ &= \int_0^t f_\xi(\xi(s), \eta(s))\eta(s)ds - \lambda \int_0^t f_\eta(\xi(s), \eta(s))\xi(s)ds \\ &\quad + \int_0^{t+} \int_{-\infty}^{\infty} \{f(\xi(s), \eta(s-) + u\xi(s)) - f(\xi(s), \eta(s-))\} N_p(ds, du) \\ &= 2(\lambda - 1)\int_0^t \frac{z(s)}{1 + z(s)^2} ds + \int_0^{t+} \int_{-\infty}^{\infty} \log \frac{1 + (z(s) - u)^2}{1 + z(s)^2} N_p(ds, du) \ . \end{aligned}$$

For simplicity put

$$f(z) = \frac{2(\lambda - 1)z}{1 + z^2}\,, \qquad g(z, u) = \log \frac{1 + (z - u)^2}{1 + z^2}\,.$$

Since there exists a suitable constant C independent of z, u such that

$$|g(z, u)| \leqq C \log (1 + |u|)$$

holds for all u, the assumption (1.4) implies

$$\int_0^{t+} ds \int_{-\infty}^{\infty} E\,|g(z(s), u)|^2\, n(du) \leqq Kt\,, \tag{1.5}$$

where K is a positive constant. This makes it possible to define

$$M(t) = \int_0^{t+} \int_{-\infty}^{\infty} g(z(s), u)\tilde{N}_p(ds, du)\,,$$

$$A(t) = \int_0^t f(z(s))ds + \int_0^t ds \int_{-\infty}^{\infty} g(z(s), u)n(du)\,,$$

where $\tilde{N}_p(ds, du) = N_p(ds, du) - dsn(du)$. The inequality (1.5) guarantees to apply Lemma 1.2 to this $M(t)$ and we obtain the following equality with probability one

$$\lim_{t \to \infty} \frac{M(t)}{t} = 0\,.$$

One the other hand, the function

$$p(z) = \int_{-\infty}^{\infty} g(z, u)n(du) + f(z)$$

is continuous in $R^1 \cup \{\infty\}$. Hence the ergodicity of $\{z(t)\}$ implies that the following equality holds with probability one.

$$\lim_{t \to \infty} \frac{A(t)}{t} = \int_{-\infty}^{\infty} p(z)T(z)dz\,.$$

Finally we express the right-hand side in terms of the Fourier transform $\phi(s)$ of $T(z)$. Observing that $T(z)$ satisfies the equation (1.1), we have

$$z\{T(z) - T(-z)\} = \frac{-i}{z^2 + \lambda}\{h(z) - h(-z)\}\,.$$

Hence the left-hand side is integrable and

(1.6) $$\mathrm{Im}\,\phi'(s) = -\int_0^\infty z\{T(z) - T(-z)\}\cos szdz\ .$$

The equation (1.1), changes of order of the integrations and (1) of Lemma 1.3 give us an identity

$$\int_{-\infty}^{\infty} p(z)T(z)dz = -2\int_0^\infty z\{T(z) - T(-z)\}dz\ .$$

This, together with (1.6), implies

$$\int_{-\infty}^{\infty} p(z)T(z)dz = 2\,\mathrm{Im}\,\phi'(0+)\ .$$

Since the function $\phi(s)$ satisfies an equation

$$\phi''(s) = \lambda\phi(s) - \frac{\psi(s)}{is}\phi(s)\ , \qquad \phi(0) = 1\ ,$$

we have

$$\frac{d}{ds}\mathrm{Im}\,\{\phi'(s)\overline{\phi(s)}\} = \mathrm{Im}\,\{\phi''(s)\overline{\phi(s)} - |\phi'(s)|^2\}$$
$$= -\frac{|\phi(s)|^2}{s}\int_{-\infty}^{\infty}(1 - \cos us)n(du)\ .$$

From (2) of Lemma 1.3 and $\phi(0) = 1$, follows the identity

$$\mathrm{Im}\,\phi'(0+) = \int_0^\infty \frac{|\phi(t)|^2}{t}dt\int_{-\infty}^{\infty}(1 - \cos ut)n(du)\ .$$

This completes the proof of Theorem 1.

§ 2. Absence of absolutely continuous spectrum

We can apply the above theorem to the proof of nonexistence of the absolutely continuous spectrum of the operator L with probability one. Our method of proof is similar to the one used by L. A. Pastur [6] in case of difference operator.

Let $q(t)$ be a right continuous function of bounded variation in each finite interval. Let $\phi(t)$ be a solution of an equation

$$d\phi^+(t) = \phi(t)dq(t)\ ,$$

for $t > 0$. Then we have the following

Lemma 2.1. *Suppose $dq(t) \geqq -q_0 dt$ for some positive constant q_0. Then for any $r < R$ we have*

$$\int_0^r \phi^+(t)^2 dt \leqq -\phi(0)\phi^+(0) + \left\{\frac{\pi^2}{4(R-r)^2} + q_0\right\} \int_0^R \phi(t)^2 dt .$$

Proof. This lemma is well-known (cf. I. M. Glazman [7] p. 176). However for completeness we give a sketch of proof. Define $\theta(t)$ by

$$\begin{aligned} \theta(t) &= 1 & 0 \leqq t \leqq r \\ &= \frac{1}{2}\left\{1 + \cos \pi \frac{t-r}{R-r}\right\} & r < t \leqq R \\ &= 0 & R < t . \end{aligned}$$

Then noting an identity

$$\int_0^R \theta(t)\phi^+(t)^2 dt = \frac{1}{2}\int_0^R \theta''(t)\phi(t)^2 dt - \int_{0+}^R \theta(t)\phi(t)^2 dq(t) - \phi(0)\phi^+(0) ,$$

we obtain the lemma.

Theorem 2. *Suppose $dq(t) \geqq 0$ a.s., and $\int_{u>1} (\log u)^2 n(du) < \infty$. Then with probability one there exists no absolutely continuous spectrum of the operator L with a fixed arbitrary boundary condition at the origin O.*

Proof. For definiteness we assume that the operator L has its boundary condition at the left end point O as $\phi^+(0) = 0$. Let $\phi_\lambda(t)$ be the solution of an equation

$$\begin{cases} L\phi_\lambda(t) = -\lambda\phi_\lambda(t) \\ \phi_\lambda^+(0) = 0 , \qquad \phi_\lambda(0) = 1 , \end{cases}$$

and $\sigma(\lambda)$ be its spectral function. Then as is well-known the fundamental solution of $\partial/\partial t - L$ can be expressed as

$$p(s, u_1, u_2) = \int_0^\infty e^{-\lambda s}\phi_\lambda(u_1)\phi_\lambda(u_2) d\sigma(\lambda) . \tag{2.1}$$

Since dq is a nonnegative measure, a comparison theorem gives us an estimate

$$p(s, u, u) \leqq C ,$$

for some constant C depending on s. Hence (2.1) implies

$$\int_0^\infty \frac{du}{1+u^\alpha} \int_0^\infty e^{-2s} \phi_\lambda(u)^2 d\sigma(\lambda) < \infty ,$$

for every fixed $\alpha > 1$ and $s > 0$, which asserts that almost everywhere with respect to the measure $d\sigma(\lambda)$

$$\int_0^\infty \frac{\phi_\lambda(u)^2}{1+u^\alpha} du < \infty \tag{2.2}$$

holds. Substituting $R/2$ for r in Lemma 2.1, we can obtain

$$\int_0^\infty \frac{\phi_\lambda^+(u)^2}{1+u^\alpha} du < \infty . \tag{2.3}$$

On the other hand, Theorem 1 says that with probability one $\phi_\lambda(t)^2 + \phi_\lambda^+(t)^2$ grows exponentially almost everywhere with respect to the Lebesgue measure $d\lambda$. This is compatible with the inequalities (2.2) and (2.3) only if with probability one the measure $d\sigma(\lambda)$ has no absolutely continuous part, which proves the theorem.

Remark. If we restrict ourselves to the case that the Lévy measure $n(du)$ is bounded and its support is contained in $(-\infty, 0)$, then under some additional conditions on $n(du)$ Theorem 1 remains valid for every $\lambda < 0$. As for Theorem 2 the author have not been able to remove the present restriction on q because it is difficult to estimate the fundamental solution $p(s, u, u)$.

References

[1] Frisch, H. L. and Lloyd, S. P., Electron levels in one-dimensional lattice, Phys. Rev., **120** (1960), 1175–1189.

[2] Matsuda, H. and Ishii, K., Localization of normal modes and energy transport in the disordered harmonic chain, Progr. Theoret. Phys. Suppl., **45** (1970), 56–86.

[3] Ishii, K., Localization of eigenstates and transport phenomena in the one-dimensional disordered system, Progr. Theoret. Phys. Suppl., **53** (1973), 77–138.

[4] Kotani, S., On asymptotic behaviours of the spectra of a one-dimensional Hamiltonian with a certain random coefficient, Publ. Res. Inst. Math. Sci., **12** (1976), 447–492.

[5] Gihman, I. I. and Skorohod, A. V., Stochastic differential equations, Springer, 1972.

[6] Pastur, L. A., On the spectrum of random Jacobian matrices and Schrö-

dinger equation with a random potential on the whole axis, 1973, preprint.
[7] Glazman, I. M., Direct methods of qualitative spectral analysis of singular differential operators, Israel Program for Scientific Translations, Jerusalem, 1965.

Department of Mathematics
Kyoto University
Kyoto 606, Japan

Proc. of Intern. Symp. SDE
Kyoto 1976, pp. 163–185

Supports of Diffusion Processes and Controllability Problems

Hiroshi KUNITA

To professor K. Itô on the occasion of his 60th birthday

§1. Introduction

Let M be a connected, paracompact C^∞-manifold of dimension d. Let $Y, X_1, \cdots, X_r$ be C^∞-vector fields on M. With local coordinate $x = (x_1, \cdots, x_d)$, these vector fields are represented as first order linear differential operators with C^∞-coefficients:

$$(1.1)\qquad Y = \sum_{i=1}^{d} Y^i(x)\frac{\partial}{\partial x_i}\,, \qquad X_j = \sum_{i=1}^{d} X_j^i(x)\frac{\partial}{\partial x_i}\,,$$
$$j = 1, \cdots, r$$

Consider a diffusion process ξ_t with state space M, governed by the stochastic differential equation

$$(1.2)\qquad d\xi_t^i = Y^i(\xi_t)dt + \sum_{j=1}^{r} X_j^i(\xi_t) \circ dB_t^j\,.$$

Here, $(\xi_t^1, \cdots, \xi_t^d)$ is the coordinate of $\xi_t, B_t = (B_t^1, \cdots, B_t^r)$ is an r-dimensional standard Brownian motion: the stochastic integral is defined in the sense of Stratonovich-Fisk. Using the standard matrix-vector notations, the equation (1.2) is written as

$$(1.2')\qquad d\xi_t = Y(\xi_t)dt + X(\xi_t) \circ dB_t$$

where $Y(x)$ is a column vector with components $Y^i(x)$ and $X(x)$ is a $d \times r$-matrix with elements $X_j^i(x)$.

Our purpose is to study the (global) geometrical property of sample paths ξ_t. For the study of ξ_t, it is convenient to approximate them by a sequence of piecewise smooth paths $\xi_t^{(n)}$, $n = 1, 2, \cdots$ such that

$$(1.3)\qquad \frac{d\xi_t^{(n)}}{dt} = Y(\xi_t^{(n)}) + X(\xi_t^{(n)})\dot{B}_t^{(n)}\,,$$

where

$$B_t^{(n)} = B_{k/n} + n\left(t - \frac{k}{n}\right)(B_{(k+1)/n} - B_{k/n}) \qquad \text{if } \frac{k}{n} \le t < \frac{k+1}{n}$$

and $\dot{B}_t^{(n)}$ is its derivative. It is known that for almost all ω, $\xi_t^{(n)}(\omega)$ converges to $\xi_t(\omega)$ uniformly on any finite subset of $[0, \infty)$ (Wong-Zakai [17], Kunita [11]).

We now replace $\dot{B}_t^{(n)}(\omega)$ by piecewise constant functions $u(t) = (u_1(t), \cdots, u_r(t))$ and obtain a system of ordinary differential equations

$$\frac{d\varphi_t}{dt} = Y(\varphi_t) + X(\varphi_t)u(t) \tag{1.4}$$

The function $u(t)$ is called a control and the equation (1.4), a control system. It is clear that almost sure property of paths $\xi_t^{(n)}$ follows from the paths φ_t of the system (1.4). More specifically, Stroock-Varadhan [15] shows that the support of the paths ξ_t coincides with the closure of all paths of the system (1.4), varying control functions. From this point of view, we shall study the control system (1.4) rather than the stochastic differential equation (1.2).

This article discusses two problems. We first establish (nearly) necessary and sufficient condition that the support of the diffusion process ξ_t is all continuous paths in M. The condition is characterized by Lie algebra generated by vector fields $X_1, \cdots, X_r$ (Section 3). We next study the support of Green measures and transition probabilities of the diffusion process. The latter problem is closely related to the controllability problem of the system (1.4). Let x and y be two points of M. The point y is called accessible (or controllable) from x at time t (> 0), if there exists a suitable control $u(s)$, $s \ge 0$ such that the corresponding solution $\varphi_t^u(x)$ starting at x satisfies $\varphi_t^u(x) = y$. The set of all points which are accessible from x at time t is denoted by $A_t(x)$. The set $A(x) = \bigcup_{t>0} A_t(x)$ is then accessible set from x at positive time. Now let $P_t(x, \cdot)$ be the transition probability of ξ_t and let $G_\alpha(x, \cdot)$ ($\alpha > 0$) be the Green measure of ξ_t. From Stroock-Varadhan's theorem cited above, it is immediate to see that

$$\overline{A_t(x)} = \text{the support of } P_t(x, \cdot)\,,$$

$$\overline{A(x)} = \text{the support of } G_\alpha(x, \cdot)\,.$$

We shall study the sets $A(x)$ and $A_t(x)$ in sections 4 and 5, respectively.

In recent years, extensive attention has been shown to the qualitative study of the accessible sets $A(x)$ and $A_t(x)$. The papers having a direct bearing on our results are the ones by Sussman-Jurdjevic [16],

Krener [10], Lobry [13] and Hirschorn [4, 5]. The thickness of the sets $A(x)$ and $A_t(x)$ (i.e., having non empty interiors) are studied in [16] and [10]. The complete controllability of the sets $A(x)$ and $A_t(x)$ (i.e., $A(x) = M$ and $A_t(x) = M$) are studied in [13] and [4]. Our results in Sections 4 and 5 are directly related to these works.

We mention finally that the problem of determining the support of the transition probabilities is also related to some problems of the second order degenerate elliptic differential operator

$$L = \frac{1}{2} \sum_{j=1}^{r} X_j^2 + Y \tag{1.5}$$

In fact, let $f(x)$ be a C^∞-function on M with compact support. Set

$$u(t, x) = \int P_t(x, dy) f(y) .$$

Then $u(t, x)$ is the solution of the parabolic equation

$$\frac{\partial u}{\partial t} = Lu$$

with the initial condition $u(0, x) = f(x)$. (Ichihara-Kunita [8]). Stroock-Varadhan [15] shows that the strong maximum principle of the operator $\partial/\partial t - L$ is characterized by the support of transition probabilities. Our goal would be to study the elliptic differential operator (1.5) via control system (1.4).

§ 2. Preliminaries

We shall introduce several notions and notations from differential geometry. The following notations will be used throughout.

$\boldsymbol{R}$ = the set of real numbers,

$\boldsymbol{R}^d$ = d-dimemsional Euclidean space,

$T_x M$ = the tangent space of the manifold M at the point x,

$\mathscr{V}(M)$ = the set of all C^∞-vector fields on M,

$C^\infty(M)$ = the set of all real C^∞-functions on M.

For X, Y of $\mathscr{V}(M)$, the Lie bracket is defined by $[X, Y] = XY - YX$ as usual. Then the space $\mathscr{V}(M)$ may be considered as a Lie algebra over $\boldsymbol{R}$. Let $\mathscr{L}$ be a Lie subalgebra of $\mathscr{V}(M)$. set

$$\mathscr{L}(x) = \{X(x) ; X \in \mathscr{L}\} , \qquad \tilde{\mathscr{L}} = \{\mathscr{L}(x) ; x \in M\} .$$

Then $\mathscr{L}(x)$ is a subspace of $T_x M$ and $\tilde{\mathscr{L}}$ is an involutive differential system. A vector field X is said to belong to $\tilde{\mathscr{L}}$ if $X(x) \in \mathscr{L}(x)$ holds for all x.

Following Hermann [3], a Lie subalgebra $\mathscr{L}$ is called *locally finitely generated*, if for any x of M, there exists an open neighborhood U of x and a finite number of elements $Z^1, \cdots, Z^n$ of $\mathscr{L}$ (depending on U) such that any X of $\mathscr{L}$ is represented as $Z = \sum_{i=1}^n f_i Z^i$ in U with $f_i \in C^\infty(U)$. The set $\{Z^i\}_{i=1}^n$ is called a base of $\mathscr{L}$ in U. If $\dim \mathscr{L}(x)$ is constant, then the Lie subalgebra $\mathscr{L}$ is locally finitely generated by Frobenius' theorem. If $\mathscr{L}$ consists of real analytic vector fields on real analytic manifold, Lobry [12] shows that $\mathscr{L}$ is locally finitely generated. Another important consequence is that if $\mathscr{L}$ is locally finitely generated, then M is written as the disjoint union of maximal connected integral manifolds of $\tilde{\mathscr{L}}$ (foliation) (See Hermann [3]).

Let D be a subset of $\mathscr{V}(M)$. We denote by $\mathscr{L}(D)$ the Lie algebra over $\boldsymbol{R}$ generated by the set D. Now, let $Y, X_1, \cdots, X_r$ be vector fields of (1.1). We shall use the following notations.

$\mathscr{B} = \mathscr{L}(X_1, \cdots, X_r)$,

$\mathscr{L} = \mathscr{L}(Y, X_1, \cdots, X_r)$,

$\mathscr{I} =$ the ideal in $\mathscr{L}$ generated by $X_1, \cdots, X_r$.

The relation $\mathscr{B} \subset \mathscr{I} \subset \mathscr{L}$ is obvious. Let $ad(Y)$ be the linear map on $\mathscr{V}(M)$ such that $ad(Y)X = [Y, X]$. Then we have

Proposition 2.1. *It holds*

$$\mathscr{I} = \mathscr{L}(ad^n(Y)X_j\,;\, j = 1, \cdots, r,\, n = 0, 1, 2, \cdots)$$

$$= \left\{\sum_{i=1}^r \lambda_i X_i + Z\,;\, \lambda_i \in \boldsymbol{R},\, Z \in [\mathscr{L}, \mathscr{L}]\right\},$$

where $[\mathscr{L}, \mathscr{L}]$ *is the derived algebra of* $\mathscr{L}$.

The proof is direct. It is omitted.

Now let X be a complete vector field. Then there exists a one parameter group of transformations φ_t, $t \in \boldsymbol{R}$ generated by X. That is to say, a) $\varphi: M \times \boldsymbol{R} \to M$ is a C^∞-map, b) for each t, φ_t is a diffeomorphism on M, c) $\varphi_t \circ \varphi_s = \varphi_{t+s}$ holds for any $t, s \in \boldsymbol{R}$, and d) $Xf(x) = (\partial/\partial t) f(\varphi_t(x))|_{t=0}$ holds for all $f \in C^\infty(M)$. The orbit $\{\varphi_t(x)\,;\, t \in \boldsymbol{R}\}$ is called the integral curve of the vector field X.

Let φ be a diffeomorphism of M. The differential $d\varphi$ is a linear map $T_x M \to T_{\varphi(x)} M$ such that $(d\varphi X)g = X(g \circ \varphi)$ holds for all $g \in C^\infty(M)$. For $X \in \mathscr{V}(M)$, we denote by $d\varphi X$ the vector field such that $(d\varphi X)(x) = (d\varphi)_{\varphi^{-1}(x)} X(\varphi^{-1}(x))$ holds for all x. Now let X be a complete vector field and let φ_t be the one parameter group of transformations generated by X.

Set $Y_t = d\varphi_t Y$, $Y \in \mathscr{V}(M)$. Then it holds

$$\frac{dY_t}{dt} = [Y_t, X] = d\varphi_t[Y, X],$$

showing the geometrical interpretation of Lie bracket.

§ 3. Support of diffusion process

In the following, we assume that all linear sums of $Y, X_1, \cdots, X_r$ are complete. We denote by φ_t^i, the one parameter group of transformations generated by X_i, and by φ_t^0 the one parameter group of transformations generated by Y.

Let $\mathscr{U}$ be the set of all piecewise constant functions with range $\boldsymbol{R}^r$. For a given u of $\mathscr{U}$, the equation (1.4) has a unique global solution $\varphi_t^u(x)$ such that $\varphi_0^u(x) = x$. We can also prove that the diffusion process ξ_t satisfying (1.2) is unique and the life time is infinite a.s.

Now let W be the set of all continuous maps φ from $[0, \infty)$ to M, equipped with compact uniform topology, and let W_x be the subset of W such that $\varphi_0 = x$. For almost all ω, the path $\xi_t(\omega)$, $t \in [0, \infty)$ may be considered as an element of W. The support of the diffusion process ξ_t is, by definition, the smallest closed subset $\mathscr{S}$ of W such that

$$P(\{\omega ; \xi_\cdot(\omega) \in \mathscr{S}\}) = 1$$

holds. Denote by $\xi_t^{(x)}$ the diffusion process with the initial condition $\xi_0^{(x)} = x$ a.s.. The support $\mathscr{S}_x$ of $\xi_t^{(x)}$ is a subset of W_x. We have

Theorem 3.1 (Stroock-Varadhan [15]). *It holds*

$$\mathscr{S}_x = \overline{\{\varphi^u(x) ; u \in \mathscr{U}\}}\ .$$

We shall obtain the condition that the support $\mathscr{S}_x$ becomes the whole space W_x.

Theorem 3.2. *Suppose* $\dim \mathscr{B}(x) = d$ *holds for all* x *of* M. *Then it holds* $\mathscr{S}_x = W_x$ *for all* x. *Conversely if* $\mathscr{S}_x = W_x$ *holds for all* x *of* M, *then the set of* x *such that* $\dim \mathscr{B}(x) = d$ *is an open dense subset of* M.

For the proof of this theorem, it is convenient to consider the control system (1.4) with a wider class of control functions. Let $\tilde{\mathscr{U}}$ be the set of all piecewise smooth functions with range $\boldsymbol{R}^r$. Then it holds $\{\varphi^u(x) ; u \in \mathscr{U}\} \subset \{\varphi^u(x) ; u \in \tilde{\mathscr{U}}\}$. However, if we take the closure of them, then both of them coincide each other. This fact can be proved easily, approximating control u of $\tilde{\mathscr{U}}$ by a sequence of controls in $\mathscr{U}$. We denote the closure of $\{\varphi^u(x)\,;\,u\in\mathscr{U}\}$ by $\mathscr{C}_x$.

We shall prepare two lemmas.

Lemma 3.3. *Assume $Y=0$. Let Z be a complete vector field belonging to $\tilde{\mathscr{B}}$ and φ_t, the one parameter group of transformations generated by Z. Then the orbit $\varphi_t(x)$, $t\geq 0$ belongs to $\mathscr{C}_x$ for every x.*

Proof. It is sufficient to show the assertion on each coordinate neighborhood. Hence we discuss the case where M is the Euclidean space $\boldsymbol{R}^d$.

a) Suppose first that Z is written as $a_1X_1+a_2X_2$ with $a_i\in C^\infty(M)$, $i=1,2$. Then it holds

$$\frac{d\varphi_t}{dt}=a_1(\varphi_t)X_1(\varphi_t)+a_2(\varphi_t)X_2(\varphi_t)\,.$$

Setting $u_1(t)=a_1(\varphi_t)$ and $u_2(t)=a_2(\varphi_t)$, we see that $\varphi_\cdot(x)\in\mathscr{C}_x$.

b) Suppose next that $Z=[X_1,X_2]$ and that $Z(x)$ satisfies the Lipschitz condition. For each positive integer n, define a piecewise smooth curve $\varphi_t^{(n)}(x)$ by

$$\varphi_t^{(n)}=\begin{cases}\varphi^1_{4\sqrt{n}(t-k/n)}\circ\varphi^{(n)}_{k/n}\,, & \text{if}\quad \dfrac{k}{n}\leq t\leq\dfrac{k+1/4}{n}\\[2ex] \varphi^2_{4\sqrt{n}(t-(k+1/4)/n)}\circ\varphi^{(n)}_{(k+1/4)/n}\,, & \text{if}\quad \dfrac{k+1/4}{n}\leq t\leq\dfrac{k+1/2}{n}\\[2ex] \varphi^1_{-4\sqrt{n}(t-(k+1/2)/n)}\circ\varphi^{(n)}_{(k+1/2)/n}\,, & \text{if}\quad \dfrac{k+1/2}{n}\leq t\leq\dfrac{k+3/4}{n}\\[2ex] \varphi^2_{-4\sqrt{n}(t-(k+3/4)/n)}\circ\varphi^{(n)}_{(k+3/4)/n}\,, & \text{if}\quad \dfrac{k+3/4}{n}\leq t\leq\dfrac{k+1}{n}\,,\end{cases}$$

for $k=0,1,\cdots$. The $\varphi_t^{(n)}$ satisfies

$$\frac{d\varphi_t^{(n)}}{dt}=X_1(\varphi_t^{(n)})u_1^{(n)}(t)+X_2(\varphi_t^{(n)})u_2^{(n)}(t)\,,$$

where

$$\begin{aligned}u_1^{(n)}(t)&=4\sqrt{n}\,, &&\text{if}\quad \frac{k}{n}\leq t<\frac{k+1/4}{n}\\ &=-4\sqrt{n}\,, &&\text{if}\quad \frac{k+1/2}{n}\leq t<\frac{k+3/4}{n}\\ &=0\,, &&\text{otherwise}\end{aligned}$$

for $k = 1, 2, \cdots$. $u_2^{(n)}(t)$ is defined similarly.

We shall prove that $\varphi_\cdot^{(n)}(x)$ converges to $\varphi_\cdot(x)$ as n tends to infinity. Note the relation

$$\frac{1}{n} Z(x) = (\varphi^2_{-1/\sqrt{n}} \circ \varphi^1_{-1/\sqrt{n}} \circ \varphi^2_{1/\sqrt{n}} \circ \varphi^1_{1/\sqrt{n}}(x) - x) + o\left(\frac{1}{n}\right).$$

Then

$$\sum_{k=0}^{[nt]} \varphi^{(n)}_{(k+1)/n}(x) - \varphi^{(n)}_{k/n}(x) = \frac{1}{n} \sum_{k=0}^{[nt]} Z(\varphi^{(n)}_{k/n}(x)) + [nt] o\left(\frac{1}{n}\right).$$

Since

$$\varphi^{(n)}_t(x) - \varphi^{(n)}_{t-(1/n)[nt]}(x) = o\left(\frac{1}{\sqrt{n}}\right),$$

we have

$$\begin{aligned} \varphi^{(n)}_t(x) - x &= \sum_{k=0}^{[nt]} \{\varphi^{(n)}_{(k+1)/n}(x) - \varphi^{(n)}_{k/n}(x)\} + \{\varphi^{(n)}_t(x) - \varphi^{(n)}_{t-(1/n)[nt]}(x)\} \\ &= \int_0^t Z(\varphi^{(n)}_s) ds + \varepsilon_n , \end{aligned}$$

where $|\varepsilon_n|$ tends to 0 as $n \to \infty$. Therefore,

$$\varphi_t(x) - \varphi^{(n)}_t(x) = \int_0^t (Z(\varphi_s) - Z(\varphi^{(n)}_s)) ds + \varepsilon_n .$$

By the Lipschitz condition of $Z(x)$, we have

$$|\varphi_t(x) - \varphi^{(n)}_t(x)| \le K \int_0^t |\varphi_s(x) - \varphi^{(n)}_s(x)|\, ds + |\varepsilon_n|$$

with a positive constant K. This implies

$$|\varphi_t(x) - \varphi^{(n)}_t(x)| \le |\varepsilon_n| \exp Kt ,$$

and $\varphi_\cdot^{(n)}(x)$ converges to $\varphi_\cdot(x)$, proving $\varphi_\cdot \in \mathscr{C}_x$.

c) Repeating the arguments (a) and (b), one can prove that any φ_t of Z belonging to $\tilde{\mathscr{B}}$ is in $\mathscr{C}_x$. We omit the detail.

Lemma 3.4. *The assertion of Lemma* 3.3 *is valid in the case where* $Y \neq 0$.

Proof. We only discuss the case where $M = R^d$ and $(Y + Z)(x)$ satisfies the Lipschitz condition. By the argument of the preceding lemma, there exists a sequence of controls $u^{(n)}(t)$, $n = 1, 2, \cdots$ such that the solution $\varphi_t^{(n)}$ of the system

$$\frac{d\varphi_t^{(n)}}{dt} = \sum_{i=1}^{r} X_i(\varphi_t^{(n)})u_i^{(n)}(t)$$

satisfies

$$\varphi_t^{(n)}(x) - \varphi_s^{(n)}(x) = \int_s^t Z(\varphi_u^{(n)}(x))\,du + (t-s)\varepsilon_n$$

where $|\varepsilon_n|$ tends to 0. Now, let $\psi_t^{(n)}(x)$ be the solution of

$$\frac{d\psi_t^{(n)}}{dt} = \begin{cases} 2Y(\psi_t^{(n)})\,, & \text{if } \dfrac{k}{n} \le t \le \dfrac{1}{n}\left(k + \dfrac{1}{2}\right) \\ 2\sum_{i=1}^{r} X_i(\psi_t^{(n)})u_i^{(n)}(t)\,, & \text{if } \dfrac{1}{n}\left(k + \dfrac{1}{2}\right) \le t \le \dfrac{k+1}{n} \end{cases}$$

Then

$$\begin{aligned} \psi_t^{(n)}(x) - x &= \sum_{k=0}^{[nt]} \{\psi_{(k+1/2)/n}^{(n)}(x) - \psi_{k/n}^{(n)}(x)\} \\ &\quad + \sum_{k=0}^{[nt]} \{\psi_{(k+1)/n}^{(n)}(x) - \psi_{(k+1/2)/n}^{(n)}(x)\} + o\left(\frac{1}{\sqrt{n}}\right) \\ &= 2\sum_{k=0}^{[nt]} \int_{k/n}^{(k+1/2)/n} Y(\psi_s^{(n)})\,ds + 2\sum_{k=0}^{[nt]} \int_{(k+1/2)/n}^{(k+1)/n} Z(\psi_s^{(n)})\,ds + \varepsilon_n' \\ &= \int_0^t (Y + Z)(\psi_s^{(n)})\,ds + \varepsilon_n''\,, \end{aligned}$$

where $|\varepsilon_n''|$ tends to 0 as $n \to \infty$. Therefore, by the same argument as Lemma 3.3, we see that $\psi_\cdot^{(n)}(x)$ converges to $\varphi_\cdot(x)$.

Proof of Theorem 3.2. For simplicity, we prove the theorem in the case where $M = \boldsymbol{R}^d$. Suppose $\dim \mathscr{B}(x) = d$ holds for all $x \in M$. Since any continuous curve is approximated by a sequence of polygons, it is enough to prove that any linear function φ_t belongs to $\mathscr{C}_x$. Now let Z be a vector field belonging to $\tilde{\mathscr{B}}$ such that

$$\frac{d\varphi_t}{dt} = \text{const} = Y + Z\,.$$

Then $\varphi_\cdot(x) \in \mathscr{C}_x$ by Lemma 3.4.

Suppose next that $\dim \mathscr{B}(x) = s\ (s < d)$ holds in some neighborhood U of x_0. Choose a local coordinate $(x_1, \cdots, x_d)$ such that $x_{s+1} = \cdots = x_d$

$= 0$ holds in the s-dimensional integral manifold of $\mathscr{B}$ containing x_0. We can assume that the d-th coordinate $Y^d(x)$ of $Y(x)$ is nonnegative in a neighborhood V of x_0, where $V \subset U$. Then the d-th component of the solution of (1.4) starting at x_0 are nonnegative until it leaves V. This shows that $\mathscr{C}_{x_0}$ does not contain paths such that the d-th component is negative in V. The set $\{x: \dim \mathscr{B}(x) = d\}$ is open, obviously. We have thus proved the latter assertion of the theorem.

§ 4. Accessible sets and support of Green measures

If $\dim \mathscr{B}(x)$ is less than d, the problem of determining the support is much more complicated. Suppose for a moment that $\mathscr{L}$ is locally finitely generated. Let $I(x)$ be the maximal connected integral manifold of $\mathscr{L}$ containing the point x. It is obvious that any orbit of (1.4) starting at x lies in $I(x)$. Hence the support $\mathscr{S}_x$ is included in the subset of continuous paths lying in $I(x)$. Thus we can restrict our attention to each maximal connected integral manifold of $\mathscr{L}$. From this point of view, we may and do assume that $\dim \mathscr{L}(x) = d$ holds for all x.

We shall turn our attention to the accessible sets and the support of Green measures. Let $A(x)$ be the set of all accessible points from x for the system (1.4). It is known that the set $A(x)$ is thick in the following sense.

Theorem 4.1 (Sussman-Jurdjevic [16], Kroner [10]). *The interior of $A(x)$ is a dense subset of $A(x)$.*

From this theorem, we have

Proposition 4.2. *It holds* $\operatorname{int} A(x) = \operatorname{int} \overline{A(x)}$.

Proof. Let $z \in \operatorname{int} \overline{A(x)}$ and let $A^*(z)$ be the set of all z' such that z is accessible from z' in positive time. Then $A^*(z)$ is the accessible set for the dual control system

$$\frac{d\varphi_t}{dt} = -Y(\varphi_t) - \sum_{i=1}^{r} X_i(\varphi_t) u_i(t) .$$

Hence $\operatorname{int} A^*(z)$ is dense in $\overline{A^*(z)}$ by the previous Theorem. Now, since $z \in \overline{A^*(z)}$, it holds $\overline{A^*(z)} \cap \operatorname{int} A(x) \neq \varnothing$, so that we have

$$\operatorname{int} A^*(z) \cap \operatorname{int} \overline{A(x)} \cap A(x) \neq \varnothing .$$

Choose a point y from the above set. Then y is accessible from x and

z is accessible from y. Therefore, z is accessible from x. This proves $\text{int}\,\overline{A(x)} \subset A(x)$. The proof is complete.

Remark. Consider the second order (degenerate) elliptic differential operator (1.5). In the case where $\dim \mathscr{L}(x) = d$, the operator L is hypoelliptic by Hörmander [6]. Let $P_t(x, dy)$ be the transition probability of the diffusion ξ_t and let

$$G_\alpha(x, dy) = \int_0^\infty e^{-\alpha t} P_t(x, dy) dt \qquad (\alpha > 0) .$$

Let $f \in C^\infty(M)$ and let

$$G_\alpha f(x) = \int G_\alpha(x, dy) f(y) .$$

Then it satisfies $(\alpha - L)G_\alpha f = f$. It is known that there exists a C^∞-function $g_\alpha(x, y)$ (except for the diagonal set) such that $G_\alpha(x, dy) = g_\alpha(x, y)m(dy)$, where m is the volume element of M (See [8]). This fact implies that $\overline{A(x)}$ contains open sets. This is an another interpretation of the "thickness" of the set $\overline{A(x)}$.

The system (1.3) is called *completely controllable* if $A(x) = M$ holds for all x. An immediate sufficient condition for the complete controll ability is

Proposition 4.3. *Suppose that* $\dim \mathscr{B}(x) = d$ *holds for all* x. *Then the system* (1.3) *is completely controllable.*

Proof. From Theorem 3.2, it is obvious that $\overline{A(x)} = M$ holds for all x. From Proposition 4.1, we have $A(x) \supset \text{int}\,A(x) = \text{int}\,\overline{A(x)} = M$.

The above condition is far from necessity for the complete controll ability. We shall obtain another sufficient condition in the case where $\dim \mathscr{B}(x) < d$.

Let X be a complete vector field and let $\varphi_t(x)$ be the one parameter group of transformations generated by X. A point x is called *Poisson stable* for φ_t, if for any neighborhood U of x and any $T > 0$, there exists t_1, t_2 greater than T such that $\varphi_{t_1}(x)$ and $\varphi_{-t_2}(x)$ are in U.

The following theorem is motivated by Lobry [12]. Recall that φ_t^0 is the one parameter group of transformations generated by Y.

Theorem 4.4. *If Poisson stable points of* φ_t^0 *is dense in* M, *then the system* (1.4) *is completely controllable.*

For the proof of this theorem, consider a slightly different control system. Let $e_0, \cdots, e_r$ be unit vectors in $\boldsymbol{R}^{r+1}$ such that $e_i = (0, \cdots, 0, 1, 0, \cdots, 0)$ (1 is in i-th coordinate). We denote by $\mathscr{U}_0$ the set of piecewise constant functions $u(t) = (u_0(t), \cdots, u_r(t))$ taking values $e_0, \pm e_1, \cdots, \pm e_r$ only. Consider the control system such that

$$\frac{d\varphi_t}{dt} = Y(\varphi_t)u_0(t) + \sum_{i=1}^{r} X_i(\varphi_t)u_i(t) \,, \tag{4.1}$$

where $u(t) = (u_0(t), \cdots, u_r(t)) \in \mathscr{U}_0$. The solution of the above is written as

$$\varphi_t = \varphi_{t_k}^{i_k} \circ \cdots \circ \varphi_{t_1}^{i_1} \,, \tag{4.2}$$

where $i_l = 0, 1, \cdots, r$, $t_l \geq 0$ if $i_l = 0$ and $t_l \in \boldsymbol{R}$ if $i_l \neq 0$, and $\sum_{l=1}^{k} |t_l| = t$. We denote the accessible set of (4.1) as $A^0(x)$. It is known that Theorem 4.1 is valid for this control system (See [16] or [10]).

We shall now introduce notations.

$\operatorname{Exp} \mathscr{L}$ = the group of diffeomorphisms generated by φ_t^i, $t \in \boldsymbol{R}$, $i = 0, \cdots, r$.

$(\operatorname{Exp} \mathscr{L})_+$ = the semigroup of diffeomorphisms generated by φ_t^0, $t \geq 0$ and φ_t^i, $t \in \boldsymbol{R}$, $i = 1, \cdots, r$.

It is known that $\operatorname{Exp} \mathscr{L}$ acts transitively on M, i.e., $\{\varphi(x)\,;\, \varphi \in \operatorname{Exp} \mathscr{L}\} = M$ holds for any x, which is known as Chow's theorem. On the other hand, we have

$$A^0(x) = \{\varphi(x)\,;\, \varphi \in (\operatorname{Exp} \mathscr{L})_+\} \,.$$

We shall prove

Lemma 4.5. $\overline{A^0(x)} = \overline{A(x)}$.

Proof. We only consider the Euclidean case. Associated with the diffeomorphism (4.2), we shall construct a sequence of controls $u^{(n)}(t)$ as follows. For each $n \geq 1$, define $0 \leq s_1^{(n)} \leq s_2^{(n)} \leq \cdots \leq s_k^{(n)}$ as

$$\begin{aligned} s_1^{(n)} &= |t_1| \,, && \text{if } i_1 = 0 \\ &= \frac{1}{n}|t_1| \,, && \text{if } i_1 \neq 0 \,. \\ s_l^{(n)} &= s_{l-1}^{(n)} + |t_l| \,, && \text{if } i_l = 0 \quad (l \geq 2) \\ &= s_{l-1}^{(n)} + \frac{1}{n}|t_l| \,, && \text{if } i_l \neq 0 \,, \end{aligned}$$

and define

$$
\begin{aligned}
u^{(n)}(t) &= n\cdot \operatorname{sign}(t_l)e_{i_l} && \text{if } s_{l-1}^{(n)} \le t \le s_l^{(n)} \text{ and } i_l \ne 0 \\
&= 0 && \text{if } s_{l-1}^{(n)} \le t \le s_l^{(n)} \text{ and } i_l = 0\,, \\
&= 0 && \text{if } t \ge s_k^{(n)}
\end{aligned}
$$

where $e_1, \cdots, e_r$ are unit vectors in $\boldsymbol{R}^r$. Let $\varphi_t^{(n)}$ be the solution of the the system (1.4) associated with the control $u^{(n)}(t)$. Then it is written as

$$
\varphi^{(n)}_{s_k^{(n)}} = \psi_{t_k}^{n,i_k} \circ \cdots \circ \psi_{t_1}^{n,i_1}\,,
$$

where ψ_t^{n,i_l} is the integral curve of Y if $i_l = 0$, and of $((1/n)Y + X_{i_l})$ if $i_l \ne 0$. Since $\psi_{t_l}^{n,i_l}(x)$ converges to $\varphi_{t_l}^{i_l}(x)$ as $n \to \infty$, we see that $\varphi^{(n)}_{s_k^{(n)}}(x)$ converges to $\varphi_{t_k}^{i_k}\circ \cdots \circ \varphi_{t_1}^{i_1}(x)$. Therefore, $\varphi_{t_k}^{i_k}\circ \cdots \circ \varphi_{t_1}^{i_1}(x) \in \overline{A(x)}$. This proves $\{\varphi(x)\,;\, \varphi \in (\operatorname{Exp} \mathscr{L})_+\} \subset \overline{A(x)}$. The converse relation can be proved similarly. We omit the detail.

Proof of Theorem 4.4. Let x and y be arbitrary points of M. By Chow's theorem, there exists $t_1, \cdots, t_k$ of $\boldsymbol{R}$ such that $y = \varphi_{t_k}^{i_k}\circ \cdots \circ \varphi_{t_1}^{i_1}(x)$ Suppose there exists $\varphi_{t_l}^{i_l}$ $(1 \le l \le k)$ such that $i_l = 0$ and $t_l < 0$. Consider the minimum l for which $i_l = 0$ and $t_l < 0$. Let U be a neighborhood of y and let $V = (\varphi_{t_k}^{i_k}\circ \cdots \circ \varphi_{t_l}^{i_l})^{-1}(U)$. Then V is a neighborhood of $z = \varphi_{t_{l-1}}^{i_{l-1}}\circ \cdots \circ \varphi_{t_1}^{i_1}(x)$. Since z belongs to $A^0(x)$, $V \cap \operatorname{int} A^0(x)$ contains a non-null open set by Theorem 4.1. By the assumption, $V \cap \operatorname{int} A^0(x)$ contains a Poisson stable point z' of φ_t^0. Then z' is written as $z' = \varphi_{s_m}^{j_m}\circ \cdots \circ \varphi_{s_1}^{j_1}$ where $s_k > 0$ if $j_k = 0$ $(1 \le k \le m)$. Also, there exists $t_l' > -t_l$ such that $\varphi_{t_l'}^0(z') \in V$. Then,

$$
\varphi_{t_k}^{i_k}\circ \cdots \circ \varphi_{t_{l+1}}^{i_{l+1}} \circ \varphi_{t_l'+t_l}^{i_l} \circ \varphi_{s_m}^{j_m} \circ \cdots \circ \varphi_{s_1}^{j_1}(x) \in U\,.
$$

Repeating this argument inductively, we can choose $s_1, \cdots, s_n$ such that $s_l > 0$ if $j_l = 0$ and $y' \equiv \varphi_{s_n}^{j_n}\circ \cdots \circ \varphi_{s_1}^{j_1}(x)$ belongs to U, i.e. $y' \in A^0(x)$. This proves $\overline{A^0(x)} = M$. The previous lemma and Proposition 4.2 implies $A(x) = M$. The proof is complete.

We shall mention a sufficient condition that the orbits φ_t have dense Poisson stable points. Suppose that there is a probability measure μ on M such that $\mu(A) = \mu(\varphi_t(A))$ holds for all $t \in \boldsymbol{R}$ and Borel set A of M: The measure μ is called an invariant measure of φ_t. It is well known that almost all x (relative to μ) are Poisson stable points. (Poincare's recurrence theorem). Hence if the support of μ is the whole space M, then the set of Poisson stable points of φ_t is dense in M.

As an example, consider an (left) invariant vector field Y on a compact Lie group. Let φ_t be the corresponding orbit. Then the Haar measure is invariant for φ_t. Hence Poisson stable points are dense in M. Therefore if we consider the control system defined on a compact connected Lie group associated with invariant vector fields $X_1, \cdots, X_r$ and Y. Then the control system is completely controllable if and only if the Lie algebra generated by $X_1, \cdots, X_r$ and Y is the full Lie algebra of the Lie group. However the assertion is not valid if M is not compact. (As a counter example, consider $X = \partial/\partial x_1$, $Y = \partial/\partial x_2$ in $\boldsymbol{R}^2$).

§ 5. Accessible sets and the supports of transition probabilities

In this section, we consider accessible sets $A_t(x)$. A result analogous to Theorem 4.1 is

Theorem 5.1 (Sussman-Jurdjevic [16], Krener [10]). *Suppose that* $\mathscr{I}$ *is locally finitely generated. Let* $I(x)$ *be the maximal connected integral manifold of* $\mathscr{I}$ *containing the point* x. *Then* $A_t(x) \subset \varphi_t^0(I(x))$. *Furthermore, the interior of* $A_t(x)$ *with respect to the topology of* $\varphi_t^0(I(x))$ *is dense in* $A_t(x)$.

Proposition 5.2. $\operatorname{int} \overline{A_t(x)} = \operatorname{int} A_t(x)$.

The proof is similar as Proposition 4.2. It is omitted.

Remark. Let $\boldsymbol{R} \times M$ be the product manifold and $\partial/\partial t$ be the invariant vector field on $\boldsymbol{R}$. Then $Y, X_1, \cdots, X_r$ and $\partial/\partial t$ are considered as vector fields on $\boldsymbol{R} \times M$ by the natural injection. Consider the Lie algebra $\hat{\mathscr{L}} = \mathscr{L}(Y + \partial/\partial t, X_1, \cdots, X_r)$. It is known that $\dim \hat{\mathscr{L}}(t, x)$ is independent of t, and that $\dim \hat{\mathscr{L}}(t, x) = d$ or $d + 1$ according as $\dim \mathscr{I}(x) = d - 1$ or d, respectively (See [8]). Therefore the operator

$$\hat{L} = \frac{1}{2} \sum_{j=1}^{r} X_j^2 + Y + \frac{\partial}{\partial t}$$

is hypoelliptic in $\boldsymbol{R} \times M$, if $\dim \mathscr{I}(x) = d$ holds for all x. Under this condition, there exists a C^∞-function $p_t(x, y)$ of (t, x, y) $(t > 0)$ such that $P_t(x, dy) = p_t(x, y)m(dy)$ with volume element m. This fact also implies the thickness of the set $\overline{A_t(x)}$.

The system (1.4) is called *strongly completely controllable* if $A_t(x) = M$ holds for all $t > 0$ and $x \in M$. A sufficient condition for this is that $\dim \mathscr{B}(x) = d$ holds for all x of M. In fact, then $\overline{A_t(x)} = M$ holds by Theorem 3.2. Hence $A_t(x) = M$ holds by Proposition 5.2. On the other hand, the condition $\dim \mathscr{I}(x) = d$, $^\forall x \in M$ is necessary for the strong controllability by the previous theorem. Hirschorn [4] obtains a sufficient condition for the strongly complete controllability in the case where dim $\mathscr{I}(x) = d$ and dim $\mathscr{B}(x) < d$. His argument is based on the Lie transformation groups generated by $\mathscr{I}$ and $\mathscr{B}$, under the assumption that $\mathscr{I}$ is of finite dimension. We shall show that his result is valid under weaker condition.

Theorem 5.3. *Suppose that* dim $\mathscr{I}(x) = d$ *holds for all* x. *Suppose that* $\mathscr{B}$ *is locally finitely generated. If* $[\mathscr{B}, \mathscr{I}](x) \subset \mathscr{B}(x)$ *holds for all* x, *then the system* (1.4) *is strongly completely controllable.*

Remark. It is immediate to see that $[\mathscr{B}, \mathscr{I}](x) \subset \mathscr{B}(x)$ holds if and only if

$$[ad^n(Y)X_i, X_j](x) \in \mathscr{B}(x)\,, \qquad n = 1, 2, \cdots, \quad i, j = 1, \cdots, r\,.$$

The proof of the theorem is divided into several steps. We first introduce

$(\operatorname{Exp}\mathscr{L})_t =$ the subset of $\varphi \in (\operatorname{Exp}\mathscr{L})_+$ which are written as

$$\varphi = \varphi_{t_k}^{i_k} \circ \cdots \circ \varphi_{t_1}^{i_1} \quad \text{with} \quad \sum_{l=1}^{k} t_l 1_{\{i_l = 0\}} = t\,.$$

Then similarly as Lemma 4.5 we have

Lemma 5.4. $A_t(x) = \{\varphi(x)\,;\, \varphi \in (\operatorname{Exp} L)_t\}$.

Now let us define three sets of diffeomorphisms.

$\operatorname{Exp}\mathscr{B} =$ the group generated by φ_t^i, $t \in \boldsymbol{R}$, $i = 1, \cdots, r$,

$$\boldsymbol{F}_t = \bigcup_{k=1}^{\infty} \{\eta_{t_k}(\psi_k) \circ \cdots \circ \eta_{t_1}(\psi_1)\,;\, t_k \leq \cdots \leq t_1 = t,\ \psi_l \in \operatorname{Exp}\mathscr{B}\}\,,$$

$$\boldsymbol{G}_t = \bigcup_{k=1}^{\infty} \{\eta_{t_k}(\psi_k) \circ \cdots \circ \eta_{t_1}(\psi_1)\,;\, \min_l t_l \geq 0,\ \max_l t_l = t,\ \psi_l \in \operatorname{Exp}\mathscr{B}\}\,,$$

where

$$\eta_t(\psi) = \varphi_t^0 \circ \psi \circ \varphi_{-t}^0\,.$$

The relation $(\operatorname{Exp}\mathscr{L})_t = \boldsymbol{F}_t \varphi_t^0$ is easily proved. We have further

Lemma 5.5. *The set* $\boldsymbol{G}_t$ *is a group. If* dim $\mathscr{I}(x) = d$ *holds for all* x, *then* $\{\varphi(x)\,;\, \varphi \in \boldsymbol{G}_t\} = M$ *holds for all* x.

Proof. If φ_1, φ_2 belong to $\boldsymbol{G}_t$, $\varphi_1 \circ \varphi_2$ belongs to $\boldsymbol{G}_t$ as is easily seen. Let $\varphi \in \boldsymbol{G}_t$ be written as $\varphi = \eta_{t_k}(\psi_k) \circ \cdots \circ \eta_{t_1}(\psi_1)$. Then $\varphi^{-1} = \eta_{t_1}(\psi_1^{-1}) \circ \cdots \circ \eta_{t_k}(\psi_k^{-1})$. Therefore, $\boldsymbol{G}_t$ is a group.

Suppose now $\dim \mathscr{I}(x) = d$ for all x. Denote the set $\{\varphi(x) ; \varphi \in G_t\}$ as $B_t(x)$. Then we show, (a) $B_t(x) = B_t(y)$ if $y \in B_t(x)$, (b) $B_t(x) \cap B_t(y) = \emptyset$ if $y \notin B_t(x)$, and (c) $x \in \operatorname{int} B_t(x)$. The properties (a) and (b) follow directly from the group property of G_t. Let $A_t^0(x)$ be the accessible set for the system (4.1). Since $B_t(x) \supset A_t^0(x)$, $B_t(x)$ contains a non-null open set U. Let $y \in U$ and choose φ of G_t such that $\varphi(x) = y$. Then $\varphi^{-1}(U)$ is an open set containing x. Obviously $\varphi^{-1}(U)$ is included in $B_t(x)$. This proves (c). The properties (a)–(c) then imply that $B_t(x)$ is open and closed. Since M is connected, we have $B_t(x) = M$ for all $t > 0$ and $x \in M$.

From Lemmas 5.4 and 5.5, one could conclude $\overline{A_t(x)} = M$, if we could prove $F_t = G_t$. In the following we shall prove this in slightly different setting. We denote by $\operatorname{Exp} \tilde{\mathscr{B}}$ the group of diffeomorphisms generated by all one parameter group of transformations generated by vector fields belonging to $\tilde{\mathscr{B}}$. The set $\tilde{F}_t$ and $\tilde{G}_t$ are defined similarly as before using $\operatorname{Exp} \tilde{\mathscr{B}}$ instead of $\operatorname{Exp} \mathscr{B}$. Then it holds $F_t \subset \tilde{F}_t$ and $G_t \subset \tilde{G}_t$. In the sequal, we shall show $\tilde{F}_t = \tilde{G}_t$.

Lemma 5.6. *Let X be a complete vector field belonging to $\tilde{\mathscr{B}}$ and let φ_t be the one parameter group of transformations generated by X. Assume the condition of Theorem 5.3. Then $d\eta_s(\varphi_t)$ is an isomorphism between $\mathscr{B}(x)$ and $\mathscr{B}(\eta_s(\varphi_t)(x))$ for each x.*

Proof. It holds $\eta_s(\varphi_{t_1}) \circ \eta_s(\varphi_{t_2}) = \eta_s(\varphi_{t_1+t_2})$ and $d\eta_s(\varphi_{t_1+t_2}) = d\eta_s(\varphi_{t_1}) \cdot d\eta_s(\varphi_{t_2})$. Hence it is enough to prove the assertion for sufficiently small $|t|$. Let Z be a vector field belonging to $\tilde{\mathscr{B}}$. Set $Y_{t,s} = d\eta_s(\varphi_t)Z$. For each s, $\eta_s(\varphi_t)$, $t \in \boldsymbol{R}$ is the one parameter group of transformations generated by $d\varphi_s^0 X$. Therefore we have $\partial Y_{t,s}/\partial t = [Y_{t,s}, d\varphi_s^0 X]$. Obviously, $d\eta_s(\varphi_t) d\varphi_s^0 X = d\varphi_s^0 X$ holds. Therefore,

$$\frac{\partial Y_{t,s}}{\partial t} = d\eta_s(\varphi_t)[Z, d\varphi_s^0 X] \; .$$

It is known that $d\varphi_s^0 X$ belongs to $\tilde{\mathscr{I}} = \{\mathscr{I}(x) ; x \in M\}$ (See ([9]). Therefore, $[Z, d\varphi_s^0 X] \in \tilde{\mathscr{B}}$ by the assumption.

We now fix a point x of M and $s \in \boldsymbol{R}$. Let $Z^1, \cdots, Z^n$ be a base of $\mathscr{B}$ in an open neighborhood U of x. Then there exists C^∞-functions f_{ij} on U such that $[Z^i, d\varphi_s^0 X] = \sum_{j=1}^n f_{ij} Z^j$ holds in U. Let ε be a positive number such that $\eta_s(\varphi_t)(x) \in U$ for $|t| < \varepsilon$. Then $d\eta_s(\varphi_t)[Z^i, d\varphi_s^0 X] = \sum_{j=1}^n f_{ij} d\eta_s(\varphi_t) Z^j$ holds for $|t| < \varepsilon$. Set $V^j(t) = d\eta_s(\varphi_t) Z^j$. Then $V^j(t)$, $|t| < \varepsilon$ satisfies the linear differenial equation

$$\frac{dV^i(t)}{dt} = \sum_{j=1}^n f_{ij} V^j(t) \; , \qquad i = 1, \cdots, n \; .$$

The solution $V^j(t)$ is written as $V^j(t) = \sum_{k=1}^n g_{jk}(t)V^k(0)$ with regular matrix (g_{jk}). Since $V^k(0) = Z^k \in \mathscr{B}$, we see that $V^k(t)$ belongs to $\tilde{\mathscr{B}}$ for $k = 1, \cdots, n$. This proves that $d\eta_s(\varphi_t)$ is the isomorphism between $\mathscr{B}(x)$ and $\mathscr{B}(\eta_s(\varphi_t)(x))$ for $|t| < \varepsilon$. The proof is complete.

The above lemma implies

Proposition 5.7. *Under the condition of Theorem* 5.3, *it holds* $\tilde{F}_t = \tilde{G}_t$ *for all* $t > 0$.

Proof. We shall first prove

$$\eta_t(\psi)\,(\mathrm{Exp}\,\tilde{\mathscr{B}})\eta_t(\psi)^{-1} \subset \mathrm{Exp}\,\tilde{\mathscr{B}} \qquad \text{for } \psi \in \tilde{\mathscr{B}}\,. \tag{5.1}$$

Let φ_t be the one parameter group of transformations of Lemma 5.6. Then $\varphi_t' \equiv \eta_s(\psi) \circ \varphi_t \cdot \eta_s(\psi)^{-1}$ (s is fixed) is a one parameter group of transformation generated by $d\eta_s(\psi)X$. The previous lemma shows that $d\eta_s(\psi)X$ belongs to $\tilde{\mathscr{B}}$. Thus φ_t', $t \in \boldsymbol{R}$ belong to $\mathrm{Exp}\,\tilde{\mathscr{B}}$. Now, $\mathrm{Exp}\,\tilde{\mathscr{B}}$ is generated by all such φ_t, $t \in \boldsymbol{R}$. Hence we have the relation (5.1).

Let φ be any element of $\tilde{G}_t$. It is written as

$$\varphi = \eta_{t_k}(\psi_{t_k}) \circ \cdots \circ \eta_{t_1}(\psi_{t_1})\,, \qquad t_l \geq 0,\ \max(t_l) = t\,.$$

We shall prove that there exists $\tilde{\psi}_1, \cdots, \tilde{\psi}_k$ of $\mathrm{Exp}\,\tilde{\mathscr{B}}$ and $0 \leq s_k \leq \cdots \leq s_1 = t$ such that

$$\eta_{t_k}(\psi_{t_k}) \circ \cdots \circ \eta_{t_1}(\psi_{t_1}) = \eta_{s_k}(\tilde{\psi}_k) \circ \cdots \circ \eta_{s_1}(\tilde{\psi}_1)\,.$$

For the simplicity, we consider the case $n = 2$. (The general case is derived easily by induction). If $t_2 \leq t_1$, there is no doubt. Suppose $t_2 > t_1$. Set $t_3 = t_2 - t_1$. Then we may write as $\eta_{t_2}(\psi_2) \circ \eta_{t_1}(\psi_1) = \eta_{t_1}(\eta_{t_3}(\psi_2) \circ \psi_1)$. By (6.1), there exists $\tilde{\psi}_1$ of $\mathrm{Exp}\,\tilde{\mathscr{B}}$ such that $\eta_{t_3}(\psi_2) \circ \psi_1 \circ \eta_{t_3}(\psi_2)^{-1} = \tilde{\psi}_1$. Then $\eta_{t_3}(\psi_2) \circ \psi_1 = \tilde{\psi}_1 \eta_{t_3}(\psi_2)$. This implies

$$\eta_{t_2}(\psi_2) \circ \eta_{t_1}(\psi_1) = \eta_{t_1}(\tilde{\psi}_1)\eta_{t_1+t_3}(\psi_2) = \eta_{t_1}(\tilde{\psi}_1)\eta_{t_2}(\psi_2)\,.$$

The proof is complete.

We are now in the position of

Proof of Theorem 5.3. Obviously the set $\{\varphi\varphi_t^0(x)\,;\, \varphi \in \boldsymbol{F}_t\}$ is included in the set $\{\varphi\varphi_t^0(x)\,;\, \varphi \in \tilde{\boldsymbol{F}}_t\}$. It is not difficult to see that the closures of the above sets coincide each other. Since $\tilde{\boldsymbol{F}}_t = \tilde{\boldsymbol{G}}_t \supset \boldsymbol{G}_t$, we see that the set $\{\varphi\varphi_t^0(x)\,;\, \varphi \in \tilde{\boldsymbol{F}}_t\}$ is the whole space M by Lemma 5.5. Therefore, we see

$\overline{A_t(x)} = M$ by Lemma 5.4. Then $A_t(x) = M$ holds by Proposition 5.2. The proof is complete.

Theorem 5.3 is generalized as follows.

Theorem 5.8. *Suppose* $\dim \mathscr{I}(x) = d$. *If there exists a finite number of locally finitely generated Lie subalgebras* $\mathscr{C}_0, \cdots, \mathscr{C}_n$ *of* $\mathscr{I}$ *with properties* (1) ~ (3) *below, then the system* (1.4) *is strongly completely controllable.*

(1) $\mathscr{C}_0 \subset \mathscr{B}$ *and* $\mathscr{I} = \mathscr{L}(\mathscr{B}, ad^n(Y)Z\,;\, n = 0, 1, \cdots, Z \in \bigcup_{i=1}^n \mathscr{C}_i)$.

(2) *For any* $\mathscr{C}_i$, $[ad^n(Y)Z, Z'](x) \in \mathscr{C}_i(x)$ *for* $Z, Z' \in \mathscr{C}_i$, $n = 1, 2, \cdots$.

(3) $\mathscr{I}_i = \mathscr{L}(ad^n(Y)Z\,;\, Z \in \mathscr{C}_i, n = 0, 1, \cdots)$ *is locally finitely generated.*

Proof. Define $\mathscr{B}_i$, $i = 1, \cdots, n$ by induction:

$$\mathscr{B}_0 = \mathscr{B}, \cdots, \mathscr{B}_i = \mathscr{L}(\mathscr{B}_{i-1}, \mathscr{I}_{i-1})\,.$$

Then it holds $\mathscr{B}_n = \mathscr{I}$ by (1). We shall prove

$$\overline{A_t(x)} \supset \{\varphi \circ \varphi_t^0(x)\,;\, \varphi \in \operatorname{Exp} \tilde{\mathscr{B}}_i\}$$

by induction. The case $i = 0$ is obvious from $\mathscr{C}_0 \subset \mathscr{B}$. Suppose the above is valid for i. Then by the same argument as the previous theorem, we have $\overline{A_t(x)} \supset \{\varphi \circ \varphi_t^0(x)\,;\, \varphi \in \operatorname{Exp} \tilde{\mathscr{I}}_i\}$. But $\operatorname{Exp} \tilde{\mathscr{B}}_{i+1}$ is generated by $\operatorname{Exp} \tilde{\mathscr{B}}_i$ and $\operatorname{Exp} \tilde{\mathscr{I}}_i$. Therefore $\varphi \circ \varphi_t^0(x) \in \overline{A_t(x)}$ holds for $\varphi \in \operatorname{Exp} \tilde{\mathscr{B}}_{i+1}$. The proof is complete.

Assertions analogous to Theorem 5.3 and 5.8 are valid in the case $\dim \mathscr{I}(x) = d - 1$. Corresponding to Theorem 5.3, we have

Theorem 5.3′. *Suppose* $\dim \mathscr{I}(x) = d - 1$ *holds for all* x. *Suppose that* $\mathscr{B}$ *is locally finitely generated and that* $[\mathscr{B}, \mathscr{I}](x) \subset \mathscr{B}(x)$ *holds for all* x. *Let* $I(x)$ *be the maximal connected integral manifold containing* x. *Then it holds* $\varphi_t^0(I(x)) = A_t(x)$.

The proof is obvious from Theorem 5.3.

§ 6. Time varying system

The controllability problem of time varying system can be reduced to the problem of autonormous one by adding the time set to the state space.

Let $Y(t), X_1(t), \cdots, X_r(t)$ be vector fields on M with parameter $t \in [0, T]$. By local coordinate $(x_1, \cdots, x_d)$, these are expressed as

$$Y(t) = \sum_{i=1}^{d} Y^i(t, x)\frac{\partial}{\partial x_i}, \qquad X_j(t) = \sum_{i=1}^{d} X_j^i(t, x)\frac{\partial}{\partial x_i}.$$

We assume that $Y^i(t, x)$ and $X_j^i(t, x)$ are C^∞-functions on $[0, T] \times M$. Consider the control system on M

$$\frac{d\varphi_t}{dt} = Y(t, \varphi_t) + \sum_{j=1}^{r} X_j(t, \varphi_t)u_j(t), \tag{6.1}$$

where $Y(t, x)$ etc. are d-vectors $(Y^1(t, x), \cdots, Y^d(t, x))$ etc. Set $\varphi_t^0 = t$, $\hat{\varphi}_t = (\varphi_t^0, \varphi_t^1, \cdots, \varphi_t^d)$ and define $d + 1$-vectors

$$\begin{aligned} \hat{Y}(t, x) &= (1, Y^1(t, x), \cdots, Y^d(t, x)), \\ \hat{X}_j(t, x) &= (0, X_j^1(t, x), \cdots, X_j^d(t, x)) \end{aligned} \tag{6.2}$$

Then $\hat{\varphi}_t$ satisfies

$$\frac{d\hat{\varphi}_t}{dt} = \hat{Y}(\hat{\varphi}_t) + \sum_{j=1}^{r} \hat{X}_j(\hat{\varphi})u_j(t). \tag{6.3}$$

The equation (6.3) may be regarded as an autonormous control system on the product manifold $[0, T] \times M$. Vector fields, on $[0, T] \times M$ corresponding to (6.2) or (6.3) are

$$\hat{Y} = \sum_{i=1}^{d} Y^i(t, x)\frac{\partial}{\partial x_i} + \frac{\partial}{\partial t}, \qquad \hat{X}_j = \sum_{i=1}^{d} X_j^i(t, x)\frac{\partial}{\partial x_i}. \tag{6.4}$$

We often identify (6.2) and (6.4), and call (6.2) themselves as vector fields.

Define now

$$\begin{aligned} \hat{\mathscr{B}} &= \mathscr{L}(\hat{X}_1, \cdots, \hat{X}_r), \\ \hat{\mathscr{L}} &= \mathscr{L}(\hat{X}_1, \cdots, \hat{X}_r, \hat{Y}), \\ \hat{\mathscr{I}} &= \text{the ideal in } \hat{\mathscr{L}} \text{ generated by } \hat{\mathscr{B}}. \end{aligned}$$

It is clear that $ad^n(\hat{Y})\hat{X}$ $(\hat{X} \in \hat{\mathscr{B}})$ do not have component of $\partial/\partial t$ i.e., they are written as $\sum_{i=1}^{d} f^i(t, x)(\partial/\partial x_i)$. Hence $\dim \hat{\mathscr{I}}(t, x) \leq d$ holds for all t and x. It holds $\dim \hat{\mathscr{I}}(t, x) = d$ if and only if $\dim \hat{\mathscr{L}}(t, x) = d + 1$.

We will assume $\dim \hat{\mathscr{I}}(t, x) = d$ holds for all (t, x). Then the maximal connected integral manifold of $\hat{\mathscr{I}}$ are the sets $(t, M) = \{(t, x); x \in M\}$. Now let $\hat{A}_t(0, x)$ be the accessible sets of (6.3) at time t starting at $(0, x)$, and $A_t(x)$, the accessible sets of (6.1) at t starting at x. Then, it holds $(t, A_t(x)) = \hat{A}_t(0, x)$, obviously. Hence Theorem 5.1 and Proposition 5.2 imply that $\operatorname{int} A_t(x)$ is dense in $A_t(x)$ and $\operatorname{int} \overline{A_t(x)} = \operatorname{int} A_t(x)$. Futhermore, from Theorem 5.3′, we have

Theorem 6.1. *Suppose that* $\dim \hat{\mathscr{I}}(t, x) = d$ *holds for all* (t, x). *Suppose* $\hat{\mathscr{B}}$ *is locally finitely generated and* $[\hat{\mathscr{B}}, \hat{\mathscr{I}}](t, x) \subset \hat{\mathscr{B}}(t, x)$ *holds for all* (t, x). *Then* $A_t(x) = M$ *holds for all* (t, x).

We shall apply the theorem to linear and bilinear system.

Example 1. Consider the linear control system on R^d:

$$\frac{d\varphi_t}{dt} = A(t)\varphi_t + F(t)u(t) \,, \qquad t \in [0, T] \tag{6.5}$$

where $A(t) = (a_{ij}(t))$ and $F(t) = (f_{ij}(t))$ are $d \times d$ and $d \times r$-matrix functions of the class C^∞, respectively. $u(t) = (u_1(t), \cdots, u_r(t))$ is an r-dimensional control function. Vector fields corresponding to (6.5) are

$$\hat{Y}(t, x) = (1, A(t)x) \,, \qquad \hat{X}_j(t, x) = (0, f_j(t)) \,,$$

where $f_j(t) = (f_{1j}(t), \cdots, f_{dj}(t))$. We shall compute $\hat{\mathscr{B}}$, $\hat{\mathscr{L}}$ and $\hat{\mathscr{I}}$.

Since $\hat{X}_i$ and $\hat{X}_j$ do not depend on x, it holds $[\hat{X}_i, \hat{X}_j] = 0$. Therefore we have

$$\hat{\mathscr{B}}(t, x) = \{(0, f_1(t)), \cdots, (0, f_r(t))\}_{LS} \,. \tag{6.6}$$

Here $\{\ \}_{LS}$ means the linear span of the elements of the set $\{\ \}$. A direct computation yields.

$$[\hat{Y}, \hat{X}_j](t, x) = (0, \dot{f}_j(t) - A(t)f_j(t)) \,,$$

where $\dot{f}_j(t) = (d/dt)f_j(t)$. Now let $f(t)$ be a d-vector function. Define a sequence of d-vector functions as

$$P_0 f = f, \cdots, P_n f = \frac{d}{dt}(P_{n-1}f) - AP_{n-1}f \,. \tag{6.7}$$

P_n, $n = 0, 1, \cdots$ are linear transformations acting on the space of smooth d-vector functions. By indnction, we have

$$ad^n(\hat{Y})\hat{X}_j = (0, P_n f_j) \,, \qquad n = 1, 2, \cdots .$$

Therefore we have

$$\hat{\mathscr{I}}(t, x) = \{(0, P_n f_j)\,;\, 1 \le j \le r,\ n = 0, 1, 2, \cdots\}_{LS} \,, \tag{6.8}$$

and

$$\hat{\mathscr{L}}(t, x) = \{(1, A(t)x),\ \hat{\mathscr{I}}(t, x)\}_{LS} \,. \tag{6.9}$$

Note that all elements of $\hat{\mathscr{B}}$ and $\hat{\mathscr{I}}$ do not depend on x. We then have $[\hat{\mathscr{I}}, \hat{\mathscr{B}}] = 0$. This shows that $\hat{\mathscr{B}}$ is an ideal in $\hat{\mathscr{I}}$. Theorem 6.1 concludes

Theorem 6.2. *Suppose that*

$$\dim \{P_n f_j(t)\,;\ 1 \leq j \leq r,\ n = 0, 1, 2, \cdots\}_{LS} = d$$

holds for all $t \in [0, T]$. *Then the system* (6.5) *is strongly completely controllable.*

In case where matrices $A(t)$ and $F(t)$ do not depend on t, it holds

$$P_n = -AP_{n-1} = \cdots = (-1)^n A^n\ .$$

Hence the above theorem implies the well known Kalman's result.

Corollary. *Suppose that* $A(t)$ *and* $F(t)$ *are constant matrices. If*

$$\text{(6.10)} \qquad \operatorname{rank}(F, AF, \cdots, A^{d-1}F) = d\ ,$$

then the system (6.1) *is strongly completely controllable.*

Example 2. Consider the homogeneous bilinear system on the punctured space $R^d - \{0\}$:

$$\text{(6.11)} \qquad \frac{d\varphi_t}{dt} = (A(t) + u_1(t)B_1(t) + \cdots + u_m(t)B_m(t))\varphi_t\ .$$

Here $A(t), B_1(t), \cdots, B_m(t)$ are $d \times d$-matrix functions of the class C^∞, $u_1(t), \cdots, u_m(t)$ are 1-dimensional control functions. Vector fields corresponding to (6.11) are

$$\hat{Y}(t, x) = (1, A(t)x)\ , \qquad \hat{X}_k(t, x) = (0, B_k(t)x)\ ,$$
$$1 \leq k \leq m\ .$$

We shall compute $\hat{\mathscr{B}}$ and $\hat{\mathscr{I}}$. For two matrices $C(t)$ and $D(t)$, we define $[C(t), B(t)] = C(t)B(t) - B(t)C(t)$ as usual. We denote by $\mathscr{L}(B_1, \cdots, B_m)$ the Lie algebra generated by matrix functions $B_1, \cdots, B_m$. A direct computation yields

$$[\hat{X}_i, \hat{X}_j](t, x) = (0, -[B_i(t), B_j(t)]x)\ .$$

Therefore we have

$$\text{(6.12)} \qquad \hat{\mathscr{B}}(t, x) = \{(0, B(t)x)\,;\ B \in \mathscr{L}(B_1, \cdots, B_m)\}_{LS}\ .$$

Let us next compute $\hat{\mathscr{I}}$. Let $B \in \mathscr{L}(B_1, \cdots, B_r)$ and $\hat{X}^B$ be the vector field corresponding to $(0, B(t)x)$. Then a direct computation induces

$$ad^n(\hat{Y})\hat{X}^B(t,x) = (0, Q_nB(t)x)$$

where $Q_n, n = 0, 1, \cdots$ are linear transformations on the space of $d \times d$-matrix functions defined by

$$(6.13) \qquad Q_0B = B, \cdots, Q_nB = \frac{d}{dt}(Q_nB) - [A, Q_{n-1}B] .$$

Set

$$\mathscr{L}(Q) = \mathscr{L}(Q_nB\,;\, B \in \mathscr{L}(B_1, \cdots, B_m),\, n = 0, 1, 2, \cdots) .$$

Then we have

$$(6.14) \qquad \hat{\mathscr{I}}(t,x) = \{(0, L(t)x)\,;\, L \in \mathscr{L}(Q)\}_{LS} .$$

$\hat{\mathscr{B}}$ is an ideal in $\hat{\mathscr{I}}$ if and only if $\mathscr{L}(B_1, \cdots, B_m)$ is an ideal in $\mathscr{L}(Q)$. Therefore we have

Theorem 6.3. *The bilinear system (6.11) is strongly completely controllable in $R^d - \{0\}$ if the following two conditions are satisfied*

a) $\dim \{L(t)x\,;\, L \in \mathscr{L}(Q)\} = d$ *for all* $x \neq 0$ *and* t.

b) $\mathscr{L}(B_1, \cdots, B_m)$ *is an ideal in* $\mathscr{L}(Q)$.

In case where $A(t), B_1(t), \cdots, B_m(t)$ do not depend on t, we have

$$Q_nB = -[A, Q_{n-1}B] = \cdots = (-1)^n ad^n(A)B .$$

Consequently, $\mathscr{L}(Q)$ coincides with the ideal in $\mathscr{L}(A, B_1, \cdots, B_m)$ generated by $B_1, \cdots, B_m$. Therefore we have

Corollary (Brockett [1]). *Let $A, B_1, \cdots, B_m$ be constant matrices. Let $\mathscr{I}(B_1, \cdots, B_m)$ be the ideal in $\mathscr{L}(A, B_1, \cdots, B_m)$ generated by $B_1, \cdots, B_m$. Let $e^{\mathscr{I}(B_1,\cdots,B_m)}$ be the group of matrices generated by e^{tB}, $B \in \mathscr{I}(B_1, \cdots, B_m)$. The system (6.11) is strongly completely controllable in $R^d - \{0\}$ if the following two conditions are satisfied.*

a) $e^{\mathscr{I}(B_1,\cdots,B_m)}$ *acts transitively on* $R^d - \{0\}$

b) $\mathscr{L}(B_1, \cdots, B_m)$ *is an ideal in* $\mathscr{I}(B_1, \cdots, B_m)$.

Example 3. Consider the inhomogeneous bilinear system on R^d:

$$(6.15) \quad \frac{d\varphi_t}{dt} = (A(t) + u_1(t)B_1(t) + \cdots + u_m(t)B_m(t))\varphi_t + F(t)v(t) .$$

Here $F(t)$ is a $d \times r$-matrix function of the class C^∞. If $B_1 = \cdots = B_m \equiv 0$, then the above system coincides with the linear system of Example 1. If $F(t) \equiv 0$, the above system equals homogeneous bilinear system of Example 2.

By the argument similar to Examples 1 and 2, Lie algebras $\hat{\mathscr{B}}$ and $\hat{\mathscr{L}}$ are computed as

$$\hat{\mathscr{B}}(t, x) = \{\{(0, B(t)x) ; B \in \mathscr{L}(B_1, \cdots, B_m)\} \cup \{(0, C(t)f_j ; C \in \boldsymbol{C},\ 1 \leq j \leq r\}\}_{LS}$$

where

$$\boldsymbol{C} = \text{semigroup generated by } \mathscr{L}(B_1, \cdots B_m)$$

and

$$\hat{\mathscr{I}}(t, x) = \{\{(0, L(t)x) ; L \in \mathscr{L}(Q)\} \cup \{(0, D(t)f_j) ; D \in \boldsymbol{D},\ 1 \leq j \leq r\}\}_{LS}$$

where

$$\boldsymbol{D} = \text{semigroup generated by } \mathscr{L}(Q) \text{ and } \{P_n ; n = 0, 1, \cdots\} .$$

From this, we have

Theorem 6.4. *The system* (6.15) *is strongly completely controllable in* $\boldsymbol{R}^d$ *if the following two conditions are satisfied.*

(a) $\dim \{D(t)f_j(t) ; D \in \boldsymbol{D},\ 1 \leq j \leq r\}_{LS} = d$ *holds for all* t,

(b) $\mathscr{L}(B_1, \cdots, B_m)$ *is an ideal in* $\mathscr{L}(Q)$.

In case where $A(t), B_1(t), \cdots, B_n(t)$ and $F(t)$ are constant matrices, it holds $\mathscr{L}(Q) = \mathscr{I}(B_1, \cdots, B_n)$ and $P_n = (-1)^n A^n$. Therefore,

$$\boldsymbol{D} = \text{semigroup generated by } \mathscr{L}(A, B_1, \cdots, B_m) .$$

Then conditions (a) and (b) are written in simpler form (Hirschorn [4]).

References

[1] Brockett, R. W., System theory on group manifolds and coset spaces. SIAM J. Control, **10** (1972), 265–284.

[2] Chow, W. L., Über systeme von linearen partiellen differentialgleichungen erster ordnung, Math. Ann., **117** (1939), 98–105.

[3] Hermann, R., The differential geometry of foliations, II, J. Math. and Mech., **11** (1962), 303–316.
[4] Hirschorn, R. M., Controllability in nonlinear systems, J. Differential Equations, **19** (1975), 46–61.
[5] ——, Topological semigroups, sets of generators, and controllability, Duke Math. J., **40** (1973), 937–947.
[6] Hörmander, L., Hypoelliptic second order differential equations, Acta Math., **119** (1967), 147–171.
[7] Ichihara, K., Diffusion processes and control systems, Master's thesis, Nagoya University, 1973.
[8] Ichihara, K. and Kunita, H., A classification of the second order degenerate elliptic operators and its probabilistic characterization, Z. Wahrscheinlichkeitstheorie und Verw. Gebiete, **30** (1974), 235–254.
[9] ——, Supplements and corrections to the above paper, Z. Wahrscheinlichkeitstheorie und Verw. Gebiete, **39** (1977), 81–84.
[10] Krener, A. J., A generalization of Chow's theorem and the bang-bang theorem to nonlinear control problems, SIAM J. Control, **12** (1974), 43–52.
[11] Kunita, H., Diffusion processes and control systems, Course at University of Paris VI, 1974.
[12] Lobry, C., Controlabilitè des systèmes non linéaires, SIAM J. Control, **8** (1970), 573–605.
[13] ——, Controllability of nonlinear systems on compact manifolds, SIAM J. Control, **12** (1974), 1–4.
[14] Silverman, L. M. and Meadows, H. E., Controllability and observability in time-variable linear systems, SIAM J. Control, **5** (1967), 64–73.
[15] Stroock, D. W. and Varadhan, S. R. S., On the support of diffusion processes with applications to the strong maximum principle, Proc. 6-th Berkeley Sympos. Math. Statist Prob., **III** (1972), 333–368.
[16] Sussman, H. J. and Jurdjevic, V., Controllability of nonlinear systems, J. Differential Equations, **12** (1972), 95–116.
[17] Wong, E. and Zakai, H., Riemann-Stieltjes approximations of stochastic integrals, Z. Wahrscheinlichkeitstheorie und Verw. Gebiete, **12** (1969), 87–97.

Department of Mathematics
Nagoya University
Nagoya 464, Japan

Present address is
Department of Applied Science
Faculty of Engineering
Kyushu University
Fukuoka 812, Japan

Proc. of Intern. Symp. SDE
Kyoto 1976, pp. 187–193

Uhlenbeck-Ornstein Process on a Riemann-Wiener Manifold

Hui-Hsiung KUO*

Abstract

It is shown that there is a diffusion process χ_μ on a Riemann-Wiener manifold $\mathscr{W}$ with Riemannian metric G corresponding to each differentiable measure μ on $\mathscr{W}$. The generator of χ_μ is

$$\mathscr{G}_\mu f(x) = \text{trace}\, \bar{G}(x)^{-1} f''(x) + \left\langle \text{sp}\,(\bar{G}^{-1})'(x) + \bar{G}(x)^{-1}\left(\frac{dD\mu}{d\mu}(x)\right),\ f'(x)\right\rangle,$$

where $\bar{G}$ is the operator associated with G and $dD\mu/d\mu$ is the logarithmic derivative of μ. Uhlenbeck-Ornstein process is the one corresponding to $\mu = q_1(x_0, \cdot)$, where x_0 is a fixed point in $\mathscr{W}$ and $q_t(x, \cdot)$ is the transition probability for a Brownian motion on $\mathscr{W}$.

§ 1. Introduction

The purpose of this paper is to construct an elliptic operator and its corresponding diffusion process on an infinite dimensional manifold. In case the manifold is flat, this elliptic operator is the so-called number operator and the corresponding process is the Uhlenbeck-Ornstein process [8]. The construction of this operator utilizes the notion of differentiable measure [7] and a formula proved in [11]. The construction of the corresponding process follows from the same procedure as in [6] which is an infinite dimensional generalization of Ito [4] and McKean [13, p. 90].

Our work is motivated by the study of potential theory over infinite dimensional manifolds and an attempt to extend Hodge's theorem to such manifolds. Let (H, B) be an abstract Wiener space [2] and p_1 the Wiener measure in B with mean 0 and variance 1. It is well-known that the number operator, i.e. the closed extension of the differential operator $Lf(x) = -\text{trace}_H f''(x) + \langle x, f'(x)\rangle$, is a self-adjoint operator acting on the Hilbert space $L^2(B, p_1)$. Thus the classical L^2-theory provides a useful tool for the study of the number operator, in particular, the potential

* Research supported by NSF Grant MPS 73-08624 A02. The author is grateful to Professor L. Gross and Professor M. Ann Piech for several stimulating conversations.

theory associated with it. On the other hand, consider a Riemann-Wiener manifold $\mathscr{W}$, i.e. a differentiable manifold modelled on (H, B). Riemannian volume element does not exist on $\mathscr{W}$ so that most of the techniques for finite dimensional Riemannian manifolds break down. The hope is to construct a Borel measure μ on $\mathscr{W}$ which is an analog of p_1 for B and serves for the same purpose as the finite dimensional volume element. From this measure μ, we then construct a differential operator N which is the analog of L and has self-adjoint extension acting on the Hilbert space $L^2(\mathscr{W}, \mu)$. The expression for N suggests to us how to define $\delta\alpha$ for a differential form α of order 1 so that $N = \delta d$, where d is the differentiation operator. We expect that it will give us a clue as how to define $\delta\alpha$ for a differential form α of higher order so that $N = \delta d + d\delta$ will serve for the same purpose as the Laplace-Beltrami operator for finite dimensional Riemannian manifolds.

We remark that Uhlenbeck-Ornstein process on an abstract Wiener space has been studied in [8; 10; 14; 15; 16].

§ 2. Differentiable measures on a Riemann-Wiener manifold

Let (H, B) be a fixed abstract Wiener space. The dual space B^* is embedded in B via $B^* \subset H^* \equiv H \subset B$. $\mathscr{L}^k(B; B^*)$ will denote the Banach space of continuous k-linear maps from B into B^*. In [5; 6], we define a $C^r (r \geq 3)$ Riemann-Wiener manifold as a C^r Banach manifold $\mathscr{W}$ modelled on B with a maximal C^r atlas $\{(U_\alpha, \varphi_\alpha)\}$ and a C^{r-1} Riemannian metric G. The atlas satisfies the following condition: for any α and β, $\varphi_\beta \circ \varphi_\alpha^{-1} = I + \rho_{\alpha,\beta}$ with $\rho_{\alpha,\beta}(x) \in H$, $\rho'_{\alpha,\beta}(x) \in \mathscr{L}(B; B^*)$ and $\rho''_{\alpha,\beta}(x) \in \mathscr{L}^2(B; B^*)$ for all x in $\varphi_\alpha(U_\alpha \cap U_\beta)$, and the maps $\rho'_{\alpha,\beta}$ and $\rho''_{\alpha,\beta}$ are continuous from $\varphi_\alpha(U_\alpha \cap U_\beta)$ into $\mathscr{L}(B; B^*)$ and $\mathscr{L}^2(B; B^*)$, respectively. The Riemannian metric G is a C^{r-1} section of the bundle $\bigcap_{p\in\mathscr{W}} \mathscr{L}^2(R_p; R) \to \mathscr{W}$ such that R_p is a Hilbert space with inner product $G(p)$, where $R_p \equiv \varphi_{*,p}^{-1}(H)$ for any chart map φ at p. For each chart (U, φ) the local expression G_φ of G is given by $G_\varphi(x)(h, k) = \langle \bar{G}_\varphi(x)h, k \rangle$, $h, k \in H$, where $\bar{G}_\varphi$ is a map from $\varphi(U)$ into $\mathscr{L}(B; B)$ satisfying the following condition: $\bar{G}_\varphi(x) - I \in \mathscr{L}(B; B^*)$, $\bar{G}'_\varphi(x) \in \mathscr{L}^2(B; B^*)$, $\bar{G}''_\varphi(x) \in \mathscr{L}^3(B; B^*)$ for all x in $\varphi(U)$, and the maps $\bar{G}'_\varphi$ and $\bar{G}''_\varphi$ are continuous from $\varphi(U)$ into $\mathscr{L}^2(B; B^*)$ and $\mathscr{L}^3(B; B^*)$, respectively.

Recall that in [7; 11] a finite positive Borel measure ν on an open subset V of B is defined to be H-differentiable if (i) for any bounded uniformly continuous function f with bounded support and dist (supp f, V^c) > 0, the function $\nu f(x) = \int_V f(x + y)\nu(dy)$ is H-differentiable at the origin and (ii) for any sequence f_n converging to zero pointwise and boundedly with

$\bigcup_n \operatorname{supp} f_n$ bounded and dist $(\bigcup_n \operatorname{supp} f_n, V^c) > 0$, $\lim_{n\to\infty} \langle D(\nu f_n)(0), h\rangle = 0$ for all h in H. If ν is H-differentiable, then there is a unique H-valued measure $D\nu$ on V such that $D(\nu f)(0) = -\int_V f(y) D\nu(dy)$ for all f satisfying the condition in (i). The logarithmic derivation $dD\nu/d\nu$ of ν, if exists, is defined to be the unique measurable function f from V into B such that $\langle D\nu(A), h\rangle = \int_A (f(x), h)\nu(dx)$ for any Borel subset A of V and any h in H. For example, $(dDp_1/dp_1)(x) = -x$. If $B = R^n$ and $\nu(dx) = w(x)dx$, then $dD\nu/d\nu = \operatorname{grad} \ln w$. (cf. [3, p. 1077]).

Definition 1. A finite positive Borel measure μ on a Riemann-Wiener manifold $\mathscr{W}$ is said to be H-differentiable if for any chart (U, φ) on $\mathscr{W}$, the measure $\mu_\varphi = \mu \circ \varphi^{-1}$ is H-differentiable on $\varphi(U)$.

Let (U, φ) and (V, ψ) be two charts on $\mathscr{W}$ with $U \cap V \neq \varnothing$ and $\theta = \psi \circ \varphi^{-1}$. For each x in $\varphi(U \cap V)$, define $T(x)\colon B \times B \to B^*$ by $T(x)(u, v) = \theta''(x)(\theta'(x)^{-1}u, \cdot)v$. By [6, Proposition II. 2, p. 443], $\sum T(x)(e_n, e_n)$ converges in H for any orthonormal basis $\{e_n\}$ of H and the limit is independent of $\{e_n\}$. Moreover, $\sum T(\cdot)(e_n, e_n)$ is a continuous function from $\varphi(U \cap V)$ into H. It is easy to check that θ satisfies the condition of [11, Theorem 1]. Therefore, by [11, Theorem 2], we have the following transformation rule for H-differentiable measures on $\mathscr{W}$.

Theorem 1. *Suppose μ is an H-differentiable measure on a Riemann-Wiener manifold $\mathscr{W}$ and, for each chart (U, φ) on $\mathscr{W}$, $\mu_\varphi = \mu \circ \varphi^{-1}$ has logarithmic derivative $dD\mu_\varphi/d\mu_\varphi$. If (U, φ) and (V, ψ) are two charts on $\mathscr{W}$ with $U \cap V \neq \varnothing$ and $\theta = \psi \circ \varphi^{-1}$, then for μ_φ-almost all x in $\varphi(U \cap V)$,*

$$\frac{dD\mu_\varphi}{d\mu_\varphi}(x) = \theta'(x)^*\left(\frac{dD\mu_\psi}{d\mu_\psi}(\theta(x))\right) + \sum_n \theta''(x)(\theta'(x)^{-1}e_n, \cdot)e_n,$$

where $\{e_n\}$ is any orthonormal basis of H.

Remark. $\theta'(x)^*$ is interpreted as follows. We know that $\theta'(x) = I + \rho'(x)$ with $\rho'(x) \in \mathscr{L}(B; B^*)$. Hence $\rho'(x)$ can be regarded as an operator of H. Let $\rho'(x)^*$ be the adjoint of $\rho'(x)$. It is easy to see that $\rho'(x)^*$ extends to an operator from B into B^*. $\theta'(x)^*$ is defined as $\theta'(x)^* = I + \rho'(x)^*$.

§ 3. Uhlenbeck-Ornstein process on a Riemann-Wiener manifold

In this section we assume that there exists a measurable norm $|||\cdot|||$ on H stronger than the B-norm such that $|||\cdot|||^2$ is twice continuously H-differentiable on the completion of H with respect to $|||\cdot|||$ and has bounded second H-derivative in the Banach space of trace class operator of H.

This assumption is weaker than the smooth assumption on B-norm in [6, p. 439]. But it is strong enough to assure that the results in [6] remain valid. We note that the assumption on the existence of Q_n [6, p. 439] is unnecessary by the remark in [9, p. 217].

In [6, p. 445], we define diffusion coefficients in $\mathscr{W}$ to be a pair $(A_\varphi, \sigma_\varphi)$ for each chart (U, φ) satisfying the transformation rules. A_φ is a map from $\varphi(U)$ into $\mathscr{L}(B; B)$ such that $A_\varphi(x) - J_\varphi \in \mathscr{L}(B; B^*)$ for some $J_\varphi \in \mathscr{L}(B, B)$ with $J_\varphi(H) \subset H$ and σ_φ is a map from $\varphi(U)$ into B. The transformation rules state that if (U, φ) and (V, ψ) are two charts with $U \cap V \neq \varnothing$ and $\bar{x} = \theta(x) = \psi \circ \varphi^{-1}(x)$ then

$$A_\psi(\bar{x})A_\psi(\bar{x})^* = \theta'(x)A_\varphi(x)A_\varphi(x)^*\theta'(x)^*$$
$$\sigma_\psi(\bar{x}) = \theta'(x)(\sigma_\varphi(x)) + \operatorname{sp}\theta''(x) \circ [A_\varphi(x) \times A_\varphi(x)] .$$

Here we have dropped the constant 1/2 in the original transformation rules [6, p. 445]. For locally Lipschitzian diffusion coefficients (A, σ) we can solve locally the following stochastic differential equation

$$dX(t) = \sqrt{2}\,A(X(t))dW(t) + \sigma(X(t))dt ,$$

where $W(t)$ is a Wiener process on B. Then we use the method of Ito [4] and McKean [13, p. 90] to construct a diffusion process on $\mathscr{W}$ with infinitesmal generator $\mathscr{G}$ [6, Theorem III. 2] given in local coordinate by

$$\mathscr{G}f(x) = \operatorname{trace} A^*(x)f''(x)A(x) + \langle \sigma(x), f'(x) \rangle .$$

A Brownian motion $\mathscr{B}_t$ on $\mathscr{W}$ has been constructed in [6] as follows. Recall that $\mathscr{W}$ has a Riemannian metric G. Let (U, φ) be a chart on $\mathscr{W}$. Define $\Gamma_\varphi : \varphi(U) \to \mathscr{L}^2(B; B^*)$ by

$$\Gamma_\varphi(x)(u, v) = \frac{1}{2}\bar{G}_\varphi(x)^{-1}\{\bar{G}'_\varphi(x)(u, v) + \bar{G}'_\varphi(x)(v, \cdot)u - \bar{G}'_\varphi(x)(\cdot, u)v\} ,$$

where $x \in \varphi(U)$ and $u, v \in B$. Then $A_\varphi(x) = (2\bar{G}_\varphi(x))^{-1/2}$ and $\sigma_\varphi(x) = -\frac{1}{2}\operatorname{sp}\Gamma_\varphi(x) \circ [\bar{G}_\varphi(x)^{-1/2} \times \bar{G}_\varphi(x)^{-1/2}]$ are locally Lipschitzian diffusion coefficients (see [6, Proposition III. 1]). They give the diffusion process $\mathscr{B}_t$ defined up to explosion time ξ.

Conjecture. For each $t > 0$ and $p \in \mathscr{W}$, the Borel measure $q_t(p, dy) = \operatorname{Prob}\{\mathscr{B}_t \in dy, t < \xi \mid \mathscr{B}_0 = p\}$ is an H-differentiable measure and its restriction to every chart has locally Lipschitzian logarithmic derivative.

Theorem 2. *Let μ be an H-differentiable measure on $\mathscr{W}$ such that* *its restriction to every chart has logarithmic derivative. Then* $A_\varphi(x) = \bar{G}_\varphi(x)^{-1/2}$ *and* $\sigma_\varphi(x) = \operatorname{sp}(\bar{G}_\varphi^{-1})'(x) + \bar{G}_\varphi(x)^{-1}((dD\mu_\varphi/d\mu_\varphi)(x))$ *are diffusion coefficients.*

Proof. It has already been proved in [6, p. 446] that $A_\varphi(x) = \bar{G}_\varphi(x)^{-1/2}$ satisfies the first transformation rule. Let (U, φ) and (V, ψ) be two charts on $\mathscr{W}$ with $U \cap V \neq \varnothing$ and $\bar{x} = \theta(x) = \psi \circ \varphi^{-1}(x)$. Then

$$\bar{G}_\psi(\bar{x})^{-1} = \theta'(x)\bar{G}_\varphi(x)^{-1}\theta'(x)^* \ .$$

Upon differentiating both sides with respect to x, we get that for u and v in B,

$$(\bar{G}_\psi^{-1})'(\bar{x})(u, v) = \theta'(x)(\bar{G}_\varphi^{-1})'(x)(\theta'(x)^{-1}u, \theta'(x)^*v)$$
$$+ \theta''(x)(\theta'(x)^{-1}u, \bar{G}_\varphi(x)^{-1}\theta'(x)^*v) + \theta'(x)\bar{G}_\varphi(x)^{-1}\theta''(x)(\theta'(x)^{-1}u, \cdot)v \ .$$

Let $\{e_n\}$ be an orthonormal basis of H. Then

$$\sum_n (\bar{G}_\varphi^{-1})'(x)(\theta'(x)^{-1}e_n, \theta'(x)^*e_n)$$
$$= \sum_n \sum_k \sum_j \langle\theta'(x)^{-1}e_n, e_k\rangle\langle\theta'(x)^*e_n, e_j\rangle(\bar{G}_\varphi^{-1})'(x)(e_k, e_j)$$
$$= \sum_k (\bar{G}_\varphi^{-1})'(x)(e_k, e_k)$$
$$= \operatorname{sp}(\bar{G}_\varphi^{-1})'(x) \ .$$

Similar argument shows that

$$\sum_k \theta''(x)(\theta'(x)^{-1}e_n, \bar{G}_\varphi(x)^{-1}\theta'(x)^*e_n)$$
$$= \sum_k \theta''(x)(e_k, \bar{G}_\varphi(x)^{-1}e_k)$$
$$= \sum_k \theta''(x)(\bar{G}_\varphi(x)^{-1/2}e_k, \bar{G}_\varphi(x)^{-1/2}e_k)$$
$$= \operatorname{sp}\theta''(x) \circ [\bar{G}_\varphi(x)^{-1/2} \times \bar{G}_\varphi(x)^{-1/2}] \ .$$

Therefore,

$$\operatorname{sp}(\bar{G}_\psi^{-1})'(\bar{x}) = \theta'(x)(\operatorname{sp}(\bar{G}_\varphi^{-1})'(x)) + \operatorname{sp}\theta''(x) \circ [\bar{G}_\varphi(x)^{-1/2} \times \bar{G}_\varphi(x)^{-1/2}]$$
$$+ \theta'(x)\bar{G}_\varphi(x)^{-1}\left(\sum_n \theta''(x)(\theta'(x)^{-1}e_n, \cdot)e_n\right) .$$

On the other hand, by Theorem 1, we have

$$\bar{G}_\psi(\bar{x})^{-1}\left(\frac{dD\mu_\psi}{d\mu_\psi}(\bar{x})\right) = \theta'(x)\bar{G}_\varphi(x)^{-1}\left(\frac{dDu_\varphi}{d\mu_\varphi}(x)\right)$$
$$- \theta'(x)\bar{G}_\varphi(x)^{-1}\left(\sum_n \theta''(x)(\theta'(x)^{-1}e_n, \cdot)e_n\right) .$$

From the last two equalities, we obtain that

$$\sigma_\psi(\bar{x}) = \theta'(x)(\sigma_\varphi(x)) + \operatorname{sp}\theta''(x) \circ [\bar{G}_\varphi(x)^{-1/2} \times \bar{G}_\varphi(x)^{-1/2}] .$$

This is the second transformation rule.

It follows from Theorem 2 that there is a diffusion process χ_μ on $\mathscr{W}$ corresponding to each H-differentiable measure μ on $\mathscr{W}$ which has locally Lipschitzian logarithmic derivative on each chart. The infinitesmal generator $\mathscr{G}_\mu$ of x_μ, in local coordinate, is given by

$$\mathscr{G}_\mu f(x) = \operatorname{trace} \bar{G}(x)^{-1} f''(x) + \left\langle \operatorname{sp}(\bar{G}^{-1})'(x) + \bar{G}(x)^{-1}\left(\frac{dD\mu}{d\mu}(x)\right), f'(x)\right\rangle .$$

The most natural and useful H-differentiable measures are the transition probabilities $q_t(p, \cdot)$ of the Brownian motion $\mathscr{B}_t$ (assuming the conjecture). In particular, take $\mu = q_1(x_0, \cdot)$ with a chosen fixed point x_0 in $\mathscr{W}$. We will call the corresponding diffusion process $\mathscr{U}_t$ the standard Uhlenbeck-Ornstein process on the point Riemann-Wiener manifold $(\mathscr{W}, x_0)$. The justification of this terminology is that when $\mathscr{W} = B$, $x_0 = 0$ and $\bar{G} \equiv I$ then then infinitesmal generator of $\mathscr{U}_t$ is $\mathscr{G}f(x) = \operatorname{trace} f''(x) - \langle x, f'(x)\rangle$.

§ 4. Comments

We point out that the construction of the generator $\mathscr{G}_\mu$ for each H-differentiable measure μ with logarithmic derivative is motivated by an integral formula in [12]. Let $N_\mu = -\mathscr{G}_\mu$. It is proved in [12] that

$$\int_{\mathscr{W}} \langle \bar{G}(x)^{-1} f'(x), g'(x)\rangle d\mu = \int_{\mathscr{W}} f(x) N_\mu g(x) d\mu$$

holds for a large core of functions f and g. N_μ has self-adjoint extension acting on the Hilbert space $L^2(\mathscr{W}, \mu)$. The expression for N_μ suggests to us to define $\delta\alpha$ for a differential form α of order 1 by

$$\delta\alpha(x) = -\operatorname{trace} \bar{G}(x)^{-1}\alpha'(x) - \left\langle \operatorname{sp}(\bar{G}^{-1})'(x) + \bar{G}(x)^{-1}\left(\frac{dD\mu}{d\mu}(x)\right), \alpha(x)\right\rangle ,$$

Then we have $N_\mu = \delta d$ acting on the 0-forms (i.e. functions defined on $\mathscr{W}$). We anticipate to define $\delta\alpha$ for a differential form α of higher order so that $N_\mu = \delta d + d\delta$ will serve as the Laplace-Beltrami operator in the Hodge theory. It seems to us that there is a theory for each H-differentiable measure μ on $\mathscr{W}$ with logarithmic derivative. However, $\mu = q_1(x_0, \cdot)$ seems to be the best measure since it carries analytic properties. But until we can prove the CONJECTURE in § 3, we can not do very much along this line. We remark that to our best knowledge this conjecture has never been considered even when the manifold $\mathscr{W}$ is finite dimensional.

Finally, we point out that if $\mathscr{W}$ is a finite dimensional Riemannian manifold and μ is the volume element $\{\det \bar{G}(x)\}^{1/2}dx$ then the operator N_μ is $-g^{ij}\nabla_i\nabla_j$. Here $g^{ij} = (\bar{G}^{-1})_{ij}$ and ∇_i is the convariant derivative. The diffusion process with infinitesmal generator $-\frac{1}{2}N_\mu = \frac{1}{2}g^{ij}\nabla_i\nabla_j$ has been constructed by Gangolli [1]. It is Brownian motion on $\mathscr{W}$.

References

[1] Gangolli, R., On the construction of certain diffusions on a differentiable manifold, Z. Wahrscheinlichkeitstheorie und Verw. Gebiete, **2** (1964), 406–419.

[2] Gross, L., Abstract Wiener spaces, Proc. 5th Berkeley Sympos. Math. Statist. and Probability, **II**, part 1 (1965), 31–42.

[3] ——, Logarithmic Sobolev inequalities, Amer. J. Math., **97** (1975), 1061–1083.

[4] Ito, K., Stochastic differential equations in a differentiable manifold, Nagoya Math. J., **1** (1950), 35–47.

[5] Kuo, H.-H., Integration theory on infinite dimensional manifolds, Trans. Amer. Math. Soc., **159** (1971), 57–78.

[6] ——, Diffusion and Brownian motion on infinite dimensional manifolds, Trans. Amer. Math. Soc., **169** (1972), 439–459.

[7] ——, Differentiable measures, Chinese J. Math., **2** (1974), 189–199.

[8] ——, Potential theory associated with Uhlenbeck-Ornstein process, J. Functional Analysis, **21** (1976), 63–75.

[9] ——, Gaussian measures in Banach spaces, Lecture Notes in Math., **463**, Springer-Verlag, 1975.

[10] ——, Distribution theory on Banach space, Proc. First International Conference on Probability in Banach Space, Obserwolfach, Lecture Notes in Math., **526**, Springer-Verlag (1976), 143–156.

[11] ——, The chain rule for differentiable measures, to appear.

[12] ——, An integral formula for Riemann-Wiener manifolds, to appear.

]13[McKean, H. P., Stochastic integrals, Academic Press, New York and London, 1969.

[14] Piech, M. Ann, Parabolic equations associated with the number operator, Trans. Amer. Math. Soc., **194** (1974), 213–222.

[15] ——, The Ornstein-Uhlenbeck semi-group in an infinite dimensional L^2 setting, J. Functional Analysis, **18** (1975), 271–285.

[16] ——, Locality of the number of particles operator, Pacific J. Math., to appear.

Department of Mathematics
State University of New York at Buffalo
Amherst, New York 14226
U.S.A.

Proc. of Intern. Symp. SDE
Kyoto 1976, pp. 195–263

Stochastic Calculus of Variation and Hypoelliptic Operators

Paul MALLIAVIN

Contents

Introduction

Given a sample path b_ω of a Brownian motion on $\boldsymbol{R}^n$, Ito's stochastic differential system

$$dX^\alpha_\omega = A^\alpha_k db^k_w + \frac{1}{2}\Big(\sum_l (\partial_\beta A^\alpha_l) A^\beta_l\Big) dt$$

defines a path $x_\omega(t)$ of the diffusion associated with the hypoelliptic operator

$$L = \frac{1}{2}\sum_{k=1}^n \mathscr{L}^2_{A_k}$$

where $\mathscr{L}_{A_k}$ denotes the Lie derivative along the vector field A^α_k, $1 \leq \alpha \leq m$. Therefore Ito's method defines a natural measure preserving map Φ_t from the probability space $\Omega_t(\boldsymbol{R}^n)$ to the probability space $\Omega_t(L)$ associated to L.

Now consider a map of $\Omega_t(L)$ to a finite dimension space, for instance, consider the *evaluation* e_t, that is, the map which associates to a path defined on $[0, t]$, its extremity, and then consider

$$g_t = e_t \circ \Phi_t .$$

Then the *fundamental solution* p_t of the heat equation for the operator L will be given as the *image of the Wiener measure* w_t *by the map* g_t.

As w_t can be considered as a "smooth" measure, its image would be smooth if the "differential" of the map g_t is a.s. surjective.

In fact g_t is not *differentiable* in the Banach space differential calculus. *Stochastic calculus of variation* means computation of stochastic differential in stochastic functional calculus. This will be done by a principle of transfer which will transfer the classical calculus of variation.

The Theorem concerning hypoellipticity appears at the beginning of Part III.

A preliminary version of this work was presented in Séminaire Jean Leray, January 1975, a resumé appeared in [25]. (See also [31], [32])

Part I. A criterion for the absolute continuity of the image of a Gaussian measure

§ 1. A lemma of harmonic analysis

Let γ be a finite measure carried by R^m. Suppose that there exists a constant c such that for every φ with compact support, we have

$$\left| \int \frac{\partial \varphi}{\partial \xi_k} d\gamma \right| \leq c \, \|\varphi\|_{L^\infty} , \qquad 1 \leq k \leq m . \tag{1.1}$$

Then

$$\gamma = k(\xi) d\xi \qquad \text{where } k \in L^1 . \tag{1.2}$$

Proof. Remark first that using Poisson's summation formula, it is sufficient to prove the lemma when the measure γ sits on a torus. We remark that the linear forms on C^∞-functions defined by

$$\varphi \to \int \frac{\partial \varphi}{\partial \xi_k} d\gamma$$

extend to those on continuous functions. Therefore there exist finite measures $\sigma_1, \cdots, \sigma_m$ such that

$$\int \frac{\partial \varphi}{\partial \xi_k} d\gamma = - \int \varphi d\sigma_k$$

which means that $d\gamma$ has for derivatives, in distribution sense, measures.

We shall use A. Calderon's technique [3] introducing the Riesz transforms which are the convolution singular integral operators defined on L^2 by

$$\mathscr{R}_k u = \overline{\mathscr{F}} \frac{\hat{\xi}_k}{|\hat{\xi}|} \mathscr{F} u \qquad (\mathscr{F} = \text{Fourier transform})$$

and the operator

$$\Lambda_\alpha = \overline{\mathscr{F}} \, |\hat{\xi}|^\alpha \mathscr{F} \, .$$

(For $\alpha < 0$, Λ_α is defined on the space of functions with mean value zero.) Then

$$\Lambda_\varepsilon \mathscr{R}_k \Lambda_{1-\varepsilon} \gamma = \sigma_k \qquad 0 < \varepsilon < 1 \, . \tag{1.3}$$

Now remark that

$$\sum \mathscr{R}_k^2 = \text{Identity},$$
$$\mathscr{R}_k \Lambda_{-\varepsilon} \text{ is a convolution kernel in } L^1 \, .$$

Then we get from (1.3)

$$\Lambda_{1-\varepsilon} \gamma = f \qquad \text{where } f \in L^1 \, . \tag{1.4}$$

Using still the fact that $\Lambda_{\varepsilon-1}$ is a convolution kernel in L^1, we get that $\gamma \in L^1$.

Remark. It is possible to get the sharper result $k \in L^{n/(n-1)}$ via the inequality of Gagliardo-Nirenberg [27],

$$\|u\|_{L^{n/(n-1)}} \leq \Pi \left\| \frac{\partial u}{\partial \xi_k} \right\|_{L^1}^{1/n} . \tag{1.5}$$

We regularize first μ, apply (1.5) and then use the weak compactness of the unit ball of $L^{n/(n-1)}$. See also [1], [29].

§ 2. An elementary study of the finite dimensional case

The result of this paragraph is obviously not sharp. Its purpose is to prepare the infinite dimensional case. We shall denote by X a finite dimensional vector space where is defined a Gaussian measure μ; its Fourier transform defined on the dual $\hat{X}$ of X is $\hat{\mu}(\hat{x}) = \exp(-\frac{1}{2}q(\hat{x}))$ where q is a positive definite quadratic form. Let $g = (g_1 \cdots g_m)$ be a map of X to R^m. We suppose that its differential $g'(x)$ exists μ-a.e.. Let σ^{ij} be the covariance matrix defined by

(2.1) $$\sigma_{i,j}(x)\eta^i\eta^j = q({}^t g'(x)\eta) .$$

We shall denote by ξ^k the first column vector of $(\sigma_{ij})^{-1}$.

If $\varphi \in \mathscr{D}(R^m)$, let us denote $\partial\varphi/\partial\zeta_s = \varphi^s$, $\tilde\varphi = \varphi \circ g$, $\tilde\varphi^s = \varphi^s \circ g$. Then let us compute

(2.2) $$I = \int_X (\xi^k g'_k \mid \varphi^s g'_s)_q \exp\left(-\frac{\|x\|^2}{2}\right) dx .$$

In this expression g'_k, g'_s are considered as elements of $\hat X$, ξ^k, $\tilde\varphi^s$ as scalars, the scalar product is associated to q, and X is equipped with the adjoint Euclidean metric of q.

Then using (2.1) we get

(2.3) $$I = \int_{R^m} \varphi^1(\zeta)\gamma(d\zeta) \qquad \text{where } \gamma = g_*\mu .$$

On the other hand, denoting by $\varDelta_q$ the elliptic operator on X which has $\frac{1}{2}q(\xi)$ for symbol and by ∇_q the gradient of a function on X, we have the identity

$$\nabla u \cdot \nabla v = \varDelta(uv) - [u\varDelta v + v\varDelta u] .$$

Consider now the Ornstein-Uhlenbeck operator L on X, defined at x by

$$L_x = \varDelta - \tfrac{1}{2}x\cdot\nabla .$$

Then we have still

(2.4) $$\nabla u \cdot \nabla v = L(uv) - [uL(v) + vL(u)] .$$

Then as

$$I = \int_X \xi^k \nabla g_k \cdot \nabla\tilde\varphi \, \mu(dx) ,$$

we get

$$I = \int_X \xi^k\{L(g_k\tilde\varphi) - g_k L(\tilde\varphi) - \tilde\varphi L(g_k)\} d\mu .$$

Now use the fact that L is selfadjoint with respect to the measure μ, we get

(2.5) $$I = \int_X \{g_k L(\xi^k) - L(g_k\xi^k) - \xi^k L(g_k)\}\tilde\varphi d\mu .$$

We have thus proved

(2.6) **Proposition.** *Suppose that every column vector* ξ^k *of the matrix* $(\sigma_{ij})^{-1}$ *satisfies*

$$\int_X |g_k L(\xi^k) - L(g_k \xi^k) - \xi^k L(g_k)|\, d\mu < +\infty .$$

Then $g_* \mu = k d\zeta$ *where* $k \in L^{n/(n-1)}$.

This result can be localized.

(2.7) **Proposition.** *Suppose that the map* g *is* C^3, *and suppose that* $\sigma_{ij}(x)$ *is* μ *a.e. invertible; then* $g_* \mu$ *is absolutely continuous.*

Proof. We shall compute the integral

$$I_\psi = \int (\xi^k \nabla g_k \cdot \nabla \varphi) \psi d\mu \tag{2.8}$$

where

$$\psi(x) = \psi_1(\|x\|)\psi_2(\det(\sigma_{ij})) , \quad \psi_1(t) \quad \text{and} \quad \psi_2(t) \quad \text{being } C^\infty$$

and $\psi_1(t) = 1$ for $|t| < R$ with compact support, $\psi_2(t) = 1$ if $t > 1/R$, $\psi_2 = 0$ on a neighbourhood of zero. Then it is possible to find ξ^k of class C^3 such that on the support of ψ we have $\xi^k \sigma_{k,j} = \delta_j^1$.

Then the same computation gives

$$I_\psi = \int_X \{g_k L(\psi \xi^k) - L(\psi g_k \xi^k) - \psi \xi^k L(g_k)\} \tilde{\varphi} du . \tag{2.9}$$

Therefore $g_*(\psi \mu)$ is absolutely continuous. As $\|\psi\mu - \mu\|$ can be made arbitrary small, we get the proposition.

§ 3. Translation of the previous paragraph in terms of stochastic differential calculus

We shall denote by x_ω the diffusion associated with the Ornstein-Uhlenbeck operator L; we shall take as starting measure the invariant measure μ, we shall denote then the expectation by E_μ. We shall suppose that the functions defined on X are sufficiently differentiable. Denote

$$\begin{aligned} B(x_0, \omega, t) = \xi^k(x_\omega(0))(g_k(x_\omega(t)) - g_k(x_\omega(0)) \\ - \int_0^t (Lg_k)(x_\omega(\xi)) d\zeta) . \end{aligned} \tag{3.1}$$

All notations of the previous paragraph will be preserved. We shall now prove the following elementary version of Girsanov's formula.

(3.2) **Lemma.**

$$\lim_{t\to 0} E_{x_0}\frac{(1+B)\tilde{\varphi}(x_\omega(t))-\tilde{\varphi}(x_\omega(0))}{t}=(L\tilde{\varphi}+\xi^k\sigma_{k,s}\tilde{\varphi}^s)(x_0)\ .$$

Proof. By Ito's stochastic calculus, denoting by $b_\omega^\cdot$ a Brownian motion on X, we have

$$x_\omega^l(t)-x_\omega^l(0)=b_\omega^l(t)-\frac{1}{2}\int_0^t x_\omega^l(\xi)d\xi\ ,$$

$$B=\xi^k(x_0)\int_0^t\frac{\partial g_k}{\partial x^l}db_\omega^l\ .$$

$$\tilde{\varphi}(x_\omega(t))=\tilde{\varphi}(x_\omega(0))+\int_0^t(L\tilde{\varphi})(x_\omega(\xi))d\xi+\int_0^t\frac{\partial g_s}{\partial x^m}\frac{\partial\varphi}{\partial\zeta_s}db_\omega^m\ .$$

We get

$$\begin{aligned}&E_{x_0}((1+B)\tilde{\varphi}(x_\omega(t))-\tilde{\varphi}(x_\omega(0)))\\ &=E_{x_0}\Big(\int_0^t L\tilde{\varphi}d\xi\Big)+\xi^k(x_0)E_{x_0}\Big(\int_0^t\sigma_{k,s}\tilde{\varphi}^s d\xi\Big)\\ &\quad+\xi^k(x_0)E_{x_0}\Big\{\Big(\int_0^t L\tilde{\varphi}\Big)\Big(\int_0^t\frac{\partial g_k}{\partial x^l}db_\omega^l\Big)\Big\}\ .\end{aligned}\tag{3.3}$$

As the last term is obviously $O(t)$, this proves (3.2).

We shall use, as in (2.7), a cut off function ψ.

(3.4) **Lemma.**

$$\int\psi(x_0)\mu(dx_0)E_{x_0}((1+B)\tilde{\varphi}(x_\omega(t)))=\int\tilde{\varphi}(y_0)\mu(dy_0)E_{y_0}((1+\tilde{B})\psi(y_\omega(t)))$$

where

$$\tilde{B}(y_0,\omega,t)=\xi^k(y_\omega(t))\Big\{g_k(y_\omega(0))-g_k(y_\omega(t))-\int_0^t(Lg_k)(y_\omega(\xi))d\xi\Big\}$$

and y_ω is a sample path of the Ornstein-Uhlenbeck process.

Proof. We shall *reverse the time* for the process L, starting with μ as initial measure. Then we apply Kolmogoroff's result [16] which gives that the reversed process has the *same probability transition* as the given process; the Kolmogoroff condition on a Riemannian manifold V for the infinitesimal generator $\Delta_V + \mathscr{L}_A$ with the starting density $\pi(V)dv$ is

$$A = \nabla(\log \pi)$$

which, for the Ornstein-Uhlenbeck process, reads as the identity $2x = \nabla \|x\|^2$.

We shall write the first member of (3.4) as

$$J = E_\mu((1 + B)\tilde{\varphi}(x_\omega(t))\psi(x_\omega(0))) .$$

Fixing t we define $\tilde{y}_\omega(\xi) = x_\omega(t - \xi)$, and get

$$J = \int \tilde{\varphi}(\tilde{y}_0)\mu(d\tilde{y}_0)E_{y_0}((1 + \tilde{B})\psi(\tilde{y}_\omega(t))) .$$

Now use the fact that the law of $(\tilde{y}_\omega(0), \tilde{y}_\omega(\tau_1), \tilde{y}_\omega(\tau_n), \tilde{y}_\omega(t))$, where $0 < \tau_1 \cdots < \tau_n < t$, is the same by Kolmogoroff's theorem as the law of $(y_{\omega'}(0), y_{\omega'}(\tau_1) \cdots y_{\omega'}(\tau_n), y_{\omega'}(t))$ where $y_{\omega'}$ is a sample of the Ornstein-Uhlenbeck process, and this proves the lemma.

(3.5) **Lemma.**

$$\lim_{t\to 0} E_{y_0} \frac{(1 + \tilde{B})\psi(y_\omega(t)) - \psi(y_0)}{t}$$

$$= \{L\psi - \xi^k \nabla g_k \cdot \nabla \psi - \psi(2\xi^k L(g_k) + \nabla \xi^k \cdot \nabla g_k)\}(y_0) .$$

Proof. It is very close to the proof of Lemma (3.2). We have

$$\xi^k(y_\omega(t)) = \xi^k(y_0) + \int_0^t L\xi^k d\xi + \int_0^t \frac{\partial \xi^k}{\partial x^l} db^l_\omega .$$

The same formula holds for $\psi(y_\omega(t))$. Finally

$$g_k(y_\omega(0)) - g_k(y_\omega(t)) = -\int_0^t Lg_k d\xi - \int_0^t \frac{\partial g_k}{\partial x^l} db^l_\omega$$

which gives the result.

Introduce now the following assumption A:

(A) Let $u(x_0, t)$ and $v(y_0, t)$ be the first members of (3.2) and (3.5) respectively. Suppose that $u(\cdot, t)$ and $v(\cdot, t)$ *converge in* $L^1(\mu)$ for the topology of the norm when $t \to 0$.

Proposition. *Under assumption* (A) *we have*

(3.6)
$$\int (L\tilde{\varphi} + \xi^k \sigma_{k,s} \tilde{\varphi}^s)\psi d\mu = \int [L\psi - \xi^k \nabla g_k \cdot \nabla \psi - \psi(2\xi^k L(g_k) + \nabla \xi^k \cdot \nabla g_k)]\tilde{\varphi} d\mu .$$

Proof. Using (3.4) we have, for fixed t,

$$E_{\mu(dx_0)}(((1 + B)\tilde{\varphi}(x_w(t)) - \tilde{\varphi}(x_0))\psi(x_0)) = E_{\mu(dy_0)}(((1 + \hat{B})\psi(y_\omega(t)) - \psi(y_0))\varphi(y_0)) .$$

Now the convergence in $L^1(\mu)$ makes possible from (3.2) and (3.5) to deduce the limit of the two integrations in $d\mu$.

Remark. As it could be expected, the formula (3.6) leads to the same result as (2.9). In fact by the selfadjointness of L

$$\int (L\tilde{\varphi})\psi = \int \tilde{\varphi} L\psi$$

and therefore (3.6) can be read

(3.7)
$$\int \tilde{\varphi}^1 d\gamma = -\int (2\xi^k \psi L(g_k) + \nabla \xi^k \psi \cdot \nabla g_k)\tilde{\varphi} d\mu .$$

Now develop the second term of the second member of (2.9), we get

$$g_k L(\psi \xi^k) - \{g_k L(\xi^k \psi) + \nabla g_k \cdot \nabla \xi^k \psi + \xi^k \psi L(g_k)\} - \xi^k \psi L(g_k)$$

which proves the identity.

§ 4. Construction of an O. U. process on an infinite dimensional space

(4.1) **Reduction.** We shall fix our attention, from now until the end of this chapter, on the Gaussian measure μ^n defined by the Wiener measure on the set trajectories of the Brownian motion on $\boldsymbol{R}^n$, starting from zero and defined on the interval of time [0, 1]. We shall use integration procedures ; therefore the Banach spaces in which this measure can be realized as a Radon measure, using the L. Gross's Theory of measurable norms [8], will not be used here. Also from this point of view of pure integration all Gaussian measures defined by their characteristic functions $\exp(-\frac{1}{2}\|\hat{x}\|^2)$ where $\hat{x}$ runs in a *separable* Hilbert space are isomorphic (cf. I. Segal [28]). Therefore we do not diminish very much the generality by restricting ourselves to μ^n. Finally μ^n being a product of n copies of the Wiener measure on $\boldsymbol{R}$, it is sufficient to construct the O. U. process for this last case, and then to deduce the general case by product of independent copies.

(4.2) **Probability space of the O.U. process**

We shall consider the white noise on the infinite strip

$$S = \{(\tau, t)\,;\, 0 < \tau < 1,\ t \in \boldsymbol{R}\}\ .$$

We shall denote by w a generic element of this white noise, by $W(S)$ its probability space. By definition if $f \in L^2(S)$ (for the $dtd\tau$ measure) then $\langle w, f\rangle$ is a Gaussian variable of mean value zero and of variance $\|f\|^2_{L^2(S)}$. The result is that if $(f, g) = 0$, then $\langle f, w\rangle$ and $\langle g, w\rangle$ are independent. In particular this is the case if f and g have disjoint supports.

We shall denote

$$S_{t_0} = \{(\tau, t) \in S\,;\, t \leq t_0\}$$

and

$$e_{t_0} \quad \text{the indicator of } S_{t_0}\ , \qquad \eta_{\tau_0} \quad \text{the indicator of } \{\tau < \tau_0\}\ .$$

We define on $W(S)$ an increasing family of σ-field $\mathscr{N}^{t_0}(-\infty < t_0 < +\infty)$ by

(4.2.1) $$\mathscr{N}^{t_0} = \sigma\text{-field generated by } \langle f, w\rangle \text{ with } f \in L^2(S)\ ,\quad f = 0, \text{ a.e. on the complement of } S_{t_0}\ .$$

We shall define

(4.2.2) $$u(w, t_0, \tau_0) = \iint e^{1/2(t-t_0)} e_{t_0}\eta_{\tau_0} dw\ ,$$

this integral is well defined as $e_{t_0}e^{t-t_0} \in L^2$. Furthermore

(4.2.3) $$E(u^2(w, t_0, \tau_0)) = \int_0^{\tau} d\tau \int_{-\infty}^{t_0} e^{(t-t_0)} dt = \tau$$

for $0 < \tau_0 < \tau_1$.

Finally

(4.2.4) $$e_{t_0}[\eta_{\tau_1} - \eta_{\tau_0}] \text{ is orthogonal to } e_{t_0}\eta_{\tau_0}\ .$$

Therefore the process in τ defined, for fixed t_0, by

(4.2.5) $$\tau \to u(w, t_0, \tau)$$

is a Brownian path starting from zero and defined on the interval of time $[0, 1]$. We shall denote by (X, μ) the *measure space* constructed by the probability space of the Brownian motion on $\boldsymbol{R}$, restricted on the time interval [0, 1]. Then the map (4.2.5) defines a process $x_w(t)$ which will be called the O. U. process. Its probability space is $W(S)$ equipped with the increasing family $\mathcal{N}^t$, $-\infty < t < +\infty$. (For another description, cf. III–5.1.)

(4.3) **Properties of the O.U. process**

(4.3.1) *The O. U. process is stationary.*

The time shift $t \to t + a$ is an isomorphism.

(4.3.2) *The O. U. process is Markovian.*

According to the stationarity it is sufficient to prove the Markov property for $\mathcal{N}^0$. Let $t_0 > 0$.

$$x_w(t_0) = \int_{-\infty}^{t_0} \int e^{1/2(t-t_0)} \eta . dw = e^{-t_0/2} \int_{-\infty}^{0} \int e^{t/2} \eta . dw + \int_0^{t_0} \int .$$

Therefore

$$x_w(t_0) \cdot = e^{-t_0/2} x_w(0) \cdot + \int_0^{t_0} \int e^{1/2(t-t_0)} \eta . dw . \tag{4.3.3}$$

The second term is independent of $\mathcal{N}^0$. Therefore, for $t_0, \cdots, t_q > 0$

$$E^{\mathcal{N}^0}[\Phi(x_w(t_0), \cdots, x_w(t_q))] = E_{x_w(0)}[\Phi(x_w(t_0), \cdots, x_w(t_q))] .$$

(4.3.4) We shall use now the vector space structure of X. Considering a dilatation of ratio λ we shall denote by μ_λ the image of μ under this dilatation.

(4.3.5) *The transition probability $p_{t_0}(x_0, d\lambda)$ of the* O. U. *process is given by*

$$\int \varphi(x) p_{t_0}(x_0, dx) = \int \varphi(x_0 e^{-t_0/2} + y) \mu_\lambda(dy) \qquad \text{where } \lambda = (1 - e^{-t_0})^{1/2} .$$

We interpret the second term of (4.3.3) as in (4.2.3) and (4.2.4) and we show that it is a dilated Brownian motion.

(4.3.6) *The measure μ is invariant under the* O. U. *process.*

$$\int \mu(dx_0) \int p_{t_0}(x_0, dx) \varphi(x) = \iint \varphi(x e^{-t_0/2} + y) \mu(dx) \mu_\lambda(dy) .$$

Let us take now two independent Brownian motions b^1 and b^2 and consider the process u defined by

$$u : \tau \to \alpha_1 b^1(\tau) + \alpha_2 b^2(\tau) \qquad \text{where } \alpha_1, \alpha_2 \text{ are two constants, } \alpha_1^2 + \alpha_2^2 = 1 .$$

Then u is a Brownian motion. Apply this to $\alpha_1 = e^{-t_0/2}$, $\alpha_2 = (1 - e^{-t_0})^{1/2}$ and we see that the last integral is equal to $\int \varphi(z)\mu(dz)$.

(4.3.7) *The* O. U. *process is invariant under the reversal of the time.*

We have first to prove the equality between the following two measures defined on $X \times X$

$$\mu(dx)p_{t_0}(x, dy) = \mu(dy)p_{t_0}(y, dx) \qquad (t_0 > 0) .$$

Integrate the first member on a test function

$$\iint \psi(x, y)\mu(dx)p_{t_0}(x, dy) = \iint \psi(x, e^{-t_0/2}x + \lambda z)\mu(dx)\mu(dz) .$$

Therefore the first measure could be considered as the image by the *linear* map:

$$\begin{pmatrix} x \\ z \end{pmatrix} \to \begin{pmatrix} x \\ e^{-t_0/2}x + \lambda z \end{pmatrix}$$

of the Gaussian measure $\mu \otimes \mu$ on X^2, and therefore is a Gaussian measure.

The same is true for the second member.

As Gaussian measures are determined by their characteristic functions, it is sufficient to show that if l_1, l_2 are linear forms on X then

$$\iint (l_1(x) + l_2(e^{-t_0/2}x + \lambda z))^2\mu(dx)\mu(dz)$$
$$= \iint (l_1(e^{-t_0/2}y + \lambda\tilde{z}) + l_2(y))^2\mu(dy)\mu(d\tilde{z}) .$$

The contributions of l_1^2, l_2^2 are the same according to (4.3.6); the cross terms which do not vanish are

$$\int l_1(x)l_2(x)e^{-t_0/2}\mu(dx) \quad \text{and} \quad \int e^{-t_0/2}l_1(y)l_2(y)\mu(dy)$$

and therefore are equal.

We shall prove now that reversed process has the Markov property; the measures $\mu(dx_0)p_{t_0}(x_0, dx_1) \cdots p_{t_n}(x_{n-1}, dx_n)$ on X^n corresponding to a sampling of the O. U. process at $t_0, t_0 + t_1, \cdots, t_0 + t_1 + \cdots + t_n$ have to satisfy that *given* x_1 *the variable* x_0 *is independent of* $x_2 \cdots x_n$. By (4.3.2) *given* $x_1, x_2, \cdots, x_n$ *are independent* of x_0, which proves the assertion by the symmetry of the independence relation.

Therefore the reversed process is Markovian and has the same probability of transition, then is isomorphic to the O. U. process.

§ 5. A criterion of absolute continuity

We shall prolongate the computation made in Section 3 to the infinite dimension.

In Section 3 we have used stochastic differential calculus on a set of scalar functions, whose cardinality is majorized by the dimension of the range, and, with the reversal of the time on the O. U. process, no other ingredients were used. As all these concepts are meaningful on an infinite dimensional O. U. process, our main task would be to define properly regularity properties such that the limits appearing in (3.2) and (3.5) exist and that the assumption (A) will be realized.

The regularity concepts will be based on the use of the Kunita-Watanabe stochastic calculus ([18]) which, being coordinate free, is well adapted to the infinite dimension.

(5.1) Some classes of regular functions

(5.1.1) We shall denote by $\mathscr{C}^0$ the class of functions u defined on X which are a.s. *continuous on the trajectories* and by $\mathscr{C}$ the class of functions *continuous in probability* a.s..

(5.1.2) We shall denote by $\mathscr{C}^1$ the functions $u \in \mathscr{C}^0_\cap L^1(\mu)$ such that there exists

(i) $v \in \mathscr{C}_\cap L^1(\mu)$ such that for $t > 0$

$$M_u(w, t) = u(x_w(t)) - u(x_w(0)) - \int_0^t v(x_w(\xi))d\xi$$

is an L^2-martingale. (i.e. $M(w, t) \in L^2(\mu)$ for every $t > 0$ and $E^{\mathscr{N}_{t'}}(M(w, t)) = M(w, t')$ for $t \geq t'$.)

(ii) There exists a function $k \in L^2(\mu)_\cap \mathscr{C}$, $k \geq 0$ such that

$$M_u^2(w, t) - \int_0^t k^2(x_w(\xi))d\xi \qquad \text{is a martingale.}$$

If $u \in \mathscr{C}^1$, we shall denote

$$v = Lu\ , \qquad k = \|\nabla u\|\ .$$

(5.1.3) *$\mathscr{C}^1$ is a vector space.*

Let $u_1, u_2 \in \mathscr{C}^1$ then $[u_1 + u_2] - \int (Lu_1 + Lu_2)$ is obviously an L^2-martingale $M_{u_1+u_2}$.

Then by Meyer's decomposition theorem there exists an increasing process $A(t, w)$ such that $M^2_{u_1+u_2}(t, w) - A(t, w)$ is a martingale and in the Kunita-Watanabe notations [18]

$$A(t, w) = \int_0^t |\nabla u_1|^2\, d\xi + 2\langle M_{u_1}, M_{u_2}\rangle + \int_0^t |\nabla u_2|^2\, d\xi\ .$$

Then, by the inequality (2.4) of [18], p. 214, we have for any non anticipating Φ,

$$E\left(\left|\int_0^t \Phi^2 d\langle M_{u_1}, M_{u_2}\rangle\right|\right) \leq \left(E\left(\int_0^t \Phi^2\, |\nabla u_1|^2\right)\right)^{1/2}\left(E\int_0^t \Phi^2\, |\nabla u_2|^2\right)^{1/2}.$$

This implies that a.s. $d\langle M_{u_1}, M_{u_2}\rangle$ is a measure absolutely continuous with respect to dt, therefore $A(t, w) = \int_0^t q(w, \xi) d\xi$.

Furthermore $q(w, s)$ is $\mathscr{N}^s$ measurable and, as it can be defined by differentiation, it depends only on the position of the process at time s. Therefore there exists $k \in L^1(\mu)$ such that $k(x_w(\xi)) = q(w, \xi)$.

(5.1.4) **Corollary.** *There exists on $\mathscr{C}^1$ a bilinear map $(u, v) \to \nabla u \cdot \nabla v$ sending $\mathscr{C}^1 \times \mathscr{C}^1 \to L^1(\mu)$ such that $\nabla u \cdot \nabla u = \|\nabla u\|^2$.*

Proof. Define $\nabla u \cdot \nabla v$ by $\int_0^t \nabla u \cdot \nabla v d\xi = \langle M_u, M_v\rangle$.

(5.1.5) *Ito's formula. Let $u_1 \cdots u_n \in \mathscr{C}^1$, consider a function $F: \boldsymbol{R}^l \to \boldsymbol{R}$ of class C^2, bounded and with bounded derivative. Then*

$$F(u_1(x_w(t)), \cdots, u_n(x_w(t))) - F(0)$$
$$= \int_0^t \frac{\partial F}{\partial \zeta_s} dM_{u_s} + \int_0^t \left(\frac{\partial F}{\partial \zeta_s} Lu_s + \frac{1}{2}\, \frac{\partial^2 F}{\partial \zeta_k \partial \zeta_l} \nabla u_k \cdot \nabla u_l\right) d\xi$$

where the first integrals are stochastic integrals in the sense of [18].

Furthermore $F(u_1, \cdots, u_n) \in \mathscr{C}^1$ and LF, $\|\nabla F\|$ can be read from the above formula.

Proof. See [18], p. 216 for the formula. To check $F \in \mathscr{C}^1$, it is easy to see, using Ito's formula, that LF and $\|\nabla F\|$ proposed by the above formula are well in fact in $L^1(\mu)$ and $L^2(\mu)$, both being in $\mathscr{C}$.

(5.2) **Integration and regularity**

Lemma. *Let $v \in L^1(\mu) \cap \mathscr{C}$, then*

(5.2.1) $$E\left|E_{x_0}\left(t^{-1}\int_0^t v(x_w(\xi))d\xi\right) - v(x_0)\right| \to 0 \qquad \textit{when } t \to 0 .$$

Proof. We majorize this expression by

$$J(t) = E\left(\int_0^1 |v(x_w(t\xi)) - v(x_w(0))|\, d\xi\right) .$$

By the invariance of the measure μ for $\varepsilon > 0$ there exists δ such that

$$E(I_K\, |v(x_w(t'))|) < \varepsilon$$

whatsoever t' and K satisfying $P(K) < \delta$. Furthermore by the continuity of v there exists t_0 such that for $t < t_0$

$$P(|v(x_w(t)) - v(x_w(0))| > \varepsilon) < \delta$$

for which we conclude that $J(t) < 3\varepsilon$ for $t < t_0$.

Lemma. *Let $u, v \in \mathscr{C}^1 \cap L^\infty(\mu)$.* Then

(5.2.2) $$\int (Lu)v\, d\mu = \int u(Lv)d\mu .$$

Proof. By (5.2.1) the first member is equal to

$$\lim_{t_0 \to 0} t_0^{-1}\{E(u(x_w(t_0))v(x_w(0))) - E(u(x_w(0))v(x_w(0)))\} .$$

Denote by y_w the reversed process $t \to t_0 - t$. Then the first term can be written

$$t_0^{-1}E(u(y_w(0))v(y_w(t_0))) .$$

Now use the isonomy of x_w and y_w and make the same computation in the reverse order, we get the second member.

(5.3) **Construction of a cut off function**

We shall resume the notations of Section 3:

g is a map of X on $\boldsymbol{R}^m$ and we suppose

(5.3.1) $$g_k \in \mathscr{C}^1 \qquad \text{for } 1 \leq k \leq m \, .$$

We introduce the covariance matrix

(5.3.2) $$\sigma_{ij}(x) = (\nabla g_i \cdot \nabla g_j)(x)$$

where the scalar product has been defined in (5.1.3).

We shall suppose that $\sigma_{ij}(x)$ satisfies the *regularity assumptions*

(5.3.3) $$\sigma_{i,j} \in \mathscr{C}^1$$

(5.3.4) $$\nabla g_k \cdot \nabla \sigma_{i,j}, \ \nabla \sigma_{i,j} \cdot \nabla \sigma_{k,l} \in \mathscr{C} \, .$$

On the other hand we shall suppose that σ_{ij} satisfies the non degeneracy condition:

(5.3.5) $$\sigma_{ij}(x) \text{ is invertible } \mu \text{ a.e. } .$$

Denote by $\mathscr{M}_n$ the set of all $m \times m$ matrices.

Lemma. *Under the assumptions* (5.3.1), (5.3.3), (5.3.4), (5.3.5), *it is possible, for every* $\varepsilon > 0$ *to construct* $\psi_1 \in \mathscr{D}(\mathscr{M}_n)$ *such that, denoting* $\psi = \psi_1 \circ \sigma$, *we have*

(5.3.6) $$\|\psi\mu - \mu\| < \varepsilon \qquad 0 \leq \psi \leq 1 \, ,$$

(5.3.7) $$\psi \in \mathscr{C}^1 \, .$$

(5.3.8) *There exists* $\xi^k \in L^\infty \cap \mathscr{C}^1$ *such that* $\xi^k \sigma_{k,i} = \delta_i^1$ if σ is in the support of ψ_1.

(5.3.9) $$\nabla g_k \cdot \nabla \psi \quad \textit{and} \quad \nabla \xi^k \cdot \nabla g_k \in \mathscr{C} \, .$$

Proof. As all the mass of $(\sigma)_* \mu$ is carried by $GL(m)$, then it exists $\psi_1 \in \mathscr{D}(GL(m))$ such that (5.3.6) is satisfied. Furthermore there exists a map $F : GL(m) \to \mathscr{M}_n$, in C^2 with bounded derivatives, such that, on the support of ψ_1, we have $F(a) = a^{-1}$.

Define ξ = first column of $F(\sigma)$. Then by Ito's formula $\psi_1(\sigma)$ and $F(\sigma)$ belongs to $\mathscr{C}^1$. Furthermore $L\psi$ can be expressed in terms of $L\sigma_{ij}$ and $\nabla \sigma_{ij} \cdot \nabla \sigma_{k,l}$ which are $\mathscr{C}$ by (5.3.4). This proves (5.3.7). We have also (5.3.9) by the identity

$$\nabla g_k \cdot \nabla \psi = \frac{\partial \psi_1}{\partial \sigma_{ij}} \nabla \sigma_{ij} \cdot \nabla g_k \, , \qquad \nabla \xi^k \cdot \nabla g_k = \frac{\partial F}{\partial \sigma_{ij}} \nabla \sigma_{ij} \cdot \nabla g_k \, .$$

(5.4) **Lemma of derivation**

Lemma. *Suppose*

(5.4.1) $$g \in \mathscr{C}^1$$

and

(5.4.2) $$Lg \in L^2(\mu)\,, \qquad \sigma_{ij} \in L^2(\mu)\,.$$

Denote $B(w, t) = \xi^k(x_w(0)) \int_0^t dM_{g_k}$ *and keep otherwise the notation of Lemma* 3.2. *Then for* $t \to 0$,

(5.4.3) $$\lim E(|t^{-1}E^{\mathscr{N}^0}((1 + B)(\tilde{\varphi}(x_w(t)) - \tilde{\varphi}(x_w(0))) - (L\tilde{\varphi} + \xi^k\sigma_{k,s}\tilde{\varphi}^s)(x_w(0))|) = 0\,.$$

Proof. Apply Ito's formula (5.1.5) to $\tilde{\varphi}$

$$\tilde{\varphi}(x_w(t)) - \tilde{\varphi}(x_w(0)) = \int_0^t \tilde{\varphi}^s dM_{g_s} + \int_0^t L\tilde{\varphi} d\xi\,.$$

Then we have to evaluate

$$E^{\mathscr{N}^0}\left(B \int_0^t \tilde{\varphi}^s dM_{g_s}\right) = \xi^k(x_0) E\left(\int_0^t \tilde{\varphi}^s \sigma_{k,s} d\xi\right),$$

$$E^{\mathscr{N}_0}\left(B \int_0^t L\tilde{\varphi}\, d\xi\right) = tE^{\mathscr{N}_0}\left(B \int_0^1 [(L\tilde{\varphi})(x_w(t\xi)) - L\tilde{\varphi}(x_w(0))] d\xi\right).$$

We majorize the expectation of the modulus of this second expression by the Cauchy-Schwarz inequality and we get

$$(E(B^2))^{1/2}\left(E\left(\left(\int_0^1\right)^2\right)\right)^{1/2} < 2t^{1/2}\, \|\xi^k\|_\infty\, \|\nabla g\|_{L^2}\, \|L\tilde{\varphi}\|_{L^2}\,.$$

(The hypothesis (5.4.2) implies $\|L\tilde{\varphi}\|_{L^2} < \infty$.)

Lemma. *Suppose* (5.3.1), (5.3.3), (5.3.4). *Suppose furthermore*

(5.4.4) $$(Lg.), (L\sigma..) \in L^2(\mu)\,, \qquad \|\nabla g.\|, \|\nabla \sigma..\| \in L^4(\mu)\,.$$

Keep the notation of Lemma 3.4 *and* 3.5 *and denote*

(5.4.5) $$G = L\psi - \xi^k \nabla g_k \cdot \nabla \psi - \psi(2\xi^k L(g_k) + \nabla \xi^k \cdot \nabla g_k)\,.$$

Then, when $t \to 0$, $t > 0$,

(5.4.6) $$\lim E(|t^{-1}E^{\mathscr{N}^0}\{(1 + \tilde{B})\psi(y_w(t)) - \psi(y_w(0))\} - G(y_w(0))|) = 0\,.$$

Proof. We remark first that the hypothesis (5.4.4) is stable by symbolic calculus with C^2 function with bounded derivative. Therefore they propagate to ψ and $\xi^{\cdot}$. We have by Ito's formula

$$-\tilde{B}(y_0, w, t) = \left(\xi_0^k + \int_0^t L\xi^k dt + \int_0^t dM_{\xi^k}\right)\left(2\int_0^t Lg_k dt + \int_0^t dM_{g_k}\right).$$

Also by (5.3.7)

$$\psi(y_w(t)) = \psi_0 + \int_0^t L\psi dt + \int_0^t dM_\psi .$$

We shall split

$$E^{\mathcal{N}_0}((1+\tilde{B})\psi(y_w(t)) - \psi(y_w(0))) = \mathrm{I} + \mathrm{II} + \mathrm{III} + \mathrm{IV} + \mathrm{V},$$

where

$$I = E_{y_0}\int_0^1 (L\psi)(y_w(t\lambda))d\lambda - 2\psi_0\xi_0^k E_{y_0}\int_0^1 (Lg_k)(y_w(t\lambda))d\lambda$$
$$- \psi_0 E_{y_0}\int_0^1 (\nabla\xi^k\cdot\nabla g_k)(y_w(t\lambda))d\lambda - \xi_0^k E_{y_0}\int_0^1 (\nabla\psi\cdot\nabla g_k)(y_w(t\lambda))d\lambda .$$

Concerning I we remark by (5.3.7), (5.3.1), (5.3.9) that the integrals satisfy the hypothesis of (5.2.1), therefore their expectation converges in $L^1(\mu(dy_0))$. As $\psi, \xi^k \in L^\infty(\mu)$, the limit holds still in $L^1(\mu(dy_0))$ and this limit is G. We have therefore to prove the L^1 convergence to zero of the other terms. We shall abbreviate

$$M_{g_k} = \int_0^t dM_{g_k}, \qquad tI_{g_k} = \int_0^t Lg_k dt$$

and so on,

$$\|\mathrm{II}\|_{L^1} \le t^{-1}E(|M_{g_k}M_\psi M_{\xi^k}|).$$

By Cauchy-Schwarz's inequality using (5.3.1), we have to prove

$$t^{-1/2}(E(|M_{g_k}M_\psi|^2))^{1/2} \to 0 .$$

Let η be an auxiliary parameter and consider the martingale $M_{g_k} + \eta M_\psi$, then its increasing process is given by

$$(\|\nabla g_k\|^2 + 2\eta\nabla g_k\cdot\nabla\psi + \eta^2\|\nabla\psi\|^2)^{1/2} \le \|\nabla g_k\| + |\eta|\,\|\nabla\psi\| .$$

Now using [9] Theorem 6, p. 26, we get, for $-1 < \eta < 1$,

$$E((M_{g_k}+\eta M_\psi)^4) \le 96t^2(\|\nabla g_k\|_{L^4}^4 + \|\nabla\psi\|_{L^4}^4) .$$

The second member is finite by (5.4.4). The first member is a polynomial of the fourth degree in η. By equivalence of the norms it is known that there exists a numerical constant c such that for all polynomial Q of degree 4

$$\left|\frac{\partial^2 Q}{\partial\eta^2}(0)\right| \le c \max_{-1<\eta<1} |Q(\eta)| .$$

And we get therefore $E(|M_{g_k}M_\psi|^2) < c't^2$ which controls II.

$$t\,\|\mathrm{III}\|_{L^1} \le E(|I_\psi M_{\xi^k} M_{g^k}|) + E(|I_{g_k} M_{\xi_k} M_\psi|) + E(|I_{\xi_k} M_\psi M_{g_k}|) .$$

Then we treat using Cauchy-Schwarz's inequality which makes appear

$$(E(|I_\psi|^2))^{1/2}(E(|M_{\xi_k}M_{g_k}|^2))^{1/2} < c'tE(|I_\psi|^2)^{1/2}$$

and

$$(5.4.7)\quad E(|I_\psi|^2) = E\left(\left|\int_0^t L_\psi dt\right|^2\right) \le tE\left(\int_0^t |L\psi|^2 dt\right) = t^2\,\|L_\psi\|_{L^2}^2 .$$

In the other terms ξ_0, ψ_0 will appear in factor but as they are in $L^\infty(\mu)$ this would not change the L^1 convergence. Then the following dominations would be done up to the multiplicative constant $(\|\xi^\cdot\|_{L^\infty} + \|\psi\|_{L^\infty} + 1)$. We have

$$t\,\|\mathrm{IV}\|_{L^1} \le E(|I_\psi M_{\xi^\cdot}|) + \cdots$$

will be treated by Cauchy-Schwarz's inequality,

$$E(|M_{\xi^\cdot}|^2) = t\,\|\nabla\xi^\cdot\|_{L^2}^2, \quad \text{and} \quad (5.4.7) .$$

Also (5.4.7) takes care of

$$t\,\|V\|_{L^1} \le E(|I_\psi I_{g.}|) + \cdots$$

and the lemma is proved.

Remark. The approach used here, very elementary and laborious, is very likely to be improved by relying on the theory of martingales and their differentiation [2], [26]. Nevertheless the hypothesis (5.4.4) is not very far from minimal hypothesis necessary to insure $G \in L^1$.

(5.5) **Theorem.** *Let μ the Wiener measure defined by the Brownian motion b on $\boldsymbol{R}^n$, running during the time* $[0,1]$.

Let g be a map defined a.e. on the probability space associated with μ, taking its values in $\boldsymbol{R}^m$.

Regularity hypothesis

(5.5.1) $$g \in \mathscr{C}^1 ,$$

(5.5.2) *the covariance matrix* $\nabla g_i \cdot \nabla g_j = \sigma_{ij} \in \mathscr{C}^1$.

(5.5.3) *All Ito's invariants* $\nabla \sigma_{ij} \cdot \nabla \sigma_{kl}, \nabla \sigma_{ij} \cdot \nabla g_k$ belong to $\mathscr{C}$.

(5.5.4) $$Lg_k, L\sigma_{ij} \in L^2(\mu) ; \|\nabla g_k\|, \|\nabla \sigma_{ij}\| \in L^4(\mu) .$$

Non degeneracy hypothesis

(5.5.5) $\sigma_{i,j}$ *is invertible* μ *a.e..*

Then the image of μ by g is absolutely continuous with respect to the Lebesgue measure of $\boldsymbol{R}^m$.

Proof. We remark first that the Lemma 3.4 based on the time reversal is valid in infinite dimension by (4.3.7).

Therefore (3.4), (5.4.3) and (5.4.6) give the identity, where G is defined in (5.4.5).

$$\int (L\tilde{\varphi})\psi d\mu + \int \tilde{\varphi}^1 \psi d\mu = \int G\tilde{\varphi} d\mu .$$

Now using (5.2.2) we get, denoting $\nu = g_*(\psi d\mu)$

$$\left| \int \frac{\partial \varphi}{\partial \zeta^1} d\nu \right| = \left| \int (G - L\psi)\tilde{\varphi} d\mu \right| \leq c \|\varphi\|_{L^\infty}$$

and now apply Section 1.

(5.5.6) *Remark.* It is of course very easy to describe in differential geometry terms the demarch we have used. We have a vector field on $\boldsymbol{R}^m$, $\partial/\partial\zeta^1$, and a map g which has a *surjective differential*. Then for all $x_0 \in X$ we can find a vector A_{x_0}

(i) $g'(x_0)A_{x_0} = \partial/\partial\zeta^1$.

Of course such choice is not unique. As we have on X a natural quadratic form, we could make the determination of A_{x_0} unique by imposing that its length would be minimum. In fact we will find

(ii) $A_{x_0} = \xi^k(x_0)\ \overrightarrow{(\text{grad } g_k)}(x_0)$
where ξ^k is as before, the first column of the inverse of the covariance matrix.

Now the vector field A generates a 1-parameter group U_t^A and we shall have

(iii) $g_*(U_t^A\mu) = U_t^{\partial/\partial\zeta}(g_*\mu)$.

Therefore the absolute continuity will result from a formula

(iv) $[(\partial/\partial t)(U_t^A\mu)]_{t=0}$ is a measure.

We can interpret (iv) as computing the *divergence* of the vector field A with respect to the base measure μ. The study of U_t^A could be attacked by the Cameron-Martin formula [4].

This approach seems nevertheless present some difficulty: the mapping g is in general not continuous for the Banach space topology and it would be the same for the vector field A. Resources of the ordinary differential calculus does not seem immediately available.

Part II. Stochastic differential geometry

Chapter I. A principle of transfer

We shall build a stochastic calculus of variation, as much as we could, relying on the ordinary calculus of variation. We shall use a transfer principle which allows to carry computation from differential geometry to stochastic geometry.

It is well known that stochastic differential equations could appear as a limit case of ordinary differential equations. We want to develop this point of view in a formalism adapted to differential geometry.

§ 1. Smoothing the Brownian motion

We shall fix a C^∞ function u, with support contained in $[0, 1]$ and with integral 1. Given $\varepsilon > 0$, we introduce the usual approximate unit $u_\varepsilon(\lambda) = \varepsilon^{-1}u(\varepsilon^{-1}\lambda)$.

Given a sample path b_ω^k of a Brownian motion on $\boldsymbol{R}^n$, we shall denote $b_{\omega,\varepsilon}^k$ the C^∞ curve defined by

(1.1) $$b_{\omega,\varepsilon}^k(\tau) = \int_0^\varepsilon b_\omega^k(\tau + \lambda)u_\varepsilon(\lambda)d\lambda\ .$$

We have the obvious property

(1.2) $$\sup_\tau \|b_{\omega,\varepsilon}(\tau) - b_\omega(\tau)\| \to 0 \quad \text{when } \varepsilon \to 0$$

(1.3) $$b_{\omega,\varepsilon}(\tau) \quad \text{is } \mathcal{N}^{\tau+\varepsilon} \text{ measurable}$$

Therefore $b_{\omega,\varepsilon}(\tau-\varepsilon)$ will be an $\mathcal{N}^{\tau}$ adapted process.

§ 2. Stratonovitch's stochastic calculus

Let V be a C^3 manifold. On a local coordinate system, we shall consider the stochastic differential system

(2.1) $$dv^i = a^i_k(v)d_S b^k_\omega + c^i(v)d\tau$$

where the d_S means that we use in the associated integral equation a Stratonovitch's stochastic integral. It is well known that the equation (2.1) is the equivalent in Ito's notation to

(2.2) $$dv^i = a^i_k(v)db^k_\omega + \left(c^i + \frac{1}{2}\sum_k \left(\frac{\partial a^i_k}{\partial v^s}\right)a^s_k\right)d\tau .$$

In fact we could have taken as *definition* of (2.1) the formula (2.2). Conversely if we have Ito's stochastic differential equation

(2.3) $$dv^i = \tilde{a}^i_k db^k_\omega + \tilde{c}^i d\tau$$

we form by method of indeterminate coefficient the (S) equation associated

(2.4) $$dv^i = \tilde{a}^i_k d_S b^k_\omega + \left(\tilde{c}^i - \frac{1}{2}\sum_k \left(\frac{\partial \tilde{a}^i_k}{\partial v^s}\right)\tilde{a}^s_k\right)d\tau .$$

Therefore (2.1) and (2.2), (2.3) and (2.4) provide a dictionary to go from the notation of Ito to the (S) notation.

(2.5) **Theorem.** *Consider the* (S) *equation* (2.1) *where the sample of the Brownian motion has been fixed in* b_{ω_0}. *Let* F *be a* C^2 *diffeomorphism* $F: u \to w$. *Then* $F(u_{\omega_0}(\tau)) = w_{\omega_0}(\tau)$ *satisfy the stochastic system*

$$dw^\alpha = A^\alpha_k d_S b^k_{\omega_0} + C^\alpha d\tau \qquad \text{(S)}$$

with

$$A^\alpha_k = \frac{\partial F^\alpha}{\partial v^s}a^s_k , \qquad C^\alpha = \frac{\partial F^\alpha}{\partial v^s}c^s .$$

Proof. Translate in Ito's formalism, apply the Ito's formula, and retranslate in (S) formalism.

Remark. Under a change of coordinate, the vectors $a_k^{\cdot}$ and $c^{\cdot}$ appearing in the differential (2.1) *transform as first order differential operators*. This fact is very remarkable because we are dealing with second order operators.

(2.6) From a differential geometry point of view $a_k^{\cdot}$, $c^{\cdot}$ can be considered as *tangent vector fields*. (I am indebted to J. Eells who has emphasized to me this important fact).

§ 3. **Limit Theorem.** *Consider the stochastic differential system defined on a bounded open set $O \subset \mathbf{R}^m$, $1 \leq k \leq n$,*

$$(3.1)\quad \begin{aligned} dv_\omega^i(\tau) &= a_k^i(v_\omega(\tau))d_S b_{\omega_1}^k(\tau) + \tilde{a}_k^i(v_\omega(\tau))db_{\omega_2}^k(\tau) + c^i(v_\omega(\tau))d\tau \\ v_\omega^i(0) &= 0 \end{aligned}$$

where $\omega = (\omega_1, \omega_2)$, $b_{\omega_1}^k$ and $b_{\omega_2}^k(t)$ being two independent Brownian motions on $\mathbf{R}^n$. We shall denote by $\mathcal{N}_s^t$ the σ-field generated by $b_{\omega_s}^{\cdot}(\xi)$, $\xi < \tau$. Then $\mathcal{N}^\tau = \mathcal{N}_1^\tau \otimes \mathcal{N}_2^\tau$.

Suppose $\tilde{a}, a \in C^2(\overline{O})$, $c \in C^1(\overline{O})$. Denote by $T(\varepsilon)$ the exit time from O. Given $\varepsilon > 0$, we consider the stochastic differential system

$$(3.2)\quad \begin{aligned} dv_{\varepsilon(\omega)}^i(\tau) &= a_k^i(v_{\varepsilon(\omega)}(\tau))db_{\varepsilon(\omega_1)}(\tau) + \tilde{a}_k^i(v_{\varepsilon(\omega)}(\tau))db_{\omega_2}^k(\tau) \\ &\quad + c^i(v_{\varepsilon(\omega)}(\tau))d\tau\ , \\ v_{\varepsilon(\omega)}(2\varepsilon) &= w\ . \end{aligned}$$

Denote by $T(\varepsilon(\omega))$ the exit time from O for $v_{\varepsilon(\omega)}$. Given a sequence $\hat{\varepsilon}_s$ we can extract a universal subsequence ε_j (see Remark 4.5) *such that* a.s. *in ω when*

$$(3.3)\quad j \to +\infty, \quad w = w_j \quad \text{with} \quad \frac{\log \|w_j\|}{\log j} \to -\infty\ ,$$

then

$$(3.4)\quad \liminf T(\varepsilon_j(\omega)) \geq T(\omega)\,;$$

for every $\tau_0 < T(\omega)$, $v_{\varepsilon_j(\omega)}(\tau)$ is defined on $[0, \tau_0]$ *for j big enough. Furthermore, under the hypothesis* (3.3), *we have*

$$(3.5)\quad \max_{0 \leq \tau \leq \tau_0} \|v_\omega(\tau) - v_{\varepsilon_j(\omega)}(\tau)\| \to 0\ .$$

Proof. We remark that $v_{\varepsilon(\omega)}(\tau)$ is $\mathcal{N}_1^{\tau+\varepsilon} \otimes \mathcal{N}_2^\tau$ measurable. Our first task would be to make appear in (3.2) a stochastic integration. We shall write

$$v^i_{\varepsilon(\omega)}(\tau_1) - w = I^i_1 + I^i_2 + I^i_3 + I^i_4\ , \quad \tau_1 > 2\varepsilon$$

$$I^i_1 = \int_{2\varepsilon}^{\tau_1} a^i_k(v_{\varepsilon(\omega)}(\tau - \varepsilon))db^k_{\varepsilon(\omega_1)}(\tau)$$

$$I^i_2 = \int_{2\varepsilon}^{\tau_1} \{a^i_k(v_{\varepsilon(\omega)}(\tau)) - a^i_k(v_{\varepsilon(\omega)}(\tau - \varepsilon))\}db^k_{\varepsilon(\omega_1)}(\tau)$$

$$I^i_3 = \int_{2\varepsilon}^{\tau_1} c^i(v_{\varepsilon(\omega)}(\tau))d\tau$$

$$I^i_4 = \int_{2\varepsilon}^{\tau_1} \tilde{a}^i_k(v_{\varepsilon(\omega)}(\tau))db^k_{\omega_2}(\tau)\ .$$

We remark, by usual properties of convolution products, that we have

$$b^l_{\varepsilon(\omega_1)}(\tau) - b^l_{\varepsilon(\omega_1)}(\tau - \varepsilon) = \int \psi_\varepsilon(\zeta - \tau)db^l_{\omega_1}(\zeta)$$

(3.6)
$$\text{where } \psi_\varepsilon(\lambda) = \int_{-\infty}^{\lambda} (u_\varepsilon(t) - u_\varepsilon(t - \varepsilon))dt\ ,$$

$$\frac{d}{d\tau}(b^k_{\varepsilon(\omega)}(\tau)) = \int u_\varepsilon(\zeta - \tau)db^k_\omega(\zeta)\ .$$

Then using this last identity I_1 can be written

$$I^i_1 = \int_{2\varepsilon}^{\tau_1} a^i_k(v_{\varepsilon(\omega)}(\tau - \varepsilon))d\tau \int u_\varepsilon(\zeta - \tau)db^k_{\omega_1}(\zeta)\ .$$

Now if we use *formally* a Fubini lemma for the "measure" $d\tau \otimes db_\omega(\xi)$, we get

$$I^i_1 = \int \psi^i_k(\omega, \varepsilon, \zeta)db^k_{\omega_1}(\zeta)$$

(3.7)
$$\text{with} \quad \psi^i_k(\omega, \varepsilon, \zeta) = \int a^i_k(v_{\varepsilon(\omega)}(\tau - \varepsilon))u_\varepsilon(\zeta - \tau)d\tau\ .$$

We remark that, by consideration of supports, $\psi(\omega, \varepsilon, \zeta) \in \mathcal{N}^\zeta$, which makes possible to justify this formal calculation. Apply to $a^i_k(v_{\varepsilon(\omega)}(\xi))$ the Ito lemma we get

$$a^i_k(v_{\varepsilon(\omega)}(\tau)) - a^i_k(v_{\varepsilon(\omega)}(\tau - \varepsilon)) = \int_{\tau-\varepsilon}^{\tau} (\partial_s a^i_k)(v_{\varepsilon(\omega)}(\xi))dv^s_{\varepsilon(\omega)}(\xi)$$
$$+ \frac{1}{2}\int_{\tau-\varepsilon}^{\tau} \tilde{L}a^i_k(v_{\varepsilon(\omega)}(\xi))d\xi$$

where $\tilde{L} = \tilde{a}^s_{\underline{k}}\tilde{a}^{s'}_{\underline{k}}\partial_s\partial_{s'}$ and where $dv_{\varepsilon(\omega)}$ is defined in (3.2). Therefore the integral I^i_2 can be written $I^i_2 = J^i_1 + J^i_2 + J^i_3$ with

$$J_1^i = \int_{2\varepsilon}^{\tau_1} db^k_{\varepsilon(\omega_1)}(\tau) \int_{\tau-\varepsilon}^{\tau} \left[\frac{1}{2}\tilde{L}a_k^i + (\partial_s a_k^i)c^s\right] d\tau$$

$$J_2^i = \int_{2\varepsilon}^{\tau_1} db^k_{\varepsilon(\omega_1)}(\tau) \int_{\tau-\varepsilon}^{\tau} (\partial_s a_k^i) a_l^s db^l_{\varepsilon(\omega_1)}(\xi) \tag{3.8}$$

$$J_3^i = \int_{2\varepsilon}^{\tau_1} db^k_{\varepsilon(\omega_1)}(\tau) \int_{\tau-\varepsilon}^{\tau} (\partial_s a_k^i) \tilde{a}_l^s db^l_{\omega_2}(\xi) .$$

Introduce

$$\delta(\varepsilon, \tau_0) = \sup_{2\varepsilon < \tau < \tau_0} \| v_{\varepsilon(\omega)}(\tau - \varepsilon) - v(\tau) \| . \tag{3.9}$$

Remark that there exists a constant γ such that

$$\| v_\omega(\tau) - v_\omega(\tau') \| \leq \gamma |\tau - \tau'|^{1/4}, \quad \tau, \tau' < T . \tag{3.10}$$

Therefore we conclude that denoting by M the C^3 norm of the coefficients (3.1)., we have, for $\tau \leqslant \tau_0$

$$\begin{aligned} M[\gamma\varepsilon^{1/4} + 2\delta] \geqslant \| \partial_s a_k^i (v_{\varepsilon(\omega)}(\tau - \varepsilon)) - \partial_s a_k^i (v_\omega(\tau)) \| \\ + \| a_l^s (v_{\varepsilon(\omega)}(\tau - \varepsilon)) - a_l^s (v_\omega(\tau)) \| . \end{aligned} \tag{3.11}$$

Denote

$$\begin{aligned} X^k_{\varepsilon,\omega_1}(\tau) &= \varepsilon^{1/2} \frac{db^k_{\varepsilon(\omega_1)}(\tau)}{d\tau} , \\ Y^l_{\varepsilon,\omega_1}(\tau) &= \varepsilon^{-1/2} (b^l_{\varepsilon(\omega_1)}(\tau) - b^l_{\varepsilon(\omega_1)}(\tau - \varepsilon)) . \end{aligned} \tag{3.12}$$

We conclude from (3.10) and (3.11) that

$$\begin{aligned} I_2^i = \int_{2\varepsilon}^{\tau_1} f^i_{k,l} X^k_{\varepsilon,\omega} Y^l_{\varepsilon,\omega} d\tau \\ + \tau_1 \theta (M + 1)^2 (\gamma\varepsilon^{1/4} + \delta(\varepsilon, \tau_0)) \sum_{k,l} \| X^k_{\varepsilon,\omega} Y^l_{\varepsilon,\omega} \|_{L^1(0,\tau_1)} + J_3^i \end{aligned} \tag{3.13}$$

if $\gamma\varepsilon^{1/4} + \delta(\varepsilon, \tau_0) < 1$, where $|\theta| < 4$, and

$$f^i_{k,l}(\tau) = a_l^s(v_\omega(\tau))(\partial_s a_k^i)(v_\omega(\tau)) . \tag{3.14}$$

§ 4. An auxiliary Gaussian process

It is clear from (3.6) that (X^k, Y^l) is the sample of a stationary vector valued Gaussian process. We shall renormalize it by introducing the action of the dilatation of ratio ε on the Brownian motion. We denote by $b^{\cdot}_{\omega_\varepsilon}$ the sample of the Brownian motion defined by

(4.1) $$b_{\omega_i^\varepsilon}(\tau) = \varepsilon^{1/2} b_{\omega_i}\left(\frac{\tau}{\varepsilon}\right) \qquad i = 1, 2 \,.$$

Define ψ as ψ_1 (ψ_ε introduced in (3.6)) and define

(4.2) $$\begin{aligned} G_\omega^k(\tau) &\equiv \int u(\lambda - \tau) db_\omega^k(\lambda) \\ H_\omega^l(\tau) &= \int \psi(\lambda - \tau) db_\omega^l(\lambda) \,. \end{aligned}$$

We remark then

(4.3) $$G_{\omega^\varepsilon}^k\left(\frac{\tau}{\varepsilon}\right) = X_{\varepsilon,\omega}^k(\tau) \,, \qquad H_{\omega^\varepsilon}\left(\frac{\tau}{\varepsilon}\right) = Y_{\varepsilon,\omega}^l(\tau) \,.$$

The covariance of the processes (4.2) is given by blocks with no correlation between k and l blocks. The elementary block is

$$\begin{bmatrix} u * \check{u} & u * \check{\psi} \\ \check{u} * \psi & \psi * \check{\psi} \end{bmatrix} .$$

Therefore as u and ψ have compact supports, the covariance has compact support, therefore the process has the *strong mixing property* and the ergodic theorem holds.

(4.4) **Lemma.** *Given a sequence $\tilde{\varepsilon}_r$, we can extract a subsequence ε_j, such that,* a.s. *for every continuous function with compact support f*

$$\lim \int f(\tau) X_{\varepsilon,\omega}^k(\tau) Y_{\varepsilon,\omega}^l(\tau) d\tau = \frac{1}{2} \delta_l^k \int f(\tau) d\tau \,.$$

Proof. Let us first control the L^1 norm

$$J_{k,l}(\varepsilon) = \int_0^1 |X_{\varepsilon,\omega}^k(\tau) Y_{\varepsilon,\omega}^l(\tau)| \, d\tau = \frac{1}{\varepsilon} \int_0^{\varepsilon^{-1}} |G_{\omega^\varepsilon}^k(\tau) H_{\omega^\varepsilon}^l(\tau)| \, d\tau \,.$$

Then by the ergodic theorem the probability that the second member is greater than $2E(|G^k(0)H^l(0)|)$ tends to zero when $\varepsilon \to 0$. Therefore *extracting a subsequence* ε_j we get that a.s.

(4.5) $$\limsup_{\varepsilon_j} |J_{k,l}(\varepsilon_j)| < 2E(|G^k(0)H^l(0)|) \,.$$

In the same way we proved that for τ', τ'' rational, extracting a subsequence we have a.s.

$$\int_{[\tau',\tau'']} X^k_{\varepsilon_j} Y^l_{\varepsilon_j} d\tau \to (\tau'' - \tau') E(X^k(0) Y^l(0)) .$$

An elementary calculus gives for the last covariance $\frac{1}{2}\delta^k_l$. We can also get, a.s.

$$\limsup |J_{k,l}(\varepsilon_j)| \leq 2 |\tau' - \tau''| E(|G^k H^l|) .$$

(4.6) *Remark.* The subsequence that we get is *universal* in the sense that it is only linked to the Gaussian process (G, H) and not at all to the geometric situation.

(4.7) We denote

$$(4.7.1) \qquad \begin{cases} J^i_{3,\varepsilon}(\tau_1) = \int_{2\varepsilon}^{\tau_1} db^l_{\varepsilon(\omega_1)}(\tau) \int_{\tau-\varepsilon}^{\tau} g^i_{k,l}(\xi) db^k_{\omega_2}(\xi) \quad \text{with} \\ g^i_{k,l}(\xi) = \tilde{a}^s_k(v_{\varepsilon(\omega)}(\xi))(\partial_s a^i_l)(v_{\varepsilon(\omega)}(\xi)) . \end{cases}$$

Lemma. *Suppose*

$$(4.7.2) \qquad \sum \varepsilon^\delta_j < +\infty \qquad \textit{for all} \quad \delta > 0 ,$$

then a.s.

$$(4.7.3) \qquad \lim_{j\to\infty} \varepsilon_j^{-\rho} \sup_{2\varepsilon_j \leq \tau_1 \leq \tau_2} \| J_{3,\varepsilon_j}(\tau_1) \| = 0 \qquad \textit{for all } \rho < \tfrac{1}{4} .$$

Proof. We shall denote $E^{\omega_1}, E^{\omega_2}$ the expectations in ω_1, ω_2; we have $E = E^{\omega_1} E^{\omega_2}$.

$$E^{\omega_2}((J^i_{3,\varepsilon}(\tau_1))^2) = \int_{2\varepsilon}^{\tau_1} \int_{2\varepsilon}^{\tau_1} db^l_{\varepsilon(\omega_1)}(\tau) db^{l'}_{\varepsilon(\omega_1)}(\tau')$$

$$E^{\omega_2}\left(\int_{\tau-\varepsilon}^{\tau} g^i_{k,l} db^k_{\omega_2} \int_{\tau'-\varepsilon}^{\tau'} g^i_{k',l'} db^{k'}_{\omega_2} \right)$$

$$\leq M \int_{2\varepsilon}^{\tau_1} \int_{2\varepsilon}^{\tau_1} \mathbf{1}(|\tau - \tau'| < \varepsilon) W_\varepsilon(\tau) W_\varepsilon(\tau') d\tau d\tau'$$

where

$$W_\varepsilon(\tau) = \sum_l \varepsilon^{1/2} \left| \frac{db^l}{d\tau} \varepsilon(\omega) \right| .$$

We have by Schwarz's inequality and (4.6)

$$E^{\omega_1}(W_\varepsilon(\tau) W_\varepsilon(\tau')) < E((W_\varepsilon(\tau))^2) < C$$

where C is a constant independent of ε. Then

(4.7.4) $$E(\|J_{3,\varepsilon}(\tau_1)\|^2) < M_1\varepsilon \quad \tau_1 < \tau_2 .$$

We shall divide $[0, \tau_2]$ in q_j equal intervals, we denote Π_j the points of this subdivision; then

$$P(\|J_{3,\varepsilon_j}(s)\| > \delta\varepsilon_j^\rho) < M_1\delta^{-2}\varepsilon_j^{1-2\rho}$$

and

(4.7.5) $$P\Big(\sup_{s\in\Pi_j}\|J_{3,\varepsilon_j}(s)\| < \delta\varepsilon_j^\rho\Big) > 1 - q_jM_1\delta^{-2}\varepsilon_j^{\prime 1-2\rho} .$$

Denote by s, s' two consecutive points of Π_j. Then by Schwarz's inequality for $s < \tau_1 < s'$.

(4.7.6) $$\|J_{3,\varepsilon_j}(\tau_1) - J_{3,\varepsilon_j}(s)\| \leqslant (\tau_2 q_j^{-1})^{1/2}(D_{\varepsilon_j}(s))^{1/2}$$

where

$$D_\varepsilon(s) = \int_s^{s'}\left[\frac{db^l}{d\tau}\varepsilon(\omega)\right]^2 d\tau \left|\int_{\tau-\varepsilon}^{\tau} g^i_{k,l}db^k_{\omega_2}\right|^2 .$$

We have

(4.7.7) $$E(D_{\varepsilon_j}(s)) = E^{\omega_1}E^{\omega_2} < M_2\tau_2 q_j^{-1} .$$

Therefore

(4.7.8) $$P\Big(\sup_{s<\tau_1<s'}\|J_{3,\varepsilon_j}(\tau_1) - J_{3,\varepsilon_j}(s)\| > \delta\varepsilon_j^\rho\Big) < \frac{M_3}{q_j^2\varepsilon_j^{2\rho}\delta^2} .$$

Now (4.7.3) will be fullfilled if

(4.7.9) $$\sum q_j\varepsilon_j^{1-2\rho} < +\infty \quad \text{and} \quad \sum q_j^{-1}\varepsilon_j^{-2\rho} < +\infty .$$

We shall realize (4.7.9) by taking $q_j = \varepsilon_j^{-\gamma}$ with $2\rho < \gamma < 1 - 2\rho$.

(4.8) **Lemma.**

(4.8.1) $$\begin{aligned} &v^i_{\varepsilon(\omega)}(\tau_1) - v^i_\omega(2\varepsilon) - \hat r^i_\varepsilon \\ &\quad = \int_{2\varepsilon}^{\tau_1} B^i_k db^k_{\omega_1} + \int_{2\varepsilon}^{\tau_1} \tilde B^i_k db^k_{\omega_2} + \theta\tau_1 M_4\delta , \end{aligned}$$

where M_4 is a constant which can be expressed in term of the C^2 norm of the coefficients of (3.1), *where $|\theta| < 1$, and where δ is defined in* (3.9); *where*

$$(4.8.2)\qquad \begin{aligned} B_k^i(\xi) &= \psi_k^i(\xi) - a_k^i(v_\omega(\xi))\,, \\ \tilde{B}_k^i(\xi) &= \tilde{a}_k^i(v_{\varepsilon(\omega)}(\xi)) - \tilde{a}_k^i(v_\omega(\xi))\,; \end{aligned}$$

where a.s.

$$(4.8.3)\qquad \sup_{2\varepsilon_j<\tau<\tau_1} \|\hat{r}_{\varepsilon_j(\tau)}\| \leqslant \eta_j \quad \textit{for } j \textit{ sufficiently big,}$$

η_j being a numerical sequence tending to zero depending only of the C^2-norm of the coefficients of (3.1) *and of the choice of ε_j.*

Proof. We have by (3.1)

$$v_\omega^i(\tau_1) - v_\omega^i(2\varepsilon) = \int_{2\varepsilon}^{\tau_1} a_k^i(v_\omega(\tau))db_{\omega_1}^k(\tau) + \frac{1}{2}\int_{2\varepsilon}^{\tau_1}\sum_k f_{k,k}^i d\tau$$
$$+ \int_{2\varepsilon}^{\tau_1} c^i d\tau + \int_{2\varepsilon}^{\tau_1} \tilde{a}_k^i db_{\omega_2}^k\,.$$

We apply then (3.13), (4.4) and (4.7.3). We remark also that the convergence in (4.4) is uniform for equicontinuous function, the speed of convergence can be controlled in term of $\frac{1}{4}$-Hölder norm of $f_{k,l}^i(v_\omega(\tau))$.

§ 5. **Proof of the theorem.** We remark first that we can extract a subsequence of ε_j universal (i.e. depending only of the C^2 norm of the coefficient of (3.1)) such that

$$\lim_{j\to\infty} j^{-1}\log\varepsilon_j = -\infty$$
$$\lim_{j\to\infty} j^{-1}\log\|\hat{r}_{\varepsilon_j}\|_{L^\infty(0,\tau_2)} = -\infty\,.$$

The last inequality is a consequence of (4.8.3).

Define

$$(5.1)\qquad \tilde{r}_{\varepsilon_j}(\tau_1) = \|\hat{r}_{\varepsilon_j}\|_{L^\infty(0,\tau_1)} + \gamma\varepsilon_j^{1/4}\,.$$

Then

$$(5.2)\qquad \lim j^{-1}\log(\tilde{r}_{\varepsilon_j}) = -\infty\,.$$

We remark that

$$(5.3)\quad |B_k^i(\xi)| + |\tilde{B}_k^i(\xi)| < 2M(\gamma\varepsilon_j^{1/4} + \delta(\tau)) \leqslant 2M_5\tilde{r}_{\varepsilon_j}(\tau) + 2M\delta(\tau)\,.$$

Then (4.8.1) can be written

$$(5.4)\quad v_{\varepsilon(\omega)}(\tau_1) - v_\omega(2\varepsilon) - \tilde{r}_\varepsilon(\tau_1)z = \int_{2\varepsilon}^{\tau_1}(B_k db_{\omega_1}^k + \tilde{B}_k db_{\omega_2}^k) + \theta\tau_1 M_4\delta$$

with $\|z\| < 1$, $\|\theta\| < 1$.

Define

$$\text{(5.5)} \qquad \begin{aligned} \tilde{\tau}_0 &= (2M_4)^{-1} \wedge T(\omega) \\ r_j &= \tilde{r}_j(2\varepsilon_j)\,. \end{aligned}$$

Introduce the stopping time $\hat{T}$ when δ reaches for the first time the value $2r_j$ and let $\tilde{T} = \hat{T} \wedge \tilde{\tau}_0$. Then (5.4) gives

$$\text{(5.6)} \qquad \frac{1}{2} r_j < \sum_{i=1}^{m} \left| \int_{2\varepsilon}^{\tilde{T}} B_k^i db_{\omega_1}^k \right| \quad \text{where } |B_k^i| < M_6 r_j \text{ by (5.3)}\,.$$

Now divide by r_j the two members of (5.6). We get, representing stochastic integral as Brownian motion with a change of time,

$$\text{(5.7)} \qquad \frac{1}{2} < \sum_{i=1}^{2m} |\hat{b}^i(T^i)| \qquad \text{where} \quad T^i < M_6^2 \tilde{T}\,.$$

Denote by F the distribution function of the first time, when the Brownian motion, b starting from zero, reaches $\pm(4m)^{-1}$. Then

$$\text{(5.8)} \qquad P(\tilde{T} < \alpha) \leqslant F(M_6^2 \alpha) \qquad \text{for all } \alpha < \tilde{\tau}_0\,.$$

It results from (5.5) and (5.8) that the law of $\tilde{T}$ is minorized uniformly, by bound which depends only upon M_6, i.e. from the coefficients of the differential system. Furthermore we can use the Markov property starting afresh form $\tilde{T}$ and getting $\tilde{T} + \tilde{T}_1, \tilde{T} + \cdots + \tilde{T}_q$. At the stage q we have for δ the estimate $2^q r_j$. By the independence of the $\tilde{T}_q$ we have by (5.8) the a priori estimate

$$P(\tilde{T} + \cdots + \tilde{T}_q < \tilde{\tau}_0) \leqslant \exp(-qc) \quad \text{with} \quad c = -\log F(M_6^2 \alpha)\,.$$

Therefore

$$P\Big(\sup_{2\varepsilon_j < \tau < \tilde{\tau}_0} \|v_{\varepsilon(\omega)}(\tau) - v_\omega(\tau)\| > 2^{q_j} r_j \Big) \leq \exp(-cq_j)\,.$$

Then take $q_j = -(2/c) \log j$, we have a.s. for j great enough

$$\text{(5.9)} \qquad \sup_{2\varepsilon_j < \tau < \tilde{\tau}_0} \|v_{\varepsilon(\omega)}(\tau) - v_\omega(\tau)\| < r_j j^{2/c}\,.$$

Remark that $r_j j^{2/c}$ tend to zero by (5.2).

Denote

$$W^1 = \|v_{\varepsilon_j(\omega)}(\tilde{\tau}_0) - v_\omega(\tilde{\tau}_0)\|\,.$$

Then we remark that

$$\frac{\log \|W_j^1\|}{\log j} \to -\infty .$$

We can therefore iterate the estimate (5.9) starting from $\tilde{\tau}_0$. We shall get in this way a majoration on $[\tilde{\tau}_0, 2\tilde{\tau}_0]$, and so on, as $\tilde{\tau}_0$ is fixed by (5.5) we have proved the theorem.

§ 6. Dependence on the initial data

(6.1) **Theorem.** *Denote by* $v_{\varepsilon(\omega)}(\tau, v_0)$ *the solution of the Cauchy problem of* (3.2) *with initial value* v_0. *Suppose* $\tilde{a}_k^i = 0$. *Then we can extract a subsequence* $\varepsilon_{j_q} = \eta_q$ *such that* a.s.

$$v_{\eta_q(\omega)}(\tau, v_0) \qquad \text{converges uniformly in } v_0 .$$

Proof. Denote by S_n the vertex of a cubic subdivision, with the mesh 2^{-n}, of the range of the local chart on which we are working. We construct the $(n+1)$th mesh by a subdivision of the n-th mesh; therefore $S_{n+1} \supset S_n$. Furthermore

$$\operatorname{card}(S_n) \leq c2^{nm} \qquad (m = \dim V) .$$

Define the set of *adjacent vertices* $\tilde{S}_n \subset S_n \times S_n$

$$\tilde{S}_n = \{(v, v') \in S_n^2 ;\ d(v, v') \leq 2^{-n+2}\} .$$

(6.1.1) Let $v_0, v_0' \in S_q$, $d(x, y) < 2^{-q+k}$ then there exists $v_1, v_2, \cdots, v_k = v_k', v_{k-1}', \cdots, v_1'$ such that (v_s, v_{s+1}) and $(v_s', v_{s+1}') \in \tilde{S}_{q-s}$.

Remark that for $\theta > 0$

$$\text{(6.1.2)} \qquad \int_\tau^{\tau+\varepsilon\theta} \left|\frac{db_{\varepsilon(\omega)}}{d\tau}\right| = \int_\tau^{\tau+\varepsilon\theta} d\tau \left|\int_0^\varepsilon u_\varepsilon'(\lambda)(b_\omega(\tau+\lambda) - b_\omega(\tau))d\lambda\right|$$

$$\leq \theta p_\omega(\varepsilon) ,$$

where

$$p_\omega(\varepsilon) = \sup_{|\tau-\tau'|<\varepsilon} \|b_\omega(\tau) - b_\omega(\tau')\| ;$$

therefore

$$\text{(6.1.3)} \qquad \|v_{\varepsilon(\omega)}(\tau) - v_{\varepsilon(\omega)}(\tau - \theta\varepsilon)\| \leq \theta M_1 p_\omega(\varepsilon) .$$

In this proof we shall denote by $M.$ constants depending on the C^3 norm of the coefficients of (3.1).

We define a subsequence $\eta_q = \varepsilon_{j_q}$ by the conditions

$$(6.1.4) \qquad \sum_q P\Big(\sup_{\substack{v_0 \in S_q \\ j > j_q}} |||v_{\varepsilon j(\omega),\, v_0} - v_{\varepsilon_{j_q}(\omega), v_0}|||_{[2\varepsilon_{j_q}, 1]} > 2^{-q}\Big) < +\infty ,$$

where $v_{\varepsilon(\omega), v_0}$ denotes the solution of (3.2) with the initial data $v_{\varepsilon(\omega)}(2\varepsilon) = v_0$, and where $|||\ \ |||_{[2\varepsilon, \tau]}$ denotes the sup of $\|\ \ \|$ on the interval $[2\varepsilon, \tau]$.

We shall split $v_{\eta_q(\omega), v_0} - v_{\eta_q(\omega), v_0'} = v^0_{\eta_q(\omega)} - v'_{\eta_q(\omega)}$ in $I_1 + I_2 + I_3$ where

$$I_1(\tau) = \int \psi_k db^k$$

where

$$\psi_k(\zeta) = \int \{a_k(v^0_{\eta(\omega)}(\tau - \eta)) - a_k(v'_{\eta(\omega)}(\tau - \eta))\} u_\eta(\zeta - \tau) d\tau .$$

We define a family of stopping times T_l by the first realizations of the following equalities

$$|||v^0 - v'|||_{[2\eta, T_l - \eta]} = 2^{l-n+3}$$

where $(v_0, v_0') \in \tilde{S}_n$. Then

$$\|\psi_k\| \leq M_2 2^{l-n} , \qquad \tau \in [T_l, T_{l+1}] .$$

Now by the calculus of variation for an ordinary differential system (cf. Chapter II) and by (6.1.2) we have

$$|||v^0 - v'|||_{[2\eta, \tau]} \leq |||v^0 - v'|||_{[\eta, \tau - \eta]} \exp(M_3 p_\omega(\eta)) .$$

Therefore

$$I_3(\tau) = \int_0^\tau c(v^0_{\eta(\omega)}) - c(v'_{\eta(\omega)}) d\tau$$

satisfies

$$\|I_3(T_{l+1}) - I_3(T_l)\| \leq (T_{l+1} - T_l) M_4 \exp(M_3 p_\omega(\eta)) 2^{l-n} .$$

Now $I_2(\tau)$ is defined by

$$I_2(\tau) = \int_{2\eta}^\tau \{[a_k(v^0_\eta(\tau)) - a_k(v'_\eta(\tau - \eta))] \\ - [a_k(v^0_\eta(\tau - \eta)) - a_k(v'_\eta(\tau - \eta))]\} d\tau$$

and satisfies

$$\|I_2(T_{l+1}) - I_2(T_l)\| \le M_5 2^{l-n} \int_{T_l}^{T_{l+1}} R_\eta(\omega, d\tau) ,$$

where

$$R_\eta(\omega, d\tau) = \sum_{k,l} |X^k_{\eta,\omega} Y^l_{\eta,\omega}| \, d\tau .$$

Consider the events

$$A(\eta_q, v_0, v_0') = \Big\{\omega \,;\, T_{[n/2]} < 1, \text{ and there exists } s_1 < s_2 \cdots < s_t < \frac{n}{2}$$

$$\text{with } l > \frac{n}{4} \text{ such that } \sup_{\tau \in [T_s, T_{s+1}]} \Big\| \int \psi_k db^k \Big\| > M_6 2^{s-n} \Big\} .$$

Now

$$P(A(\eta_q, v_0, v_0')) \le \exp(-M_7 n^2) .$$

Define now the events

$$B(\eta_q, n) = \bigcup_{v_0 \in \tilde{S}_n} A(\eta_q, v_0, v_0') .$$

Then

$$P(B(\eta_q, n)) \le \exp(-M_7 n^2 + M_8 n)$$

Define finally $C_q = \bigcup_{n \ge q} B(\eta_q, n)$. Then

$$\sum P(C_q) < +\infty$$

and almost surely only a finite number of C_q is realized. Denote by $q(\omega)$ the last q for which C_q happens.

It results that from a.s. estimate on R, obtained in (4.5) and of a.s. modulus of continuity for p_ω, there exists $q_1(\omega) > q(\omega)$ such that

$$|||v^0_{\eta_q} - v'_{\eta_q}|||_{[\eta_q, 1]} < 2^{-n/2} \qquad \text{for } q > q_1(\omega),\ v_0, v_0' \in \tilde{S}_n \text{ and } n \ge q_1(\omega) .$$

It results from (6.1.1) that we have

$$|||v^0_{\eta_q} - v'_{\eta_q}||| \le 2(d(v_0, v_0'))^{1/2} \qquad \text{for } q > q_1(\omega) \text{ and } d(v_0, v_0') \le 2^{-q} .$$

Use now (6.1.4) and the almost sure uniform convergence on S_q; we get

$$(6.1.5) \qquad |||v_{\eta_q} - v'_{\eta_q}||| \leq 2(d(v_0, v'_0))^{1/2} \quad \text{for all } q > q_2(\omega) \text{ and } d(v_0, v'_0) < 2^{-q_2(\omega)}$$

which implies the *equicontinuity* in v_0 of the functions $v_{\eta_{q(\omega)}}(\cdot, v_0)$. As by Theorem 3.1 we have the convergence on $\cup S_n$, then the Ascoli's theorem implies the uniform convergence.

(6.1.6) *Remark.* The inequality is a temporary result. In fact as we shall see in the next chapter, we have a C^1 convergence of $v_{\eta_q}(\tau, \cdot) \to v_\omega(\tau, \cdot)$.

(6.2) **Theorem.** *Given a sample path ω of the Brownian motion, then almost surely the Cauchy problem for* (3.1) *has a solution for all initial data v_0.*

Proof. We define the solution as the uniform limit of $v_{\eta(\omega)}(\tau, v_0)$.

Chapter II. Stochastic calculus of variation

The semi elliptic operator

$$(0.1) \qquad L = \frac{1}{2} \sum_{k=1}^{n} \mathscr{L}^2_{A_k}$$

has been studied recently by Stroock-Varadhan [30], and Kunita [17] using the family of difierential equations

$$(0.2) \qquad \frac{dv}{d\tau} = A_k \frac{df^k}{d\tau} = A_f$$

where $\tau \to f(\tau)$ is a smooth curve of R^n. We shall first write the well known variational theory of (0.2), and then transfer it to the stochastic case.

§ 1. Variation of a differential system

Given a vector field B on V, depending on time we shall consider the differential system

$$(1.0) \qquad \frac{dv}{d\tau} = B_{v(\tau)} \, .$$

Given (v_0, τ_0), we denote by $U_{B,\tau}(v_0, \tau_0)$ the value at instant $\tau_0 + \tau$ of the solution of (1.0), with the Cauchy data

(1.1) $$v(\tau_0) = v_0 .$$

Then

(1.2) $$U_{B,\tau} \circ U_{B,\tau'} = U_{B,\tau+\tau'}$$

and we have a flow on $V \times \boldsymbol{R}$.

Bundle of linear frames

We shall denote by $GL(V)$ the *principal bundle of linear frames* of V (cf. [20]). A frame r will be by definition a linear isomorphism,

$$r: R^m \to T_{v_1}(V) .$$

We have the mapping $r \to v_1$ which is the projection $\pi: GL(V) \to V$. Furthermore if V is of class C^k, $GL(V)$ is of class C^{k-1}.

(1.2.1) The linear group $G = GL(m, \boldsymbol{R})$ operates on $GL(V)$ via

$$a_g(r) = r \circ g .$$

This action satisfies

$$a_{g_1} \circ a_{g_2} = a_{g_2 g_1} .$$

Therefore

(1.2.2) there is a natural isomorphism of $T_r(\pi^{-1}(v_0))$ with the *right* Lie algebra of $GL(m, R) \simeq \mathcal{M}_m$.

(1.3) **First prolongation of a vector field**

Let B be a vector field on V depending on time, $U_{B,\tau}$ the flow associated on $V \times \boldsymbol{R}$. We shall define a prolongation $\tilde{U}_{B,\tau}$ of $U_{B,\tau}$ to $GL(V) \times \boldsymbol{R}$ by the formula

(1.3.1) $$\tilde{U}_\tau(r_0, \tau_0) = [U'_{B,\tau}(\pi(r_0), \tau_0)] \circ r_0 .$$

Then $\tilde{U}_\tau$ is a map of $GL(V) \times \boldsymbol{R}$ into itself, which satisfies the semi group property (1.2). Therefore it is the flow associated with a vector field, as its time component is known we shall write this vector field on the form

(1.3.2) $$\left(\tilde{B}, \frac{\partial}{\partial \tau}\right)$$

where $\tilde{B}$ is a time dependent tangent vector field on $GL(V)$. We shall call $\tilde{B}$ the first prolongation of B. As $\pi \circ \tilde{U}_\tau = U_{B,\tau} \circ \pi$ it results that $\pi'(r_0)\tilde{B} = B_{\pi(r_0)}$; therefore $\tilde{B}$ is a lifting of B to $GL(V)$.

(1.3.3) *Scholia.* In order to compute the differential of $U_{B,\tau}$ at (v_0, τ_0), we choose a frame $r_0 \in \pi^{-1}(v_0)$, integrate the vector field $\tilde{B}$, starting at r_0, then obtain $r(\tau_0 + \tau)$ and the differential is given by

(1.3.4) $$U'_{B,\tau}(v_0, \tau_0) = r(\tau_0 + \tau) \circ r_0^{-1} = \tilde{U}_{\tau_0}(r_0) \circ r_0^{-1} \, .$$

We can deduce therefore the differential by *integration of the prolongated vector field*. This scholia is only a rephrasing of (1.3.1).

(1.3.5) Lifting a local chart

Let $\varphi = (v^1, \cdots, v^m)$ be a local chart of V, with value in R^m, denote by J the differential φ'. Then $v \to J^{-1}(v)$ defines a section of $GL(V)$ which realizes a local chart $\tilde{\varphi}$ of $GL(V)$ into $R^m \times GL(m, R)$. Then

(1.3.6) **Proposition.** *Given the chart φ of V, let $\tilde{\varphi}$ its lifting to $GL(V)$, then the prolongation $\tilde{B}$ of B is read*

$$(\tilde{\varphi})_* \tilde{B} = (B^i, \partial_k B^i) \, .$$

Remark. In this formula $(B^i) = \varphi_* B$, ∂_k is the kth partial derivative in R^m, $\partial_k B^i$ is the matrix where k is the indices of columns, i the indices of rows. Furthermore the tangent space $T_{r_0}(\pi^{-1}(x_0))$ has been identified (cf. (1.2.2)) to $\mathcal{M}_n$ by the *right* invariant Maurer Cartan form (cf. [20]).

Proof. In coordinate we have

$$\frac{dv^i}{d\tau} = B^i(v(\tau)) \, ,$$

we have the linearized variational equation

(1.3.7) $$\frac{d}{d\tau}\delta v^i = (\partial_k B^i)\delta v^k \, .$$

We have to solve this linear equation with initial value $(\delta v^i)(\tau_0)$. On the other hand, we consider the *right* invariant vector field on $GL(m, R)$, depending on time, defined by $(\partial_k B^i)(v(\tau))$. Then the integration of this vector field is equivalent to the integration of the matrix equation associated with (1.3.7)

$$\frac{dX}{d\tau} = (\partial B^{\cdot})X$$

and this proves the proposition.

(1.4) The first variation

We resume the study of (0.2) and we suppose that f has a variation δf. We want to compute the variation of the first order

$$(1.4.1) \qquad \delta v = U_{f+\delta f,\tau}(v_0, \tau_0) - U_{f,\tau}(v_0, \tau_0) ,$$

which is the variation of the integral curve of the flow A_f, when the origin is fixed at time τ_0.

$$(1.4.2) \qquad A_{\delta f} = A_k \frac{d}{d\tau}(\delta f^k) .$$

The variational equation

Using the same line as in (1.3.7) we get in a chart the variational equation

$$(1.4.3) \qquad \frac{d}{d\tau}(\delta v^i) = (\partial_k A_f^i)\delta v^k + A_{\delta f}^i .$$

Proposition. *The first variation δv is equal to*

$$(1.4.4) \qquad \delta v(\tau_1) = \int_{\tau_0}^{\tau_1} U'_{A_f,\tau_1-\tau} \cdot A_{\delta f(\tau)} d\tau .$$

Proof. Integrate (1.4.3) via Lagrange's method and then translate in intrinsic term via (1.3.7), (1.3.6), (1.3.3).

Variation reduced at the initial value

We define the *reduced* variation δ^* via

$$(1.4.5) \qquad (\delta^* v)(\tau_1) = \int_{\tau_0}^{\tau_1} (U'_{A_f,\tau-\tau_0})^{-1} A_{\delta f(\tau)} d\tau .$$

An advantage of reduced variation is that it does not involve "anticipating" integral as (1.4.4). The relation between δ and δ^* is obviously

$$(1.4.6) \qquad (\delta v)(\tau_1) = U'_{A_f,\tau_1-\tau_0} \cdot (\delta^* v)(\tau_1)$$

or formally

(1.4.7) $$(v + \delta v)(\tau_1) = U_{A_f,\tau_1-\tau_0}(v(\tau_0) + (\delta^* v)(\tau_1)) .$$

§ 2. The semigroup $U_{\omega,\tau}$

We shall consider the semi elliptic operator (0.1). In order to simplify our statement we shall suppose

(2.0) The life time of the associated diffusion is infinite. This is the case when V is compact. Now, given a sample b^k_ω of a Brownian motion on R^n, we define for $\tau > 0$, $U_{\omega,\tau}(v_0)$ as $v_{\omega,v_0}(\tau)$, where v_{ω,v_0} is the solution of the system

(2.1) $$\begin{cases} dv_{\omega,v_0} = A_k d_S b^k_\omega , & \tau > 0 \\ v_{\omega,v_0}(0) = v_0 . \end{cases}$$

We shall denote by ω^τ the time shift by τ on the probability space of the Brownian motion; then we have by Markov's property

(2.2) $$U_{\omega,\tau_1+\tau_2} = U_{\omega^{\tau_1},\tau_2} \circ U_{\omega,\tau_1} .$$

Theorem. *For almost all $\omega \in \Omega(R^n)$, $U_{\omega,\tau}$ is defined on $V \times R^+$. Supposing A_k of class C^3, then for every fixed $\tau \geq 0$, the map $U_{\omega,\tau}: V \to V$ is of class C^2.*

Denote by $\hat{A}_k$ the prolongation of A_k to $GL(V)$, then

(2.3.1) $$U'_{\omega,\tau} = r_\omega(\tau) \circ r_0^{-1}$$

where

(2.3.2) $$dr_\omega(\tau) = \tilde{A}_k d_S b^k_\omega .$$

Proof. We regularize the Brownian motion b_ω and get $b_{\varepsilon(\omega)}$. We shall denote by $U_{\varepsilon(\omega),\tau}$ the flow associated with the ordinary differential equation

$$dv = A_k db^k_{\varepsilon(\omega)} .$$

Now by the limit theorem, a.s., in ω

$$U_{\varepsilon_j(\omega),\tau}(v_0) \to U_{\omega,\tau}(v_0)$$

locally uniformly (I.6). We have by (1.3.3)

$$U'_{\varepsilon(\omega),\tau} = r_{\varepsilon(\omega)}(\tau) \circ r_0^{-1}$$

where $r_\omega(\tau)$ is a solution of

$$dr_{\varepsilon(\omega)} = \tilde{A}_k db^k_{\varepsilon(\omega)} .$$

By the limit theorem we know that $U'_{\varepsilon_j(\omega)}$ converges locally uniformly to the second member of (2.3.1). Therefore $U_{\omega,\tau}$ is of class C^1 and its differential is given by (2.3.1). For the second derivatives we use the fact that the second variation $\delta^2 v$ for the equation $dv = Bd\tau$ is given by

$$\frac{d}{d\tau}(\delta^2 v^i) - (\partial_k B^i)(\delta^2 v^k) = (\partial^2_{k,l} B^i)(\delta v^k, \delta v^l)$$

(2.3.3) $$\frac{d}{d\tau}(\delta v^i) - (\partial_k B^i)(\delta v) = 0$$

$$\delta v(0) = a , \qquad (\delta^2 v)(0) = 0 .$$

Apply this to $B = A_k db^k_{\varepsilon(\omega)}$, we get $(U''_{\omega,\tau}(v_0)(a, a))^i = G^i$.

(2.3.4) $$\begin{aligned} dG^i - (\partial_q A^i_k) G^q d_S b^k_\omega &= (\partial^2_{q,l} A^i_k) z^q z^l d_S b^k_\omega \\ dz^q - (\partial_l A^q_k) z^l d_S b^k_\omega &= 0 , \qquad z^q(0) = a^q . \end{aligned}$$

(2.3.5) **Remark.** We have here a differential geometry approach to Part II, Section 8 of [9].

§ 3. Elementary stochastic variation

Given the semi elliptic operator (0.1), we denote by X the probability space of the Brownian motion on R^n defined on [0, 1] as in Part I, Section 4. We fix once and for all $v_0 \in V$, and we want to study *the stochastic differential* of the map

(3.0) $$g_\tau : \omega \to U_{\omega,\tau}(v_0) , \qquad 0 \leq \tau \leq 1$$

when X is equipped with the O. U. process (Part I, Section 4). (We denote now by ω what we have denoted by x in Part I).

(3.1) **Lemma.** *The stochastic differential of* $b^{\cdot}_\omega$ *in the* O. U. *process is*

$$b^{\cdot}_{\omega+\delta\omega} - b^{\cdot}_\omega = b^{\cdot}_{\delta\omega} - \frac{\delta t}{2} b^{\cdot}_\omega$$

where δt *is the increment of time on the trajectory of the* O. U. *process, and where* $b^{\cdot}_{\delta\omega} = (\delta t)^{1/2} b^{\cdot}_{\omega'}$, $b_{\omega'}$ *being a Brownian motion, defined on* [0, 1], *independent of* $b^{\cdot}_\omega$.

Proof. Part I, formula (4.3.3).

(3.2) Computation of the Drift

We shall evaluate the effect of the variation $-(\delta t/2)b^{\cdot}_{\omega}$ on g_τ. Using the ε approximation and denoting $y^i = \delta v^i$ we have in a local chart for $T_j < \tau < T_{j+1}$

$$dv^i_\varepsilon = A^i_k db^k_{\varepsilon(\omega)}$$

$$dy^i_\varepsilon = (\partial_s A^i_k) y^s_\varepsilon db^k_{\varepsilon(\omega)} - \frac{\delta t}{2} A^i_k db^k_{\varepsilon(\omega)}$$

$$y^i_\varepsilon(T_j) = c^i_{j,\varepsilon}$$

where $T_1 < T_j < T_{j+1} <$ is the sequence of stopping time describing the change of local chart and where $y^i(0) = 0$.

Now by the limit theorem the solution of this system when $\varepsilon_j \to 0$ and δt is fixed tends to the solution of

$$dv^i = A^i_k d_S b^k_\omega$$

$$dy^i = (\partial_s A^i_k) y^s d_S b^k_\omega - \frac{\delta t}{2} A^i_k d_S b^k_\omega$$

which will be translated in Ito's notation. (In writing the formula we shall make the convention to underline subscript on which the summation has to be made to realize the stochastic contraction).

(3.2.1) $$dv^i_k = A^i_k db^k_\omega + \tfrac{1}{2}(\partial_q A^i_{\underline{k}}) A^q_{\underline{k}} d\tau$$

(3.2.2) $$d\tilde{y}^i = ((\partial_s A^i_k)\tilde{y}^s - \tfrac{1}{2}A^i_k) db^k_\omega + \tfrac{1}{2}\{(\partial^2_{r,s} A^i_{\underline{k}}) A^r_{\underline{k}} \tilde{y}^s + (\partial_s A^i_{\underline{k}})[(\partial_q A^s_{\underline{k}})\tilde{y}^q - \tfrac{1}{2}A^s_{\underline{k}}]\} d\tau \, .$$

In (3.2.2) we denote $\tilde{y}^i = (\delta t)^{-1} y^i$. Now we have proved

(3.2.3) **Lemma.** *The drift of the* O. U. *process gives as contribution to the differential of g_τ the vector $\tilde{y}(\tau)$ computed from* (3.2.1), (3.2.2) *with* $\tilde{y}(0) = 0$.

§ 4. Computation of the diffusion term

We shall work on reduced variation in the sense of (1.4.5). For this we would need the inverse of the map $U_{\omega,\tau}$.

(4.1) **Proposition.** *The map $U_{\omega,\tau}$ is a diffeomorphim of V.*

Proof. This fact, which is familiar for flows, need some explanation here. We use the theorem that Stratanovitch's integrals are invariant under the reversal of the time, fact which is proved in [14] with some important geometric consequences [13]. Fixing τ_0 we reverse the Brownian motion from time τ_0 defining

$$b_{\hat{\omega}}(\tau) = b^{\cdot}_{\omega}(\tau_0) - b^{\cdot}_{\omega}(\tau_0 - \tau)\,, \qquad 0 \le \tau \le \tau_0\,.$$

Now by [14] we have

$$U_{\hat{\omega},\tau_0} \circ U_{\omega,\tau_0} = \text{identity} \quad \text{and} \quad U_{\omega,\tau_0} \circ U_{\hat{\omega},\tau_0} = \text{identity}\,.$$

Furthermore $U_{\hat{\omega},\tau}$ is of class C^1, therefore $U_{\omega,\tau}$ is a diffeomorphism. We shall denote by $(U_{\omega,\tau})^{-1}$ its inverse diffeomorphism. We shall now work under assumption (A)

(A) $\begin{cases} \omega \text{ is fixed, } \delta t \text{ is fixed, } b_{\delta\omega} \text{ is an independent Brownian motion,} \\ \mathcal{N}^{\tau} \text{ the } \sigma\text{-field generated by } \{b_{\delta\omega}(\xi),\, \xi < \tau\}. \text{ All expectation will} \\ \text{be taken relatively to } \delta\omega. \end{cases}$

We shall denote by $\eta(\delta\omega)$ a smoothing of the Brownian motion $b_{\delta\omega}$ in the sense of Chapter I. We denote by $v_{\omega}, v_{\omega+\delta\omega}, v_{\omega+\eta(\delta\omega)}$ the solution with same initial values v_0 of

(4.2.1) $$dv^i_{\omega} = A^i_k d_S b^k_{\omega}$$

(4.2.2) $$dv^i_{\omega+\delta\omega} = A^i_k d_S b^k_{\omega} + A^i_k d_S b^k_{\delta\omega}$$

(4.2.3) $$dv^i_{\omega+\eta(\delta\omega)} = A^i_k d_S b^k_{\omega} + A^i_k db^k_{\eta(\delta\omega)}\,.$$

The last term of (4.2.3) is an ordinary drift. Furthermore the limit theorem applies and shows for $\eta_j \to 0$, the solution of (4.2.3) tends to the solution of (4.2.2).

Define the *reduced variations*

(4.2.4) $$\delta^*_{\delta\omega}(\tau) = (U_{\omega,\tau})^{-1}(v_{\omega+\delta\omega}(\tau))$$

(4.2.5) $$\delta^*_{\eta(\delta\omega)}(\tau) = (U_{\omega,\tau})^{-1}(v_{\omega+\eta(\delta\omega)}(\tau))\,.$$

We remark that these definitions do not suppose δt "small", they are valid on the manifold V itself and not on some tangent bundle.

(4.3) **Main Lemma.** *Under assumption* (A), $\delta^*_{\delta\omega}$ *has a stochastic differential with respect to* $\mathcal{N}^{\tau}$ *and this differential satisfies the stochastic differential equation*

(4.3.1) $$d\delta^*_{\delta\omega}(\tau) = X_k(\tau, \delta^*_{\delta\omega}(\tau))d_S b^k_{\delta\omega}(\tau)$$

where

(4.3.2) $$X_k(\tau, v) = [U'_{\omega,\tau}(U_{\omega,\tau}(v))]^{-1} \cdot A_k(U_{\omega,\tau}(v)) .$$

The equation (4.3.1) *with the limit condition*

(4.3.3) $$\delta^*_{\delta\omega}(0) = v_0$$

determines $\delta^*_{\delta\omega}$.

(4.3.4) *Remark.* The inverse diffeomorphism does not appear in (4.3.1) and (4.3.2). Its existence was only needed in order to define δ^*.

Proof. Regularize now the Brownian motion $b_{\varepsilon(\omega)}$. Define $\delta^*_{\varepsilon(\omega),\eta(\delta\omega)}$ by the equation

$$U_{\varepsilon(\omega),\tau}(\delta^*_{\varepsilon(\omega),\eta(\delta\omega)}(\tau)) = v_{\varepsilon(\omega)+\eta(\delta\omega)}(\tau) .$$

All the functions appearing here are smooth; we differentiate in τ

$$A_k(v_{\varepsilon(\omega)+\eta(\delta\omega)})db^k_{\varepsilon(\omega)} + U'_{\varepsilon(\omega),\tau} d\delta^*_{\varepsilon(\omega),\eta(\omega)} = A_k(v_{\varepsilon(\omega)+\eta(\delta\omega)})(db^k_{\varepsilon(\omega)} + db^k_{\eta(\delta\omega)})$$

which means that we get the differential equation

(4.3.5) $$\frac{d}{d\tau}\delta^*_{\varepsilon(\omega),\eta(\omega)} = X_k(v_{\varepsilon(\omega)+\eta(\delta\omega)})db^k_{\eta(\delta\omega)} .$$

We remark now that *the differentials* $db^{\cdot}_{\varepsilon(\omega)}$ *have disappeared* in (4.3.5). Therefore when $\varepsilon_j \to 0$ we get at the limit the following ordinary differential equation for $\delta^*_{\eta(\delta\omega)}$

$$d\delta^*_{\eta(\delta\omega)}(\tau) = X_k(v_{\omega+\eta(\delta\omega)}(\tau))db^k_{\eta(\delta\omega)} .$$

Now let $\eta \to 0$ and apply again the limit theorem. We get (4.3.2).

(4.4) **Theorem.** *Consider* g_{τ_0}, *defined by* (3.0). *Then* g_{τ_0} *has a stochastic differential with respect to the* O. U. *process, in the sense that the following limits exist* a.s. *in* ω:

(4.4.1) $$\lim_{t\to 0} t^{-1}E_\omega(g_{\tau_0}(x_w(t)) - g_{\tau_0}(\omega)) = (Qg_{\tau_0})(\omega)$$

(4.4.2) $$\lim_{t\to 0} t^{-1}E_\omega(l^2(g_{\tau_0}(x_w(t)) - g_{\tau_0}(\omega))) = q_{\omega,\tau}(l)$$

with

$$q_{\omega,\tau_0}(l) = \int_0^{\tau_0} \sum_k \{l(U'_{\tau_0-\tau}(v_\omega(\tau)))\cdot A_k(v_\omega(\tau))\}^2 d\tau$$

where

$$l \in T^*_{g_{\tau_0}(\omega)}(V) .$$

Proof. Let $A_{k,t} = e^{-t/2}A_k$, U^t_ω the associated semigroup and the corresponding $X_{k,t}$ (cf. (4.3.2)) then by continuous dependence on the parameter t ([9] p. 276) of (2.1), (2.3.2), (2.3.3), $U^t_{\omega,\tau} \to U_{\omega,\tau}$ in C^2 and

(4.4.3) $$\sup_{\tau,v} \|X_{k,t} - X_k\| \to 0 \qquad \text{when } t \to 0 .$$

Let $\lambda = (1 - e^{-t})^{1/2}$, $\omega' \in \Omega(R^n)$, independent of ω. Define $u_{\omega',t}$ by $u_{\omega',t}(0) = v_0$ and

(4.4.4) $$du_{\omega',t} = X_{k,t}(\tau, \lambda u_{\omega',t})db^k_{\omega'} + \frac{\lambda}{2}(\partial_q X_{\underline{k},t})X^q_{\underline{k},t}d\tau .$$

The law of g_τ, by I-(4.3.5) and (4.3), is given by

(4.4.5) $$E_{\mathscr{W}}(\varphi(g_{\tau_0}(x_w(t)))) = E^{\omega'}(\varphi(U^t_{\omega,\tau_0}(\lambda u_{\omega',t}(\tau_0)))) .$$

We will take φ of class C^2; if we replace in the second member U^t by U we still bring to (4.4) the contribution

$$\varphi'\left(\left(\frac{\partial}{\partial t}U^t(v_0)\right)_{t=0}\right) = \varphi'(\tilde{y}(t_0)) \qquad (\tilde{y} \text{ defined in (3.2.2)}) .$$

As U_{ω,τ_0} is a diffeomorphism of class C^2, it is sufficient to evaluate (4.4.1), (4.4.2) for $\lambda u_{\omega,\tau}(\tau_0)$. Define

$$R_{\omega',t}(\tau) = \int_0^\tau (X_{k,t}(\tau, \lambda u_\tau) - X_k(\tau, 0))db^k_{\omega'}(\tau) .$$

Remark by (4.4.3)

(4.4.6) $$\sup_{t<1} P^{\omega'}\left(\sup_\tau \|u_{\omega',t}(\tau)\| > M\right) \to 0, \qquad M \to \infty .$$

Hence

(4.4.7) $$E^{\omega'}(\|R_{\omega'}\|^2) \to 0 ,$$
$$E^{\omega'}\left(\int_0^{\tau_0} X^q_{\underline{k},t}\partial_q X_{\underline{k},t}d\tau\right) \to I = \int_0^{\tau_0} X_{\underline{k}}\partial_q X_{\underline{k}}d\tau .$$

Remark now that $R_{\omega'}$ is a martingale and so will not contribute to the *drift*, remark that by (4.4.7) R_ω will not contribute to the gradient, therefore

$$\lim t^{-1}E^{\omega'}(\lambda u_{\omega',t} - v_0) = \tfrac{1}{2}I$$

$$\lim t^{-1}E^{\omega'}(l^2(\lambda u_{\omega',t} - v_0)) = E^{\omega'}\left(l^2\left(\int_0^{\tau_0} X_k db^k\right)\right) = \sum_k \int_0^{\tau_0} l^2(X_k)d\tau \ ,$$

which proves (4.4.2), and (4.4.1) with

(4.4.8) $$Qg_{\tau_0} = \tilde{y}(\tau_0) + U'_{\omega,\tau_0}I + \tfrac{1}{2}\,\mathrm{Trace}_q\, U''_{\omega,\tau_0} \ .$$

(4.5) Summary of computations in a local chart

For later convenience we summarize here the formulae which give in the local chart x^α the stochastic variation of the equation

(4.5.1) $$dx^\alpha = A_k^\alpha d_S b_\omega^k \ , \qquad 1 \le k \le n,\ 1 \le \alpha \le N \ .$$

The local chart x^α will be used between two stopping times, T_0, $T_0 + T_1$. Using the time shift, we shall take $T_0 = 0$ and we will define $\tilde{\tau}_0 = T_1 \wedge \tau_0$.

Define $\Gamma^\alpha_{\beta,k} = \partial_\beta A_k^\alpha$, $\Gamma^\alpha_{\beta,\gamma,k} = \partial^2_{\beta,\gamma}A_k^\alpha$. Define $Z_\cdot^\cdot$, $G_{\cdot\cdot}^\cdot$ by

(4.5.2) $$\begin{aligned} &dZ_\beta^\alpha - \Gamma^\alpha_{\lambda,k}Z_\beta^\lambda d_S b_\omega^k = 0 \ , \\ &Z_\beta^\alpha(0) = \delta_\beta^\alpha\ ;\ (\text{Let } \check{Z}\text{: the inverse matrix of } Z\text{:}) \end{aligned}$$

(4.5.3) $$dG^\alpha_{\beta,\gamma} - \Gamma^\alpha_{\lambda,k}G^\lambda_{\beta,\gamma}d_S b_\omega^k = \Gamma^\alpha_{\lambda,\mu,k}Z_\beta^\lambda Z_\gamma^\mu d_S b_\omega^k \ , \qquad G^\alpha_{\beta,\gamma}(0) = 0 \ .$$

(4.5.4) $$\begin{aligned} &\delta^* x^\alpha(\tilde{\tau}_0) - \delta^* x^\alpha(0) \\ &= \int_0^{\tilde{\tau}_0} \check{Z}_\beta^\alpha A_k^\beta db_{\delta\omega}^k + \frac{\delta t}{2}\int_0^{\tilde{\tau}_0} (-\check{Z}_\rho^\mu G^\gamma_{\lambda,\mu}A^\lambda_{\underline{k}} + \Gamma^\gamma_{\rho,\underline{k}})\check{Z}_\gamma^\alpha A^\rho_{\underline{k}} d\tau \end{aligned}$$

(4.5.5) $$H^\alpha(\tilde{\tau}_0) = (\tilde{y}^\alpha(\tilde{\tau}_0) - \tilde{y}^\alpha(0)$$

is defined by integrating (3.2.2).

(4.5.6) $$(U'_{\omega,\tau}(x_0))^\alpha_\beta = Z_\beta^\alpha \ , \qquad (U''_{\omega,\tau}(x_0))^\alpha_{\beta,\gamma} = G^\alpha_{\beta,\gamma}$$

(4.5.7) $$\begin{aligned} \delta x^\alpha(\tilde{\tau}_0) &= \delta t(H^\alpha(\tilde{\tau}_0) + Z_\beta^\alpha(\tilde{\tau}_0)y^\beta(0)) \\ &\quad + Z_\beta^\alpha(\tilde{\tau}_0)\delta^* x^\alpha(\tilde{\tau}_0) + \frac{\delta t}{2}G^\alpha_{\beta,\gamma}(\tilde{\tau}_0)\int_0^{\tau_0} \check{Z}_\rho^\beta A^\rho_{\underline{k}}\check{Z}_\mu^\gamma A^\mu_{\underline{k}} d\tau \ . \end{aligned}$$

(4.5.8) When we go from the chart α to the chart $\tilde{\alpha}$ we have to multiply $\delta x(T)$ by the Jacobian φ' of the mapping describing the change of chart and by φ'' according to Ito's rule.

(4.5.9) *Remark.* Suppose the lifetime of the equation (4.5.1) is infinite, it is not clear that the other stochastic equation defines processes with infinite lifetime. This point will be cleared in the case of compact base manifold in Part III, Section 4. In any case all this formula holds locally, in the sense that we can introduce new stopping time when the matrices $Z{:}, \check{Z}{:}, G{:}.$ become too big, and the differentiability would be obtained for time τ sufficiently small.

Part III. Hypoellipticity with degeneracy

§ 1. We shall consider a C^∞-compact manifold and the semi elliptic operator

(1.1) $$L = \frac{1}{2}\sum_{k=1}^{n} \mathscr{L}^2_{A_k}\,.$$

We shall denote by $\mathscr{E}$ the exceptional set

(1.2) $\mathscr{E} = \{v_0 : \{A_k\}_{v_0}$ *and all their brackets do not generate* $T_{v_0}(V)\}$.

A theorem by Hörmander [10] says that if $\mathscr{E}$ is empty then the operator L is hypoelliptic.

(1.3) **Fine interior**

Given a Borelian subset of V and $k_0 \in K$ we shall say that k_0 is L-finely interior ([6]) to K if, denoting by T^K the exit time of K, we have

(1.3.1) $$P_{k_0}(T^K > 0) = 1\,.$$

(1.3.2) This probability is always equal to 0 or 1 by the zero-one law.

We shall denote by K' the L-fine interior of K. Obviously $(K_1 \cap K_2)' = K_1' \cap K_2'$.

(1.4) **Covering process and fine interior**

Let π be a surjective map of $V \to \hat{V}$. Suppose that there exists on $\hat{V}$ a semi elliptic operator of the form (1.1) which is compatible with π (i.e. $(\hat{L}\hat{f}) \circ \pi = L(\hat{f} \circ \pi)$). Let $K = \pi^{-1}(\pi(K))$, now $k_0 \in K'$ if and only if $\pi(k_0) \in (\pi(K))'$. For instance if $\hat{L}$ is elliptic we have then a very explicit characterization of the $\hat{L}$-fine interior, the Wiener criterion [12]; $\tilde{v}_0$ is finely interior to $\tilde{K}$ if and only if

(1.4.1) $$\sum_k 2^{k(\tilde{m}-2)} \operatorname{cap}(\tilde{K}^c \cap \{\tilde{v}, 2^{-k} < d(\tilde{v}_0, \tilde{v}) \le 2^{-k+1}\}) < +\infty$$

where $\tilde{m} = \dim \tilde{V}$ and cap $(\cdot)$ is the Newtonian capacity.

(1.4.2) A geometric example of covering is a case in which $\hat{V}$ is a Riemannian manifold, $\hat{L}$ its Laplace Beltrami operator, V the bundle of orthonormal frames of $\tilde{V}$ and L its Bochner's Laplacian (see [20], [24]). Then $\pi(\mathscr{E})$ is contained in the set where the curvature of V degenerates.

(1.5) **Main Theorem.** *Denote by $p_\tau(v_0, dv)$ the transition probability of the process associated with* (1.1), *with the above notations and hypothesis.*

(1.5.1) *Suppose $v_0 \notin \mathscr{E}'$, then*

(1.5.2) *$p_\tau(v_0, dv)$ is absolutely continuous with respect to the Lebesgue measure, for every $\tau > 0$.*

The proof will be based on Part I and will be independent of the methods of Hörmander; our approach is related to the point of view of [30], [17]. At the end of this part a short remark will be made about a reciprocal statement.

§ 2. Bundle of linear frames

We shall denote by $GL(V)$ the bundle of linear frames, $\tilde{A}_k$ the prolongation of A_k to $GL(V)$. Given a vector field Z on V we associate

$$f_Z(r) = r^{-1}(Z_v)\,, \qquad \pi(r) = v\,.$$

Then f_Z is a map of $GL(V)$ in $\boldsymbol{R}^n$, equivariant under the $GL(m, \boldsymbol{R})$ action.

(2.1) **Lemma.** $f_{[A,Z]} = \mathscr{L}_{\tilde{A}} f_Z$.

Proof. By definition of the prolongation

(2.2) $U'_{\tau,A} = r_{\tilde{A}}(\tau) \circ r_{\tilde{A}}^{-1}(0)$, where $r(\tau)$ is an integral line of $\tilde{A}$, therefore

$$f_{(U_{-\tau,A})_* Z}(r) = f_Z(U_{\tau,\tilde{A}}(r))$$

which gives (2.1) by derivation.

We shall introduce

$$\tilde{L} = \frac{1}{2} \sum_{k=1}^{n} \mathscr{L}^2_{\tilde{A}_k} \tag{2.3}$$

and the associated process

$$dr = \tilde{A}_k db^k_\omega\,.$$

Now

(2.4) $$(\pi^{-1}(\mathscr{E}))' = \pi^{-1}(\mathscr{E}')$$

by (1.4).

Furthermore by virtue of (2.1) we have the following equivalence

(2.5) $$r_0 \in \pi^{-1}(\mathscr{E})$$

is *equivalent* to the existence of

$$l \in (\boldsymbol{R}^m)^* , \qquad l \neq 0$$

such that

$$\langle l, f_{A_k}(r_0)\rangle = 0 , \quad k = 1, 2, \cdots, n$$

and such that, for every $s_1, \cdots, s_q, k$, *we have*

$$\langle l, \mathscr{L}_{\tilde{A}_{s_1}} \mathscr{L}_{\tilde{A}_{s_2}} \cdots \mathscr{L}_{\tilde{A}_{s_q}} f_{A_k}(r_0)\rangle = 0 .$$

Proof. (2.1).

(2.6) **Proposition.** *Let* W *be an* $\tilde{L}$*-fine open set. Let* f *be a* C^∞ *function which is constant on* W. *Then*

$$(\mathscr{L}_{\tilde{A}_{s_1}} \cdots \mathscr{L}_{\tilde{A}_{s_q}} f)(w) = 0 \qquad \text{for every } s_1 \cdots s_q, \text{ and } w \in W .$$

Proof. By induction it is sufficient to prove it for the first derivatives. Now write Ito's formula, for the r_ω defined in (2.3) starting from r_0 and denoting by T the exit time from W

$$0 = \int_0^{\tau \wedge T} (\mathscr{L}_{A_k} f)(r_\omega(\xi)) db_\omega^k(\xi) + \int_0^{\tau \wedge T} (Lf)(r_\omega(\xi)) d\xi .$$

If for some k_0, $(\mathscr{L}_{A_{k_0}} f)(w_0) \neq 0$, then the stochastic integral term would dominate for $\tau \to 0$ which leads to a contradiction.

§ 3. Verification of the non degeneracy

(3.1) **Proposition.** *With the notations* (2.3), *and* $q_{\omega,\tau}$, *being defined in* II–2–(4.4.2), *we have*

(3.1.1) $$q_{\omega,\tau_1}(z) = \sum_{k=1}^n \int_0^{\tau_1} [\langle l, f_{A_k}(r_\omega(\tau))\rangle]^2 d\tau$$

where

$$z \in T^*_{v_\omega(\tau_1)}(V) , \qquad l = r^*_\omega(\tau_1) z .$$

Proof. We use II–2–(2.3.1)

$$r^{-1}_{v_\omega(\tau_1)} \circ U'_{\omega,\tau_1-\tau} \cdot A_k(v(\tau)) = r^{-1}_\omega(\tau) A_k(v(\tau)) = f_{A_k}(r_\omega(\tau)) .$$

(3.2) *Rescaling.* Making a dilation $A_k \to \lambda A_k$, $\lambda \in R^+$, we can reduce the Theorem 1.5 to the case $\tau = 1$.

(3.3) **Lemma.** *When* $\pi(r_0) \notin \mathscr{E}'$, *then* a.s. *in* ω, $q_{\omega,1}(l)$ *defined in* (3.1.1) *is a positive definite quadratic form.*

Proof. As $r_\omega(1)$ is an isomorphism we can restrict to consider the second member of (3.1.1). As the function considered is continuous in τ the negation of (3.3) is that there will exist $l(\omega) \in (\boldsymbol{R}^m)^*$ such that

(3.3.1) $P_{r_0}\{\langle l, f_{A_k}(r_\omega(\tau))\rangle = 0,\ 0 \le \tau \le 1,\ 1 \le k \le n\} > 0 .$

$$H(\omega) = \lim_{\tau \to 0} \bigcap_{0 \le \tau' \le \tau} \{f_{A_k}(r_\omega(\tau')),\ 1 \le k \le n\}^\perp$$

is *independent* of ω .

(3.3.2) $K = \{r ; \langle l_0, f_{A_k}(r)\rangle = 0,\ 1 \le k \le n\}$, $\quad l_0 \in H,\ l_0 \ne 0$.

Then according to (1.3.2), we deduce from (3.3.2)

(3.3.3) $$r_0 \in K' .$$

Now apply (2.5) and (2.6) and we get

$$K' = (\pi^{-1}(\mathscr{E}))' .$$

Use now (2.4)

$$K' = \pi^{-1}(\mathscr{E}') .$$

Therefore $\pi(r_0) \in \mathscr{E}'$, and we get a contradiction.

§ 4. Verification of the regularity hypothesis

We want to verify the hypothesis of the Theorem 5.5 in Part I. We remark that g_τ has a stochastic differential written in Part II, Chapter II, formula (4.5.7), where the indices α are roman letters i, j, $1 \le i, j \le m$. We shall keep with this convention all the formulae of Part II, Chapter II, (4.5) in action.

(4.1) **Proposition.** *The covariance matrix* σ^{ij} *has a stochastic differential.*

Proof. Using Part II, Chapter II, (4.4.1) and (4.5.6), we get

$$\sigma^{i,j}(\omega, \tau_0) = \int_0^{\tau_0} Z^i_l(\tau_0)\check{Z}^l_m(\tau) Z^j_n(\tau_0)\check{Z}^n_p(\tau) A^m_k A^p_k d\tau \,. \tag{4.1.1}$$

To compute $\delta\sigma$ we can derivate under the integral sign, and then use Ito's rule to compute the 6-linear integrand. Therefore the computation is reduced to the calculation of

$$\delta A^m_k = \Gamma^m_{i,k}\delta v^i + \Gamma^m_{ijk}\delta v^i \cdot \delta v^j \tag{4.1.2}$$

and of δZ.

We use now the mechanism of *prolongation*; we can define Z as the solution of the stochastic system consisting of II–2–(4.5.1) and (4.5.2) together, which can be written

$$dr^\alpha = \tilde{A}^\alpha_k d_S b^k_\omega \,. \tag{4.1.3}$$

Now α range in two sets of indices: α take latin letters i, $1 \le i \le m$, or α take ordered couple of two latin letters (j, j'), such couple will be denoted by $\mathbf{j}$. With this convention (4.1.3) will be read as

$$r^i = v^i \,, \qquad r^{\mathbf{j}} = Z^j_{j'}$$
$$\tilde{A}^i_k = A^i_k \,, \qquad \tilde{A}^{\mathbf{j}}_k = \Gamma^j_{q,k} Z^q_{j'} \,.$$

We can apply therefore the formulae of (4.5), Part II and then we get the existence of δZ which proves the lemma.

(4.2) We want to remark for later use, that $\tilde{\Gamma}^{\mathbf{j}}_{\cdot,k}$ splits in two species

$$\tilde{\Gamma}^{\mathbf{j}}_{p,k} = \Gamma^j_{p,q,k} Z^q_{j'} \,, \qquad \tilde{\Gamma}^{\mathbf{j}}_{\mathbf{l},k} = \Gamma^j_{l,k}\delta^{l'}_{j'} \,. \tag{4.2.1}$$
$$(\delta^{l'}_{j} = \text{Kronecker's symbol}) \,.$$

In the same way the $\tilde{\Gamma}^{\mathbf{j}}_{\cdot,k}$ splits in three species

$$\tilde{\Gamma}^{\mathbf{j}}_{\mathbf{i},\mathbf{p},k} = 0 \,, \quad \tilde{\Gamma}^{\mathbf{j}}_{i,p,k} = \Gamma^j_{i,p,q,k} Z^q_{j'} \,, \quad \tilde{\Gamma}^{\mathbf{j}}_{\mathbf{i},p,k} = \Gamma^j_{i,p,k}\delta^{i'}_{j'} \,. \tag{4.2.2}$$

We know by prolongation

$$\tilde{Z}^i_j = Z^i_j \,, \quad \tilde{G}^i_{p,q} = G^i_{p,q} \,, \quad \tilde{Z}^i_{\mathbf{j}} = 0 \,, \quad \tilde{G}^i_{\mathbf{j},p} = 0 \,, \quad \tilde{G}^i_{\mathbf{j},\mathbf{p}} = 0 \,.$$

Now (4.5.2) gives

$$\begin{cases} d\tilde{Z}^{\mathbf{j}}_i - \Gamma^j_{l,k}\delta^{l'}_{j'}\tilde{Z}^{\mathbf{l}}_i d_S b^k_\omega = M_k d_S b^k_\omega \\ d\tilde{Z}^{\mathbf{j}}_{\mathbf{i}} - \Gamma^j_{l,k}\delta^{l'}_{j'}\tilde{Z}^{\mathbf{l}}_{\mathbf{i}} = 0 \\ d\tilde{G}^{\mathbf{j}}_{p,q} - \Gamma^j_{l,k}\delta^{l'}_{j'}\tilde{G}^{\mathbf{l}}_{p,q} d_S b^k_\omega = P_k d_S b^k_\omega \\ dG^{\mathbf{j}}_{\mathbf{p},q} - \Gamma^j_{l,k}\delta^{l'}_{j'} G^{\mathbf{l}}_{\mathbf{p},q} db^k_\omega = Q_k d_S b^k_\omega \\ dG^{\mathbf{j}}_{\mathbf{p},\mathbf{q}} - \Gamma^j_{l,k}\delta^{l'}_{j'} G^{\mathbf{l}}_{\mathbf{p},\mathbf{q}} db^k_\omega = R_k d_S b^k_\omega \,. \end{cases} \tag{4.2.3}$$

In these equations, M_k, P_k, Q_k, R_k denote polynomials in the $A, \Gamma :, \Gamma :.$, and functions calculated at a previous stage when we integrate the system step by step, starting by the non prolongated system, and after proceeding in the order of the lines.

All these equations are of the type

$$dx - B_k x d_S b^k_\omega = c_k d_S b^k_\omega \tag{4.2.4}$$

where x is an unknown vector, B_k are matrix functions which can be expressed in terms of the A: and their derivatives. In the second member, the c_k are polynomials in the previously computed functions, with coefficients depending only on $A : (v_\omega(\tau)), \Gamma :. (v_\omega(\tau)), \Gamma :.., \Gamma :...$. The coefficients of these polynomials are therefore uniformly bounded.

We shall integrate (4.2.4) using the $b_{\varepsilon(\omega)}$ approximation

$$dx - B_k x db^k_{\varepsilon(\omega)} = c_k db^k_{\varepsilon(\omega)} . \tag{4.2.5}$$

We shall use Lagrange's method and let $\varepsilon_j \to 0$. We get

$$\begin{aligned} dX &= (B_k X) d_S b^k_\omega \\ d\tilde{x} &= (X^{-1} c_k) d_S b^k_\omega \quad \text{and} \quad x(\tau) = X(\tau)\tilde{x}(\tau) . \end{aligned} \tag{4.2.6}$$

Now write (4.2.6) in Ito's formalism

$$dX = B_k X db^k_\omega + \tfrac{1}{2}(B_{\underline{k}} B_{\underline{k}} + \partial_{A_{\underline{k}}} B_{\underline{k}}) X d\tau , \tag{4.2.7}$$

$$dx = X^{-1}\{c_k db^k_\omega - \tfrac{1}{2}(B_{\underline{k}} c_{\underline{k}} + \tilde{c}_k) d\tau\} , \tag{4.2.8}$$

where $\tilde{c}_k$ is, as c_k, a polynomial in the previously computed functions.

Denoting $X^{-1} = Y$, we have

$$dY = -Y B_k db^k_\omega + \tfrac{1}{2} Y (B_{\underline{k}} B_{\underline{k}} - \partial_{A_{\underline{k}}} B_{\underline{k}}) d\tau . \tag{4.2.9}$$

(4.3) Majorization of the solution of matrix valued equations

We are willing to have majorization of X and Y, not only in a chart, but on the whole interval of time $[0, 1]$, starting at zero with the identity. We shall denote by

$$0 < T_1 < T_2 , \quad T_q < 1 < T_{q+1} , \quad q = q(\omega) , \tag{4.3.1}$$

the stopping time describing the change of charts.

We shall put subscript $X^s, B^s, s = s(\omega)$, to denote that we integrate (4.2.7) in the chart s where X^s stays for $\tau \in (T_s, T_{s+1})$. We have then

$$X^s(T_{s+0}) = J^s_{s-1}(v(T_j)) X^{s-1}(T_{s-0}) , \tag{4.3.2}$$

where J is a Jacobian of first or higher derivatives associated to the change of chart. We have uniform estimate on the norm of this Jacobians and therefore we get

(4.3.3) $$\max_{0<\tau<1} \|X(\tau)\| \leq \gamma_0^{q(\omega)} \prod_s \max_{T_s \leq \tau \leq T_{s+1}} \|X^s(\tau)(X^s(T_s))^{-1}\| .$$

Introduce the class L^π of functions

(4.3.4) $$L^\pi = \left\{X(\omega, \tau); E\left(\left(\sup_{0\leq\tau\leq 1} \|X(\omega, \tau)\|\right)^p\right) < +\infty ,\right.$$
$$\left. \text{for all } p < +\infty\right\} .$$

(4.3.5) For scalar valued functions L^π is an algebra. (There is obvious generalization for vector valued functions).

Lemma. *Let γ_0 be a fixed constant, $q(\omega)$ defined in* (4.3.1). *Then* $(\gamma_0^{q(\omega)}) \in L^\pi$.

Proof. Choose on V a Riemannian metric of reference. Then there is a constant $\rho > 0$ such that

(4.3.6) $$\text{dist}\, l(v_\omega(T_s), v_\omega(T_{s+1})) \geq \rho .$$

Use now the method of comparison equation of Debiard-Gaveau-Mazet [5]; denote by T'_s the first time where

$$\text{dist}\,(v_\omega(T_s), v_\omega(T'_s)) = \tfrac{1}{2}\rho .$$

Then there exists by [5] a constant γ_1 and an abstract scalar Brownian motion $\tilde{b}$, which is $\mathcal{N}^\tau$ adapted, such that

$$\tilde{b}(T_{s+1}) - \tilde{b}(T'_s) \geq \frac{\rho}{2} - \gamma_1(T_{s+1} - T'_s) ,$$

therefore

$$P(T_{s+1} - T_s < \alpha) < \exp\left(-\frac{\rho^2}{10\alpha}\right) \qquad \text{for } \alpha < \frac{\rho}{4\gamma_1} ,$$

(4.3.7) $P(q(\omega) > N) < \exp(-\gamma_3 N^2)$, where γ_3 is a constant > 0.

(4.3.8) **Proposition.**

$$X \in L^\pi , \qquad (X)^{-1} \in L^\pi .$$

Proof. We remark first that $(X)^{-1}$ satisfies (4.2.9) and therefore it is sufficient to prove the proposition for X only. Now use (4.3.3), according to (4.3.5), (4.3.6) it is sufficient to majorize

$$I_s = \sup_{T_s<\tau<T_s+1} \|X^s(\tau)\| \quad \text{with} \quad X^s(0) = I\ .$$

We denote by G the linear group of the appropriate dimension such that the matrix B_k could be considered as element of the right Lie algebra g. Then $X^s(\tau)$ can be written as the following right multiplicative stochastic integral in the McKean's sense ([19], [11])

(4.3.9) $$X^s(\tau) = \exp\left(* \int_{[\tau,0]} B_k db^k_\omega + c d\tau \right)$$

where c is bounded. Take on G a *right invariant* Riemannian metric of reference. Let $\rho_\omega(\tau) = \text{dist}\,(X^s(\tau), \text{identity})$ for this metric. As the curvature tensor is constant we have [5],

(4.3.10) $$\rho_\omega(\tau) < 1 + M\tau + b(\gamma_4 \tau)$$

where b is an abstract Brownian motion, γ_4 is the constant $\gamma_4 = \sum_k \|B_k\|^2$ and M is bounded in terms of bounds on B_k and c. Therefore

$$P\Big(\sup_{0<\tau<1} \rho_\omega(\tau) > R + M + 2\Big) \le \exp\,(-\gamma_5 R^2)\ .$$

As

$$\|X^s(\tau)\| < \exp\,(\gamma_6(\rho(\tau) + 1))$$

we get

$$P(I_s > e^{M+R}) < \exp\,(-\gamma_7 R^2)\ ,$$

therefore

$$E(I_s^p) < e^{Mp} \exp\,(\gamma_8 p^2) = a_p\ .$$

Now by (4.3.7)

$$E\Big(\Big(\sup_{0<\tau<1} \|X_\omega(\tau)\|\Big)^p\Big) \le \sum\,(a_p)^N \exp\,(-\gamma_3 N^2) < +\infty\ .$$

(4.4) **Proposition.** g_τ, σ^{ij}, $Lg_\tau \in L^\pi$, $L\sigma^{ij}$, $\nabla\sigma^{ij} \cdot \nabla g_\tau \in L^\pi$.

(4.4.1) **Lemma.** *Consider the stochastic system*

$$dx = u_k db^k_\omega + c d\tau\ , \quad x(0) = 0\ , \quad \textit{where } u_k \in L^\pi,\ c \in L^\pi\ ,$$

then

$$x \in L^\pi .$$

Proof of (4.4.1). [9] Theorem 6, p. 26, written with a stopping time T shows that for every integer m there exists a constant γ_m such that

$$P(\sup \|x(\tau)\| > a) \leq \gamma_m a^{-2m}\ .$$

Proof of (4.4). It results from (4.3.8), (4.4.1) and from the procedure of integration in (4.2.3) that $Z, G, \tilde{Z}, \tilde{G} \in L^\pi$. Now the drifts coming from Part II, Chapter II, (3.3.2) are given by stochastic integral where only appear the $A:, \Gamma:.., \Gamma:...$, therefore are in L^π. By (4.4.1), $\tilde{y}^\cdot$ also is in L^π.

§ 5. Continuity on the trajectories

We have to prove the continuity of g_τ and all its stochastic differential invariants used on the trajectories of the O. U. process. Our approach will consist in proving that the limit theorem II–1–3 *holds uniformly* on a trajectory of the O. U. process. The estimates will be made with the tools of Fourier's analysis. In order to avoid unnecessary complication of indices we shall work as much as possible in dimension 1.

(5.1) A Fourier series representation of the O. U. process

Proposition. *Let* $u(t, \tau)$ *be defined in* I–(4.2.2), *then*

$$u(t, \tau) = \tau G_0(t) + \sum_{q=1}^{+\infty} G_q(t) \frac{\sin \pi q \tau}{q\pi} (2)^{1/2} \tag{5.1.1}$$

where $G_0, \cdots, G_q, \cdots$ *are independent samples of the* O. U. *process on* $\boldsymbol{R}$, *which has for infinitesimal generator* $\frac{1}{2}(d^2/dx^2) - \frac{1}{2}x(d/dx)$.

Proof. For fixed t we get Wiener's representation of the Brownian path. We have only to check that probability of transition given in I–(4.3.5) is the same for (5.1.1), looked upon as a process in t with values in the continuous functions in τ, result which is obvious by looking at the $G_0, G_1, \cdots, G_q$ components.

(5.2) **Lemma.** *Let* G *be a* 1-*dimension* O. U. *process, then there exists a constant* $R_0 > 0$ *such that for* $R > R_0$

$$P\left(\sup_{0<t<1} |G(t)| > R\right) \le \exp\left(-\frac{R^2}{6}\right).$$

Proof. We denote T the stopping time where $|G|$ is greater or equal to $R(T \ge 0)$. Using the Markov's property from T we have for $\xi > T$

$$x(\xi) - x(T) = -\frac{1}{2}\int_T^\xi x(\lambda)d\lambda + b(\xi) - b(T).$$

As

$$P\left(\sup_{0<\eta<1} |b(\eta) - b(0)| > 4\right) < \frac{1}{2}$$

we have

$$P\left(x(1-T) > \frac{R}{\sqrt{e}} - 4\right) \ge \frac{1}{2}.$$

Then

$$P(T \le R) \le 2\int_{Re^{-1/2}-4}^{+\infty} \exp\left(-\frac{x^2}{2}\right)dx.$$

(5.3) **Lemma.** *Let $G_0, \cdots, G_q$ be independent samples of the* O. U. *process and let c_k a sequence of constants such that*

$$\sigma = \sum_k (c_k)^2 < +\infty.$$

Then

$$Y = \sum c_k G_k$$

converge a.s. *for all t and $\sigma^{-1}Y$ is governed by an* O. U. *process.*

Proof. If all the c_k are zeroes, except a finite number, then $(G_0 \cdots G_q)$ span the O. U. process on $\boldsymbol{R}^q$ and Ito's lemma applied to a linear form defined on $\boldsymbol{R}^q$ gives the answer. Furthermore by (5.2)

$$\text{(5.3.1)} \qquad P\left\{\max_{0\le t\le 1}\left|\sum_{q_0}^{q_0+h} c_q G_q\right| > \left(\sum_{q>q_0}^{+\infty}(c_q)^2\right)R\right\} \le \exp\left(-\frac{R^2}{6}\right)$$

which makes it possible to pass to the limit.

(5.4) **Theorem.** *The application $t \to u(t, \cdot)$ is* a.s. *a continuous map from $\boldsymbol{R}^+$ to the Banach space $C^r([0, 1])$ where $C^r([0, 1])$ denotes the set* *of continuous functions on* [0, 1] *which satisfy a Hölder condition of order* γ ($\gamma < \frac{1}{2}$).

Proof. We shall split the γ fractional derivative of (5.1.1) in the following blocks

(5.4.1) $$Q_m(t, \tau) = \sum_{2^m \leq q < 2^{m+1}} G_q(t) q^{\gamma-1} \sin \pi q \tau \, .$$

The advantage of this splitting is, according to [15], that Q_m, being a polynomial of degree 2^{m+1}, we have

(5.4.2) $$\max_{\tau} |Q_m(t, \tau)| \leq 2 \max_{\tau \in U_m} |Q_m(\tau)|$$

where U_m denotes the 2^{m+1} roots of the unity.

On the other hand, by (5.2) and (5.3) for any τ fixed

(5.4.3) $$P\Big(\max_{0 \leq t \leq 1} |Q_m(t, \tau)| > c_\gamma R (2^m)^{\gamma - 1/2}\Big) < \exp\left(-\frac{R^2}{6}\right).$$

We choose $R = m^{-2}(2m)^{1/2-\gamma}$ and we get by (5.4.2), that there exists $\varepsilon > 0$ such that

$$P(\|Q_m\|_{L^\infty} > c_\gamma m^{-2}) \leq 2^{m+1} \exp(-2^{m\varepsilon})$$

which by Borel-Cantelli implies the convergence a.s. of $\sum Q_m$, uniformly on $[0, 1] \times [0, 1]$.

If we apply this result with $\gamma = 0$, we get that a.s. $u(t, \tau)$ is a continuous function in the two variables (t, τ).

Now use the same inequality for $\gamma' < \frac{1}{2}$, we get that the partial sums

$$\sum_{m=1}^{m_0} Q_m(t, \cdot)$$

stay in a bounded set of $C^{\gamma'}[0, 1]$ whatsoever is m_0 or $t \in [0, 1]$. As a bounded set of $C^{\gamma'}[0, 1]$ is a compact set in $C^{\gamma}[0, 1]$ for $\gamma < \gamma'$, Ascoli's Theorem finishes the proof.

(5.4.4) *Remark.* The $G^0_\cdot$ provide a *coordinate system* for the stochastic calculus on the O. U. process. We can denote by $d_M G_\cdot$ the martingale part of the stochastic differential of $dG_\cdot$. Then the theorem II–2–(4.4) can be written

$$dg_{\tau_0}(\omega) = (\partial_{k,q} g_{\tau_0})(\omega) d_M G_{k,q} + \mathcal{Q} g_{\tau_0}(\omega) d\tau$$

where

$$(2)^{-1/2}(\partial_{k,q}g_{\tau_0})(\omega) = \int_0^{\tau_0} U'_{\omega,\tau_0-\tau}(v_\omega(\tau))A_k \frac{\sin q\pi\tau}{q\pi} d\tau \, .$$

(5.5) Let $\tilde{F}$ be a C^2 function on V. Fix τ_0 and define $F = \tilde{F} \circ g_{\tau_0}$. Define

(5.5.1) $$QF = F'Qg_{\tau_0} + \tfrac{1}{2}\,\mathrm{Trace}_q\, F''$$

where q, Qg_{τ_0} have been defined in II–2–(4.4.2) and (4.4.8).

(5.6) **Theorem.** *On almost all trajectories of the* O. U. *process, the following integral is absolutely convergent*

(5.6.1) $$\int_0^t (QF)(x_w(\xi))d\xi = Y(w, t) \, ,$$

furthermore

(5.6.2) $$Y(w, t) \in L^2(W) \quad \text{a.s.}$$

and

(5.6.3) $$F(x_w(t)) - Y(w, t)$$

is an L^2 *martingale with continuous trajectories and* g_τ *is continuous on almost all paths.*

Proof. We remark first that the O. U. process is strongly mixing, therefore the ergodic theorem holds. As $QF \in L^2(X)$ by (4.4), then a direct application of the ergodic theorem gives (5.6.1) and (5.6.2).

Define

$$g^\varepsilon_{\tau_0}(\omega) = v_{\varepsilon(\omega)}(\tau_0)$$

and

(5.6.4) $$F^\varepsilon(\omega) = \tilde{F} \circ g^\varepsilon_{\tau_0} \, .$$

Then, for fixed ε, the C^1 norm of $A_k db^k_{\varepsilon(\omega)}$ will be controlled by the sup norm on b_ω. Therefore by Theorem 5.4 and by the classical theory of the variation of an ordinary differential system

(5.6.5) $F^\varepsilon(x_w(t))$ is, a.s., a continuous function of t.

Furthermore, F^ε is $\mathscr{N}^t$ adapted.

Define

$$(Qg^\varepsilon_\tau)(\omega), \; q^\varepsilon_\omega(l)$$

as in (5.5), replacing the stochastic differential equation appearing for the construction done in II–2 $d_S b_\omega$ by $db_{\varepsilon(\omega)}$. Define QF^ε as in (5.5). Now consider the following assumption (A):

(A) *There exists a subsequence ε'_j of ε_j for which*

$$\sup \|Qg_\tau^{\varepsilon'_j}\|_{L^2(X)} < +\infty\,, \qquad \sup \|q^{\varepsilon'_j}\|_{L^2(X)} < \infty\,.$$

We shall finish the proof under this assumption. Remark first that, as in (5.6.5),

$$|QF^\varepsilon(\omega)| \le c_\varepsilon \max_\tau |b_\omega(\tau)|\,.$$

Therefore $QF^\varepsilon \in L^2(X)$: we define

$$Y^\varepsilon(w, t) = \int_0^t (QF^\varepsilon)(x_w(\xi))d\xi \tag{5.6.6}$$

which clearly belongs to $L^2(W)$. Then

$$F^\varepsilon(x_w(t)) - Y^\varepsilon(w, t) = M^\varepsilon(w, t) \in \mathcal{M}_c \tag{5.6.7}$$

where $\mathcal{M}_c$ denotes the *Hilbert space of L^2 martingale with continuous paths* (cf. [18]).

As its L^2 norm is bounded by (A) we can extract a subsequence which is weakly convergent:

$$M_{\varepsilon''_j} \to M_0$$

and M_0 will be still an *L^2 martingale with continuous path* (cf. [18]).

We could have chosen also ε''_j such that $QF^{\varepsilon''}$ will converge weakly in $L^2(X)$. We get in this way a weakly convergent sequence in the Hilbert direct sum $\mathcal{H} = L^2(X) \oplus \mathcal{M}_c$. Then it is possible by *finite convex combination* to build another sequence which is strongly convergent in $\mathcal{H}$, that is we have found a triangular infinite matrix β_j^q, $\beta_j^q \ge 0$, $\sum_j \beta_j^q = 1$, such that defining

$$F^{\beta q} = \beta_j^q F^{\varepsilon''_j}\,, \quad QF^{\beta q} = \beta_j^q QF^{\varepsilon''_j}\,, \quad M^{\beta q} = \beta_j^q M^{\varepsilon''_j}\,,$$

then the two last functions converge in L^2 strongly; let $(QF)^\infty$, $(M)^\infty$ be their limits. By the ergodic theorem we know that

$$a_{k,q} = P\left(\int_0^1 [(QF^{\beta q} - (QF)^\infty)(x_w(t))]^2 dt > k^{-2}\right) \to 0\,, \tag{5.6.8}$$

when $q \to \infty$. Then extracting a subsequence q' which realizes $\sum a_{k,q'} < +\infty$ we have that a.s.

$$(5.6,9) \qquad \| QF^{\beta q'}(x_w(t)) - QF(x_w(t)) \|_{L^2([0,1],dt)} \to 0 \ .$$

Now (5.6.9) shows that $Y^{\beta q}$ defined via (5.6.6) converges to Y^∞ in L^2. As $M^{\beta q}$ does equally, then (5.6.7) shows that $F^{\beta q}$ converges in L^2. As by II–1–3 $g_\tau^{\varepsilon j}$ converges to g_τ a.s.,

$$F^{\beta q} \to F \qquad \text{in } L^2 \ .$$

Then we have

$$(5.6.10) \qquad F - (Y)^\infty = M^\infty \qquad \text{with } M^\infty \in \mathscr{M}_c \ .$$

Furthermore by (5.6.9), Y^∞ is pathcontinuous as being the indefinite integral of an L^2 function; then (5.6.10) proves that $F(x_w(t))$ is a continuous function of t. As this is true for every C^2 function $\tilde{F}$, defined on V, we have proved the theorem.

(5.7) **Proof of assumption** (A)

We remark that, by the limit theorem I–(1.3), the stochastic differential of dF^ε with respect to τ can be obtained by I–(2.4.5) replacing in the stochastic differential equations in τ $d_S b_\omega^k$ by $db_{\varepsilon(\omega)}^k$. Then we have to prove the analogue of Proposition 4.4, for the ordinary differential system. The recursive procedure of integration (4.2.3) holds without modification. We shall use (4.2.6) instead of (4.2.7), (4.2.8). Denote by L^π the class of function $u(x, \tau, \varepsilon)$ such that for a *universal* sequence ε_r

$$(5.7.1) \qquad \sup_r E((\sup_\tau \| u(x, \tau, \varepsilon_r) \|)^p) < +\infty \ , \qquad \text{for every } p < +\infty \ .$$

Let X_ε be the matrix function defined by the differential equation

$$dX_\varepsilon = B_k X db_{\varepsilon(\omega)}^k \ , \qquad X_\varepsilon(0) = I$$

where $B_.$ is the same as in (4.3.8), *then*

$$(5.7.2) \qquad X_. \in L^\pi \ .$$

Consider the ordinary differential system

$$dx_\varepsilon = u_{k,\varepsilon} db_{\varepsilon(\omega)}^k + c_\varepsilon d\tau \ , \qquad x(0) = 0$$

where

$$u_{k,\cdot}\ ,\quad (\partial_y u_{k,\cdot})\ ,\quad c_{\cdot} \in L^{\pi}\ ,$$

then

(5.7.3) $$x_{\cdot} \in L^{\pi}\ .$$

In order to prove these both statements we remark that the limit for $\varepsilon_j \to 0$ is in L^{π}. Therefore we have only to have a domination of the remainder term which appears in the limit theorem. We have, by known result on the modulus of continuity of the Brownian motion:

$$\gamma = \sup_{\tau,\tau'} \frac{\|v_{\omega}(\tau) - v_{\omega}(\tau')\|}{|\tau - \tau'|^{1/4}}\ ,$$

then

$$\gamma \in L^{\pi}\ .$$

As u, ψ appearing in II–1–(4.2) are bounded, $G^k, H^l \in L^{\pi}$. We can extract a subsequence ε_j'' as in (5.6.8) such that the ergodic theorem will imply

$$\|X^k_{\varepsilon,\omega}\|_{L^p([0,1],d\tau)} \in L^{\pi}$$

and the same is true for Y^l_{ε}. Finally we have to estimate

$$Z_{\varepsilon,\omega} = \sup_{\tau} \left| \int_0^{\tau} \psi(\omega, \varepsilon, \zeta) db_{\omega}(\zeta) \right|$$

where ψ is defined in II–1–(3.7).

We use again [9] Theorem 6, p. 26,

$$P(Z_{\varepsilon,\omega} > a) \leq (2p)!\ \|\psi\|^{2p}_{L^{2p}} a^{-2p}\ .$$

As $\psi \in L^{\pi}$ by the recursive integration procedure, we get $Z_{\varepsilon,\omega} \in L^{\pi}$.

(5.8) Final regularity requirement

As the σ^{ij} have a stochastic differential which has the same regularity properties as the differential of g_{τ} the method of Theorem 5.6 gives the path continuity of σ^{ij}.

Now $Qg_{\tau}, \nabla g\cdot\nabla\sigma^{ij}, Q\sigma^{ij}$ are *continuous in probability*. In fact these invariants come from the integration of stochastic differential system. The law of their increments is known by I–(4.3.5) and the exponential martingale inequality on the term in $b_{\delta\omega}$ will give the continuity in probability.

§ 6. Conclusion. The proof of the main Theorem will result in I–(5.5), combined with the Sections 3, 4, 5.

Remark. The proof of the necessity of the non degeneracy hypothesis in the main Theorem, when it is looked upon by a direct approach, leads to an elementary, but not obvious, monodromy problem. We shall describe it on a very concrete example.

Let D be the unit disk of the complex plane $\boldsymbol{C}$, D_n a family of open disks, all disjoints, contained in D. Let $K = (UD_n)^c$ and suppose the fine interior K' of K for the logarithm potential is not empty.

Now construct $\varphi_n \in C^\infty(\bar{D}_n)$, $\varphi_n \geq 0$, such that $h = \sum \varphi_n \in C^\infty(\bar{D})$. Consider now

$$u(z) = -\int_D g(z, z')h(z')dz' ,$$

where g is the Green function of U.

We shall put on D the Riemannian metric

$$ds^2 = e^u \, |dz|^2 .$$

We know that its Gaussian curvature is h.

Let $O(D)$ be the circle bundle of the orthonormal frames. If $b_\omega(\tau)$ is a sample of the diffusion on D, we lift it on $O(M)$, computing

$$I(\omega, \tau_0) = \int_0^{\tau_0} \frac{\partial u}{\partial x} db_\omega^2(\tau) - \frac{\partial u}{\partial y} db_\omega^1(\tau) .$$

Now consider the map $g_{\tau_0} : \Omega(R^2) \to \bar{D} \times R$ defined by

$$\omega \to b_\omega(\tau_0), \, I(\omega, \tau_0)$$

and let ρ be the law on $D \times \boldsymbol{R}$ so obtained. Then it is clear by the infinite connectivity of K^c that the support of ρ is always $D \times \boldsymbol{R}$. Now does $b_\omega(0) \notin K'$ imply that ρ is not absolutely continuous?

Part IV. Simplified version of the calculus of variation on a Lie group

Chapter I. Some elliptic operator associated with hypoelliptic left invariant operator on a Lie group

Let G be a Lie group. For convenience we shall suppose furthermore that G is a group of matrix. We denote by $\mathfrak{G}$ the Lie algebra of left invariant vector field on G. Let $e_1, \cdots, e_q \in \mathfrak{G}$, denote by ∂_{e_r} the derivative associated with e_r and let

$$L = \sum_{r=1}^{q} \partial_{e_r}^2 . \tag{1.1}$$

Even if L is not hypoelliptic, a left invariant diffusion can be associated with L by the multiplicative stochastic integral [19],

$$g(t) = \exp\left(* \int_{[0,t]} e_k db_\omega^k \right) \tag{1.2}$$

where $b_\omega^1, \cdots, b_\omega^q$ is a sample of the Brownian motion on R^q. Let $p_t(dg)$ denote the law of $g_\omega(t)$. Suppose L is *hypoelliptic*, then $p_t(dy)$ has a C^∞ density $p_t'(g)$ with respect to a left invariant Haar measure dg and p_t' satisfies the differential equation

$$2\frac{\partial p_t'}{\partial t} = L{*}p_t' \qquad \text{(where } L{*} \text{ denotes the adjoint of } L\text{)} . \tag{1.3}$$

Denote by Ad (γ) the adjoint action of G on $\mathfrak{G}$: Ad $(\gamma)z = \gamma z \gamma^{-1}$.

Then Ad acts on L by the formula

$$L^\gamma = \sum \partial_{e_k^\gamma}^2 \qquad \text{where} \qquad e^\gamma = \mathrm{Ad}\,(\gamma^{-1})e . \tag{1.4}$$

Define a differential operator P on G

$$(P\varphi)(g) = (L^g\varphi)(g) . \tag{1.5}$$

It is to be remarked that P is no more a left invariant differential operator.

(1.6) **Theorem.** *For every C^∞ function with compact support we have*

$$4\frac{\partial}{\partial t}\int p_t(dg)\varphi(g) = \int p_t(dg)\{P\varphi + L\varphi\} .$$

Proof. The semigroup property implies

$$p_{t+2\varepsilon} = p_\varepsilon * p_t * p_\varepsilon . \tag{1.7}$$

This can be written

$$\int p_{t+2\varepsilon}(dg')\varphi(g') = \int p_t(dg) \iint \varphi(g_1 g g_2) p_\varepsilon(dg_1) p_\varepsilon(dg_2) .$$

Now g *being fixed*, the double integral can be written

$$E\left(\varphi\left(\exp\left(*\int_{[0,\varepsilon]} e_k db_\omega^k\right) g \exp\left(*\int_{[0,\varepsilon]} e_k db_{\omega'}^k\right)\right)\right)$$
$$= E\left(\varphi\left(g \exp\left(*\int_{[0,\varepsilon]} e_k^g db^k\right) \exp\left(*\int_{[0,\varepsilon]} e_k db_{\omega'}^k\right)\right)\right).$$

Using the formula

$$\exp(a)\exp(b) = \exp(a + b + \tfrac{1}{2}[a, b] + 0(\|a\|^2 + \|b\|^2))$$

and the fact that b_ω and $b_{\omega'}$ are independent we get

$$2\frac{\partial}{\partial t}\int p_t(dg)\varphi(g)$$
$$= \lim_{\varepsilon\to 0} \varepsilon^{-1} E\left\{\varphi\left(g \exp\left(*\int_{[0,\varepsilon]} e_k^g db_\omega^k + e_k db_{\omega'}^k\right)\right) - \varphi(g)\right\}$$

which, using again the independence of b_ω and $b_{\omega'}$, proves the theorem.

(1.8) **Corollary.** *Suppose that L is hypoelliptic then*

$$\frac{\partial p_t'}{\partial t} = \frac{1}{4}(L* + P*)p_t' \cdot$$

Proof. By the theorem

$$\int\left\{\frac{\partial}{\partial t}p_t' - \frac{1}{2}(L* + P*)p_t'\right\}\varphi dg = 0$$

for any test function φ.

Obviously any convex combination of $L*$ and $P*$ will give a heat equation for p_t. The interesting fact is that, in some case $P* + L*$ can be elliptic (cf. Chapter 2).

(1.9) Interpretation as a stochastic dilatation

Let $m_\omega(t)$ be a diffusion associated to a hypoelliptic operator L, defined on a manifold M. We shall denote by $\Omega(M, L, m)$ the space of probability where m_0 denotes the starting point of $m_\omega(t)$. We shall fix now m_0. We shall call a *stochastic dilatation* of $\Omega(M, L, m_0)$ a Markov process, with state space M', and $(\Omega', \mathscr{F}_t', P')$ probability space, such that there exists an $\mathscr{F}_t'$ measurable map from Ω' to M for which the image of P' is $p_t(m_0, dm)$. Therefore evaluation of $p_t(m_0, dm)$ could be carried on the Markov process $(\Omega', \mathscr{F}_t', P')$.

Let us show that the Theorem 3.6 can be interpreted in this formalism. Using (1.2) there is a map of $\Omega(\boldsymbol{R}^q)$ on $\Omega(G, L, e)$. We shall consider

$\Omega(\boldsymbol{R}^q)$ in the non integrated form, that is as the set of q independent random variables, each of which is *white noise* on $\boldsymbol{R}^+$. The white noise is preserved by any measure preserving map, in particular by a map $t \to c - t$. Take two independent copies of $\Omega(\boldsymbol{R}^q)$ and define a map

$$f_t : \Omega(\boldsymbol{R}^q) \times \Omega(\boldsymbol{R}^q) \to \Omega(\boldsymbol{R}^q)$$

associating to the white noises (w_1^k, w_2^k), the white noise w_3^k defined on $[0, 2t]$ by

$$\begin{aligned} w_3^k(\xi) &= w_1^k(t-\xi) \qquad 0 < \xi < t \\ w_3^k(\xi) &= w_2(\xi - t) \qquad t < \xi < 2t\,. \end{aligned}$$

This transformation corresponds as the level of stochastic integral and for small ε at

(1.10) $$\exp\left(*\int_{[0,\varepsilon]} e_k w_1^k\right) g_\omega(t) \exp\left(*\int_{[0,\varepsilon]} e_k w_2^k\right)$$

and therefore to (1.7).

We also remark that the multiplicative integral (1.2) will make possible to estimate via the approach of Part I the derivative of p_t for t large.

Chapter II. Case of $SL(2, R)$. Factorization and limit law for the corresponding diffusion.

We shall make explicit now the preceding computation on an example. We shall use on $SL(2)$ the Cartan's coordinates

$$g = k'_\theta,\ ak_{-\theta}$$

where

$$k_\theta = \begin{bmatrix} \cos\theta & -\sin\theta \\ \sin\theta & \cos\theta \end{bmatrix}, \qquad a = \begin{bmatrix} e^{\tilde{a}} & 0 \\ 0 & e^{-\tilde{a}} \end{bmatrix}.$$

Take on the Lie algebra $sl(2)$ the basis

$$e_1 = \begin{bmatrix} 1 & 0 \\ 0 & -1 \end{bmatrix}, \qquad e_2 = \begin{bmatrix} 0 & 1 \\ 1 & 0 \end{bmatrix}, \qquad e_3 = \begin{bmatrix} 0 & 1 \\ -1 & 0 \end{bmatrix}$$

then denoting by $\sigma(g)$ the adjoint representation it is defined in this basis by the following 3×3 matrices

(2.1) $$\sigma(k_\theta) = \begin{bmatrix} k_\theta & 0 \\ 0 & 1 \end{bmatrix}, \qquad \sigma(a) = \begin{bmatrix} 1 & 0 \\ 0 & r(\tilde{a}) \end{bmatrix}$$

where

$$r(\tilde{a}) = \begin{bmatrix} \operatorname{ch} \tilde{a} & -\operatorname{sh} \tilde{a} \\ -\operatorname{sh} \tilde{a} & \operatorname{ch} \tilde{a} \end{bmatrix} .$$

We shall work with the hypoelliptic operator

$$L = \partial_{e_1}^2 + \partial_{e_2}^2 . \tag{2.2}$$

(2.3) **Proposition.** *The operator $L + L^g$ is elliptic on an open dense subset of $SL(2)$.*

Proof. Let (ξ_1, ξ_2, ξ_3) be the coordinates on the dual of $sl(2)$ equipped with the dual basis. Then the symbol of L is the quadratic form

$$q(\xi) = \xi_1^2 + \xi_2^2 .$$

The symbol of L^g is the quadratic form

$$q^g(\xi) = q(\sigma(g^{-1})\xi) .$$

Therefore the symbol of $L + L^g$ is the quadratic form

$$\begin{aligned} l(g, \xi) = {} & (\xi_1 \cos\theta + \xi_2 \sin\theta)^2 \\ & + (-\xi_1 \operatorname{ch} \tilde{a} \sin\theta + \xi_2 \operatorname{ch} \tilde{a} \cos\theta + \xi_3 \operatorname{sh} \tilde{a})^2 \\ & + \xi_1^2 + \xi_2^2 . \end{aligned} \tag{2.4}$$

As long as $\tilde{a} \neq 0$, $l(g, \xi)$ is a positive definite quadratic form and then the proposition is proved.

Remark. If $l(g, \xi) = a^{ij}\xi_i\xi_j$ then

$$L + L^g = \tfrac{1}{2}a^{ij}(g)(\partial_{e_i}\partial_{e_j} + \partial_{e_j}\partial_{e_i}) . \tag{2.5}$$

This comes from the fact that we compute in terms of left invariant differential operators.

(2.6) **Theorem.** *Consider the diffusion $g_\omega(t)$ associated to the heat equation*

$$\frac{\partial}{\partial t} - \frac{1}{4}(L + L^g)$$

and starting from the identity, then

$$g_\omega(t) = k_{\theta'\omega_2} a_{\omega_1} k_{-\theta\omega_2}$$

where (ω_1, ω_2) are independent variables; the radial motion $\tilde{a}_{\omega_1}(t)$ is governed by the equation

(2.6.1)
$$\frac{d^2}{d\tilde{a}^2} + \coth \tilde{a}\, \frac{d}{d\tilde{a}^2} .$$

We denote by ω_2 the data of a Brownian motion b_{ω_2} on $\boldsymbol{R}^2$ which determine, when ω_1 is known, the motion of θ', θ via the stochastic differential equations

(2.6.2)
$$d\theta' = \frac{1}{2 \operatorname{sh} \tilde{a}}[db^1_{\omega_2} + \operatorname{ch} \tilde{a}\, db^2_{\omega_2}]$$
$$d\theta = \frac{1}{2 \operatorname{sh} \tilde{a}}[db^2_{\omega_2} + \operatorname{ch} \tilde{a}\, db^1_{\omega_2}] .$$

(2.7) **Corollary.** *There exists a Brownian motion b_{ω_3} defined on $\boldsymbol{R}$, independent of the motion of θ', and a function $\varphi(\tilde{a})$, such that*

$$2d\theta = \frac{\operatorname{sh} \tilde{a}}{(2 + \operatorname{sh}^2 \tilde{a})^{1/2}} db_{\omega_3} + \varphi(\tilde{a}) d\theta' .$$

Remark. This factorization can be compared to the factorization obtained in [21] for L. Then the stochastic differentials of $d\theta$ and $d\theta'$ were exactly coupled, and a phenomena like (2.7) did not happen. Furthermore when $t \to \infty$ θ' tends to a limit when by (2.6.2) θ' turns at an increasing speed when $t \to \infty$. The qualitative aspect of the two diffusions is then fundamentally different.

Proof of the Theorem. We shall follow the method of [21]. Denote $b_1, b_2, \hat{b}_1, \hat{b}_2$ the Brownian motion on $\boldsymbol{R}^2$, we have to identify

(2.7.1)
$$k'ak^{-1} \exp(e_1 \Delta b_1 + e_1^g \Delta \hat{b}_1 + e_2 \Delta b_2 + e_2^g \Delta \hat{b}_2)$$
$$= k' \exp(\Delta k') a \exp(e_1 \Delta \tilde{a}) \exp(-\Delta k) k^{-1} .$$

We shall first move k^{-1} to the right in the first member. It appears first $\operatorname{Ad}(k^{-1})e_1$, $\operatorname{Ad}(k^{-1})e_2$, but $\operatorname{Ad}(k^{-1})$ acts as an orthogonal transformation on (e_1, e_2), therefore define a probability preserving isomorphism of the Brownian motion on $\boldsymbol{R}^2$ (b_1, b_2) and can be forgotten. Secondly appears

$$\operatorname{Ad}(k^{-1})e_1^g = \operatorname{Ad}(k^{-1}) \operatorname{Ad}(g^{-1})e_1$$
$$= \operatorname{Ad}(k^{-1}) \operatorname{Ad}(k) \operatorname{Ad}(a^{-1}) \operatorname{Ad}((k')^{-1})e_1 .$$

The two left terms compensated, now $\operatorname{Ad}((k')^{-1})$ define as before an isomorphism on $(\Delta\hat{b}_1, \Delta\hat{b}_2)$ and can be forgotten. Finally, up to stochastic isomorphisms (2.7.1) is equivalent to

$$\text{(2.7.2)} \quad \begin{aligned} &\exp(e_1 \Delta b_1 + e_2 \Delta b_2 + e_1^a \Delta \hat{b}_1 + e_2^a \Delta \hat{b}_2) \\ &\quad = \exp(e_3^a \Delta\theta') \exp(e_1 \Delta\tilde{a}) \exp(-e_3 \Delta\theta) . \end{aligned}$$

We shall solve (2.7.2) using the two steps of the Ito's differential calculus, first order meaning martingale terms, second order meaning drift terms computed in the system of coordinates $(\theta', \tilde{a}, \theta)$. At the *first order* (2.7.2) implies

$$\text{(2.7.3)} \quad e_1 \Delta b_1 + e_2 \Delta b_2 + e_1^a \Delta \hat{b}_1 + e_2^a \Delta \hat{b}_2 = e_3^a \Delta\theta' + e_1 \Delta\tilde{a} - e_3 \Delta\theta .$$

We have, according to (2.1),

$$\text{(2.7.4)} \quad \begin{aligned} e_1^a &= e_1 \\ e_2^a &= \operatorname{ch} \tilde{a}\, e_2 + \operatorname{sh} \tilde{a}\, e_3 \\ e_3^a &= \operatorname{sh} \tilde{a}\, e_2 + \operatorname{ch} \tilde{a}\, e_3 . \end{aligned}$$

Therefore (2.7.3) implies

$$\text{(2.7.5)} \quad \begin{cases} \Delta b_1 + \Delta \hat{b}_1 = \Delta \tilde{a} \\ \Delta\theta' \quad = \dfrac{1}{\operatorname{sh} \tilde{a}} [\Delta b_2 + (\operatorname{ch} \tilde{a}) \Delta \hat{b}_2] \\ \Delta\theta \quad = \dfrac{1}{\operatorname{sh} \tilde{a}} [\Delta \hat{b}_2 + (\operatorname{ch} \tilde{a}) \Delta b_2] . \end{cases}$$

Now to compute the drift term we shall use

$$\exp(u) \exp(v) = \exp(u + v + \tfrac{1}{2}[u, v] + o(\|u\| + \|v\|)) .$$

Applying this identity to (2.7.2), we see that the drifts $Z', \tilde{Z}, Z$ on the variable $(\theta', \tilde{a}, \theta)$ will satisfy

$$\text{(2.7.6)} \quad \begin{aligned} 0 = &\tfrac{1}{2}[e_3^a, e_1] \langle \Delta\theta', \Delta\tilde{a} \rangle - \tfrac{1}{2}[e_3, e_1] \langle \Delta\tilde{a}, \Delta\theta \rangle \\ &+ \tfrac{1}{2}[e_3, e_3^a] \langle \Delta\theta', \Delta\theta \rangle + e_3^a Z' + e_1 \tilde{Z} - e_3 Z . \end{aligned}$$

We substitute in this formula $\Delta\theta, \Delta\tilde{a}, \Delta\theta'$ by their first order approximation (2.7.6), then

$$\langle \Delta\theta, \Delta\tilde{a} \rangle = \langle \Delta\theta', \Delta\tilde{a} \rangle = 0$$

and

$$\langle \Delta\theta, \Delta\theta' \rangle = \frac{2 \operatorname{ch} \tilde{a}}{(\operatorname{sh} \tilde{a})^2} .$$

Then (2.7.6) implies, using (2.7.4)

$$Z' = Z = 0\,,$$
$$\tilde{Z} = 2\coth\tilde{a}$$

and the Theorem is proved.

(2.8) **Application to the equivariant spectrum**

The operator L commutes with the right action of the compact subgroup $K = \{\exp(te_3)\}$. Given an integer n, we consider $L^2_n(G)$ the subspace of function of $L^2(G)$ satisfying

(2.8.1) $$f(gk_\theta) = e^{-in\theta}f(g)\,.$$

Then $L(L^2_n(G) \cap \mathscr{D}(G)) \subset L^2_n(G) \cap \mathscr{D}(G)$; We shall denote by $-\rho(n)$ the sup of the spectrum of $-L$ restricted to $L^2_n(G)$. In the case of $G = SL(2)$ this spectrum can be computed from the knowledge of the Plancherel formula. However, the problem of determining such equivariant spectrum is a serious question in the general theory of representations of semisimple Lie groups. (cf. [22]).

Define the *stochastic holonomy* as

(2.8.2) $$\lim_{t\to\infty} t^{-1}\log\|p_{t,n}(g)\|_{L^2(G)} = \tilde{\rho}(n)$$

where

$$p_{t,n}(g) = \int_0^{2\pi} p_{t,n}(gk_\theta)e^{in\theta}\frac{d\theta}{2\pi}\,.$$

Then a natural conjecture is that $\tilde{\rho}(n) = \rho(n)$ and this has been verified on the nilpotent groups of order 2 [7]. Then we have

Proposition.

(2.8.3) $$|p_{t,n}(g)| \le E_0\left(\exp\left(-n^2\int_0^t q(\tilde{a}_\omega(\xi))d\xi\right)\right)$$

where $g = k'\exp(\tilde{a}_\omega(t))k^{-1}$, $\tilde{a}_\omega$ is the radial process governed by

$$\frac{1}{2}\frac{d^2}{d\tilde{a}^2} + \frac{1}{2}\coth\tilde{a}\frac{d}{d\tilde{a}}$$

and starting from zero and

$$8q(\tilde{a}) = \frac{\operatorname{sh}^2 \tilde{a}}{2 + \operatorname{sh}^2 \tilde{a}} .$$

Remark. The interest of (2.8.3) is to replace the evaluation of an oscillating integral by a minoration of a positive integral.

Proof. We shall use the expression of $p_t(k'a)$ as the conditional expectation

$$p_t(k'a) = E_e(e^{in\theta_\omega(t)} \mid k'_\omega(t) = k', a_\omega(t) = a) . \tag{2.8.4}$$

Let $\mathscr{F}_t$ be the σ field generalized by $(k'_\omega(\xi), a_\omega(\xi))$, $0 \leq \xi \leq t$. Then

$$p_t(k'a) = E(E^{\mathscr{F}_t}(e^{in\theta_\omega(t)}) \mid k'_\omega(t) = k', a_\omega(t) = a) .$$

Furthermore by the Corollary 4.7

$$2\theta(t) = \int_0^t \frac{\operatorname{sh} \tilde{a}}{(2 + \operatorname{sh}^2 \tilde{a})^{1/2}} db + \psi(\omega, t) \tag{2.8.5}$$

where ψ is $\mathscr{F}_t$-measurable. Then

$$|E^{\mathscr{F}_t}(e^{in\theta_\omega(t)})| = \exp\left(-n^2 \int_0^t q(\tilde{a}_\omega(\xi)) d\xi\right)$$

and (2.8.3) is proved.

Remark. This chapter could be developped above an arbitrary symmetric space (cf. [21]). In this setting there is even an exact formula of the type (2.8.3) (cf. [23]). The interest of the preceding approach is that it could perhaps give some light on the G/H case (where H is maximal torus), case where the method used in [23] fails.

Remark. Of course the calculus of variation used in Chapter II could have been used on a Lie group. The covariance matrix σ_{ij} reads on the *right* invariant Lie algebra $\mathfrak{g}$ as

$$q_\omega(l) = \int_0^t \sum_{k=1}^n l^2(\operatorname{Ad}(g_\omega(\xi))e_k) d\xi = \sigma_{i,j}(\omega) l^i l^j \tag{2.8.6}$$

where $l \in \mathfrak{g}^*$ and $g_\omega(t)$ is the *left* invariant process on G associated to

$$\frac{1}{2} \sum_{k=1}^n \partial^2_{e_k} .$$

The formula (2.8.6) corresponds to a variation of the path around each of its points; the operator $L + L^g$ corresponds to a variation at the extremity and at the origin; it is then clear that the introduction of $L + L^g$ is a simplified approach to the theory developped in Part II.

Bibliography

[1] Adams, A., Sobolev spaces, Academic Press, 1975.

[2] Airault, H. and Föllmer, H., Relative densities of semi-martingales, Invent. Math., **27** (1974), 299–327.

[3] Calderon, A., Lebesgue spaces of differentiable functions and distributions, Proc. Symp. Pure Math., **IV** (1964), 33–49.

[4] Cameron, R. H. and Martin, W. T., Transformation of Wiener integrals by non linear transformation, Trans. Amer. Math. Soc., **66** (1949), 253–283.

[5] Debiard, A., Gaveau, B. and Mazet, E., Théorèmes de comparaison, Publ. Res. Inst. Math. Sci., **12** (1976), 391–425.

[6] Fuglede, B., Finely harmonic functions, Lecture notes in Mathematics, **289**, Springer, 1972.

[7] Gaveau, B., Estimées hypoelliptiques sur certains groupes nilpotents, Acta Math., 1977.

[8] Gross, L., Potential theory in Hilbert space, J. Functional Analysis, **1** (1967), 123–181.

[9] Gihman, I. I. and Skorohod, A. V., Stochastic differential equations, Ergebnisse der Mathematik, **72**, Springer, 1972.

[10] Hörmander, L., Hypoelliptic second order differential equations, Acta Math., **119** (1967), 147–171.

[11] Ibero, M., Intégrales stochastiques multiplicatives, Bull. Sci. Math., 1976.

[12] Ito, K. and McKean, H. P., Stochastic process and their sample paths, Grundlehren der mathematischen Wissenschaften, **125**, Springer, 1965.

[13] Ito, K., Stochastic parallel displacement, Proceedings of the Victoria Conference on probabilistic methods in differential equations 1974, Lecture Notes in Mathematics, **451**, Springer, 1975.

[14] ——, Extension of stochastic integrals, These Proceedings, 95–109.

[15] Kahane, J. P., Propriétés locales des fonctions à séries de Fourier aléatoires, Studia Math., **19** (1960), 1–25.

[16] Kolmogoroff, A., Umkehrbarkeit der statistischen Naturgesetze, Math. Ann., **133** (1937), 766–772.

[17] Kunita, H., Support of a diffusion and controllability, These Proceedings, 163–185.

[18] Kunita, H. and Watanabe, S., On square integrable martingales, Nagoya Math. J., **30** (1967), 209–245.

[19] McKean, H. P., Jr., Stochastic integrals, Academic Press, 1969.

[20] Kobayashi, S. and Nomizu, K., Foundations of differential geometry, 1963.

[21] Malliavin, M. P. and Malliavin, P., Factorisations et lois limites de la diffusion horizontale au dessus d'un espace riemannien symétrique, Lecture Notes in Mathematics, **404**, Springer, 1974, 166–217.

[22] ——Diagonalisation du système de de Rham-Hodge au dessus d'un espace riemannien homogène, Lecture Notes in Mathematics, **466**, Springer, 1975, 135–146.

[23] ——, Holonomie stochastique au dessus d'un espace riemannien symétrique, C. R. Acad. Sci. Paris, **280** (1975), 793–795.

[24] Malliavin, P., Formule de la moyenne pour les formes harmoniques, J. Functional Analysis, **17** (1974), 274–291.

[25] ——, Sur un principe de transfert en géométrie stochastique, C. R. Acad. Sci. Paris, Nov., 1976.

[26] Meyer, P. A., Cours sur les intégrales stochastiques, Lecture Notes in Mathematics, **511**, Springer, 1976, 261–400.

[27] Nirenberg, L., On elliptic partial differential equations, Ann. Scuola Norm. Sup. Pisa, **13** (1959), 115–162.

[28] Segal, I., Tensor algebras over Hilbert space I, Trans. Amer. Math. Soc., 81 (1956), 106–134; II, Ann. of Math., **63** (1956), 160–175.

[29] Stein, E. M., Singular integrals and differentiability of functions, Princeton, 1970.

[30] Stroock, D. W. and Varadhan, S. R. S., On the support of diffusion processes with application to the strong maximum principle, Proc. 6th Berkeley Symposium, **III** (1972), 333–368.

[31] Michel, D., Formules de Stokes stochastiques, Comptes Rendus, Paris, Mars 1978 et Bulletin des Sciences Mathématiques, 1978.

[32] Malliavin, P., Stochastic Jacobi field. Proceedings of the Conference on Partial Differential Equations on Manifolds, Salt Lake City, 1977, to appear.

Université de Paris VI
Mathématiques
4, Place Jussieu
Paris (5e), France

Proc. of Intern. Symp. SDE
Kyoto 1976, pp. 265-281

Periodic Boundary Problems of the Two Dimensional Brownian Motion on Upperhalf Plane

Minoru MOTOO

Introduction

We shall investigate a class of conservative Feller processes with continuous path functions on the upperhalf plane, which are extensions of the absorbing Brownian motion. We also require them to be periodic with period 2π, which means that the probability laws of the processes are invariant under the translation of length 2π parallel to the x-axis. Since the absorbing Brownian motion on upperhalf plane can be transformed to that on the unit disk, by a conformal mapping and a time change, our investigation is essentially equivalent to the investigation of processes on the unit disk.

In this paper, it will be shown that each process in the class satisfies a certain boundary condition, which are generalization of Wentzell's condition [1] in the continuous case. Using this fact, every process in the class can be characterized by two periodic measures on the real line and two constants. Roughly speaking, these measures and constants have the following probabilistic meanings for the process. One measure is the supporting measure of the "density of the invariant measure" of the process, which is a positive harmonic function, and the other one is the supporting measure of the "x-derivative of the harmonic scale" of the process, which is also a positive harmonic function. One constant represents the "trend of right sided shift", while the second constant is non-negative and represents the "fluctuation on the boundary".

In the paper we shall only sketch the results.

Notations

$D = \{\text{open upperhalf plane}\} = \{z = (x, y) : y > 0\}$
$\bar{D} = \{z = (x, y) : y \geqq 0\}$
$D^a = \{z = (x, y) : 0 < y < a\}\ a > 0$
$R = \{\text{the whole real line}\}$
$C(D) = \{\text{the set of all bounded continuous functions on } D\}$
$C_0(D) = \{\text{the set of all continuous functions with compact support on } D\}$

$C(\bar{D}), C_0(\bar{D}), C(R)$ and $C_0(R)$ are defined in the similar way.

A function f on R (or D) is called periodic of period $2N\pi$ if $f(x+2N\pi) = f(x)$ for any $x \in R$, (or $f(z+2N\pi) = f(z)$ for any $z \in D$, where $z + 2N\pi = (x+2N\pi, y)$ if $z = (x, y)$.) Functions of period 2π are simply called periodic.

$\mathscr{B}(D)$, $\mathscr{B}(\bar{D})$ or $\mathscr{B}(R)$ are topological Borel fields of $D, \bar{D}$ or R.

$D^* = D \cup \{\partial\}$ is the topological space constructed by identifying the set $\{y = 0\}$ as a single point ∂. More precisely, let f be the mapping from $\bar{D}$ into D^* such that

$$f(z) = \begin{cases} z & \text{if } y > 0 \\ \partial & \text{if } y = 0 \end{cases} \tag{0.1}$$

then the topology of D^* is the strongest topology which makes f to be continuous.

§ 1. Processes in class $\mathscr{P}$ and harmonic measures

1.1 Set

$W = \{$the set of all continuous mappings from $[0.\infty)$ into $D^*\}$. For $w \in W$, we shall write $z_t(w) = w(t)$. $\boldsymbol{B}_t$ is the Borel field of subsets of W generated by $\{z_s(w) : s \leq t\}$, and $\boldsymbol{B} = \bigvee_t \boldsymbol{B}_t$.

We shall consider the following class of Markov processes.

Definition. $\boldsymbol{P} = \{P_z(\mathscr{A}) : z \in D, \mathscr{A} \in \boldsymbol{B}\}$ is in class $\mathscr{P}$ if and only if

(P.0) $P_z(\mathscr{A})$ is a probability kernel on $D \times \boldsymbol{B}$.

(P.1) $P_z(z_t \in D) = 1$.

(P.2) $P_z(z_{t+s} \in A \mid \boldsymbol{B}_t) = P_{z_t}(z_s \in A)$ a.e. P_z for $A \in \mathscr{B}(D)$

(P.3) Let $\sigma = \inf\{t : z_t = \partial\}$, then

$$P_z(z_t \in A : t < \sigma) = \tilde{P}_z(z_t \in A : t < \sigma) \quad \text{for } A \in \mathscr{B}(D),$$

where $\tilde{P}_z(\cdot)$ is the probability measure of the absorbing Brownian motion on D starting from z.

(P.4) $$P_{z+2\pi}(z_t \in A + 2\pi) = P_z(z_t \in A) \quad \text{for } A \in \mathscr{B}(D),$$

where $A + 2\pi = \{(x + 2\pi, y) : (x, y) \in A\}$.

For a measurable function $F(w)$ on W, we shall write $E_z(F(w))$

$$= \int F(w)P_z(dw).$$

(P.1), (P.2) and (P.3) show that the process $\boldsymbol{P}$ has no sojourn on the boundary, it is a Markov process on D, and it is an extension of the absorbing Brownian motion. While (P.4) means periodicity of the process. Since $\boldsymbol{P}$ is defined on W, path functions of $\boldsymbol{P}$ are continuous in D^* and have no jump from the boundary into D. We can also prove $\boldsymbol{P}$ has strong Markov property in the following sense.

Proposition 1.1. *Let σ be a Markov time, that is, $\{\sigma < t\} \in \boldsymbol{B}_t$ for any t, and set*

$\boldsymbol{B}_t = \{\mathscr{A} \in \boldsymbol{B} : \mathscr{A} \cap \{\sigma < t\} \in \boldsymbol{B}_t\}$, *then*

$$P_z(z_{\sigma+t} \in A \mid \boldsymbol{B}_t) = P_{z_\sigma}(z_t \in A) \qquad \text{a.e. } P_z$$

on the set $\{w : z_\sigma \in D\}$ for any $A \in \mathscr{B}(D)$ and $t > 0$.

1.2. Now, we shall define continuous process on $\bar{D}$. Let

$\bar{W}$ = {the set of all continuous mappings from $[0, \infty)$ into $\bar{D}$}

$\bar{\boldsymbol{B}}_t$ and $\bar{\boldsymbol{B}}$ are defined in the similar way as $\boldsymbol{B}_t$ and $\boldsymbol{B}$ are defined.

Definition. $\bar{\boldsymbol{P}} = \{\bar{P}_z(\bar{\mathscr{A}}) : z \in D, \bar{\mathscr{A}} \in \bar{\boldsymbol{B}}\}$ is in class $\bar{\mathscr{P}}$ if and only if $\bar{\boldsymbol{P}}$ satisfies (P.0), (P.1), (P.2), (P.3) and (P.4), where $\boldsymbol{B}$, $\boldsymbol{B}_t$ and ∂ should be replaced by $\bar{\boldsymbol{B}}$, $\bar{\boldsymbol{B}}_t$ and $\{y = 0\}$ respectively.

Let f be the mapping defined in (0.1), defining the measurable mapping f^* from $\bar{W}$ into W by $z_t(f^*(\bar{w})) = f(z_t(\bar{w}))$ for $\bar{w} \in \bar{W}$, we can induce the mapping $\bar{f}$ from $\bar{\mathscr{P}}$ into $\mathscr{P}$ by $\bar{f} \circ \bar{P}_z(\mathscr{A}) = \bar{P}_z(f^{*-1}(\mathscr{A}))$ for $\mathscr{A} \in \boldsymbol{B}$. It is easily seen $\bar{f}$ is an injection, and we shall identify the image of $\bar{\mathscr{P}}$ by $\bar{f}$ with $\bar{\mathscr{P}}$. Thus $\bar{\mathscr{P}}$ can be embedded in $\mathscr{P}$.

For $\boldsymbol{P}$ in $\mathscr{P}$, set $T_t\phi(z) = E_z(\phi(z_t))$ for $\phi \in C(D)$. It is easily seen $T_t\phi$ is continuous in D by (P.3).

Definition. For $\boldsymbol{P}$ in $\mathscr{P}$, $\boldsymbol{P}$ is in $\mathscr{P}_f$ if and only if (1) for any $\phi \in C_0(D)$, there exists the (unique) continuous extension $\bar{T}_t\phi(z)$ to $\bar{D}$ of $T_t\phi(z)$, and (2) if $\phi_n \in C_0(D)$ and $0 \leq \phi_n \uparrow 1$ $(n \to \infty)$, then $\bar{T}_t\phi_n \uparrow 1$ $(n \to \infty)$ in $\bar{D}$. We set $\bar{\mathscr{P}}_f = \bar{\mathscr{P}} \cap \mathscr{P}_f$.

Processes in $\mathscr{P}_f$ are Feller processes, and $\bar{\mathscr{P}}_f$ is the class which we have discussed in introduction.

1.3. Let $\boldsymbol{P}$ be in $\mathscr{P}$. For any $a > 0$ and $B \in \mathscr{B}(R)$, set

(1.1) $$H_z^a(B) = P_z(x_{\sigma_a} \in B) \qquad \text{for } z \in D^a ,$$

where $\sigma_a = \inf\{t : y_t \geqq a\}$ and $z_t(w) = (x_t, y_t)$.

It is easily seen:

(H.0) For a fixed $a > 0$, $H_z^a(B)$ is a probability kernel on $D^a \times \mathscr{B}(R)$.

(H.1) For a fixed $a > 0$ and a fixed $B \in \mathscr{B}(R)$, $H_z^a(B)$ is a harmonic function in D^a.

(H.2) For $0 < b < a$ and $z \in D^b$

$$H_z^a(B) = \int H_{(\xi,b)}^a(B) H_z^b(d\xi) .$$

(H.3) $$H_{z+2\pi}^a(B + 2\pi) = H_z^a(B) \qquad \text{for } z \in D^a .$$

We shall write $H^a f(z) = \int H_z^a(d\xi) f(\xi)$ for $f \in C(R)$

Definition. $H = \{H_z^a(B) : a > 0, z \in D^a, B \in \mathscr{B}(R)\}$ is in class $\mathscr{H}$. if and only if H satisfies (H.0), (H.1), (H.2) and (H.3).

We have the following correspondence between $\mathscr{P}$ and $\mathscr{H}$.

Theorem 1.2. *For any* H *in* $\mathscr{H}$, *there exists the unique* P *in* $\mathscr{P}$ *which satisfies* (1.1). *Therefore, there is the bijection from* $\mathscr{P}$ *onto* $\mathscr{H}$ *by the mapping* (1.1).

Feller property of the process can be stated in terms of H:

Proposition 1.3. *Let* P *be in* $\mathscr{P}$. *Then* P *is in* $\mathscr{P}_f$ *if and only if* (1) $H^a f(z)$ $(z \in D^a)$ *has the continuous extension* $\overline{H}^a f(z)$ *to* $\overline{D}^a = \{z : 0 \leq y \leq a\}$ *for any* $a > 0$ *and* $f \in C_0(R)$, *and* (2) *if* $f_n \in C_0(R)$ *and* $0 \leq f_n \uparrow 1$ $(M \to \infty)$, *then* $\overline{H}^a f_n \uparrow 1$ $(n \to \infty)$ *in* $\overline{D}^a$.

§ 2. Processes with smooth boundary conditions

Let $\alpha(x)$ and $\beta(x)$ be smooth periodic functions on R and $\alpha(x) > 0$. For $f \in C(R)$ and $\alpha > 0$, we shall consider the following boundary value problem.

(2.1) $$\begin{cases} \phi = f & \text{on } y = a \\ \Delta\phi = 0 & \text{in } D^a \\ \alpha(x)\dfrac{\partial^2\phi}{\partial x^2} + \beta(x)\dfrac{\partial\phi}{\partial x} + \dfrac{\partial\phi}{\partial y} = 0 & \text{on } y = 0 . \end{cases}$$

Then we can show that (2.1) has the unique smooth solution $\phi(z) \equiv H^{\alpha}f(z) = \int H_z^{\alpha}(d\xi)f(\xi)$, and $H = \{H_z^{\alpha}(B)\}$ is in class $\mathscr{H}$. We shall denote the process corresponding to H by $P_{\alpha,\beta}$ (c.f. theorem 1.2). We can also show that $P_{\alpha,\beta}$ is in $\overline{\mathscr{P}}_f$.

Let us consider;

$$(2.2)\qquad \begin{cases} \Delta u = 0 & \text{in } D \\ \alpha(x)\dfrac{\partial^2 u}{\partial x^2} + \beta(x)\dfrac{\partial u}{\partial x} + \dfrac{\partial u}{\partial y} = 0 & \text{on } y = 0 \\ \dfrac{\partial u}{\partial x} \text{ is periodic and } \lim\limits_{y\to\infty} \dfrac{\partial u}{\partial x} = 1 \\ u(0, 1) = 0 \end{cases}$$

and

$$(2.3)\qquad \begin{cases} \Delta m = 0 & \text{in } D \\ \dfrac{\partial^2}{\partial x^2}(\alpha(x)m) - \dfrac{\partial}{\partial x}(\beta(x)m) + \dfrac{\partial m}{\partial y} = 0 & \text{on } y = 0 \\ m \text{ is periodic and } \lim\limits_{y\to\infty} m = 1\,. \end{cases}$$

Then it can be shown that (2.2) and (2.3) have the unique solutions u and m respectively, and $s = \partial u/\partial x$ and m are positive, periodic harmonic functions on D and smooth in $\bar{D}$. We can also show that $u(z)$ is harmonic with respect to $P_{\alpha,\beta}$ and $m(z)dz$ is a periodic invariant measure of $P_{\alpha,\beta}$.

Integrating the boundary condition in (2.3) we have;

$$(2.3)'\qquad \frac{\partial}{\partial x}(\alpha m) - \beta m + \int^{x} \frac{\partial m}{\partial y} = \text{const.} \qquad \text{on } y = 0\,.$$

Set,

$$(2.4)\qquad 2\pi k = \int_0^{2\pi} \beta(x)m(x, 0)dx\,,$$

and take the harmonic conjugate l of m such that

$$(2.5)\qquad 2\pi k = -\int_0^{2\pi} l(x, y)dx\,.$$

Note that l is periodic harmonic in D and the right side of (2.5) is independent of y. Now, (2.3)′ becomes;

$$\frac{\partial}{\partial x}(\alpha m) - \beta m - l = 0 \qquad \text{on } y = 0 \tag{2.6}$$

Eliminating β from (2.6) and the boundary condition in (2.2), it holds that

$$\frac{\partial}{\partial x}(\alpha m s) - ls - mt = 0 \qquad \text{on } y = 0 \,. \tag{2.7}$$

where $s = \partial u/\partial x$ and $t = -\partial u/\partial y$. We can prove t is a periodic harmonic function with $2\pi k = \int_0^{2\pi} t(x, y)dx$.

Let $U(z)$ $(z \in D)$ be a solution of

$$\begin{cases} \dfrac{\partial U}{\partial x} = ls + mt \\ \dfrac{\partial U}{\partial y} = -lt + ms \end{cases} \qquad \text{in } D \,, \tag{2.8}$$

then U is periodic harmonic in D and smooth in $\bar{D}$. Taking a suitable additive constant, we can assume

$$U = \alpha m s \qquad \text{on } y = 0 \tag{2.9}$$

by (2.7).

In the similar way, starting from (2.1) and (2.6), we can show that there exist a harmonic function $U_\phi(z)$ in D^a such that

$$\begin{cases} \dfrac{\partial U_\phi}{\partial x} = l\dfrac{\partial \phi}{\partial x} - m\dfrac{\partial \phi}{\partial y} \\ \dfrac{\partial U_\phi}{\partial y} = l\dfrac{\partial \phi}{\partial y} + m\dfrac{\partial \phi}{\partial x} \,, \end{cases} \tag{2.10}$$

and

$$U_\phi = \alpha m \frac{\partial \phi}{\partial x} \qquad \text{on } y = 0 \,. \tag{2.11}$$

Eliminating α from (2.9) and (2.11), (2.1) can be written in the form:

$$\begin{cases} \phi = f \qquad \text{on } y = a \\ \Delta \phi = 0 \qquad \text{in } D^a \\ U_\phi s = U \dfrac{\partial \phi}{\partial x} \qquad \text{on } y = 0 \\ \text{where } U_\phi \text{ satisfies (2.10)} \,. \end{cases} \tag{2.12}$$

Conversely, from (2.12) we can easily go back to (2.1).

Proposition 2.1. (2.1) *is equivalent to* (2.12). *In other words, ϕ is a solution of* (2.1) *if and only if ϕ is a solution of* (2.12).

In § 3, we shall generalize boundary problems in the form (2.12). For given $a > 0$ and $f \in C(R)$, l, m and U determine the problem (2.12). Note that l, m and U are determined by s, m, k and p such that

$$(2.13) \qquad 2\pi p = \int_0^{2\pi} U(x, 0)dx = \int_0^{2\pi} \alpha(x)\left(\frac{\partial u}{\partial x}(x, 0)\right)^2 m(x, 0)dx .$$

In § 5, we shall use this fact to characterize processes in $\bar{\mathscr{P}}_f$.

§ 3. Boundary conditions in general case

Retaining proposition 2.1 in mind, we shall define the general boundary condition.

3.1.

[I] Let $\sigma(dx)$ and $\mu(dx)$ be a positive periodic measures on R such that

$$(3.1) \qquad 2\pi = \int_0^{2\pi} \sigma(dx) = \int_0^{2\pi} \mu(dx) .$$

Define two harmonic functions $s(z)$ and $m(z)$ on D by

$$(3.2) \qquad \begin{cases} s(z) = \dfrac{1}{\pi}\displaystyle\int_R \frac{y\sigma(d\xi)}{y^2 + (\xi - x)^2} \\ m(z) = \dfrac{1}{\pi}\displaystyle\int_R \frac{y\mu(d\xi)}{y^2 + (\xi - x)^2} . \end{cases}$$

[II] Let k be any constant.
Denote by $t(z)$ and $l(z)$ the harmonic conjugates of $s(z)$ and $m(z)$, respectively, such that

$$(3.3) \qquad 2\pi k = \int_0^{2\pi} t(x, y)dx = -\int_0^{2\pi} l(x, y)dx .$$

Note that s, m, t and l are periodic harmonic, and s and m are positive in D.

Let U be a function in D which satisfies

$$(3.4) \quad \begin{cases} \dfrac{\partial U}{\partial x} = ls + mt \\ \dfrac{\partial U}{\partial y} = -lt + ms \end{cases} \quad \text{in } D$$

Then, U is a periodic harmonic function in D and determined except for an additive constant.

We note the following theorem:

Theorem 3.1. (3.4) *has at least one non-negative solution if and only if*

[P] *σ and μ have no common discrete mass, and* $\int_0^{2\pi}\int_0^{2\pi} |\cot(\xi - n)/2|$ $\times\, s(x, \xi, \eta)\sigma(d\xi)\mu(d\eta)$ *is bounded in x, where*

$$s(x, \xi, \eta) = \begin{cases} 1 & \text{if } (x, \xi, \eta) \in \left(\bigcup_{n,m} A_{n,m}\right) \cup \left(\bigcup_{n,m} B_{n,m}\right) \\ \dfrac{1}{2} & \text{if } (x, \xi, \eta) \in \left(\bigcup_n C_n\right) \cup \left(\bigcup_n D_n\right) \\ 0 & \text{otherwise} \end{cases}$$

and

$$\begin{aligned} A_{n,m} &= \{2(n-1)\pi < \eta - \xi < (2n+1)\pi,\ \eta < x + 2m\pi < \xi\} \\ B_{n,m} &= \{2(n-1)\pi < \eta - \xi < (2n+1)\pi,\ \xi < x + 2m\pi < \eta\} \\ C_n &= \{\xi = x + 2n\pi\} \\ D^n &= \{\eta = x + 2n\pi\}\,. \end{aligned}$$

In this section, we shall assume the condition [P]. So, there exists the minimum non-negative solution U_0 of (3.4). Set,

$$(3.5) \qquad p_0 \equiv p_0(\sigma, \mu, k) = \inf_y \int_0^{2\pi} U_0(x, y)s(x, y)dx \geqq 0\,.$$

[III] Let p be any constant that $p \geqq p_0$. Then, it is easily seen that there exists the unique non-negative solution U of (3.4) in D such that

$$(3.6) \qquad p = \inf_y \int_0^{2\pi} U(x, y)s(x, y)dx\,.$$

Actually, we can show that $p(y) = \int_0^{2\pi} U(x, y)s(x, y)dx$ is increasing in y.

3.2. Let a system $B \equiv \{\sigma, \mu, k, p\}$, satisfying [I], [II], [P] and [III], be given. Then, m, l, t, s and U are defined as in 3.1.

Definition. Let $B \equiv \{\sigma, \mu, k, p\}$ be given. For $a > 0$ and $f \in C(R)$, a function $\phi \in C(\{z: 0 < y \leqq a\}) \cap C^2(D^a)$ is a B-solution of f in D^a if and only if

$$\begin{cases} \phi = f & \text{on } y = a \\ \Delta\phi = 0 & \text{in } D^a \end{cases}$$

and measures $\nu_y(dx) = (U(x, y)(\partial\phi/\partial x)(x, y) - U_\phi(x, y)s(x, y))dx \ (y > 0)$ converge to 0 weak* on R as $y \to 0$. Here U_ϕ is a function in D^a which satisfies;

$$(3.7) \qquad \begin{cases} \dfrac{\partial U_\phi}{\partial x} = l\dfrac{\partial \phi}{\partial x} - m\dfrac{\partial \phi}{\partial y} \\ \dfrac{\partial U_\phi}{\partial y} = -l\dfrac{\partial \phi}{\partial y} + m\dfrac{\partial \phi}{\partial x} \end{cases} \text{in } D^a .$$

Then, we can prove the following theorem.

Theorem 3.2. *Let $B \equiv \{\sigma, \mu, k, p\}$ satisfying* [I], [II], [P] *and* [III] *be given. Then, there exists one and only one process $\boldsymbol{P}_B$ in $\mathscr{P}$ such that, for any $a > 0$ and $f \in C(R)$ of period $2N\pi$ (N is any positive integer) $\phi = H^a f$ is a B-solution of f in D^a.*

We shall call $\boldsymbol{P}_B$ in the theorem B-process, and set

$$\mathscr{P}_B = \{\text{the set of all } B\text{-processes}\}$$

§ 4. Probabilistic interpretation of $B \equiv \{\sigma, \mu, k, p\}$

4.1. In this section, we shall fix a process $\boldsymbol{P}$ in $\mathscr{P}$ and assume the following two conditions:

[V] $$r(y, a) = \sup_x \int_R H^a_{(x,y)}(d\xi)(\xi - x)^2 < \infty \qquad \text{for any } a > y > 0 .$$

[M] If f is increasing function in $C(R)$, $H^a f(x, y)$ is also increasing in x for any $a > y > 0$.

The "horizontal harmonic scale" $u(z)$ can be defined by the following theorem.

Theorem 4.1. *If $\boldsymbol{P}$ in $\mathscr{P}$ satisfies* [M] *and* [V], *then there exists the unique function $u(z)$ in D such that*

(1) $u(z) = \int H_z^a(d\xi)u(\xi, a)$ *for any* $a > 0$ *and* $z \in D^a$

(2) $\frac{\partial u}{\partial x}(z) \geq 0$ *and* $u(z + 2\pi) - u(z) = 2\pi$ *for any* $z \in D$,

(3) $u(0, 1) = 0$.

The function u is harmonic in D. $s = \partial u/\partial x$ is positive, periodic and harmonic in D, which satisfies $\lim_{y\to\infty} s(z) = 1$. $t = -\partial u/\partial y$ is a periodic conjugate of s in D.

We shall set

$$2\pi k_P = \int_0^{2\pi} t(x, y)dx ; \tag{4.1}$$

note that the right side of (2.4) is independent of y. Using Martin's representation theorem for positive harmonic functions in D, we can define the periodic measure σ_P on R by,

$$s(z) = \frac{1}{\pi}\int \frac{y\sigma_P(d\xi)}{y^2 + (\xi - x)^2} . \tag{4.2}$$

About the invariant measure for P, we have the following theorem.

Theorem 4.2. *If P is in $\mathscr{P}$, then P has an invariant measure $m(dz)$ which is unique except multiplicative constant. $m(dz)$ has a density function $m(z)$ which is positive, periodic and harmonic in D.*

We define the periodic measure μ_P on R by

$$m(z) = \frac{1}{\pi}\int_{-\infty}^{\infty} \frac{y\mu_P(d\xi)}{y^2 + (\xi - x)^2} \tag{4.3}$$

with the normalizing condition

$$2\pi = \int_0^{2\pi} \mu_P(d\xi) . \tag{4.4}$$

Note that (4.4) means

$$2\pi = \int_0^{2\pi} m(x, y)dx . \tag{4.4$'$}$$

The function $l(z)$ is defined as a periodic harmonic conjugate of m with

(4.5) $$2\pi k_P = \int_0^{2\pi} l(x, y)dx\,.$$

4.2. Now, we shall define the energy integral on the boundary.

Proposition 4.3. *Let* P *in* $\mathscr{P}$ *satisfy* [M] *and* [V], *for* $f \in C^2(R)$ *we can define*

(4.6) $$B^a f(x) = \lim_{y \to a} \frac{1}{a - y} \int H^a_{(x,y)}(d\xi)(f(\xi) - f(x))$$

for $a > 0$ *and* $x \in R$. $B^a f(x)$ *can be represented by*

$$B^a f(x) = \lim_{\varepsilon \downarrow 0} \int_{|\xi - x| > \varepsilon} B^a(x, d\xi)(f(\xi) - f(x))\,,$$

where $B^a(x, A)$ *is a positive singular kernel on* $R \times \mathscr{B}(R)$ *which satisfies*

$$B^a(x + 2\pi, A + 2\pi) = B^a(x, A) \quad \text{and}$$

$$\int B^a(x, d\xi)(\xi - x)^2 < \infty\,.$$

Note that we can prove

(4.7) $$2\pi k_P = \int_0^{2\pi} m(x, a)dx\left(\lim_{\varepsilon \downarrow 0} \int_{|\xi - x| > \varepsilon} B^a(x, d\xi)(\xi - x)\right).$$

For the smooth processes in § 2, we can show

(4.8) $$\frac{1}{2}\int_0^{2\pi} m(x, a)dx \int B^a(x, d\xi)(u(\xi, a) - u(x, a))^2$$
$$= \frac{1}{2}\iint_{D^a \cap \{0 \leq x < 2\pi\}} m\left\{\left(\frac{\partial u}{\partial x}\right)^2 + \left(\frac{\partial u}{\partial y}\right)^2\right\}dxdy$$
$$+ \int_0^{2\pi} \alpha(x)\left(\frac{\partial u}{\partial x}(x, 0)\right)^2 dx\,,$$

and the right side converges to $2\pi p$ $(a \to 0)$ defined by (2.13). In general, if P in $\mathscr{P}$ satisfies [M] and [V], set

$$p(a) = \frac{1}{2}\int_0^{2\pi} m(x, a)dx \int B^a(x, d\xi)(u(\xi, a) - u(x, a))^2\,,$$

then $p(a)$ decreases as a decreases. We define,

(4.9) $$2\pi p_P = \lim_{a \to 0} p(a)\,.$$

4.3. Let f and g be twice differentiable functions on R such that df/dx, d^2f/dx^2, dg/dx, $d^2g/dx^2 \in C(R)$. Set

$$\rho_{f,g}(x, \xi) = \int_x^\xi g'(t)(f(t) - f(x))dt = \int_x^\xi (g(\xi) - g(t))f'(t)dt$$

and

$$B^a(f, g) = \int_0^{2\pi} m(x, a)dx \int B^a(x, d\xi)\rho_{f,g}(x, \xi) ,$$

then

$$B^a(f, f) = \frac{1}{2}\int_0^{2\pi} m(x, a)dx \int B^a(x, d\xi)(f(\xi) - f(x))^2 .$$

Theorem 4.4. *Let* P *in* $\mathscr{P}$ *satisfy* [M] *and* [V], *then there exists the unique function* U_P *in* D *such that*

$$B^a(f, u(\cdot a)) = \int_0^{2\pi} f'(x)U_P(x, a)dx \tag{4.10}$$

for any a *and any twice differentiable function* f *on* R *with* $f(x + 2\pi) - f(x) =$ const. *Then* U_P *is positive, periodic and harmonic in* D, *and satisfies*

$$\begin{cases} \dfrac{\partial U_P}{\partial x} = ls + mt \\ \dfrac{\partial U_P}{\partial y} = -lt + ms . \end{cases} \tag{4.11}$$

Then, we have

$$p_P = \lim_{a \to 0} B^a(u(\cdot, a), u(\cdot, a)) = \lim_{a \to 0} \int_0^{2\pi} s(x, a)U_P(x, a)dx \tag{4.12}$$

Since U_P is non-negative, by theorem 3.1, σ_P and μ_P satisfy the condition [P] and $p_P \geqq p_0 (\sigma_P, \mu_P, k_P)$.

Proposition 4.5. *Let* P *in* $\mathscr{P}$ *satisfy* [M] *and* [V], *and* $B(P) = \{\sigma_P, \mu_P, k_P, p_P\}$ *be defined by* (4.2), (4.3), (4.1) *and* (4.9). *Then* $B(P)$ *satisfies conditions* [I], [II], [P] *and* [III] *in* § 3.

The next theorem shows that, for B-process P, $B(P)$ defined in this section coinsides with B.

Theorem 4.6. *If* $\boldsymbol{P}$ *is B-process for* $B = \{\sigma, \mu, k, p\}$, *then* $\boldsymbol{P}$ *satisfies* [M] *and* [V], *and* $B = B(\boldsymbol{P})$. *That is,* $\sigma = \sigma_{\boldsymbol{P}}$, $\mu = \mu_{\boldsymbol{P}}$, $k = k_{\boldsymbol{P}}$ *and* $p = p_{\boldsymbol{P}}$. *Especially, if* $\boldsymbol{P}$ *is in* $\mathscr{P}_B$, *then* $\boldsymbol{P}$ *is* $B(\boldsymbol{P})$*-process.*

For any $\boldsymbol{P}$ in $\mathscr{P}$ satisfying [M] and [V], we can define $B(\boldsymbol{P})$. But, in general, $\boldsymbol{P}$ is not necessarily $B(\boldsymbol{P})$-process. There are many processes in $\mathscr{P}$ which have the same $B(\boldsymbol{P})$. Example 1 in § 6 shows the situation.

§ 5. Local condition and processes with continuous path function

As a sufficient condition for processes in $\mathscr{P}$ to be B-process, the following one is useful.

Proposition 5.1. *If* $\boldsymbol{P}$ *in* $\mathscr{P}$ *satisfies* [M], [V] *and*

$$\text{[L]} \qquad \lim_{a \to 0} \int_0^{2\pi} m(x, a)dx \int_{|\xi - x| > \varepsilon} B^a(x, d\xi)(u(\xi, a) - u(\xi, x))^2 = 0$$

for any $\varepsilon > 0$,
then $\boldsymbol{P}$ *is* $B(\boldsymbol{P})$*-process, where* $B(\boldsymbol{P}) = \{\sigma_{\boldsymbol{P}}, \mu_{\boldsymbol{P}}, k_{\boldsymbol{P}}, p_{\boldsymbol{P}}\}$ *is defined as in* § 4.

The condition [L] also has a close connection with continuity of path functions of the process.

Theorem 5.2. *If* $\boldsymbol{P}$ *in* $\mathscr{P}$ *satisfies* [M] *and* [V], *then* $\boldsymbol{P}$ *is in* $\overline{\mathscr{P}}$ *if and only if* $\boldsymbol{P}$ *satisfies* [L], *and* $\sigma_{\boldsymbol{P}}$ *and* $\mu_{\boldsymbol{P}}$ *are positive for any open non empty set in* R.

Finally, we can discuss processes in $\overline{\mathscr{P}}_f$ (Feller processes with continuous path functions) which are main objects of the paper.

Theorem 5.3. *Every process* $\boldsymbol{P}$ *in* $\overline{\mathscr{P}}_f$ *satisfies* [M], [V] *and* [L]. *Therefore,* $\boldsymbol{P}$ *is* $B(\boldsymbol{P})$*-process. Moreover,*

(1) $\sigma_{\boldsymbol{P}}$ *and* $\mu_{\boldsymbol{P}}$ *are positive for any non empty open interval in* R, *and*
(2) $\sigma_{\boldsymbol{P}}$ *is a continuous (atomless) measure.*

Conversely, we have the following theorem.

Theorem 5.4. *If* $B = B(\sigma, \mu, k, p)$ *satisfies* [I], [II], [P] *and* [III] *in* § 3, *and* σ *and* μ *satisfy conditions* (1) *and* (2) *of theorem* 5.3, *then* B*-process* $\boldsymbol{P}_B$ *is in* $\overline{\mathscr{P}}_f$.

In conclusion, set

$$V = \left\{ \begin{array}{l} B \equiv \{\sigma, \mu, k, p\}: \\ \sigma \text{ and } \mu \text{ satisfy [I] and [P] in § 3} \\ \text{and (1) and (2) in theorem 5.3,} \\ k \text{ is any constant} \\ p \text{ is non negative constant with} \\ p \geqq p_0\ (\sigma, \mu, k, p) \end{array} \right\},$$

then, by the correspondence of B in V with B-process $\boldsymbol{P}_B$, we have a bijection from V onto $\overline{\mathscr{P}}_f$.

§ 6. Examples

Example 1. Translation invariant processes.
The class $\mathscr{P}_t$ of translation invariant processes are defined by

$$\mathscr{P}_t = \{\boldsymbol{P} \in \mathscr{P}: P_{z+c}(z_t \in A + c) = P_z(z_t \in A) \text{ for any } c > 0\}$$

By the methods which are similar to those used in calculating infinite divisible laws, we can get the following proposition.

Proposition 6.1. *Let $\boldsymbol{P}$ be in $\mathscr{P}_t$, then for any $a > 0$ and $f \in C(R)$, $\phi = H^a f$ satisfies;*

$$(6.1) \quad \begin{cases} \phi = f & \text{on } y = a \\ \Delta\phi = 0 & \text{in } D^a \\ \alpha \dfrac{\partial^2 \phi}{\partial x^2} + \beta \dfrac{\partial \phi}{\partial x} + \displaystyle\int \Pi(d\xi)\Big(\phi(x+\xi, 0) - \phi(x, 0) & \\ \qquad\qquad - \dfrac{\xi}{1+\xi^2}\dfrac{\partial \phi}{\partial x}(x, 0)\Big) + \dfrac{\partial \phi}{\partial y} = 0\,, & \end{cases}$$

where,

$$(6.2) \quad \begin{cases} \alpha \text{ and } \beta \text{ are constants and } \alpha \geqq 0, \text{ and } \Pi \text{ is a measure on } R \text{ such} \\ \text{that } \displaystyle\int \Pi(d\xi)\frac{\xi^2}{1+\xi^2} < \infty\,. \end{cases}$$

Conversely, α, β and Π satisfying (6.2) be given, then (6.1) has the unique solution ϕ;

$$\phi(z) = \int H_z^a(d\xi) f(\xi)$$

and $H = \{H_z^a(A)\}$ *determines a process in* $\mathscr{P}_t$.

Every process in $\mathscr{P}_t$ satisfies [M], and it satisfies [V] if and only if

$$[V_t] \qquad \int \Pi(d\xi)\xi^2 < \infty \ .$$

And under $[V_t]$, we can get

$$\mu_P = \sigma_P = dx \qquad \text{(Lebesque measure on } R)$$

$$k_P = \beta + \int \Pi(d\xi)\frac{\xi^3}{1+\xi^2}$$

$$p_P = \alpha + \frac{1}{2}\int \Pi(d\xi)\xi^2 \ , \quad \text{and}$$

$$\lim_{a\to 0}\int_{|\xi-x|>\varepsilon} B^a(x, d\xi)f(\xi) = \int_{|\xi-x|>\varepsilon} \Pi(d\xi)f(x+\xi) \qquad \text{for any } \varepsilon > 0$$

for f in $C(R)$. Especially, P is not determined by $B(P) = \{\mu_P, \sigma_P, k_P, p_P\}$.

Proposition 6.2. *Let* P *be in* $\mathscr{P}$, *then the following five conditions are equivalent.*

(1) $P \in \mathscr{P}_t \cap \overline{\mathscr{P}}_f$

(2) $P \in \mathscr{P}_t \cap \overline{\mathscr{P}}_B$

(3) $P \in \mathscr{P}_B$ *and* $P = P_B$ *for* $B = \{dx, dx, \beta, \alpha\}$ *where* $\alpha \geqq 0$.

(4) $P \in \mathscr{P}_t$ *and* $\Pi \equiv 0$ *in* (6.1)

(5) $P \in \mathscr{P}_t$, P *satisfies* $[V_t]$ *and*

$$\lim_{a\to 0}\int_{|\xi-x|>\varepsilon} B^a(x, d\xi)(\xi-x)^2 = 0 \ .$$

Example 2. Processes with inclined reflections.

These processes are discussed in [2]. Now, we shall consider them in our general scheme.

Let μ be a periodic measure on R with $2\pi = \int_0^{2\pi} \mu(dx)$. m be the harmonic function in D defined by

$$m(z) = \frac{1}{\pi}\int \frac{y\mu(d\xi)}{y^2+(\xi-x)^2} \ ,$$

and l be the harmonic conjugate of m with $2\pi k = -\int_0^{2\pi} l(x, y)dx$. Then, the real part s of the analytic function

$$F(z) = \frac{1 + k^2}{m(z) + il(z)}$$

is positive, periodic and harmonic and satisfies $2\pi = \int_0^{2\pi} s(x, y)dx$. We define $\sigma \equiv \sigma(\mu, k)$ by

(6.3) $$s(z) = \frac{1}{\pi} \int \frac{y\sigma(d\xi)}{y^2 + (\xi - x)^2}.$$

and t be the harmonic conjugate of s with $2k\pi = \int t(x, y)dx$. Then, since $V_0(z) = (1 + k^2)y$ is the minimum non negative solution of (3.4) for s, t, m, and l defined above, $\sigma(\mu, k)$ and μ satisfy conditions [I] and [P] and $p_0(\sigma(\mu, k), \mu, k) = 0$ (c.f (3.5)).

Therefore, for $B = \{\sigma(\mu, k), \mu, k, 0\}$, we can define B-process $\boldsymbol{P}_{\mu,k}$. Let $\mathscr{P}_d$ be the set of all such processes, then we have the following propositions.

Proposition 6.3. *Let $\boldsymbol{P}_{\mu,k}$ be in $\mathscr{P}_d$, then for any $a > 0$ and any f which is periodic of period $2N\pi$ ($N = 1, 2, \cdots$) in $C(R)$, $\phi = H^a f$ is a solution of*

(6.4) $$\begin{cases} \phi = f & \text{on } y = a \\ \Delta\phi = 0 & \text{in } D^a \\ m\dfrac{\partial\phi}{\partial y} - l\dfrac{\partial\phi}{\partial x} \to 0 & \text{(boundedly) as } y \to 0 . \end{cases}$$

$\phi = H^a f$ is the unique solution, if we restrict ϕ in the set:

(6.5) $$\begin{cases} \phi \text{ is periodic of period } 2N\pi , \\ \phi \text{ is continuous in } \{0 < y \leqq a\} , \quad \text{and} \\ \phi \text{ is twice differentiable in } D^a . \end{cases}$$

Proposition 6.4. *Let $\boldsymbol{P}$ be in $\mathscr{P}$ and satisfy* [M] *and* [V], *then $\boldsymbol{P}$ is in $\mathscr{P}_d$ if and only if*

[D] $$\lim_{a\to 0} \int_0^{2\pi} m(x, a)dx \int B^a(x, d\xi)(\xi - x)(u(\xi, a) - u(\xi, x)) = 0$$

where u is the harmonic function such that $\partial u/\partial x = s$ and $\partial u/\partial y = -t$.

Proposition 6.5. *For $\boldsymbol{P}_{\mu,k}$ in $\mathscr{P}_d$, $\boldsymbol{P}_{\mu,k}$ belongs $\bar{\mathscr{P}}$ if and only if μ is positive for non empty open sets in R.*

§7. Symmetric processes

Let $\boldsymbol{P}$ be in $\mathscr{P}$, we say $\boldsymbol{P}$ is in $\mathscr{P}_s$ ($\boldsymbol{P}$ is in symmetric class) if and only if, there exists a function $P_t(z, \zeta)$ in $C(D \times D)$ with $P_t(z, \zeta) = P_t(\zeta, z)$ and

$$T_t f(z) = E_z(f(z_t)) = \int P_t(z, \zeta) f(\zeta) d\zeta$$

for f in $C(D)$, where $d\zeta$ is Lebesgue measure on D. Then, we can prove:

Proposition 7.1. *Let $\boldsymbol{P} = \boldsymbol{P}_B$ be B-process with $B = \{\sigma, \mu, k, p\}$, then $\boldsymbol{P}$ is in $\mathscr{P}_s$ if and only if $\mu = dx$ (Lebesgue measure on R) and $k = 0$.*

For $\mu = dx$ the condition [P] can be reduced to

$$[P_s] \qquad \int_0^{2\pi} |\log(1 - \cos(\xi - x))| \sigma(d\xi) \text{ is bounded in } x .$$

and we can show

$$(7.1) \qquad \begin{aligned} p_0 &= p_0(\sigma, dx, 0) \\ &= \sup_x \frac{1}{\pi} \int_0^{2\pi} |\log(1 - \cos(\xi - x))| \sigma(d\xi) \\ &\quad - \frac{1}{2\pi^2} \int_0^{2\pi} \int_0^{2\pi} |\log(1 - \cos(\xi - \eta))| \sigma(d\xi)\sigma(d\eta) , \end{aligned}$$

and σ is continuous (atomless measure) under $[P_s]$.

Therefore, we have the following correspondence. Set,

$$V_s = \left\{ \begin{array}{l} \{\sigma, p\} : \sigma \text{ satisfies } [P_s] \text{ and is positive for non empty open} \\ \text{set, and } p \geqq p_0 \text{ where } p_0 \text{ is given in (7.1).} \end{array} \right\} ,$$

then, by the correspondence of $\{\sigma, p\}$ with the $\{\sigma, dx, 0, p\}$-process, we have a bijection from V_s onto $\mathscr{P}_s \cap \overline{\mathscr{P}}_f$.

References

[1] Wentzell, A. D., On lateral conditions for multi-dimensional diffusion processes, Theor. of Probability Appl., 4 (1959), 172–185.

[2] Motoo, M., Brownian motion in the halfplane with singular inclined periodic boundary conditions, Topics in Probability theory, 163–179, New York Univ., 1973.

Department of Applied Physics
Tokyo Institute of Technology
O-okayama Tokyo 152, Japan

Proc. of Intern. Symp. SDE
Kyoto 1976, pp. 283–296

Approximation Theorem on Stochastic Differential Equations

Shintaro NAKAO[1] and Yuiti YAMATO

§ 1. Introduction

In this paper we shall discuss a relation between ordinary and stochastic differential equations. Let $Q(t) = (Q^1(t), \cdots, Q^r(t))$ be a system of quasi-martingales and let $\sigma(t, x, \xi)$ be a (d, r)-matrix valued function defined on $[0, \infty) \times R^d \times R^r$. We shall consider the solution $X(t)$ of the following stochastic differential equation:

$$(1.1) \qquad \begin{cases} dX(t) = \sigma(t, X(t), Q(t)) \circ dQ(t) \\ X(0) = x_0 \in R^d \,, \end{cases}$$

where the small circle $\circ$ denotes the symmetric multiplication of stochastic differentials (K. Itô [3]). Let $Q_n(t) = (Q_n^1(t), \cdots, Q_n^r(t))$ $(n = 1, 2, \cdots)$ be a sequence of approximations to $Q(t)$. For each n, $Q_n(t)$ is a process with piecewise smooth sample functions. The precise explanation of $Q_n(t)$ will be given in section 2. We shall consider the solution $X_n(t)$ of the following ordinary differential equation:

$$(1.2) \qquad \begin{cases} dX_n(t) = \sigma(t, X_n(t), Q_n(t)) dQ_n(t) \\ X_n(0) = x_0 \in R^d \,. \end{cases}$$

Our aim is to establish a convergence theorem of $X_n(t)$ to $X(t)$.

In this connection, E. Wong and M. Zakai [8] first showed that $X_n(t)$ converges in the mean to $X(t)$ when $Q(t)$ is a one-dimensional Brownian motion. When $Q(t)$ is a multi-dimensional Brownian motion and $Q_n(t)$ is its polygonal approximation, D. W. Stroock and S. R. S. Varadhan [7] proved the convergence in law and utilized it to decide the support of a diffusion process. In this paper we shall treat a much more general case that $Q(t)$ is a system of continuous quasi-martingales satisfying a certain integrability condition. Our theorem will be stated in section 2. In particular, it generalizes a pathwise approximation theorem in H. Kunita [5]. The proof of our theorem will be given in section 3. Section 4 is devoted to its application to stochastic parallel displacement (K. Itô [2] [4], E. B. Dynkin [1]).

§ 2. Approximation theorem

Let $(\Omega, \mathcal{F}, P)$ be the basic probability space and let $\{\mathcal{F}_t\}_{t\geq 0}$ be an increasing family of sub-σ-algebras of $\mathcal{F}$. We shall consider a system of $\mathcal{F}_t$-quasi-martingales $Q(t)=(Q^1(t)=M^1(t)+V^1(t), \cdots, Q^r(t)=M^r(t)+V^r(t))$ ($M^i(t)$ is a martingale and $V^i(t)$ is a process whose sample functions are of bounded variation) defined on $(\Omega, \mathcal{F}, P)$ satisfying the following conditions:

Condition A

(i) $M^i(0)=V^i(0)=0$ for $1\leqq i\leqq r$.

(ii) $M^i(t)$ is a continuous square-integrable $\mathcal{F}_t$-martingale for $1\leqq i\leqq r$.

(iii) $V^i(t)$ is a continuous $\mathcal{F}_t$-adapted process and $|||V^i|||(t)$[2] has a finite expectation for $1\leqq i\leqq r$ and $t>0$.

(iv) $\sup_{s\leqq t}\|Q(s)\|^6$[3] has a finite expectation for $t>0$.

$P([0,\infty))$ denotes the space of countable partitions of $[0,\infty)$ with finite mesh. For $\Pi\in P([0,\infty))$, *mesh*(Π) is the length of the largest subinterval of Π.

Definition 2.1. Let κ be a positive constant. For $\Pi=(0=t_0<t_1<t_2<\cdots)\in P([0,\infty))$, we will say that $\tilde{Q}(t)=(\tilde{Q}^1(t),\cdots,\tilde{Q}^r(t))\in C(\Pi,\kappa)$ if $\tilde{Q}(t)$ satisfies the following conditions:

(i) $\tilde{Q}(t)$ has continuous and piecewise smooth sample functions and $\tilde{Q}(t)=Q(t)$ for every division points t of Π.

(ii) $\tilde{Q}(t)$ is $\mathcal{F}_{t_\nu}$-measurable for $t\leqq t_\nu$ and $\nu=0,1,2,\cdots$.

(iii)
$$\sum_{i=1}^{r}\sup_{t_\nu<t<t_{\nu+1}}\left|\frac{\partial}{\partial t}\tilde{Q}^i(t)-\frac{Q^i(t_{\nu+1})-Q^i(t_\nu)}{t_{\nu+1}-t_\nu}\right|\leqq\kappa\frac{\|Q(t_{\nu+1})-Q(t_\nu)\|^2}{t_{\nu+1}-t_\nu}\quad\text{for }\nu=0,1,2,\cdots.$$

Let $\sigma(t,x,\xi)=(\sigma_i^\alpha(t,x,\xi))$ $(1\leqq\alpha\leqq d,\ 1\leqq i\leqq r)$ be a (d,r)-matrix valued function defined on $[0,\infty)\times R^d\times R^r$. We assume that each component has first order partial derivatives in (x,ξ). Further we assume that $\sigma(t,x,\xi)$ satisfies the following condition:

Condition B. For $1\leqq\alpha,\beta\leqq d$ and $1\leqq i,j\leqq r$, σ_i^α, $(\partial/\partial x^\beta)\sigma_i^\alpha$ and

[1] The auther was supported in part by the Sakkokai Foundation.

[2] Let $f(t)$ be a function on $[0,\infty)$. $|||f|||(t)$ denotes the total variation of f on $[0,t]$.

[3] For $y=(y^1,\cdots,y^m)\in R^m$, $\|y\|=\{\sum_{i=1}^m(y^i)^2\}^{1/2}$.

$(\partial/\partial\xi^j)\sigma_i^\alpha$ are bounded and Lipschitz continuous in (t, x, ξ).

Let $\{\tau(s); s \geqq 0\}$ be a family of $\mathscr{F}_t$-stopping times satisfying the following conditions:

(2.1) $$\tau(0) = 0 \quad \text{a.s.} \ .$$

(2.2) There exists a subset $\Omega' \subset \Omega$ with $P(\Omega') = 1$ such that $\tau(u, \omega) - \tau(v, \omega) \leqq u - v$, $|||V^i|||(\tau(u, \omega), \omega) - |||V^i|||(\tau(v, \omega), \omega) \leqq u - v$ and $|||\varphi^{ij}|||(\tau(u, \omega), \omega) - |||\varphi^{ij}|||(\tau(v, \omega), \omega) \leqq u - v$ for $\omega \in \Omega'$, $0 \leqq v < u$ and $1 \leqq i, j \leqq r$, where $\varphi^{ij}(t)$ is the continuous $\mathscr{F}_t$-adapted process of bounded variation such that $M^i(t)M^j(t) - \varphi^{ij}(t)$ is a $\mathscr{F}_t$-martingale (H. Kunita and S. Watanabe [6]).

Consider a sequence $\Pi_n \in P([0, \infty))$ with $\lim_{n\to\infty} mesh(\Pi_n) = 0$ and a sequence $Q_n(t) \in C(\Pi_n, \kappa)$. Let $X(t)$ and $X_n(t)$ be the unique solutions of (1.1) and (1.2) respectively. Our main result can now be stated as

Theorem. *Let $\{\tau(s); s \geqq 0\}$ be a family of $\mathscr{F}_t$-stopping times satisfying the conditions* (2.1) *and* (2.2). *If Condition A and Condition B hold, then we have, for each $T > 0$,*

$$\lim_{n\to\infty} E\,[\sup_{s\leqq T} \|X_n(\tau(s)) - X(\tau(s))\|^2] = 0 \ .$$

The proof of the above theorem will be given in section 3. By theorem, we can obtain a pathwise approximation theorem under more general conditions.

Definition 2.2. Let $\{\kappa_N\}$ be a positive sequence. For $\Pi = (0 = t_0 < t_1 < t_2 < \cdots) \in P([0, \infty))$, we will say that $\tilde{Q}(t) = (\tilde{Q}^1(t), \cdots, \tilde{Q}^r(t)) \in C^{loc}(\Pi, \{\kappa_N\})$ if $\tilde{Q}(t)$ satisfies (i), (ii) in Definition 2.1 and moreover satisfies that, for each N,

$$\sum_{i=1}^{r} \sup_{t_\nu < t < t_{\nu+1}} \left| \frac{\partial}{\partial t}\tilde{Q}^i(t) - \frac{Q^i(t_{\nu+1}) - Q^i(t_\nu)}{t_{\nu+1} - t_\nu} \right| \leqq \kappa_N \frac{\|Q(t_{\nu+1}) - Q(t_\nu)\|^2}{t_{\nu+1} - t_\nu}$$

for ν: $t_{\nu+1} \leqq \inf\{t; \|Q(t)\| = N\}$.

Condition B'. $\sigma(t, x, \xi)$ satisfies the following hypotheses:

(i) For each compact subset Γ_1 of $[0, \infty) \times R^d \times R^r$, there exists a constant L_1 such that

$$|\sigma_i^\alpha(t, x, \xi) - \sigma_i^\alpha(s, y, \eta)| \leqq L_1(|t - s| + \|x - y\| + \|\xi - \eta\|) \ ,$$

$$\left| \frac{\partial}{\partial x^\beta}\sigma_i^\alpha(t, x, \xi) - \frac{\partial}{\partial x^\beta}\sigma_i^\alpha(s, y, \eta) \right| \leqq L_1(|t - s| + \|x - y\| + \|\xi - \eta\|)$$

and

$$\left|\frac{\partial}{\partial \xi^j}\sigma_i^\alpha(t, x, \xi) - \frac{\partial}{\partial \xi^j}\sigma_i^\alpha(s, y, \eta)\right| \leqq L_1(|t-s| + \|x-y\| + \|\xi-\eta\|)$$

for any (t, x, ξ), $(s, y, \eta) \in \Gamma_1$, $1 \leqq \alpha, \beta \leqq d$ and $1 \leqq i, j \leqq r$.

(ii) For each compact subset Γ_2 of $[0, \infty) \times R^r$, there exists a constant L_2 such that

$$|\sigma_i^\alpha(t, x, \xi) - \sigma_i^\alpha(t, y, \xi)| \leqq L_2\|x-y\| ,$$

$$\left|\frac{\partial}{\partial x^\beta}\sigma_i^\alpha(t, x, \xi) - \frac{\partial}{\partial x^\beta}\sigma_i^\alpha(t, y, \xi)\right| \leqq L_2\|x-y\|$$

and

$$\left|\frac{\partial}{\partial \xi^j}\sigma_i^\alpha(t, x, \xi) - \frac{\partial}{\partial \xi^j}\sigma_i^\alpha(t, y, \xi)\right| \leqq L_2\|x-y\|$$

for any $(t, \xi) \in \Gamma_2$, $x, y \in R^d$, $1 \leqq \alpha, \beta \leqq d$ and $1 \leqq i, j \leqq r$.

Consider a sequence $\Pi_n \in P([0, \infty))$ with $\lim_{n\to\infty} mesh(\Pi_n) = 0$ and a sequence $Q_n(t) \in C^{loc}(\Pi_n, \{\kappa_N\})$. Let $X(t)$ and $X_n(t)$ be the unique solutions of (1.1) and (1.2) respectively. We have

Corollary. *Under Condition A and Condition B′, there exists a subsequence $\{n_j\}$ such that $X_{n_j}(t)$ converges uniformly on any compact time interval of $[0, \infty)$ to $X(t)$ a.s.(P).*

§ 3. Proof of Theorem

Before the proof, we note that (1.1) is also expressed as

$$(1.1)' \quad \begin{cases} dX^\alpha(t) = \sum_{i=1}^r \sigma_i^\alpha(t, X(t), Q(t))dQ^i(t) \\ \qquad + \dfrac{1}{2}\sum_{i,j=1}^r \left(\sum_{\beta=1}^d \sigma_j^\beta \dfrac{\partial}{\partial x^\beta}\sigma_i^\alpha + \dfrac{\partial}{\partial \xi^j}\sigma_i^\alpha\right)(t, X(t), Q(t))d\varphi^{ij}(t) , \\ \qquad\qquad 1 \leqq \alpha \leqq d , \\ X(0) = x_0 . \end{cases}$$

Throughout this section we assume that $Q(t)$ satisfies Condition A. For $1 \leqq p \leqq 6$, $t > 0$ and $\delta > 0$, set

$$\gamma_p(t, \delta) = E[\sup_{\substack{0 \leqq u, v \leqq t \\ |u-v| < \delta}} \|Q(u) - Q(v)\|^p] .$$

The following lemma is an immediate consequence of Lebesgue dominated convergence theorem.

Lemma 3.1. *We have, for* $1 \leqq p \leqq 6$ *and* $t > 0$, $\lim_{\delta \to 0} \gamma_p(t, \delta) = 0$.

Fix $\Pi = (0 = t_0 < t_1 < \cdots) \in P([0, \infty))$ with $mesh(\Pi) \leqq 1$ and $\{\tau(s) : s \geqq 0\}$ satisfying the conditions (2.1) and (2.2). For $t \geqq 0$, we define $[t] = \max\{t_\nu \leqq t; \nu = 0, 1, 2, \cdots\}$, $[t]^+ = \min\{t_\nu > t; \nu = 0, 1, 2, \cdots\}$ and $\tau_\nu(t) = \min\{t_\nu, \tau(t)\}$. Further we introduce the notations, for any process $Z(t)$, $\Delta^\nu Z(\tau(t)) = Z(\tau_{\nu+1}(t)) - Z(\tau_\nu(t))$, $\Delta Z(\tau(t)) = Z(\tau(t)) - Z([\tau(t)])$ and $\Delta^+ Z(\tau(t)) = Z([\tau(t)]^+) - Z([\tau(t)])$. In case that $Z(t) = t$, we denote them simply by $\Delta^\nu(\tau(t))$, $\Delta(\tau(t))$ and $\Delta^+(\tau(t))$ respectively.

Lemma 3.2. *We have, for* $1 \leqq p \leqq 6$ *and* $T > 0$,

$$E[\sup_{t \leqq T} \| \Delta Q(\tau(t)) \|^p] \leqq \gamma_p(T, mesh(\Pi))$$

and

$$E[\sup_{t \leqq T} \| \Delta^+ Q(\tau(t)) \|^p] \leqq \gamma_p(T + 1, mesh(\Pi)) .$$

Proof. Noting

$$\sup_{t \leqq T} \| \Delta Q(\tau(t)) \|^p \leqq \sup_{\substack{0 \leqq u, v \leqq T \\ |u - v| \leqq \delta}} \| Q(u) - Q(v) \|^p \quad a.s.$$

and

$$\sup_{t \leqq T} \| \Delta^+ Q(\tau(t)) \|^p \leqq \sup_{\substack{0 \leqq u, v \leqq T+1 \\ |u - v| \leqq \delta}} \| Q(u) - Q(v) \|^p \quad a.s.,$$

we obtain Lemma 3.2. Q.E.D.

Lemma 3.3. *Let* $Y(t, \omega)$ *be a bounded* $\mathscr{F}_t$*-adapted process defined on* $(\Omega, \mathscr{F}, P)$ *and set* $C = \sup_{t, \omega} |Y(t, \omega)|$. *Let p be a constant such as* $1 \leqq p \leqq 3$. *If a continuous function* $g(x)$ *satifies* $|g(x)| \leqq |x|^p$ $(x \in R^1)$, *then we have*

$$E\left[\sup_{t \leqq T} \left| \int_0^{\tau(t)} Y(s) g(Q^i(s) - Q^i([s])) dQ^j(s) \right|^2 \right] \leqq 2C^2 T(4 + T) \gamma_{2p}(T, mesh(\Pi)) ,$$

for $1 \leqq i, j \leqq r$ *and* $T > 0$.

Proof. It follows from martingale inequality and Schwarz inequality that

$$E\left[\sup_{t\leq T}\left|\int_0^{\tau(t)} Y(s)g(Q^i(s)-Q^i([s]))dQ^j(s)\right|^2\right]$$
$$\leq 2E\left[\sup_{t\leq T}\left|\int_0^{\tau(t)} Y(s)g(Q^i(s)-Q^i([s]))dM^j(s)\right|^2\right.$$
$$\left.+\sup_{t\leq T}\left|\int_0^{\tau(t)} Y(s)g(Q^i(s)-Q^i([s]))dV^j(s)\right|^2\right]$$
$$\leq 2C^2\left\{4E\left[\int_0^{\tau(T)}|Q^i(s)-Q^i([s])|^{2p}d\varphi^{jj}(s)\right]\right.$$
$$\left.+TE\left[\int_0^{\tau(T)}|Q^i(s)-Q^i([s])|^{2p}d\,|||V^j|||\,(s)\right]\right\}.$$

Hence Lemma 3.2 leads us to Lemma 3.3. Q.E.D.

Lemma 3.4. *For each $T>0$, there exist positive constants C_1, C_2, C_3 and C_4 depending only on r and T such that*

$$E\left[\sup_{t\leq T}\left\{\sum_{\nu=0}^{\infty}\Delta^\nu(\tau(t))\,\|\Delta^\nu Q(\tau(t))\|\right\}^2\right] \tag{3.1}$$
$$\leq C_1 mesh(\Pi)\{1+\gamma_1(T, mesh(\Pi))+\gamma_2(T, mesh(\Pi))\},$$

$$E\left[\sup_{t\leq T}\left\{\sum_{\nu=0}^{\infty}\Delta^\nu(\tau(t))\,\|\Delta^\nu Q(\tau(t))\|^2\right\}^2\right] \tag{3.2}$$
$$\leq C_2 mesh(\Pi)^2\{1+\gamma_2(T, mesh(\Pi))\},$$

$$E\left[\sup_{t\leq T}\left\{\sum_{\nu=0}^{\infty}\|\Delta^\nu Q(\tau(t))\|^3\right\}^2\right] \tag{3.3}$$
$$\leq C_3\{\gamma_2(T, mesh(\Pi))+\gamma_4(T, mesh(\Pi))\},$$

$$E\left[\sup_{t\leq T}\left\{\sum_{\nu=0}^{\infty}\|\Delta^\nu Q(\tau(t))\|^4\right\}^2\right] \tag{3.4}$$
$$\leq C_4\{\gamma_4(T, mesh(\Pi))+\gamma_6(T, mesh(\Pi))\},$$

Proof. By Schwarz inequality, we have

$$\sup_{t\leq T}\left\{\sum_{\nu=0}^{\infty}\Delta^\nu(\tau(t))\,|\Delta^\nu Q^i(\tau(t))|\right\}^2$$
$$\leq \sup_{t\leq T}\left\{\sum_{\nu=0}^{\infty}\Delta^\nu(\tau(t))^2\sum_{\nu=0}^{\infty}|\Delta^\nu Q^i(\tau(t))|^2\right\}$$
$$\leq mesh(\Pi)T\left\{\sum_{\nu=0}^{\infty}|\Delta^\nu Q^i(\tau(T))|^2+\sup_{t\leq T}|\Delta Q^i(\tau(t))|^2\right\} \quad a.s..$$

Itô's formula implies that

(3.5) $$\sum_{\nu=0}^{\infty} |\Delta^\nu Q^i(\tau(T))|^2 = 2\int_0^{\tau(T)} \{Q^i(s) - Q^i([s])\} dQ^i(s) + \varphi^{ii}(\tau(T)) \ a.s.,$$

(H. Kunita and S. Watanabe [6]). Hence we get the inequality (3.1).

We will show (3.2). First we note that

$$\left\{\sum_{\nu=0}^{\infty} \Delta^\nu(\tau(t))\, |\Delta^\nu Q^i(\tau(t))|^2\right\}^2 \leqq \{mesh(\Pi)\}^2 \left\{\sum_{\nu=0}^{\infty} |\Delta^\nu Q^i(\tau(t))|^2\right\}^2 .$$

Therefore we have (3.2) by Lemma 3.3 and (3.5).

(3.3) and (3.4) can be obtained in the same way as (3.2). Q.E.D.

From now on we assume that $\sigma(t, x, \xi)$ satisfies Condition B. Set

$$b^\alpha_{ij}(t, x, \xi) = \left(\sum_{\beta=1}^{d} \sigma^\beta_j \frac{\partial}{\partial x^\beta} \sigma^\alpha_i + \frac{\partial}{\partial \xi^j} \sigma^\alpha_i\right)(t, x, \xi) .$$

The common constant of boundedness (Lipschitz constant) of $\sigma(t, x, \xi)$ and the first order derivatives of $\sigma(t, x, \xi)$ in (x, ξ) is denoted by B (L) respectively. Let $\xi(u) = (\xi^1(u), \cdots, \xi^r(u))$ be an R^r-valued smooth function on $[s, t]$. We shall consider a solution $\eta(u) = (\eta^1(u), \cdots, \eta^d(u))$ of the equation

(3.6) $$d\eta(u) = \sigma(u, \eta(u), \xi(u)) d\xi(u) .$$

Now we shall prepare a lemma which plays a fundamental role in the future proof. We define

$$F = \sum_{i=1}^{r} \sup_{s<u<t} \left| \frac{d}{dv} \xi^i(v) - \frac{\xi^i(t) - \xi^i(s)}{t - s} \right|$$

and we define, for $s < u \leqq t$,

$$\Delta = u - s , \qquad \Delta\xi = \xi(u) - \xi(s)$$

and

$$G = \sum_{i=1}^{r} \sup_{s<v<u} \left| \frac{d}{dv} \xi^i(v) - \frac{\Delta\xi^i}{\Delta} \right| .$$

Lemma 3.5. *Suppose that $\eta(u)$ satisfies (3.6). Then there exist positive constants C_1 depending only on d, r and B and C_2 depending only on d, r, B and L such that, for $s < u \leqq t$ and $1 \leqq \alpha \leqq d$,*

(3.7) $$\|\eta(u) - \eta(s)\| \leqq C_1\{(t - s)F + \|\xi(t) - \xi(s)\|\} ,$$

$$\left|\eta^\alpha(u) - \eta^\alpha(s) - \sum_{i=1}^r \sigma_i^\alpha(s, \eta(s), \xi(s))\Delta\xi^i \right.$$

$$(3.8) \qquad \left. - \frac{1}{2}\sum_{i,j=1}^r b_{ij}^\alpha(s, \eta(s), \xi(s))\Delta\xi^i\Delta\xi^j \right|$$

$$\leqq C_2\{\Delta\|\Delta\xi\| + \Delta\|\Delta\xi\|^2 + \|\Delta\xi\|^3 + \Delta^2 G + \Delta^3 G^2$$
$$+ \Delta\|\Delta\xi\| G + \Delta^2\|\Delta\xi\| G + \Delta^2 G^2 + \Delta\|\Delta\xi\|^2 G\}\,.$$

Proof. (3.7) is obvious. We shall show (3.8). In this proof K_i ($i = 1, 2, 3, 4$) denote positive constants depending only on d, r, B and L. Set

$$\eta^\alpha(u) - \eta^\alpha(s) - \sum_{i=1}^r \sigma_i^\alpha(s, \eta(s), \xi(s))\Delta\xi^i$$
$$= \sum_{i=1}^r \int_s^u \{\sigma_i^\alpha(s, \eta(v), \xi(v)) - \sigma_i^\alpha(s, \eta(s), \xi(s))\}d\xi^i(v)$$
$$+ \sum_{i=1}^r \int_s^u \{\sigma_i^\alpha(v, \eta(v), \xi(v)) - \sigma_i^\alpha(s, \eta(v), \xi(v))\}d\xi^i(v)$$
$$= I_1 + \varepsilon_1\,.$$

Since σ is Lipschitz continuous, we have

$$|\varepsilon_1| \leqq K_1\{\Delta^2 G + \Delta\|\Delta\xi\|\}\,.$$

Denoting that

$$I_1 = \sum_{i,j=1}^r \int_s^u d\xi^i(v)\int_s^v b_{ij}^\alpha(s, \eta(w), \xi(w))d\xi^j(w)$$
$$+ \sum_{\beta=1}^d \sum_{i,j=1}^r \int_s^u d\xi^i(v)\int_s^v \frac{\partial}{\partial x^\beta}\sigma_i^\alpha(s, \eta(w), \xi(w))$$
$$\times \{\sigma_j^\beta(w, \eta(w), \xi(w)) - \sigma_j^\beta(s, \eta(w), \xi(w))\}d\xi^j(w)$$
$$= I_2 + \varepsilon_2\,,$$

we find that $|\varepsilon_2| \leqq K_2\{\Delta^3 G^2 + \Delta^2\|\Delta\xi\| G + \Delta\|\Delta\xi\|^2\}$.

$$\varepsilon_3 = I_2 - \sum_{i,j=1}^r \int_s^u dv \int_s^v b_{ij}^\alpha(s, \eta(w), \xi(w))\frac{\Delta\xi^i\Delta\xi^j}{\Delta^2}dw$$

has an estimate $|\varepsilon_3| \leqq K_3\{\Delta\|\Delta\xi\| G + \Delta^2 G^2\}$. Since

$$\varepsilon_4 = \sum_{i,j=1}^r \int_s^u dv \int_s^v b_{ij}^\alpha(s, \eta(w), \xi(w))\frac{\Delta\xi^i\Delta\xi^j}{\Delta^2}dw$$

$$- \frac{1}{2} \sum_{i,j=1}^{r} b_{ij}^{\alpha}(s, \eta(s), \xi(s)) \Delta\xi^i \Delta\xi^j$$

$$= \sum_{i,j=1}^{r} \int_s^u dv \int_s^v \{b_{ij}^{\alpha}(s, \eta(w), \xi(w)) - b_{ij}^{\alpha}(s, \eta(s), \xi(s))\} \frac{\Delta\xi^i \Delta\xi^j}{\Delta^2} dw ,$$

(3.7) implies that $|\varepsilon_4| \leqq K_4\{\Delta \|\Delta\xi\|^2 G + \|\Delta\xi\|^3\}$. Q.E.D.

Let $\Pi = (0 = t_0 < t_1 < \cdots) \in P([0, \infty))$ with $mesh(\Pi) \leqq 1$ and let $\tilde{Q}(t) \in C(\Pi, \kappa)$. We shall consider the unique solution $\tilde{X}(t)$ of the equation:

$$\begin{cases} d\tilde{X}(t) = \sigma(t, \tilde{X}(t), \tilde{Q}(t)) d\tilde{Q}(t) \\ \tilde{X}(0) = x_0 . \end{cases} \tag{3.9}$$

Lemma 3.6. *For $T > 0$, there exists a positive constant C depending only on d, r, κ and B such that*

$$E\Big[\sup_{t \leqq T} \|\tilde{X}(\tau(t)) - \tilde{X}([\tau(t)])\|^2\Big]$$
$$\leqq C\{\gamma_2(T + 1, mesh(\Pi)) + \gamma_4(T + 1, mesh(\Pi))\} .$$

Proof. From (3.7), there exists a positive constant K depending only on d, r, κ and B such that, for $t > 0$,

$$\|\tilde{X}(\tau(t)) - \tilde{X}([\tau(t)])\| \leqq K\{\|\Delta^+ Q(\tau(t))\| + \|\Delta^+ Q(\tau(t))\|^2\} .$$

Combining the above inequality with Lemma 3.2, we obtain Lemma 3.6. Q.E.D.

Lemma 3.7. *For $T > 0$, there exists a positive constant C depending only on d, r, T, κ, B and L such that*

$$E\Big[\sup_{t \leqq T} \Big| \tilde{X}^{\alpha}(\tau(t)) - x_0^{\alpha} - \sum_{i=1}^{r} \int_0^{\tau(t)} \sigma_i^{\alpha}([s], \tilde{X}([s]), \tilde{Q}([s])) dQ^i(s)$$
$$- \frac{1}{2} \sum_{i,j=1}^{r} \int_0^{\tau(t)} b_{ij}^{\alpha}([s], \tilde{X}([s]), \tilde{Q}([s])) d\varphi^{ij}(s) \Big|^2\Big]$$
$$\leqq C\{mesh(\Pi) + mesh(\Pi)\gamma_1(T, mesh(\Pi)) + \gamma_2(T + 1, mesh(\Pi))$$
$$+ \gamma_4(T + 1, mesh(\Pi)) + \gamma_6(T, mesh(\Pi))\} ,$$

for $1 \leqq \alpha \leqq d$.

Proof. Noting Lemma 3.5, we have

$$\Big| \tilde{X}^{\alpha}(\tau(t)) - x_0^{\alpha} - \sum_{i=1}^{r} \int_0^{\tau(t)} \sigma_i^{\alpha}([s], \tilde{X}([s]), \tilde{Q}([s])) dQ^i(s)$$

$$
\begin{aligned}
&-\frac{1}{2}\sum_{i,j=1}^{r}\int_0^{\tau(t)} b_{ij}^{\alpha}([s],\tilde{X}([s]),\tilde{Q}([s]))d\varphi^{ij}(s)\Big| \\
&\leqq |\tilde{X}^{\alpha}(\tau(t))-\tilde{X}^{\alpha}([\tau(t)])| \\
&\quad + K_1\Big(\sum_{\nu=0}^{\infty}\{\Delta^{\nu}(\tau(t))\,\|\Delta^{\nu}Q(\tau(t))\| + \Delta^{\nu}(\tau(t))\,\|\Delta^{\nu}Q(\tau(t))\|^2 \\
&\quad + \|\Delta^{\nu}Q(\tau(t))\|^3 + \|\Delta^{\nu}Q(\tau(t))\|^4\}\Big) \\
&\quad + K_2\{\|\Delta Q(\tau(t))\| + \|\Delta Q(\tau(t))\|^2\} \\
&\quad + \frac{1}{2}\sum_{i,j=1}^{r}\Big|\int_0^{\tau(t)} b_{ij}^{\alpha}([s],\tilde{X}([s]),\tilde{Q}([s]))d\varphi^{ij}(s) \\
&\quad - \sum_{\nu=0}^{\infty} b_{ij}^{\alpha}(\tau_\nu(t),\tilde{X}(\tau_\nu(t)),\tilde{Q}(\tau_\nu(t)))\Delta^{\nu}Q^i(\tau(t))\Delta^{\nu}Q^j(\tau(t))\Big|,
\end{aligned}
$$

where K_1 (K_2) is a positive constant depending only on d, r, κ, B and L (d, r and B) respectively. Itô's formula leads us to

$$
\begin{aligned}
&\int_0^{\tau(t)} b_{ij}^{\alpha}([s],\tilde{X}([s]),\tilde{Q}([s]))d\varphi^{ij}(s) \\
&\qquad - \sum_{\nu=0}^{\infty} b_{ij}^{\alpha}(\tau_\nu(t),\tilde{X}(\tau_\nu(t)),\tilde{Q}(\tau_\nu(t)))\Delta^{\nu}Q^i(\tau(t))\Delta^{\nu}Q^j(\tau(t)) \\
&= -\int_0^{\tau(t)} b_{ij}^{\alpha}([s],\tilde{X}([s]),\tilde{Q}([s]))\{Q^i(s)-Q^i([s])\}dQ^j(s) \\
&\quad - \int_0^{\tau(t)} b_{ij}^{\alpha}([s],\tilde{X}([s]),\tilde{Q}([s]))\{Q^j(s)-Q^j([s])\}dQ^i(s) \qquad a.s.\,.
\end{aligned}
$$

Therefore Lemma 3.3, Lemma 3.4 and Lemma 3.6 imply Lemma 3.7.

Q.E.D.

Let $X(t)$ and $\tilde{X}(t)$ be the unique solutions of the equation (1.1) and (3.9) respectively.

Proposition 3.1. *For $T>0$, there exist a positive constant C_1 depending only on d, r, T, B and L and a constant C_2 depending only on d, r, T, κ, B and L such that, for $t \leqq T$,*

$$
\begin{aligned}
&E\Big[\sup_{u\leqq t}\|X(\tau(u))-\tilde{X}(\tau(u))\|^2\Big] \\
&\leqq C_1\int_0^t E\Big[\sup_{u\leqq s}\|X(\tau(u))-\tilde{X}(\tau(u))\|^2\Big]ds \\
&\quad + C_2\{mesh(\Pi)(1+\gamma_1(T, mesh(\Pi))) + \gamma_2(T+1, mesh(\Pi)) \\
&\quad + \gamma_4(T+1, mesh(\Pi)) + \gamma_6(T, mesh(\Pi))\}\,.
\end{aligned}
$$

Proof. By martingale inequality and Schwarz inequality, we have, for $t \leqq T$,

$$E\left[\sup_{u \leqq t}\left|\int_0^{\tau(u)} \{\sigma_i^\alpha(v, X(v), Q(v)) - \sigma_i^\alpha([v], \tilde{X}([v]), \tilde{Q}([v]))\} dQ^i(v)\right.\right.$$

$$+ \frac{1}{2}\sum_{j=1}^r \int_0^{\tau(u)} \{b_{ij}^\alpha(v, X(v), Q(v))$$

$$\left.\left. - b_{ij}^\alpha([v], \tilde{X}([v]), \tilde{Q}([v]))\} d\varphi^{ij}(v)\right|^2\right]$$

$$\leqq KE\left[\sum_{j=1}^r \int_0^{\tau(t)} (|v - [v]|^2 + \|X(v) - \tilde{X}(v)\|^2 + \|\tilde{X}(v) - \tilde{X}([v])\|^2\right.$$

$$\left. + \|Q(v) - Q([v])\|^2)(d\varphi^{ii}(v) + d\,|||V^i|||\,(v) + d\,|||\varphi^{ij}|||\,(v))\right],$$

where K is a positive constant depending only on d, r, T, B and L. Therefore Lemma 3.6 and Lemma 3.7 leads us to Proposition 3.1.

Q.E.D.

Finally, using the above proposition, we shall prove Theorem in section 2.

Proof of Theorem. By Lemma 3.1 and Proposition 3.1, there exists a positive sequence $\{\varepsilon_n\}$ with $\lim_{n\to\infty} \varepsilon_n = 0$ such that, for $t \leqq T$,

$$E\left[\sup_{u \leqq t} \|X(\tau(u)) - X_n(\tau(u))\|^2\right]$$

$$\leqq C_1 \int_0^t E\left[\sup_{u \leqq s} \|X(\tau(u)) - X_n(\tau(u))\|^2\right] ds + \varepsilon_n ,$$

where C_1 is a constant in Proposition 3.1. The above functional inequality implies that, for $t \leqq T$

$$E\left[\sup_{u \leqq t} \|X(\tau(u)) - X_n(\tau(u))\|^2\right] \leqq \varepsilon_n \exp C_1 t .$$

This proves

$$\lim_{n\to\infty} E\left[\sup_{u \leqq t} \|X(\tau(u)) - X_n(\tau(u))\|^2\right] = 0 . \qquad \text{Q.E.D.}$$

§4. Application to the stochastic parallel displacement

Let M be a simply connected Riemannian manifold of dimension r

with non-positive sectional curvature. We fix a point m_0 in M, then we can take a global coordinate ψ induced by $\exp_{m_0}$. We consider an M-valued process $\{Y(t)\}_{t\geq 0}$ defined on $(\Omega, \mathcal{F}, P)$. Suppose that $Q(t)=\psi(Y(t))$ satisfies Condition A in section 2.

Let $T_p^q=(T_p^q(m), m\in M)$ be the bundle of tensors of type (p, q) and let $(\Gamma_{ij}^k)_{1\leq i,j,k\leq r}$ be the Riemannian connection of M. Set $I=\{\binom{j_1\cdots j_q}{i_1\cdots i_p}; 1 \leq i_1, \cdots, i_p, j_1, \cdots, j_q \leq r\}$ and set

$$
(4.1)\qquad \begin{aligned}\sigma_i^\alpha(x, \xi) = & \sum_{\mu=1}^{p}\sum_{k=1}^{r} \Gamma_{ii_\mu}^k(\psi^{-1}(\xi))x_{i_1\cdots i_{\mu-1}k\, i_{\mu+1}\cdots i_p}^{j_1\cdots j_q} \\ & -\sum_{\nu=1}^{q}\sum_{k=1}^{r} \Gamma_{ik}^{j_\nu}(\psi^{-1}(\xi))x_{i_1\cdots i_p}^{j_1\cdots j_{\nu-1}k\, j_{\nu+1}\cdots j_q}\end{aligned}
$$

for $\alpha=\binom{j_1\cdots j_q}{i_1\cdots i_p}\in I$, $1\leq i\leq r$, $x\in R^I$ and $\xi\in R^r$. We denote by Ψ_m the coordinate of $T_p^q(m)$ induced by ψ. Following K. Itô [4], the stochastic parallel displacement of the tensor $\Psi_{m_0}^{-1}(x_0)$ along $Y(t)$ is defined as $\Psi_{Y(t)}^{-1}(X(t))$, where $X(t)$ is the solution of the stochastic differential equation:

$$
(4.2)\qquad \begin{cases} dX(t)=\sigma(X(t), Q(t))\circ dQ(t) \\ X(0)=x_0\,, \end{cases}
$$

where $\sigma(x, \xi)=(\sigma_i^\alpha(x, \xi))_{\alpha\in I, 1\leq i\leq r}$.

For any partition $\Pi=(0=t_0<t_1<t_2<\cdots)\in P([0, \infty))$, we connect $Y(t_\nu)$ with $Y(t_{\nu+1})$ by the geodesic for $\nu=0, 1, \cdots$. Then we obtain a piecewise smooth curve $\{Y_\Pi(t)\}_{t\geq 0}$. $\psi(Y_\Pi(t))$ is denoted by $Q_\Pi(t)=(Q_\Pi^1(t), \cdots, Q_\Pi^r(t))$. The parallel displacement of the tensor $\Psi_{m_0}^{-1}(x_0)$ along $Y_\Pi(t)$ is defined as $\Psi_{Y_\Pi(t)}^{-1}(X_\Pi(t))$, where $X_\Pi(t)$ is the solution of the ordinary differential equation:

$$
(4.3)\qquad \begin{cases} dX_\Pi(t)=\sigma(X_\Pi(t), Q_\Pi(t))dQ_\Pi(t) \\ X_\Pi(0)=x_0\,. \end{cases}
$$

Consider a sequence $\Pi_n\in P([0, \infty))$ with $\lim_{n\to\infty} mesh(\Pi_n)=0$ and let $X(t)$ and $X_{\Pi_n}(t)$ be the solution of (4.2) and (4.3) for $\Pi=\Pi_n$ respectively.

Proposition 4.1. *There exists a subsequence $\{n_j\}$ such that $\Psi_{Y_{\Pi_{n_j}}(t)}^{-1}(X_{\Pi_{n_j}}(t))$ converges uniformly on any compact time interval of $[0, \infty)$ to $\Psi_{Y(t)}^{-1}(X(t))$ in T_p^q a.s..*

Proof. Noting the following lemma, the approximate sequence $Q_{\Pi_n}(t)$ belongs to $C^{loc}(\Pi_n, \{\kappa_N\})$ for some sequence $\{\kappa_N\}$. Therefore Corollary gives us the above proposition.

Lemma 4.1. *Let $y_{m_1,m_2}(u)$, $s \leqq u \leqq t$ be the geodesic such that $y_{m_1,m_2}(s) = m_1$, $y_{m_1,m_2}(t) = m_2$. Then for each compact subset Γ of M, there exists a constant κ such that*

$$\sup_{s \leqq u \leqq t} \left\| \frac{d}{du}\psi(y_{m_1,m_2}(u)) - \frac{\psi(m_2) - \psi(m_1)}{t-s} \right\| \leqq \kappa \frac{\|\psi(m_2) - \psi(m_1)\|^2}{t-s} \tag{4.4}$$

for every $0 \leqq s < t$ and $m_1, m_2 \in \Gamma$.

Proof. We denote by φ_{m_1} the coordinate of $T_{m_1}M$ induced by ψ. Set $f = \psi \circ \exp_{m_1} \circ \varphi_{m_1}^{-1}$, $g = f^{-1}$, $y = \psi(m_1)$, $\Delta y = \psi(m_2) - \psi(m_1)$, $\Delta = t - s$, $\Gamma_1 = \{\lambda\psi(m_1) + (1-\lambda)\psi(m_2)\,;\, 0 \leqq \lambda \leqq 1, m_1, m_2 \in \Gamma\}$ and $\Gamma_2 = \{\lambda g(\Gamma_1)\,;\, 0 \leqq \lambda \leqq 1\}$. Since

$$(y^i(v))_{1 \leqq i \leqq r} \equiv \psi(y_{m_1,m_2}(s+v)) = f(v\Delta^{-1}g(y + \Delta y)) ,$$

we have

$$\frac{d}{dv}y^i(v) = \sum_j \partial_j f^i(v\Delta^{-1}g(y+\Delta y))\Delta^{-1}g^j(y+\Delta y) .$$

We note that $\partial_j f^i(0) = \partial_j g^i(0) = \delta^i_j$ and $g_j(y) = 0$. It is easy to see that

$$\left| \frac{d}{dv}y^i(v) - \sum_j \partial_j f^i(v\Delta^{-1}g(y+\Delta y))\Delta^{-1}\Delta y^j \right|$$
$$\leqq \sum_j \max_{\xi \in \Gamma_2} |\partial_j f^i(\xi)| \max_{\xi \in \Gamma_1} \|\operatorname{grad} g^j(\xi)\| v\Delta^{-1} \|\Delta y\|^2 ,$$

if we note that

$$|g^j(y + \Delta y) - \Delta y^j| \leqq \max_{\xi \in \Gamma_1} \|\operatorname{grad} g^j(\xi)\| \|\Delta y\|^2 .$$

Using

$$|\partial_j f^i(v\Delta^{-1}g(y+\Delta y)) - \partial_j f^i(0)|$$
$$\leqq \max_{\xi \in \Gamma_2} \|\operatorname{grad} \partial_j f^i(\xi)\| v\Delta^{-1} \|g(y+\Delta y)\|$$
$$\leqq \max_{\xi \in \Gamma_2} \|\operatorname{grad} \partial_j f^i(\xi)\| \max_{\xi \in \Gamma_1} \sum_k \|\operatorname{grad} g^k(\xi)\| v\Delta^{-1} \|\Delta y\| ,$$

we have

$$\left|\sum_j \partial_j f^i(v\Delta^{-1}g(y+\Delta y))\Delta^{-1}\Delta y^j - \Delta^{-1}\Delta y^i\right|$$
$$= \left|\sum_j \{\partial_j f^i(v\Delta^{-1}g(y+\Delta y)) - \partial_j f^i(0)\}\Delta^{-1}\Delta y^j\right|$$
$$\leqq \sum_j \max_{\xi\in\Gamma_2} \|\operatorname{grad}\partial_j f^i(\xi)\| \max_{\xi\in\Gamma_1} \sum_k \|\operatorname{grad} g^k(\xi)\| v\Delta^{-2}\|\Delta y\|^2 .$$

Hence we obtain (4.4). Q.E.D.

References

[1] Dynkin, E. B., Diffusion of tensors, Dokl. Akad. Nauk SSSR, **179** (1968), 532–535.

[2] Itô, K., The Brownian motion and tensor fields on Riemannian manifold, Proc. Int. Congress Math., 1962 (Stockholm), 536–539.

[3] ——, Stochastic differentials, Appl. Math. Optimization, **1** (1975), 374–381.

[4] ——, Stochastic parallel displacement, Probabilistic Methods in Differential Equations, Lecture Notes in Math., **451**, Springer. (1975), 1–7.

[5] Kunita, H., Diffusion processes and control systems, Lecture Note Paris VI, 1973–1974.

[6] Kunita, H. and Watanabe, S., On square integrable martingales, Nagoya Math. J., **30** (1967), 209–245.

[7] Stroock, D. W. and Varadhan, S. R. S., On the support of diffusion processes with applications to strong maximum principle, Proc. 6-th Berkeley Symp. on Math. Stat. and Prob., **3** (1970), 333–360.

[8] Wong, E. and Zakai, M., On the relation between ordinary and stochastic differential equations, Internat. J. Engrg. Sci., **3** (1965), 213–229.

Shintaro Nakao
DEPARTMENT OF MATHEMATICS
NARA WOMEN'S UNIVERSITY
NARA 630, JAPAN

Yuiti YAMATO
DEPARTMENT OF MATHEMATICS
HIROSHIMA UNIVERSITY
HIROSHIMA 730, JAPAN

Proc. of Intern. Symp. SDE
Kyoto 1976, pp. 297–325

On Stochastic Optimal Controls and Envelope of Markovian Semi-Groups

Makiko NISIO

§ 1. Introduction

In [8] we introduced a non-linear semi-group associated with the so-called Bellman principle in stochastic optimal controls of diffusion type. Let Γ be a convex σ-compact subset of R^k, called a control region. Let a triple (Ω, B, U) be an admissible system where Ω is a probability space, B is an n-dimensional Brownian motion on Ω and U is a Γ-valued B-non-anticipative bounded process on Ω. For an admissible system (Ω, B, U) we consider the following n-dimensional stochastic differential equation.

$$dX(t) = \alpha(X(t), U(t))dB(t) + \gamma(X(t), U(t))dt \tag{1}$$

where $\alpha(x, u)$ is a symmetric $n \times n$ matrix and $\gamma(x, u)$ an n-vector. We assume the following conditions of boundedness and smoothness,

$$|F(x, u)| \leq b\ , \qquad F = \alpha, \gamma \tag{2}$$

and

$$|F(x, u) - F(y, v)| \leq \mu|x - y| + \rho(|u - v|)\ , \qquad F = \alpha, \gamma\ , \tag{3}$$

where ρ is concave, strictly increasing and continuous on $[0, \infty)$ and $\rho(0) = 0$ and μ is a positive constant. Under these conditions the stochastic differential equation (1) has a unique solution X, which is called the response for (Ω, B, U).

By C we denote the Banach lattice of all bounded and uniformly continuous functions on R^n, endowed with the usual supremum norm and the usual order. Let $c(x, u)$ be non-negative and $f(x, u)$ real. We assume both c and f to satisfy the conditions (2) and (3). For any $\phi \in C$ we define Q_t by

$$\begin{aligned} Q_t\phi(x) = \sup_{\text{adm. syst.}} \ E_x \int_0^t \exp\left(-\int_0^s c(X(\theta), U(\theta))d\theta\right) f(X(s), U(s))ds \\ + \exp\left(-\int_0^t c(X(\theta), U(\theta))d\theta\right)\phi(X(t))\ , \end{aligned} \tag{4}$$

where X is the response for (Ω, B, U), starting at $X(0) = x$. Then by virtue of (2) and (3) Q_t is a strongly continuous non-linear semi-group on C, which is contractive and monotone. Moreover the generator G of Q_t is given by

$$G\phi = \sup_{u \in \Gamma} [A^u\phi + f^u] \tag{5}$$

$$A^u\phi(x) = \frac{1}{2} \sum_{ij} \alpha^2(x, u)_{ij} \frac{\partial^2 \phi}{\partial x_i \partial x_j}(x) + \sum_i \gamma_i(x, u) \frac{\partial \phi}{\partial x_i}(x) - c(x, u)\phi(x) \tag{6}$$

for every ϕ whose first and second derivatives are in C. The right side of (5) can be found in the famous Bellman equation, [2], [5]. Furthermore the least Q_t-excessive majorant is closely related to the optimal stopping problem, [3], [5].

In this note we shall discuss a similar problem in a more general set-up. Let A^u denote the generator of a Markov process. We seek a semi-group of operators acting on C whose generator is an extension of $G\phi = \sup_u (A^u\phi + f^u)$. Such a semi-group (with generator G) will be obtained as the envelope of the semi-groups

$$T_t^u\phi = P_t^u\phi + \int_0^t P_\theta^u f^u d\theta \ , \qquad u \in \Gamma$$

whose generators are

$$G^u\phi = A^u\phi + f^u \ , \qquad u \in \Gamma$$

respectively, as we can imagine from the fact that G is the envelope of $G^u, u \in \Gamma$. In fact we will prove the following theorem in § 3.

Theorem 1. *Let Γ be a parameter set. Let A^u be the generator of positive contractive and strongly continuous linear semi-group P_t^u on C. We assume the following conditions* (A1) ~ (A4).

(A1) If $\phi_n \in C$ is an increasing sequence tending to $\phi \in C$ at each point, then $P_t^u\phi_n$ increases and tends to $P_t^u\phi$ at each point, for every $u \in \Gamma$ and every $t \geq 0$.

(A2) Let $\mathscr{D}(A^u)$ denote the domain of the generator A^u. The subset C^2 of C defined by

$$C^2 = \left\{\phi \in C, \frac{\partial \phi}{\partial x_i} \text{ and } \frac{\partial^2 \phi}{\partial x_i \partial x_j} \in C, \ i, j = 1, \cdots, n\right\}$$

is contained in $\bigcap_u \mathscr{D}(A^u)$. Moreover

$$(7) \qquad \sup_u \|A^u\phi\| < \infty \qquad \text{for } \phi \in C^2 .$$

(A3) For a positive constant h

$$(8) \qquad \sup_u \|f^u\| \leq h \quad \text{and} \quad \sup_u |f^u(x) - f^u(y)| \leq h|x-y| .$$

(A4) For any positive T there exists a constant $\lambda = \lambda(T)$ such that

$$(9) \qquad \sup_u |P_t^u\phi(x) - P_t^u\phi(y)| \leq e^{\lambda t}|x-y| \qquad \forall t \leq T ,$$

if $|\phi(x) - \phi(y)| \leq |x-y|$ and $\|\phi\| \leq 1$. Then there exists a unique non-linear semi-group S_t on C satisfying the following conditions (0) ~ (vi),

(0) semi-group property: $S_0 =$ identity, $S_{t+\theta}\phi = S_\theta(S_\theta\phi) = S_\theta(S_t\phi)$,
(i) monotone: $S_t\phi \leq S_t\psi$ whenever $\phi \leq \psi$,
(ii) contractive: $\|S_t\phi - S_t\psi\| \leq \|\phi - \psi\|$,
(iii) strongly continuous: $\|S_t\phi - S_\theta\phi\| \to 0$ as $t \to \theta$,

$$\text{(iv)} \qquad P_t^u\phi + \int_0^t P_\theta^u f^u d\theta \leq S_t\phi , \qquad \text{for } \forall t \text{ and } u ,$$

where the integral stands for the Bochner integral,
(v) the generator G of S_t is expressed by

$$(10) \qquad G\phi = \sup_u [A^u\phi + f^u] \qquad \text{for } \phi \in \mathscr{D}(G) \cap C^2 .$$

Moreover if the following conditions (11) and (12) hold

$$(11) \qquad \sup_u \left\| \frac{1}{t}(P_t^u\phi - \phi) - A^u\phi \right\| \to 0 \qquad \text{as } t \to 0 \qquad \text{for } \phi \in C^2$$

and

$$(12) \qquad \sup_u \|P_t^u f^u - f^u\| \to 0 \qquad \text{as } t \to 0 ,$$

then

$$\mathscr{D}(G) \supset C^2 ,$$

(vi) minimum: if $\tilde{S}_t$ is a non-linear semi-group with (i) ~ (iv), then

$$S_t\phi \leq \tilde{S}_t\phi .$$

When A^u is the second order elliptic differential operator expressed by (6), P_t^u is the transition semi-group of diffusion expressed by stochastic differential equation with the killing rate $c(x, u)$. The following theorem will be proved in § 4.

Theorem 2. *Let Γ be a σ-compact convex subset of R^k. Let A^u be the following elliptic differential operator*

$$A^u\phi(x) = \frac{1}{2}\sum_{ij}\alpha^2(x,u)_{ij}\frac{\partial^2\phi}{\partial x_i \partial x_j}(x) + \sum_i \gamma_i(x,u)\frac{\partial\phi}{\partial x_i}(x) - c(x,u)\phi(x) . \tag{6}$$

Suppose α, γ, c and f satisfy the conditions (2) *and* (3). *Then the conditions* (A1) $\sim$ (A4) *are valid and*

$$Q_t = S_t , \qquad \forall t \geq 0 .$$

The S_t-excessive majorant and the optimal stopping problem will be treated elsewhere.

§ 2. Preliminaries

Let C denote the set of all bounded and uniformly continuous functions on R^n. C becomes a Banach lattice by the usual norm and partial order, i.e.

$$\|\phi\| \equiv \sup_{x\in R^n} |\phi(x)|$$

and "$\phi \leq \psi$" is defined by "$\phi(x) \leq \psi(x), \forall x$". When $\phi_n \in C$ is increasing to $\phi \in C$ at each point, we say $\phi = 0_i\text{-}\lim \phi_n$. If a subset $\{\phi_\alpha\}$ of C is uniformly bounded and equi-uniformly continuous then $\sup \phi_\alpha$ and $\inf \phi_\alpha$ exist in C. For any positive M and K the set $\sum(M, K)$ defined by

$$\sum(M, K) = \{\phi \in C,\ \|\phi\| \leq M,\ |\phi(x) - \phi(y)| \leq K|x - y|,\ \forall x, y\}$$

is a complete lattice. But C is not complete. According to the difinitions of $\sup \phi_\alpha$ and $\inf \phi_\alpha$ the following inequalities are clear

$$\inf(\phi_\alpha - \psi_\alpha) \leq \sup\phi_\alpha - \sup\psi_\alpha \leq \sup(\phi_\alpha - \psi_\alpha) \tag{1}$$

$$\|\sup\phi_\alpha - \sup\psi_\alpha\| \leq \sup\|\phi_\alpha - \psi_\alpha\| . \tag{2}$$

Let P_t be a positive, contractive and strongly continuous linear semigroup on C. Define T_t by, for $f \in C$,

$$(3)\qquad T_t\phi = P_t\phi + \int_0^t P_\theta f d\theta\,, \qquad \phi \in C\,.$$

Proposition 1. *T_t is a mapping from C into C and has the following properties*

(T0) *semi-group property*: $T_0\phi = \phi$, $T_{t+\theta}\phi = T_t(T_\theta\phi) = T_\theta(T_t\phi)$,

(T1) *monotone*: $T_t\phi \le T_t\psi$ *whenever* $\phi \le \psi$,

(T2) *contractive*: $\|T_t\phi - T_t\psi\| \le \|\phi - \psi\|$

(T3) *strongly continuous*: $\|T_t\phi - T_\theta\phi\| \to 0$ *as* $t \to \theta$

(T4) *the generator G of* T_t: *Let A be the generator of* P_t. *Then* $\mathscr{D}(G) = \mathscr{D}(A)$ *and* $G\phi = A\phi + f$

(T5) $T_t\phi - \phi = \int_0^t P_\theta G\phi d\theta, \qquad \phi \in \mathscr{D}(G).$

Proposition 2. *Suppose* (A1), (A3) *and* (A4). *If*

$$\phi = 0_i\text{-lim}\,\phi_n \quad \textit{and} \quad \phi_n \in \sum(K, K)(\textit{say} \sum(K))\,, \qquad n = 1, 2, \cdots$$

then

$$(4)\qquad \sup T_t^u\phi\,, \qquad \sup T_t^u\phi_n \in \sum(K + ht,\ e^{\lambda t}(K + ht))$$

where $\lambda = \lambda(T)$ *and* $t \le T$ *and*

$$(5)\qquad \sup T_t^u\phi = 0_i\text{-lim}\sup T_t^u\phi_n\,.$$

Proof. By the linearity of P_t^u, (9) of (A4) means the following inequality

$$(6)\qquad |P_t^u\phi(x) - P_t^u\phi(y)| \le e^{\lambda t}K|x - y|\,, \qquad t \le T,\ u \in \Gamma,\ \phi \in \sum(K)\,.$$

We fix T and t moves on $[0, T]$. For $\psi \in \sum(K)$ we have

$$\begin{aligned}
|T_t^u\psi(x) - T_t^u\psi(y)| &\le |P_t^u\psi(x) - P_t^u\psi(y)| \\
&\quad + \int_0^t |P_\theta^u f^u(x) - P_\theta^u f^u(y)|\, d\theta \\
(7)\qquad &\le e^{\lambda t}K|x - y| + \int_0^t e^{\lambda\theta}h|x - y|\, d\theta \\
&\le e^{\lambda t}|x - y|\left(K + \frac{h}{\lambda}(1 - e^{-\lambda t})\right) \\
&\le e^{\lambda t}|x - y|(K + ht)\,.
\end{aligned}$$

$$(8)\qquad \|T_t^u\psi\| \le \|\psi\| + \int_0^t \|f^u\|\, d\theta \le K + ht\,.$$

Hence $\sup T_t^u\psi$ exists and belongs to $\sum(K+ht, e^{\lambda t}(K+ht))$. This means (4).

Next we shall prove (5). Since T_t^u is monotone,

$$T_t^u\phi_n \leq T_t^u\phi_{n+1} \leq T_\theta^u\phi$$

Therefore

$$\sup T_t^u\phi_n \leq \sup T_t^u\phi_{n+1} \leq \sup T_t^u\phi\,. \tag{9}$$

Recalling (4) we can see that $0_i\text{-}\lim_n \sup T_t^u\phi_n$ exists and

$$0_i\text{-}\lim \sup T_t^u\phi_n \leq \sup T_t^u\phi\,. \tag{10}$$

On the other hand we have, by (A1)

$$T_t^u\phi = P_t^u\phi + \int_0^t P_\theta^u f^u d\theta = 0_i\text{-}\lim_n P_t^u\phi_n + \int_0^t P_\theta^u f^u d\theta$$
$$= 0_i\text{-}\lim_n \left(P_t^u\phi_n + \int_0^t P_\theta^u f^u d\theta\right) = 0_i\text{-}\lim_n T_t^u\phi_n\,.$$

Thus

$$T_t^u\phi = 0_i\text{-}\lim_n T_t^u\phi_n \leq 0_i\text{-}\lim_n \sup T_t^u\phi_n\,, \qquad \forall u\in\Gamma\,.$$

Taking the supremum *w.r.* to u we get

$$\sup T_t^u\phi \leq 0_i\text{-}\lim_n \sup T_t^u\phi_n\,. \tag{11}$$

From (10) and (11) we derive (5).

§ 3. Proof of Theorem 1

In this section we fix T arbitrarily and t moves on $[0, T]$. The method of proof is nearly the same as one in [9]. Define $J = J(N)$ by

$$J\phi = \sup T_{1/2^N}^u\phi\,, \qquad \phi\in\bigcup_{K>0}\sum(K)\,(\equiv\sum)\,. \tag{1}$$

Recalling Proposition 2, J is a mapping from $\sum$ into $\sum$. So we can define J^{k+1} by

$$J^{k+1}\phi = J(J^k\phi)\,, \qquad \phi\in\sum\,. \tag{2}$$

Lemma 1. *J^k has the following properties,*

(J0) $J^{k+l}\phi = J^k(J^l\phi) = J^l(J^k\phi)$, $J^0\phi \equiv \phi$,

(J1) *monotone*: $J^k\phi \leq J^k\psi$ *whenever* $\phi \leq \psi$,

(J2) *contractive*: $\|J^k\phi - J^k\psi\| \leq \|\phi - \psi\|$,

(J3) $\|J^k\phi - \phi\| \leq \dfrac{k}{2^N}(\sup \|A^u\phi\| + \sup \|f^u\|)$, $\phi \in C^2$,

(J4) $T^u_{k/2^N}\phi \leq J^k\phi$,

(J5) $J^k\phi = 0_i\text{-}\lim J^k\phi_n$, if $\phi = 0_i\text{-}\lim \phi_n$ and $\phi_n \in \sum(K)$,

(J6) $|J^k\phi(x) - J^k\phi(y)| \leq |x - y|\, e^{\lambda k/2^N}\left[K + \dfrac{h}{\lambda}(1 - e^{-\lambda k/2^N})\right]$, $\phi \in \sum(K)$.

Proof. (J1) Since T^u_t is monotone we have

$$J\phi \leq J\psi \qquad \text{whenever } \phi \leq \psi .$$

Hence we derive (J1) by induction.

(J2) Putting $\Delta = 1/2^N$ we have the following evaluation,

$$\begin{aligned}\|J\phi - J\psi\| &= \|\sup T^u_\Delta\phi - \sup T^u_\Delta\psi\| \\ &\leq \sup \|T^u_\Delta\phi - T^u_\Delta\psi\| \leq \|\phi - \psi\| .\end{aligned}$$

Hence if we assume (J2) for k, then

$$\begin{aligned}\|J^{k+1}\phi - J^{k+1}\psi\| &= \|J(J^k\phi) - J(J^k\psi)\| \\ &\leq \|J^k\phi - J^k\psi\| \leq \|\phi - \psi\| ,\end{aligned}$$

namely (J2) holds for $k + 1$.

(J3) Putting $K(\phi) = \sup\limits_n \|A^u\phi\| + \sup\limits_n \|f^u\|$ for $\phi \in C^2$, we can see

$$T^u\phi - \phi = \int_0^\Delta P^u_\theta A^u\phi d\theta + \int_0^\Delta P^u_\theta f^u d\theta$$

So

$$\|J\phi - \phi\| \leq \sup \|T^u_\Delta\phi - \phi\| \leq \Delta K(\phi) .$$

Therefore by (J2) we have

$$\begin{aligned}\|J^k\phi - \phi\| &\leq \sum_{l=1}^k \|J^l\phi - J^{l-1}\phi\| \\ &= \sum_{l=1}^k \|J^{l-1}(J\phi) - J^{l-1}\phi\| \leq k\,\|J\phi - \phi\| \leq k\Delta K(\phi) .\end{aligned}$$

This completes the proof of (J3).

(J4) By the definition of J we get

$$T_\Delta^u \psi \le J\psi \qquad \forall \psi \in \Sigma$$

Hence if we assume (J4) for k, then

$$(3) \qquad T^u_{(k+1)\Delta}\phi = T^u_\Delta(T^u_{k\Delta}\phi) \le T^u_\Delta(J^k\phi) \le J(J^k\phi) = J^{k+1}\phi\,.$$

namely (J4) holds for $k+1$.

(J5) For $k = 1$ (J5) is Proposition 2. If (J5) holds for k, then

$$J^{k+1}\phi = J(J^k\phi) = J(0_i\text{-}\lim_n J^k\phi_n) = 0_i\text{-}\lim_n J(J^k\phi) = 0_i\text{-}\lim_n J^{k+1}\phi_n\,.$$

Therefore we get (J5).

(J6) For $k = 1$ (J6) comes from (7) of Proposition 2. Suppose (J6) holds for k. Appealing to the following inequality

$$\|J^k\phi\| \le K + hk\Delta \le e^{\lambda k\Delta}K + h(e^{\lambda k\Delta} - 1)/\lambda$$

we have $J^k\phi \in \sum(e^{\lambda k\Delta}[K + h(1 - e^{-\lambda k\Delta})/\lambda])$. Again using the similar evalution as (7) of Proposition 2, we get

$$\begin{aligned}
|J^{k+1}\phi(x) - J^{k+1}\phi(y)| &\le \sup |T^u_\Delta(J^k\phi)(x) - T^u_\Delta(J^k\phi)(y)| \\
&= \sup_u \left[|P^u_\Delta(J^k\phi)(x) - P^u_\Delta(J^k\phi)(y)| + \int_0^\Delta |P^u_\theta f^u(x) - P^u_\theta f^u(y)|\, d\theta\right] \\
(4) \qquad &\le e^{\lambda\Delta}|x-y|\, e^{\lambda k\Delta}\left(K + \frac{h}{\lambda}(1 - e^{-\lambda k\Delta})\right) + \frac{h}{\lambda}|x-y|\,(e^{\lambda\Delta} - 1) \\
&= e^{\lambda(k+1)\Delta}|x-y|\left|K + \frac{h}{\lambda}(1 - e^{-\lambda(k+1)\Delta})\right|,
\end{aligned}$$

namely (J6) holds for $k+1$.

Put $S_t^{(N)}\phi = J^k(N)\phi$ for $t = k/2^N$ and $\phi \in \sum$.

Lemma 2. *$S_t^{(N)}\phi$ is increasing as $N \to \infty$, i.e.*

$$(5) \qquad S_t^{(N)}\phi \le S_t^{(N+1)}\phi \qquad \textit{for } t = k/2^N \textit{ and } \phi \in \textstyle\sum\,.$$

Moreover 0_i-$\lim_N S_t^{(N)}\phi$ exists in $\sum$ for any binary $t = j/2^l$.

Proof. Put $\Delta = 1/2^{N+1}$. Recalling (T0) and (T1) we have

$$(6) \qquad T^u_{2\Delta}\phi = T^u_\Delta(T^u_\Delta\phi) \le T^u_\Delta(S_\Delta^{(N+1)}\phi)$$

Taking the supremum of both sides we get

$$(7) \qquad S_{2\Delta}^{(N)}\phi \le S_\Delta^{(N+1)}(S_\Delta^{(N+1)}\phi) = S_{2\Delta}^{(N+1)}\phi\,,$$

namely (5) is valid for $k = 1$. If (5) holds for k then

$$(8)\qquad \begin{aligned} S^{(N)}_{2(k+1)\Delta}\phi &= S^{(N)}_{2\Delta}(S^{(N)}_{2k\Delta}\phi) \le S^{(N)}_{2\Delta}(S^{(N+1)}_{2k\Delta}\phi) \\ &\le S^{(N+1)}_{2\Delta}(S^{(N+1)}_{2k\Delta}\phi) = S^{(N+1)}_{2(k+1)\Delta}\phi\ . \end{aligned}$$

namely (5) holds for $k+1$.

By virtue of (J6) we can see

$$S^{(N)}_t\phi \in \sum(e^{\lambda t}[K + h(1-e^{-\lambda t})/\lambda]) \qquad \text{for } \phi \in \sum(K),\ N \ge l\ .$$

Since $\sum(p)$ $(p = e^{\lambda t}[K + h(1-e^{-\lambda t})/\lambda])$ is a complete lattice, (5) implies the existence of $0_i\text{-}\lim_N S^{(N)}_t\phi$ in $\sum(p)$. This completes the proof of Lemma 2.

Let us define S_t by

$$(9)\qquad S_t\phi = 0_i\text{-}\lim_N S^{(N)}_t\phi$$

for binary t and $\phi \in \sum$. Then the following Lemma 3 is clear

Lemma 3. *For binary t and θ and $\phi, \psi \in \sum$*

(S0) $S_0\phi = \phi$

(S1) *monotone*: $S_t\phi \le S_t\psi$ *whenever* $\phi \le \psi$

(S2) *contractive*: $\|S_t\phi - S_\theta\psi\| \le \|\phi - \psi\|$

(S3) $\|S_t\phi - S_\theta\phi\| \le |t-\theta|K(\phi)$, *for* $\phi \in C^2$

(S4) $T^u_t\phi \le S_t\phi$, $\forall u \in \Gamma$

(S5) $\|S_t\phi\| \le \|\phi\| + th$

(S6) $|S_t\phi(x) - S_t\phi(y)| \le |x-y|e^{\lambda t}(K + h(1-e^{-t\lambda})/\lambda)$, $\phi \in \sum(K)$.

Since C^2 is dense in C, we can extends S_t by the following way,

$$(10)\qquad S_t\phi = \lim S_{t_l}\phi_l\ , \qquad \text{for } t \ge 0 \text{ and } \phi \in C\ ,$$

where t_l is a binary approximation to t and $\phi_l \in C^2$ an approximation to ϕ. From (S2) and (S3) we can see

$$(11)\qquad \begin{aligned} \|S_{t_l}\phi_l - S_{t_k}\phi_k\| &\le \|S_{t_l}\phi_l - S_{t_l}\phi_N\| \\ &\quad + \|S_{t_l}\phi_N - S_{t_k}\phi_N\| + \|S_{t_k}\phi_N - S_{t_k}\phi_k\| \\ &\le \|\phi_l - \phi_N\| + |t_l - t_k|\,K(\phi_N) + \|\phi_N - \phi_k\|\ . \end{aligned}$$

For any $\varepsilon > 0$, we take a large N such that

$$|\phi_i - \phi_j| < \varepsilon \qquad \text{for } i, i \ge N\ .$$

Putting $\delta = \varepsilon/(K(\phi_N)+1)$ we see

$$\|S_{t_l}\phi_l - S_{t_k}\phi_k\| < 3\varepsilon \qquad \text{for } |t_l - t_k| < \delta \text{ and } l, k > N\ .$$

Hence $S_{t_l}\phi_l$ is a Cauchy sequence in C. Moreover the left side of (10) does not depend on the special choice of approximation (t_l, ϕ_l).

Proposition 3. *S_t has the following properties*
(0) *S_t maps $\sum$ into $\sum$,*
(i) *monotone: $S_t\phi \leq S_t\psi$ whenever $\phi \leq \psi$,*
(ii) *contractive: $\|S_t\phi - S_t\psi\| \leq \|\phi - \psi\|$,*
(iii) *strongly continuous: $\|S_t\phi - S_\theta\phi\| \to 0$ as $t \to \theta$,*
(iv) *$T_t^u\phi \leq S_t\phi$,*
(v) *$\|S_t\phi\| \leq \|\phi\| + th$.*

Proof. First we shall show (ii). Let $\phi_l \in C^2$ and $\psi_l \in C^2$ be approximations to ϕ and ψ respectively. Hence for a binary approximation t_l to t, we have

$$\|S_t\phi - S_t\psi\| = \lim_l \|S_{t_l}\phi_l - S_{t_l}\psi_l\| \leq \lim \|\phi_l - \psi_l\| = \|\phi - \psi\| .$$

(i) For $\varepsilon > 0$ we take an approximation $\phi_l(\varepsilon) \in C^2$ to $\phi - \varepsilon$. Let $\psi_n \in C^2$ be an approximation to ψ. Then

$$\phi_l(\varepsilon) \leq \psi_l \qquad \text{for large } l .$$

Hence by (S1)

$$S_{t_l}\phi_l(\varepsilon) \leq S_{t_l}\psi_l \qquad \text{for large } l .$$

Therefore tending l to ∞ we have

$$S_t(\phi - \varepsilon) \leq S_t\psi .$$

On the other hand $\phi - \varepsilon$ tends to ϕ, so (ii) imples

$$S_t\phi = \lim_{\varepsilon \downarrow 0} S_t(\phi - \varepsilon) .$$

Hence we have (i).

(iii) By ($S3$) and (10) we have, for $\phi \in C^2$,

$$\|S_t\phi - S_\theta\phi\| = \lim \|S_{t_l}\phi - S_{\theta_l}\phi\| \leq \lim \|t_l - \theta_l| K(\phi) = |t - \theta| K(\phi)$$

where t_l and θ_l are binary approximations to t and θ respectively. For $\phi \in C$ and $\varepsilon > 0$ we take $\psi \in C^2$ such that $\|\phi - \psi\| < \varepsilon$. Then we have

$$\begin{aligned}\|S_t\phi - S_\theta\phi\| &\leq \|S_t\phi - S_t\psi\| + \|S_t\psi - S_\theta\psi\| + \|S_\theta\psi - S_\theta\phi\| \\ &\leq 2\varepsilon + |t - \theta| K(\psi) ,\end{aligned}$$

Hence there exists a small positive $\delta = \delta(\phi, \varepsilon)$ such that $\|S_t\phi - S_\theta\phi\| < 3\varepsilon$ whenever $|t - \theta| < \delta$.

(iv) and (v) are clear from (S4) and (S5).

(0) Let ϕ be in $\sum(K)$. Let t_l be a binary approximation to t. Then we have, by (iii) and (S6),

$$\begin{aligned}
&|S_t\phi(x) - S_t\phi(y)| \\
&\quad \leq \lim_l |S_t\phi(x) - S_{t_l}\phi(x)| + |S_{t_l}\phi(x) - S_{t_l}\phi(y)| + |S_{t_l}\phi(y) - S_t\phi(y)| \\
&\quad \leq \lim_l 2\|S_t\phi - S_{t_l}\phi\| + |x - y|\, e^{\lambda t_l}\Big(K + \frac{h}{\lambda}(1 - e^{-t_l\lambda})\Big) \\
&\quad = |x - y|\, e^{\lambda t}\Big(K + \frac{h}{\lambda}(1 - e^{-\lambda t})\Big).
\end{aligned}$$

Recalling (v) we can derive (0).

Proposition 4. *S_t is a semi-group on C.*

Proof. Let t and θ be binary, say $t = i/2^l$ and $\theta = j/2^l$. For $N \geq l$ and $\phi \in C^2 (\subset \sum)$ we have

$$S_{t+\theta}^{(N)}\phi = S_\theta^{(N)}(S_t^{(N)}\phi) \leq S_\theta^{(N)}(S_t\phi) \leq S_\theta(S_t\phi) \tag{12}$$

$$S_\theta(S_t\phi) = 0_i\text{-lim}\, S_\theta^{(N)}(S_t\phi) \tag{13}$$

and

$$S_{\theta+t}\phi = 0_i\text{-lim}\, S_{\theta+t}^{(N)}\phi\,.$$

Hence

$$S_{\theta+t}\phi \leq S_\theta(S_t\phi)\,. \tag{14}$$

On the other hand, for $l \leq n \leq N$,

$$S_\theta^{(n)}(S_t^{(N)}\phi) \leq S_\theta^{(N)}(S^{(N)}) = S_{\theta+t}^{(N)}\phi \leq S_{\theta+t}\phi\,. \tag{15}$$

Moreover, (J5) of Lemma 1 means

$$S_\theta^{(n)}(S_t\phi) = 0_i\text{-}\lim_N S_\theta^{(n)}(S_t^{(N)}\phi)\,.$$

Hence, by (15), we have

$$S_\theta^{(n)}(S_t\phi) \leq S_{\theta+t}\phi\,, \qquad \text{for } n \geq l\,.$$

Tending n to ∞, we can see

(16) $$S_\theta(S_t\phi) \le S_{\theta+t}\phi \,.$$

From (14) and (16) we get

$$S_\theta(S_t\phi) = S_{\theta+t}\phi \,.$$

Let t_n and θ_n be binary approximations to t and θ respectively. For $\varepsilon > 0$ we take ψ and ζ in C^2 such that $\|\phi - \psi\| < \varepsilon$ and $\|S_t\psi - \zeta\| < \varepsilon$. Then we get

(17) $$\|S_{t_n+\theta_n}\phi - S_{t_n+\theta_n}\psi\| \le \|\phi - \psi\| < \varepsilon \,.$$

(18) $$\begin{aligned} \|S_\theta(S_t\phi) - S_{\theta_n+t_n}\psi\| &= \|S_\theta(S_t\phi) - S_{\theta_n}(S_{t_n}\psi)\| \\ &\le \|S_\theta(S_t\phi) - S_\theta(S_t\psi)\| + \|S_\theta(S_t\psi) - S_\theta\zeta\| \\ &\quad + \|S_\theta\zeta - S_{\theta_n}\zeta\| + \|S_{\theta_n}\zeta - S_{\theta_n}(S_{t_n}\psi)\| \\ &\le \|\phi - \psi\| + \|S_t\psi - \zeta\| + |\theta - \theta_n| K(\zeta) + \|\zeta - S_{t_n}\psi\| \\ &\le 2\varepsilon + |\theta - \theta_n| K(\zeta) + \|\zeta - S_t\psi\| + \|S_t\psi - S_{t_n}\psi\| \\ &\le 3\varepsilon + |\theta - \theta_n| K(\zeta) + |t - t_n| K(\psi) \,. \end{aligned}$$

Hence by (17) and (18) we have

$$\|S_\theta(S_t\phi) - S_{\theta_n+t_n}\phi\| \le 4\varepsilon + |\theta - \theta_n| K(\zeta) + |t - t_n| K(\psi) \,.$$

Therefore $S_{\theta_n+t_n}\phi$ converges to $S_\theta(S_t\phi)$. On the other hand $S_{\theta_n+t_n}\phi$ tends to $S_{\theta+t}\phi$ by (iii) of Proposition 3. Hence

(19) $$S_{\theta+t}\phi = S_\theta(S_t\phi) \,.$$

This means the semi-group property of S_t.

Let G be the generator of S_t, namely

(20) $$G\phi = \lim_{t\downarrow 0} \frac{1}{t}(S_t\phi - \phi)$$

and

$$\mathscr{D}(G) = \left\{\phi \in C, \lim_{t\downarrow 0} \frac{1}{t}\,(S_t\phi - \phi) \text{ exists}\right\} .$$

Proposition 5.

$$G\phi = \sup_u \,(A^u\phi + f^u) \,, \qquad \textit{for } \phi \in C^2 \cap \mathscr{D}(G) \,.$$

Moreover if the conditions (11) *and* (12) *of Theorem* 1 *hold, i.e.*

$$\sup_u \left\| \frac{1}{t}(P_t^u \phi - \phi) - A^u \phi \right\| \xrightarrow[t \downarrow 0]{} 0 \qquad \text{for } \phi \in C^2 \tag{11}$$

and

$$\sup_u \| P_t^u f^u - f^u \| \xrightarrow[t \downarrow 0]{} 0 \,, \tag{12}$$

then

$$\mathscr{D}(G) \supset C^2 \,.$$

Proof. In the case $f^u \equiv 0$ for any u, we denote S_t by Λ_t. Put $A\phi = \sup_u G^u \phi = \sup_u (A^u \phi + f^u)$ and $\Delta = 1/2^N$. Recalling (T5) of Proposition 1 we have for $\phi \in C^2$.

$$\begin{aligned} S_\Delta^{(N)} \phi - \phi &= \sup_u (T_\Delta^u \phi - \phi) = \sup_u \int_0^\Delta P_\theta^u G^u \phi d\theta \\ &\le \sup_u \int_0^\Delta P_\theta^u A \phi d\theta \le \int_0^\Delta \Lambda_\theta A \phi d\theta \,. \end{aligned} \tag{21}$$

Moreover

$$\begin{aligned} S_{2\Delta}^{(N)} \phi - S_\Delta^{(N)} \phi &= \sup_u T_\Delta^u (S_\Delta^{(N)} \phi) - \sup_u T_\Delta^u \phi \\ &\le \sup_u [T_\Delta^u (S_\Delta^{(N)} \phi) - T_\Delta^u \phi] = \sup_u [P_\Delta^u (S_\Delta^{(N)} \phi) - P_\Delta^u \phi] \\ &= \sup_u [P_\Delta^u (S_\Delta^{(N)} \phi - \phi)] \le \Lambda_\Delta (S_\Delta^{(N)} \phi - \phi) \\ &\le \Lambda_\Delta \Big(\int_0^\Delta \Lambda_\theta A \phi d\theta \Big) \le \int_0^\Delta \Lambda_{\Delta + \theta} A \phi d\theta = \int_\Delta^{2\Delta} \Lambda_\theta A \phi d\theta \,. \end{aligned} \tag{22}$$

Suppose

$$S_{k\Delta}^{(N)} \phi - S_{(k-1)\Delta}^{(N)} \phi \le \int_{(k-1)\Delta}^{k\Delta} \Lambda_\theta A \phi d\theta \,.$$

Then by the similar calculation we see

$$S_{(k+1)\Delta}^{(N)} \phi - S_{k\Delta}^{(N)} \phi \le \Lambda_\Delta (S_{k\Delta}^{(N)} \phi - S_{(k-1)\Delta}^{(N)} \phi) < \int_{k\Delta}^{(k+1)\Delta} \Lambda_\theta A \phi d\theta \,.$$

Hence taking the summation for k we get

$$S_t^{(N)} \phi - \phi \le \int_0^t \Lambda_\theta A \phi d\theta \qquad \text{for } t = \frac{i}{2^N} \,. \tag{23}$$

Tending N to ∞ we have

(24) $$S_t\phi - \phi \leq \int_0^t \Lambda_\theta A\phi d\theta$$

for binary t and $\phi \in C^2$. Since the both sides of (24) are continuous in t, (24) holds for any $t \geq 0$. Furthermore,

(25) $$\frac{1}{t}(S_t\phi - \phi) \leq \frac{1}{t}\int_0^t \Lambda_\theta A\phi d\theta .$$

So we get

(26) $$\overline{\lim_{t\downarrow 0}} \frac{1}{t}(S_t\phi(x) - \phi(x)) \leq A\phi(x) , \qquad \forall x .$$

On the other hand by virtue of (iv) of Proposition 3 we have

(27) $$\frac{1}{t}(S_t\phi - \phi) \geq \frac{1}{t}(T_t^u\phi - \phi) .$$

So we get

$$\underline{\lim_{t\downarrow 0}} \frac{1}{t}(S_t\phi(x) - \phi(x)) \geq G^u\phi(x) = A^u\phi(x) + f^u(x) , \qquad \forall u \in \Gamma .$$

Therefore we have the following evaluation

(28) $$\underline{\lim_{t\downarrow 0}} \frac{1}{t}(S_t\phi(x) - \phi(x)) \geq \sup_u G^u\phi(x) = A\phi(x) .$$

Thus from (26) and (28) we obtain

$$\lim_{t\downarrow 0} \frac{1}{t}(S_t\phi(x) - \phi(x)) = A\phi(x) \qquad \forall x .$$

This means that the domain of weak generator G_w of S_t contains C^2 and

$$G_w\phi = A\phi .$$

Hence we have

$$G\phi = A\phi , \qquad \text{for } \phi \in \mathscr{D}(G) \cap C^2 .$$

Next we shall show the latter half. From (25) we can see

(29) $$\frac{1}{t}(S_t\phi - \phi) - A\phi \leq \frac{1}{t}\int_0^t (\Lambda_\theta A\phi - A\phi)d\theta , \qquad \phi \in C^2 .$$

Since the right side converges to 0 as $t \to 0$, for $\varepsilon > 0$ there exists a positive δ such that

$$\sup_x \left[\frac{1}{t}(S_t\phi(x) - \phi(x)) - A\phi(x)\right] < \varepsilon\,, \qquad t < \delta\,. \tag{30}$$

On the other hand by (27) we have

$$\begin{aligned} \frac{1}{t}(S_t\phi - \phi) - A\phi &\geq \sup_u \frac{1}{t}(T_t^u\phi - \phi) - \sup_u G^u\phi \\ &\geq \inf_u \left[\frac{1}{t}(T_t^u\phi - \phi) - G^u\phi\right]. \end{aligned} \tag{31}$$

On the other hand we can see, from the conditions (11) and (12) of Theorem 1,

$$\begin{aligned} \left|\frac{1}{t}(T_t^u\phi - \phi) - G^u\phi\right| &\leq \left|\frac{1}{t}(P_t^u\phi - \phi) - A^u\phi\right| \\ &\quad + \frac{1}{t}\int_0^t |P_\theta^u f^u - f^u|\, d\theta \to 0 \end{aligned}$$

uniformly in u, as $t \downarrow 0$. Thus for $\varepsilon > 0$ there exists a positive δ, such that

$$\inf_x \left[\frac{1}{t}(S_t\phi(x) - \phi(x)) - A\phi(x)\right] \geq -\varepsilon\,, \qquad \text{for } t < \delta\,. \tag{32}$$

Hence from (30) and (32) we get for $\phi \in C^2$

$$\left\|\frac{1}{t}(S_t\phi - \phi) - A\phi\right\| \to 0\,, \qquad \text{as } t \downarrow 0 \tag{33}$$

namely $\phi \in \mathscr{D}(G)$.

Proposition 6. *If $\tilde{S}_t$ is a semi-group on C satisfying the conditions* (i) $\sim$ (iv), *then we have for $t \geq 0$ and $\phi \in C$*

$$S_t\phi \leq \tilde{S}_t\phi\,. \tag{34}$$

Proof. Putting $\Delta = 1/2^N$ we have

$$S_\Delta^{(N)}\phi = \sup T_\Delta^u\phi \leq \tilde{S}_\Delta\phi\,, \qquad \phi \in \Sigma\,. \tag{35}$$

Suppose

$$S_{k\Delta}^{(N)}\phi \leq \tilde{S}_{k\Delta}\phi\,, \qquad \phi \in \Sigma\,, \tag{36}$$

then we have

$$\begin{aligned} S^{(N)}_{(k+1)\Delta}\phi &= S^{(N)}_{\Delta}(S^{(N)}_{k}\phi) \leq \tilde{S}_{\Delta}(S^{(N)}_{k}\phi) \\ &\leq \tilde{S}_{\Delta}(\tilde{S}_{k\Delta}\phi) = \tilde{S}_{(k+1)\Delta}\phi\,, \qquad \phi \in \textstyle\sum\,. \end{aligned}$$

Hence (36) holds for any k. This means for any binary t and $\phi \in \sum$

(37) $$S^{(N)}_{t}\phi \leq \tilde{S}_{t}\phi \qquad \text{for large } N\,.$$

Therefore for binary t

(38) $$S_{t}\phi \leq \tilde{S}_{t}\phi\,, \qquad \phi \in \textstyle\sum\,.$$

Since the both sides of (38) are continuous in t and contractive in $\phi \in C$, the inequality (38) holds for any $t \geq 0$ and $\phi \in C$. This completes the proof.

We assume Γ is a σ-compact subset of R^{k}. Let W be the path space, namely the set of all right continuous R^{n}-valued functions on $[0, \infty)$ with left limits. Let $\mathscr{F}_{t}$ and $\mathscr{F}$ be σ-algebras on W generated by $\{w(s), s \leq t\}$ and $\{w(s), s < \infty\}$ respectively. By P^{u}_{x} we denote the probability law of the path starting at x with transition probability $P^{u}(t, x, B)$, which is a substochastic measure on R^{n} for any u, t and x and a Borel measurable functions of (u, t, x) for any B. $X(t, w)$ denotes the t-th coordinate of $w \in W$ and the system $(X(t), W, \mathscr{F}_{t}, \mathscr{F}, P^{u}_{x}, x \in R^{n})$ is called a Markov process with transition probability P^{u}. Define the transition operator P^{u}_{t} by

$$P^{u}_{t}\phi(x) = \int_{R^{n}} \phi(y)P^{u}(t, x, dy) = E^{u}_{x}\phi(X(t))$$

for a bounded Borel function ϕ. We assume P^{u}_{t} is a semi-group satisfied the conditions of Theorem 1, when we restrict P^{u}_{t} on C.

Let d be a non-anticipative step function on $[0, \infty) \times W$, i.e.

(39) $$d(t, w) = d(k\Delta, w) \qquad \text{for } t \in [k\Delta, (k+1)\Delta)$$

where $\Delta = 2^{-N}$ and $d(s, \cdot)$ is $\mathscr{F}_{s}$-measurable. By $\mathfrak{A}_{N}$ we denote of all functions expressed by (39). Put $\mathfrak{A} = \bigcup_{N=1}^{\infty} \mathfrak{A}_{N}$ and an element of $\mathfrak{A}$ is called an admissible control, precisely speaking control of measures of switchings at binary times, [11]. An admissible control d written by (39) define the new measure Q^{d}_{x} on $(W, \mathscr{F})$, which satisfies the following conditions

(40) $$\begin{aligned} &Q^{d}_{x}(A) = P^{d(0)}_{x}(A)\,, \qquad A \in \mathscr{F}_{\Delta} \\ &Q^{d}_{x}(X(t_i) \in B_i,\ i = 1 \cdots k / \mathscr{F}_{\Delta}) = P^{d(\Delta)}_{X(\Delta)}(X(t_i - \Delta) \in B_i,\ i = 1 \cdots k)\,, \\ &\hspace{20em} \Delta < t_i \leq 2\Delta\,, \end{aligned}$$

$$Q_x^d(X(t_i) \in B_i,\ i = 1 \cdots k / F_{l\Delta})$$
$$= P_{X(l\Delta)}^{d(l\Delta)}(X(t_i - \Delta) \in B_i,\ i = 1 \cdots k)\ ,$$
$$l\Delta < t_i \leq (l+1)\Delta\ .$$

We define $V_t^d\phi$, $v_k^{(N)}\phi$ and $V_t\phi$, $\phi \in C$, by

(41) $$V_t^d\phi(x) = E_x^d\left|\int_0^t f^{d(s)}(X(s))ds + \phi(X(t))\right|$$

where E_x^d means the expectation w.r. to Q_x^d,

(42) $$v_k^{(N)}\phi(x) = \sup_{d \in \mathfrak{A}_N} V_{k\Delta}^d\phi(x)$$

and

(43) $$V_t\phi(x) = \sup_{d \in \mathfrak{A}} V_t^d\phi(x)$$

respectively. According to the definition of Q_x^d we can see

$$E_x^d\left(\int_\Delta^t f^{d(s)}(X(s))ds + \phi(X(t)) / \mathscr{F}_\Delta\right) = V_{t-\Delta}^{\tilde{d}}\phi(X(\Delta))$$

where $\tilde{d}(s, w) = d(s + \Delta, \theta_\Delta w)$ and $\theta_\Delta w(s) = w(s + \Delta)$, and

(44) $$V_{(k+1)\Delta}^d\phi(x) = E_x^{d(0)}\int_0^\Delta f^{d(0)}(X(s))ds + V_{k\Delta}^{\tilde{d}}\phi(X(\Delta))\ .$$

Hence we have

(45) $$J^k(N)\phi = v_k^{(N)}\phi\ , \qquad \phi \in C\ .$$

Proof of (45). Since $u \in \Gamma$ can be regarded as a constant admissible control $d(0) = u$, we have

$$J\phi(x) = \sup_{u \in \Gamma} T_\Delta^u\phi(x) = \sup_u\left(\int_0^\Delta P_\theta^u f^u(x)d\theta + P_\Delta^u\phi(x)\right)$$
$$= \sup_u E_x^u\int_0^\Delta f^u(X(s))ds + \phi(X(\Delta))$$
$$\leq \sup_{d \in \mathfrak{A}_N} E_x^d\int_0^\Delta f^{d(0)}(X(s))ds + \phi(X(\Delta)) = v_1^{(N)}\phi(x)\ .$$

For any $d \in \mathfrak{A}_N$ we get

$$E_x^d \int_0^\Delta f^{d(0)}(X(s))ds + \phi(X(\Delta))$$
$$\leq \sup_{u \in \Gamma} E_x^u \int_0^\Delta f^u(X(s))ds + \phi(X(\Delta)) = J\phi(x) .$$

Hence $v_1^{(N)}\phi(x) \leq J\phi(x)$. So (45) holds for $k = 1$. Suppose (45) holds for k. Thus $v_k^{(N)}\phi$ is in C. Using (44) we have

$$V^d_{(k+1)\Delta}\phi(x) \leq E_x^{d(0)}\left(\int_0^\Delta f^{d(0)}(X(s))ds + v_k^{(N)}\phi(X(\Delta))\right) \tag{46}$$
$$\leq v_1^{(N)}(v_k^{(N)}\phi)(x) , \qquad d \in \mathfrak{A}_N .$$

Therefore

$$v_{k+1}^{(N)}\phi \leq v_1^{(N)}(v_k^{(N)}\phi) = J(J^k\phi) = J^{k+1}\phi . \tag{47}$$

Let Γ_l be a sequence of compact sets which increases to Γ as $l \uparrow \infty$. Replacing Γ by Γ_l we define $J_l(N)$ in the same way. Put $F(x, u) = T_\Delta^u\psi(x)$, $\psi \in C$. Since $F(x, u)$ is continuous in (x, u) and $J_l\psi(x) \in F(x, \Gamma_l)$, the implicit function theorem tells us the existence of a Borel function $m\colon R^n \to \Gamma_l$ such that

$$F(x, m(x)) = J_l\psi(x) .$$

Hence $F(x, m(x))$ is in C. So there exist Γ_l-valued Borel functions m_i, $i = 0, 1, \cdots, k$ such that

$$T_\Delta^{m_k(x)}\phi(x) = J_l\phi(x) , \qquad T_\Delta^{m_{k-1}(x)}(J_l\phi)(x) = J_l^2\phi(x) ,$$
$$T_\Delta^{m_0(x)}(J_l^k\phi)(x) = J_l^{k+1}\phi(x) .$$

Define $d \in \mathfrak{A}_N$ by

$$d(t, w) = m_j(w(j\Delta)) , \quad j\Delta \leq t < (j+1)\Delta , \qquad j = 0, \cdots, k . \tag{48}$$

An admissible control, like (48), is called a Markovian step control. Then

$$v_{k+1}^{(N)}\phi(x) \geq E_x^d \int_0^{(k+1)\Delta} f^{d(s)}(X(s))ds + \phi(X((k+1)\Delta)) \tag{49}$$
$$= T_\Delta^{m_0} \cdots T_\Delta^{m_k}\phi(x) = J_l^{k+1}\phi(x) , \qquad \forall l .$$

We show the following convergence (50).

$$0_i\text{-}\lim_l J_l^k\phi = J^k\phi . \tag{50}$$

Since $T_\Delta^u\phi \le J_l\phi$ for $u \in \Gamma_l$ and $J_l\phi(x)$ increase as $l \uparrow \infty$, we have

$$T_\Delta^u\phi(x) \le \lim_l J_l\phi(x) , \qquad (u, x) \in \Gamma \times R^n .$$

Hence taking the supremum w.r. to u, we can see

$$J\phi(x) \le \lim_l J_l\phi(x) .$$

Since $J_l\phi \le J\phi$, we get (50) for $k = 1$. Suppose (50) holds for k. Then we have

$$T_\Delta^u(J^k\phi) = T_\Delta^u(0_i\text{-}\lim_l J_l^k\phi) = 0_i\text{-}\lim_l T_\Delta^u(J_l^k\phi) .$$

Because

$$T_\Delta^u(J_l^k\phi) \le J_l(J_l^k\phi) = J_l^{k+1}\phi \qquad \text{for } u \in \Gamma_l ,$$

we have

$$T_\Delta^u(J^k\phi)(x) \le \lim_l J_l^{k+1}\phi(x) .$$

This means $J^{k+1}\phi(x) \le \lim_l J_l^{k+1}\phi(x)$. So (50) holds for $k + 1$.

Using (49) and (50) we have the converse inequality of (47). This completes the proof of (45).

Since $v_{p2^N-l}^{(N)}\phi$ increases to $V_{p2^{-l}}\phi$ as $N \uparrow \infty$, we have

(51) $$S_t\phi = V_t\phi \qquad \text{for binary } t .$$

For binary t the continuous function $S_t^{(N)}\phi$ increases to a continuous function $S_t\phi$ as $N \uparrow \infty$. Hence $S_t^{(N)}\phi$ converges to $S_t\phi$ uniformly on any compact set A of R^n. Moreover by (50) $J_l^k(N)\phi$ increases to $J^k(N)\phi$ as $l \uparrow \infty$. Therefore $J_l^k(N)\phi$ converges to $J^k(N)\phi$ uniformly on A. Since $J_l^k(N)\phi$ is attained by a Markovian step control d of (48), (51) implies the following proposition.

Proposition. *For any compact set A of R^n and any binary t, there exists an ε-optimal Markovian step control $\tilde{d}$, namely*

$$\sup_{d \in \mathfrak{A}} V_t^d\phi(x) < V_t^{\tilde{d}}\phi(x) + \varepsilon , \qquad {}^\forall x \in A .$$

§ 4. Proof of Theorem 2

First we show that the assumption of Theorem 1 is satisfied. Let X and Y be the solutions for $(\Omega, \mathscr{B}, u) \in \mathfrak{A}$, starting at x and y respectively. Then by (2) and (3) in § 1 the following evaluation is well-known.

$$(1) \qquad E|X(t) - Y(t)|^2 \le |x-y|^2 e^{2vt} \qquad t \le T$$

with a positive $v = v(T)$. Hence we have for $\phi \in \sum_K$

$$\begin{aligned} &|P_t^u\phi(x) - P_t^u\phi(y)| \\ &\quad = \left| E \exp\left\{-\int_0^t c(X(\theta), u)\, d\theta\right\}\phi(X(t)) \right. \\ (2) &\quad \left. - \exp\left\{-\int_0^t c(Y(\theta), u)\, d\theta\right\}\phi(Y(t)) \right| \\ &\quad \le E|\phi(X(t)) - \phi(Y(t))| \\ &\quad\quad + \|\phi\| \int_0^t E|c(X(\theta), u) - c(Y(\theta), u)|\, d\theta\,, \end{aligned}$$

Using the inequality

$$|e^\xi - e^\eta| \le e^{\xi\vee\eta}(1 - e^{-(\xi\vee\eta - \xi\wedge\eta)}) \le e^{\xi\vee\eta}|\xi - \eta|\,,$$

$$\begin{aligned} \|P_t^u\phi(x) - P_t^u\phi(y)| &\le KE|X(t) - Y(t)| + K\mu\int_0^t E|X(\theta) - Y(\theta)|\, d\theta \\ &\le K|x-y|e^{vt}(1+\mu t) \le K|x-y|\, e^{(v+\mu)t}\,, \qquad t \le T\,. \end{aligned}$$

Moreover we have for

$$(3) \qquad \|A^u\phi\| \le \left(\sum_{ij}\left\|\frac{\partial^2\phi}{\partial x_i\partial x_j}\right\| + \sum_i\left\|\frac{\partial\phi}{\partial x_i}\right\| + \|\phi\|\right)b\,.$$

Therefore the assumptions of Theorem 1 are satisfied. Thus we have two semi-groups, Q_t and S_t. Since Q_t satisfies (i) ~ (iv) of Theorem 1,

$$(4) \qquad S_t\phi \le Q_t\phi\,.$$

Now we shall show the converse inequalily. Define $\mathfrak{A}_N$ by

$$(5) \qquad \mathfrak{A}_N = \left\{(\Omega, B, U) \in \mathfrak{A}\,;\, U(t) = U\left(\frac{k}{2^N}\right),\ \frac{k}{2^N} \le t < \frac{k+1}{2^N}\right\}$$

and V by

$$\begin{aligned} &V(t, x, \Omega, B, U) \\ &\quad = E_x\int_0^t \exp\left\{-\int_0^s c(X(\theta), U(\theta))d\theta\right\} f(X(s), U(s))ds \\ &\quad\quad + \exp\left\{-\int_0^t c(X(\theta), U(\theta))d\theta\right\}\phi(X(t)) \end{aligned}$$

where X is the solution for $(\Omega, B, U) \in \mathfrak{A}$. For simplicity we denote the inside of expectation by $I(t, U, \phi)$, i.e.

$$I(t, U, \phi) = \int_0^t \exp\left\{-\int_0^s c(X(\theta), U(\theta))d\theta\right\} f(X(s), U(s))ds$$
$$+ \exp\left\{-\int_0^t c(X(\theta), U(\theta))d\theta\right\} \phi(X(t)) .$$

Putting $\Delta = 2^{-N}$ we get

$$V(\Delta, x, \Omega, B, U) = E_x E(I(\Delta, U, \phi)/U(0)) \tag{6}$$

and

$$E(I(\Delta, U, \phi)/U(0)) = T_\Delta^{U(0)}\phi(X(0)) \leq J\phi(X(0)) \tag{7}$$
$$\text{for } (\Omega, B, U) \in \mathfrak{A}_N$$

we have

$$V(\Delta, x, \Omega, B, U) \leq J\phi(x) . \tag{8}$$

Suppose for any $\phi \in C$ and $(\Omega, B, U) \in \mathfrak{A}_N$.

$$V(k\Delta, x, \Omega, B, U) = E_x I(k\Delta, U, \phi) \leq J^k\phi(x) . \tag{9}$$

Then, using the following equality, we can show that (9) is valid for $k + 1$.

$$E_x I((k + 1)\Delta, U, \phi) = E_x E(I((k + 1)\Delta, U, \phi)/\sigma_{k\Delta}(B, U)) . \tag{10}$$

On the conditional probability field $P(/\sigma_{k\Delta}(U, B))$, $U(t)$, $k\Delta \leq t < (k + 1)\Delta$, can be regarded as constant and $\{dB(t), k\Delta \leq t\}$ is a Brownian motion. Hence the solution $X(t)$ of the following stochastic equation

$$X(t) = X(k\Delta) + \int_{k\Delta}^t \alpha(X(s), U(k\Delta))dB(s)$$
$$+ \int_{k\Delta}^t \gamma(X(s), U(k\Delta))ds , \qquad k\Delta \leq t < (k + 1)\Delta$$

is a diffusion. Hence we have

$$E(I((k + 1)\Delta, U, \phi)/\sigma_{k\Delta}(B, U))$$
$$= \int_0^{k\Delta} \exp\left\{-\int_0^s c(X, U)\right\} f(X(s), U(s))ds \tag{11}$$

$$+ \exp\left\{-\int_0^{k\Delta} c(X, U)\right\} E\left(\int_{k\Delta}^{(k+1)\Delta} \exp\left\{-\int_{k\Delta}^{s} c(X, U)\right\}\right.$$

$$\times f(X(s), U(k\Delta))ds + \exp\left\{-\int_{k\Delta}^{(k+1)} c(X, U)\right\}$$

$$\left.\times \phi(X(k+1)\Delta)/\sigma_{k\Delta}(B, U)\right)$$

$$= \int_0^{k\Delta} \exp\left\{-\int_0^s c(X, U)\right\} f(X, U)ds$$

$$+ \exp\left\{-\int_0^{k\Delta} c(X, U)\right\} T_\Delta^{U(k\Delta)} \phi(X(k\Delta))$$

$$\leq \text{1st term} + \exp\left\{-\int_0^{k\Delta} c(X, U) J\phi(X(k\Delta)) = I(k\Delta, U, J\phi)\right. .$$

So from (10) and (11) we can see

$$V((k+1)\Delta, x, \Omega, B, U)$$
$$= E_x I((k+1)\Delta, U, \phi) \leq E_x I(k\Delta, U, J\phi)$$
$$\leq J^k(J\phi)(x) = J^{k+1}\phi(x) .$$

This is our wanted inequality.

Put

$$V_N(t, x) = \sup_{\mathfrak{A}_N} V(t, x, \Omega, B, U) \qquad \text{for } t = k/2^N .$$

Thus for any binary t and $x \in R^N$, $V_N(t, x)$ is increasing. Denote $\lim_{N\uparrow\infty} V_N(t, x)$ by $\tilde{V}(t, x)$. Since (9) implies the following inequalities

$$V_N(t, x) \leq S_t^{(N)}\phi(x) \leq S_t\phi(x) \qquad \text{for } t = k/2^N ,$$

we have

(12) $$\tilde{V}(t, x) \leq S_t\phi(x) , \qquad \text{for binary } t .$$

In order to obtain the converse inequality of (12), we use the following approximation lemma.

Lemma. *For any* $(\Omega, B, U) \in \mathfrak{A}$ *there exists* $(\Omega, B, U_k) \in \bigcup \mathfrak{A}_N$, *such that*

(13) $$V(t, x, \Omega, B, U_k) \to V(t, x, \Omega, B, U) , \qquad {}^\forall t, x .$$

Proof. Define W_k and W_{kl} as follows, [7],

(14) $$W_k(t, U) = 2^k \int_{t-1/2^k}^{t} U(s)ds \quad \text{for } t \geq 0 ,$$

where $U(s) = U(0)$ for $s \leq 0$, for convenience, and

(15) $$W_{kl}(t, U) = W_k(2^{-l}[2^l t], U)$$

where $[\cdot]$ means the integer part of $\cdot$. Then

(16) $$W_{kl}(t, U) \in F$$

where F is a convex σ-compact subset of Γ such that $U(t, \omega) \in F$, $^\forall t, \omega$. Moreover

(17) $$\lim_{k\uparrow\infty} \lim_{l\uparrow\infty} E \int_0^T |W_{kl}(t, U) - U(t)|^2 dt = 0 , \qquad ^\forall T > 0 .$$

Hence we can successively take two increasing integer k_m and l_m as follows, (k_1, l_1) is chosen so that $k_1 \leq l_1$ and

$$E \int_0^1 |W_{kl}(t, U) - U(t)|^2 dt < 2^{-1} .$$

When $(k_1, l_1), \cdots, (k_{m-1}, l_{m-1})$ are determined, we take (k_m, l_m) such that $k_m \leq l_m$,

(18) $$E \int_0^m |W_{kl}(t, U) - U(t)|^2 dt < 2^{-m}$$

and

(19) $$E \int_0^m |W_{kl}(t, U_j) - U_j(t)|^2 dt < 2^{-m} , \qquad j = 1, \cdots, m-1$$

where $U_j(t) = W_{k_j l_j}(t, U)$. Hence U_m converges to U in L^2(loc) with probability 1 and

(20) $$(\Omega, B, U_m) \in \mathfrak{A}_{l_m} \subset \bigcup_N \mathfrak{A}_N$$

For any fixed i, j $(i \leq j)$ and an integer T,

(21) $$\int_0^T |W_{ij}(t, U_p) - W_{ij}(t, U)|^2 dt = \int_0^T |W_i(\Pi_j(t), U_p) - W_i(\Pi_j(t), U)|^2 dt$$

where $\Pi_j(t) = 2^{-j}[2^j t]$. From the definition of W_i we have

$$(22)\qquad |W_i(s, U_p) - W_i(s, U)| \le 2^i \int_{s-2^{-i}}^{s} |U_p(\theta) - U(\theta)|^2\, d\theta$$

Hence by (21) and (22) we can see

$$(23)\qquad \begin{aligned} &\int_0^T |W_{ij}(t, U_p) - W_{ij}(t, U)|^2 \\ &\le 2^i \sum_{n=0}^{T2^j} \int_{n2^{-j}}^{(n+1)2^{-j}} dt \int_{n2^{-j}-2^{-i}}^{n2^{-j}} |U_p(\theta) - U(\theta)|^2\, d\theta \\ &= 2^{i-j} \sum_{n=0}^{T2^j} \int_{n2^{-j}-2^{-i}}^{n2^{-j}} |U_p(\theta) - U(\theta)|^2\, d\theta \\ &\le 2^{i-j}2^{j-i} \int_0^T |U_p(\theta) - U(\theta)|^2 d\theta = \int_0^T |U_p(\theta) - U(\theta)|^2\, d\theta \end{aligned}$$

Thus we have for $k = k_m$ and $l = l_m$

$$(24)\qquad \begin{aligned} &E\int_0^m |W_{kl}(t, U_p) - U_p(t)|^2\, dt \\ &= E\int_0^m |W_{kl}(t, U_p) - W_{kl}(t, U) + W_{kl}(t, U) \\ &\qquad\qquad - U(t) + U(t) - U_p(t)|^2\, dt \\ &\le 3E\int_0^m |W_{kl}(t, U) - U(t)|^2\, dt + 6E\int_0^m |U(t) - U_p(t)|^2\, dt \\ &= 3E\int_0^m |U_m(t) - U(t)|^2\, dt + 6E\int_0^p |U(t) - U_p(t)|^2\, dt\,. \end{aligned}$$

$$E\int_0^m |U(t) - U_p(t)|^2\, dt \le E\int_0^p |U(t) - U_p(t)|^2\, dt < 2^{-p} \qquad \text{for } p \ge m\,.$$

Therefore

$$(25)\qquad E\int_0^m |W_{kl}(t, U_p) - U_p(t)|^2\, dt < 92^{-m}\,, \qquad \text{for } p \ge m\,.$$

Recalling (19) we can see that (25) is valid for $p \le m - 1$.

Let X_m and X be solutions for U_m and U respectively. For any $T > 0$, $\{U_m(t), t \le T\}$ can be regarded as an $L^2[0, T]$-valued random variable, and B and X_m as $C_n[0, T]$-valued random variables where $C_n[0, T]$ is the Banach space of the set of all R^n-valued continuous functions on $[0, T]$. Since U_m converges to U in $L^2[0, T]$ with probability 1, U_m converges to U in Prohorov topology on $L^2[0, T]$. On the other hand $\{X_m, m = 1, 2, \cdots\}$ is totally bounded in Prohorov topology on $C_n[0, T]$, by the conditions (2) and (3) in § 1. Hence $\{(B, U_m, X_m), m = 1, 2, \cdots$

is also totally bounded in Prohorov topology and (B, U_m) converges to (B, U) in Prohovov topology. Therefore on a probability space $\tilde{\Omega}$we can construct $(\tilde{B}, \tilde{U}, \tilde{X})$ and a version of (B, U_m, X_m), say $(\tilde{B}_m, \tilde{U}_m, \tilde{X}_m)$, such that for some subsequence m'

$$\begin{aligned} &\tilde{B}_{m'}(t) \to \tilde{B}(t) && \text{uniformly on } [0, T] \\ &\tilde{X}_{m'}(t) \to \tilde{X}(t) && \text{''} \\ &\tilde{U}_{m'}(t) \to U(t) && \text{in } L^2[0, T] \end{aligned} \tag{26}$$

with probability 1. Thus $(\tilde{B}, \tilde{U})$ has the same law as (B, U) on $C_n[0, T]$. Moreover we can take an integer $p = p(m')$ such that

$$\int_0^T E\,|\tilde{X}(\Pi_p(t)) - \tilde{X}(t)|^2\, dt < 2^{-m'} \tag{27}$$

and

$$\int_0^T E\,|\tilde{X}_{j'}(\Pi_p(t)) - \tilde{X}_{j'}(t)|^2\, dt < 2^{-m'}\,, \qquad j = 1, 2, \cdots \tag{28}$$

Recalling the conditions (3) in § 1 we have

$$\begin{aligned} E&\left|\int_0^T \alpha(\tilde{X}(s),\, U(s))dB(s) - \int_0^T \alpha(\tilde{X}(\Pi_p(s), W_{kl}(s,\, U))dB(s)\right|^2 \\ &\le 2\mu^2 \int_0^T E\,|\tilde{X}(s) - \tilde{X}(\Pi_p(s))|^2\, ds \\ &\quad + 2b \int_0^T E\rho(|\tilde{U}(s) - W_{kl}(s,\, U)|)ds \end{aligned} \tag{29}$$

Since ρ^{-1} is convex and increasing, we can see

$$\begin{aligned} \rho^{-1}&\left(\frac{1}{t}E\int_0^t \rho(|\tilde{U}(s) - W_{kl}(s, \tilde{U})|)ds\right) \\ &\le \frac{1}{t}E\int_0^t \rho^{-1}(\rho(|\tilde{U}(s) - W_{kl}(s, \tilde{U})|)ds \\ &= \frac{1}{t}E\int_0^t |\tilde{U}(s) - W_{kl}(s, \tilde{U})|\, ds \\ &= \frac{1}{t}E\int_0^t |U(s) - W_{kl}(s, U)|\, ds < \frac{2^{-m'}}{t}\,, \\ &\qquad\qquad \text{for } k = k_{m'},\ l = l_m\,. \end{aligned} \tag{30}$$

Therefore by (27) and (30)

$$E\left|\int_0^T \alpha(\tilde{X}(s),\tilde{U}(s))d\tilde{B}(s)-\int_0^T \alpha(\tilde{X}(\Pi_p(s),W_{kl}(s,\tilde{U}))d\tilde{B}(s)\right|^2 \le 2\mu^2 2^{-m'}+2bT\rho\left(\frac{2^{-m'}}{T}\right). \tag{31}$$

For j' we have the same evaluation, namely

$$E\left|\int_0^T \alpha(\tilde{X}_{j'}(s),\tilde{U}_{j'}(s))d\tilde{B}_{j'}(s)-\int_0^T \alpha(\tilde{X}_{j'}(\Pi_p(s),W_{kl}(s,\tilde{U}_{j'}))dB_{j'}\right|^2 \le 2\mu^2 2^{-m'}+18bT\rho\left(\frac{2^{-m'}}{T}\right) \tag{32}$$

by (25) and (28). On the other hand, by (26),

$$\begin{aligned} W_k(s,\tilde{U}) &= 2^k\int_{s-2^{-k}}^{s}\tilde{U}(\theta)d\theta=\lim_j 2^k\int_{s-2^{-k}}^{s}\tilde{U}_{j'}(\theta)d\theta \\ &= \lim_j W_k(s,\tilde{U}_{j'})\,, \qquad \text{for any } s\,. \end{aligned} \tag{33}$$

From (26) and (33) we see

$$\begin{aligned} &\int_0^t \alpha(\tilde{X}(\Pi_p(s),W_{kl}(s,\tilde{U})))d\tilde{B}(s) \\ &\quad=\lim_j\int_0^t \alpha(\tilde{X}_{j'}(\Pi_p(s),W_{kl}(s,\tilde{U}_{j'})))d\tilde{B}_{j'} \end{aligned} \tag{34}$$

with probability 1.

Combining above evaluations we have

$$\int_0^t \alpha(X_{j'}(s),\tilde{U}_{j'}(s))d\tilde{B}_{j'}(s)\to\int_0^t \alpha(\tilde{X}(s),\tilde{U}(s))d\tilde{B}(s)\,, \tag{35}$$

in proba. Applying the similar evaluation for the drift term,

$$\int_0^T \gamma(\tilde{X}_{j'}(s),\tilde{U}_{j'}(s))d\tilde{B}_{j'}(s)\to\int_0^t \gamma(\tilde{X}(s),\tilde{U}(s))d\tilde{B}(s)\,, \tag{36}$$

in proba. Therefore $\tilde{X}$ is a solution for $(\tilde{B},\tilde{U})$.

Again recalling (26), we have

$$V(t,x,\tilde{\Omega},\tilde{B}_{j'},\tilde{U}_{j'})\to V(t,x,\tilde{\Omega},\tilde{B},\tilde{U}) \qquad \forall x,t\le T\,.$$

Since $(\tilde{B}_{j'},\tilde{U}_{j'})$ and $(\tilde{B},\tilde{U})$ have the same laws as $(B,U_{j'})$ and (B,U) respectively,

$$V(t,x,\tilde{\Omega},\tilde{B}_j,\tilde{U}_j)=V(t,x,\Omega,B_j,U_{j'})$$

and

$$V(t, x, \tilde{\Omega}, \tilde{B}, \tilde{U}) = V(t, x, \Omega, B, U) .$$

Since T is arbitrary, $U_{j'}$ is our wanted one. This completes the proof of Lemma.

Lemma implies that $V(t, x, \Omega, B, U) \leq \tilde{V}(t, x)$ for any $(\Omega, B, U) \in \mathfrak{A}$. Hence $S_t\phi(x) \leq \tilde{V}(t, x)$. Therefore we have

(37) $$S_t\phi(x) = \tilde{V}(t, x) \qquad \text{for binary } t .$$

Since both sides of (37) are continuous in t, (37) holds for any t. This completes the proof of Theorem 2.

When $\alpha(x, u)$ is uniformly positive definite and α, γ, c, f and ϕ are in C^2 uniformly in u, the following Bellman equation of parabolic type has a unique solution in $W^{1,2}_{p,\mathrm{loc}}$ [6]

$$\begin{cases} \dfrac{\partial V}{\partial t}(t, x) = \sup\limits_{u \in \Gamma} [A^u V(t, x) + f^u(x)] = GV(t, x) , \\ \qquad\qquad\qquad\qquad\qquad\qquad\qquad \text{a.e. in } (0, \infty) \times R^n \\ V(0, x) = \phi(x) \qquad \text{on } R^n . \end{cases}$$

Hence S_t is e^{tG} in the sense of [1].

§ 5. Examples

Let X^u be an 1-dimensional homogeneous Levy process of pure jump type, i.e.

$$X^u(t) = x + \lim_{m \to \infty} \int_{|z| > 1/m} \int_0^t zN^u(ds, dz) - \frac{z}{1 + z^2} dsn^u(dz) \equiv x + Y^u(t)$$

where $N(ds, dz)$ is Poisson random measure with $EN^u(ds, dz) = dsn^u(dz)$ and we assume

(1) $$\sup_u \int_{|z| \leq 1} z^2 n^u(dz) < \infty \quad \text{and} \quad \sup_u \int_{|z| > 1} n^u(dz) < \infty .$$

Let P^u_t be the transition semi-group of X^u on C, namely

(2) $$P^u_t\phi(x) = \int \phi(x + y)\nu^u(t, dy)$$

where $\nu^u(t, \cdot)$ is the probability law of $Y^u(t)$. Then we have

$$(3) \qquad A^u\phi(x) = \int\left(\phi(x+z) - \phi(x) - \frac{z}{1+z^2}\phi'(x)\right)n^u(dz) \qquad \text{for } \phi \in C^2$$

and

$$\Phi(x, z) \equiv \phi(x+z) - \phi(x) - \frac{z}{1+z^2}\phi'(x)$$

$$(4) \qquad = \phi'(x)z + \frac{1}{2}\phi''(\xi_x)z^2 - \frac{z}{1+z^2}\phi'(x)$$

$$= \frac{z^3}{1+z^2}\phi'(x) + \frac{1}{2}z^2\phi''(\xi_x)$$

with some $\xi_x \in (x, x+z)$. Hence for $\phi \in C^2$ we have

$$(5) \qquad \int_{|z|\leq 1} \Phi(x, z)n^u(dz) \leq h_\phi \int_{|z|\leq 1} z^2 n^u(dz)$$

and

$$(6) \qquad \int_{|z|\geq 1} \Phi(x, z)n^u(dz) \leq h'_\phi \int_{|z|\geq 1} n^u(dz)$$

with positive constants h_ϕ and h'_ϕ. Thus, recalling (1), we have

$$(7) \qquad \sup_u |A^u\phi| < \infty\,, \qquad \text{for } \phi \in C^2\,.$$

$$(8) \qquad |P^u_t\phi(x) - P^u_t\phi(x')| \leq \int |\phi(x+y) - \phi(x'+y)|\,\nu^u(t, dy)$$

$$\leq K|x - x'|\,, \qquad \text{for } \phi \in \textstyle\sum(K)\,.$$

Therefore the conditions of Theorem 1 holds if f^u satisfies (A3).

References

[1] Crandall, M. G. and Liggett, T. M., Generation of semi-groups of non-linear transformations on general Banach spaces, Amer. J. Math., 18 (1971), 265–298.

[2] Fleming, W. H. and Rishel, R. W., Deterministic and stochastic optimal control, Appl. of Math., 1, Springer-Verlag, 1975.

[3] Grigelionis, B. I. and Shiryaev, A. N., On Stefan problem and optimal stopping rules for Markov processes, Theor. Probability Appl., **11** (1966), 541–558.
[4] Ito, K., Stochastic processes, Lecture Notes, **16**, Aarhus Univ., 1969.
[5] Krylov, N. V., Control of a solution of a stochastic integral equation, Theor. Probability Appl., **17** (1972), 114–131.
[6] ——, On Bellman's equation, Proc. School-Seminar on the theory of random processes, Part 1, Vilnius, 1974, 203–235.
[7] McKean, H. P., Stochastic integrals, Acad. Press, 1969.
[8] Nisio, M., Some remarks on stochastic optimal controls, 3rd USSR-Japan Symp. Prob. Th., 1975, Lecture Notes in Math., Springer-Verlag.
[9] ——, On a non-linear semi-group attached to stochastic optimal control, Publ. Res. Inst. Math. Sci., **12** (1976), 513–537.
[10] Yosida, K., Functional analysis, Springer-Verlag, 1968.
[11] Zabczyk, J., Optimal control by means of switchings, Studia Math., **45**, (1973), 161–171.

Department of Mathematics
Kobe University
Rokko, Kobe 657, Japan

Proc. of Intern. Symp. SDE
Kyoto 1976, pp. 327-339

Remarks on the *B*-shifts of Generalized Random Processes

Shigeyoshi OGAWA

1. Given the R^1-valued Brownian motion process $(\Omega, \mathcal{F}, P, B_t)$ we consider the formal quadratic functional of the white noise $\dot{B}_s = (d/ds)B_s$ as follows,

$$(1) \qquad A_t(k) = \exp\left[k\int_0^t (\dot{B}_s)^2 ds\right], \qquad \text{where } k \text{ is a number} .$$

The functional appears in the probabilistic study of the Feynman path integral (cf. [1], [10]): In his famous paper [1], R. P. Feynman proposed an attempt to get the wave function, a solution of the Schrödinger equation, by means of a function space integral. The principal interest of his study was to preserve the concepts in classical mechanics such as the path, the action integral or the principle of minimum action, etc. But the argument was soon proved to be mathematically inexact. It is well known that the measure realizing Feynman's path integral can not exist but as a cylindrical measure. Thus we were led far from the beautiful dream of preserving in non-relativistic quantum mechanics the classical concepts, especially the notion of path of a particle.

However, in 1976 the author showed in [10] the possibility of justifying Feynman's idea as a true function space integral under the condition that the mass of a particle is a purely imaginary quantity. The author's idea was to start with the hypothesis that the path of a particle in a free space exhibits a Brownian motion process. In such an interpretation the functional $k\int (\dot{B}_s)^2 ds$ now turns out to be the action integral of this dynamical system and the limit procedure in Feynman's argument is understood as taking the expectation of the functional $A_t(k)$ with respect to the Wiener measure P. The guiding motivation of the present study is to find a reasonable, at least mathematical, interpretation of this formal functional $A_t(k)$. The best possible way may be to define it as an operator mapping a distribution $\psi_0(x)$ into a generalized random process $A_t(k)\psi_0(x) = \psi(t, x)$. To be more precise, as it is expected for the Feynman path integral to provide us with a solution of the Schrödinger equation, the operator must be such that the expectation $\bar{\psi}(t, x) = E[A_t(k)\psi_0(x)]$ solves the following Cauchy problem;

$$(2) \qquad \frac{d}{dt}\bar{\psi} = \frac{k^2}{2}\left(\frac{\partial}{\partial x}\right)^2\bar{\psi}, \qquad \bar{\psi}(0, x) = \psi_0(x).$$

On the other hand, we already know that the expectation of a generalized random process ψ satisfying the following equations (3)–(4) is a solution of the problem (2);

$$(3) \qquad \frac{d}{dt}\psi = 2(\dot{B})^2\psi, \qquad \psi(0, x) = \psi_0(x)$$

$$(4) \qquad \hat{\psi} + k\frac{\partial}{\partial x}\psi = 0, \qquad \text{where } \hat{\psi} \text{ is the } B\text{-derivative of } \psi.$$

Therefore it is natural to define the $A_t(k)$ as such operator that acts on a certain space of generalized random processes and for a distribution $\psi_0(x)$, $A_t(k)\psi_0(x)$ provides us with a solution of the problem (3)–(4). In order that the $A_t(k)$ can be well defined in this way, we have to examine the uniqueness property of solutions for the problem (3)–(4) and this is the subject of the paper.

For the simplicity of discussions, we shall limit ourselves to the case that k is a real number and B_t is one-dimensional. But, as we shall see, the arguments and the calculus developed here will also apply to other cases with suitable modifications and extensions. The discussions on it will be published elsewhere.

2. In this paragraph we shall give some comments on the stochastic calculus concerning the white noise. For the detailes on the subject we shall refer to the papers [3], [7], [8], [9]. In what follows we always suppose that the random functions $f(t, \omega)$ satisfy the conditions: (1) $\omega \to f(t, \omega)$ is adapted to the σ-field $\sigma(B_s : s \leq t)$ for each t, and (2) $E\left\{\int_0^T |f(t, \omega)|^2\, dt\right\} < \infty \quad (T < \infty)$.

A random function $f(t, \omega)$ is said to be uniformly $B(M_\alpha)$-differentiable ($\alpha \geq 1$) if there exists a random function $g(t, \omega)$ such that

(5)

$$\lim_{h \downarrow 0} \sup_t E\left\{\left|\frac{1}{\sqrt{h}}(f(t+h, \omega) - f(t, \omega) - g(t, \omega)(B_{t+h} - B_t))\right|^\alpha\right\} = 0.$$

We called such random function $g(t, \omega)$, which is uniquely determined, the B-derivative of $f(t, \omega)$ and used the notation $g(t, \omega) = \hat{f}(t, \omega)$. It is proved that for a uniformly $B(M_4)$-differentiable function $f(t, \omega)$ the sequence of symmetrized Riemann sums $\{J_n(f)\}_n$ given below converges in the $L^2(\Omega)$-sense to the limit $\mathscr{I}_{1/2}(f)$ that we called the $\mathscr{I}_{1/2}$-integral of $f(t, \omega)$.

$$(6) \qquad J_n(f) = \sum_i f\left(\frac{t_{i+1}^n + t_i^n}{2}, \omega\right)(B_{t_{i+1}^n} - B_{t_i^n})$$

where $\{t_i^n; 1 \leq i \leq n\}$ is a canonical family of partitions of $[0, T]$. As for this integral we obtained the following relation

$$(7) \qquad \mathscr{I}_{1/2}(f) \equiv \int_0^T f(t, \omega)dB_t = \frac{1}{2}\int_0^T \dot{f}(t, \omega)dt + \int_0^T f(t, \omega)d^\circ B_t$$

where $\int f(t, \omega)d^\circ B_t$ means the Ito-integral of f.

From the relation (7) we see that the $\mathscr{I}_{1/2}$-integral is a generalization of the so-called Stratonovich integral and therefore the calculus based on our integral conserves the rules in classical calculus of integrations. It is not difficult to extend the $\mathscr{I}_{1/2}$-integral and the notion of B-differentiability into the case of multi-dimensional Brownian motion process or, possibly to the case of square-integrable martingales instead of the Brownian motion, (cf. [3], [12]). As for the generalization of the Stratonovich integral, there is another integral know by the name "Fisk integral" ([2]). The integrable class of functions consists of continuous quasi-martingales. G. Kallianpur and C. Striebel obtained the concrete form of the Fisk integral when applied to the case of second order continuous quasi-martingales $M(t)$ of the form $dM(t) = a(t)dt + b(t)d^\circ B_t$ ([4]). Their result shows that the values of two integrals, namely the $\mathscr{I}_{1/2}$-integral and the Fisk integral of a random function $f(t, \omega)$, coincide with each other when they exist. However it should be remarked that the two are not the same. For there are uniformly $B(M_4)$-differentiable functions which are not continuous quasi-martingales and, on the contrary, the quasi-martingales which are not uniformly $B(M_4)$-differentiable. The disadvantage of our calculus was that even a simple function like $\int^t f(t, \omega)d^\circ B_t$ may fail to be uniformly $B(M_4)$-differentiable. Therefore we think it necessary to ameliorate the discussions so as to make the $\mathscr{I}_{1/2}$-integral be free from the disadvantage.

We begin with the following

Definition 1. A random function $f(t, \omega)$ is said to be B-differentiable if there exists a random function $g(t, \omega)$ such that

$$(5)' \quad \lim_{h \downarrow 0} \sup_t E\left\{\left|\frac{1}{\sqrt{h}}(f(t+h,\omega) - f(t,\omega)) - \int_t^{t+h} g(s,\omega)d^\circ B_s)\right|^2\right\} = 0\,.$$

Also in this case we can verify the uniqueness of the random function $g(t, \omega)$ in the definition; let $h(t, \omega)$ be another such function then we have $P[h(t, \omega) = g(t, \omega)] = 1$ for almost all t. It is clear from the definition itself of the Ito-integral that a uniformely $B(M_2)$-differentiable function is B-differentiable in the sense of the new definition. Conversely, a B-differentiable function $f(t, \omega)$ is so in the old sense if the B-derivative $\hat{f}(t, \omega)$ admits a version which is continuous in the L^2-sense, that is: $E[|\hat{f}(t,\omega) - \hat{f}(s,\omega)|^2] \to 0$ as $|t-s| \to 0$. Based on this modification of B-differentiability we set the

Definition 2. The $\mathscr{I}_{1/2}$-integral is defined for B-differentiable random functions $f(t, \omega)$ by the following relation

$$(7)' \quad \int_0^T f(t,\omega)dB_t \equiv \frac{1}{2}\int_0^T \hat{f}(t,\omega)dt + \int_0^T f(t,\omega)d^\circ B_t\,.$$

Remark 1. We can prove that the $\mathscr{I}_{1/2}$-integral defined above becomes the limit in probability of the sequence $\{S_n(f)\}_n$.

$$(6)' \quad S_n(f) = \sum_i \frac{1}{2}[f(t_{i+1}^n, \omega) + f(t_i^n, \omega)](B_{t_{i+1}^n} - B_{t_i^n})$$

where $\{t_i^n;\, 1 \le i \le n\}$ is a family of partitions of $[0, T]$ such that $\lim_{n\to\infty} \max_i (t_{i+1}^n - t_i^n) = 0$. But this fact will not play a significant role in the calculus.

We confirme that the $\mathscr{I}_{1/2}$-integral is now free from the disadvantage mentioned above since all the square-integrable quasi-martingales are B-differentiable. Moreover with our integral we need not suppose the path-wise continuity of the integrands, nor that they are quasi-martingales. This is certainly the advantage of our integral. We finish the paragraph by the following

Lemma 1. (i) *Let $f(t, \omega)$, $g(t, \omega)$ be B-differentiable functions such that*

$$(8) \quad f(t,\omega)\int^t g(s,\omega)dB_s \text{ and } g(t,\omega)\int^t f(s,\omega)dB_s \text{ are B-differentiable.}$$

Then the following relation holds.

$$(9) \quad \int_0^t f(s,\omega)dB_s \int_0^t g(s,\omega)dB_s$$

$$= \int_0^t \left[f(s,\omega) \int_0^s g(r,\omega)dB_r + g(s,\omega) \int_0^s f(r,\omega)dB_r \right] dB_s$$

(ii) *The condition* (8) *is fulfiled when* $f(t,\omega)$ *and* $g(t,\omega)$ *are continuous quasi-martingales.*

(iii) *As a special case of* (i), *we have the following relation for any bounded smooth function* $u(x)$ *with bounded derivatives* u'_x, u''_{xx};

$$(9)' \quad u(B_t) \int_0^t g(s,\omega)dB_s$$

$$= \int_0^t \left[u'_x(B_s) \int_0^s g(r,\omega)dB_r + u(B_s)g(s,\omega) \right] dB_s \,.$$

Proof. The assertions follow immediately from the definition 2 and the Ito's chain rule.

3. Let $K(G)(G = [0,T] \times R^1,\ [0,T]$ or $R^1)$ be the space of indefinitely differentiable functions with compact supports in G, endowed with the usual topology of L. Schwartz. Especially we shall suppose that $K([0,T] \times R^1)$ (or $K([0,T])$ resp.) consists of such elements $w(t,x)$ (or, $v(t)$) that $w(0,x) = w(T,x) = 0$ ($v(0) = v(T) = 0$ resp.). We shall denote by $K'(\Omega)$ and $K'_{[0,T]}(\Omega)$ the set of all linear continuous applications from $K([0,T] \times R^1)$, and from $K([0,T])$ respectively, into the Hilbert space $L^2(\Omega)$ and call their elements the generalized random processes. In what follows we shall employ the notations, $u(x)$ and $v(t)$, to represent the arbitrary elements of $K(R^1)$ and $K([0,T])$. For a fixed $u(x)$, we associate to $\psi \in K'(\Omega)$ the generalized random process $\psi.(u) \in K'_{[0,T]}(\Omega)$ such that $\langle \psi, u \cdot v \rangle = \langle \psi.(u), v \rangle$ for all $v(t)$. We call $\psi.(u)$ the temporal cut of ψ. If the temporal cut $\psi.(u)$ of ψ can be identified with a usual stochastic process for each fixed $u(x)$, then we shall call such ψ a regular process.

Definition 3. A regular process ψ is called B-differentiable provided that; for each fixed $u(x)$, (i) the temporal cut $\psi_t(u)$ is B-differentiable as a random function, and (ii) the B-derivative satisfies the condition ess $\sup_t E[|\widehat{\psi_t(u)}|^2] < \infty$.

It follows immediately from this definition that the temporal cut $\psi_t(u)$ of a B-differentiable process ψ is continuous in t in the L^2-sense. As for the continuity in $u(x)$ of $\widehat{\psi_t(u)}$ we have the next statement that can be verified by a discussion similar to that given in [10] (page 180).

Lemma 2. *The bilinear form $u, v \to \langle \widehat{\psi.(u)}, v \rangle$ is separately continuous.*

Consequently the kernel theorem (Theorem 1 of [10]) assures the existence of an element $\hat{\psi}$ of $K'(\Omega)$ that we call the B-derivative of ψ. It is important to notice that for a B-differentiable process ψ, it holds the relation $A\hat{\psi} = \widehat{(A\psi)}$ where $A = \Sigma a_j(x)(\partial/\partial x)^j$ with sufficiently smooth coefficients $a_j(x)$. Therefore if a B-differentiable process ψ satisfies the wave condition (4) then ψ is indefinitely B-differentiable.

Given a B-differentiable process ψ we want to introduce the products $\psi(\dot{B})^n$ $(n \geq 1)$ as the elements of $K'(\Omega)$. For the case $n = 1$, there is no problem in defining it as being the distribution derivative of the element $\int \psi dB$; $\psi\dot{B} = D_t\left(\int \psi dB\right)$ where $\int \psi dB$ is such that $\left\langle \int \psi dB, u \cdot v \right\rangle = \int v(t)dt \int^t \psi_s(u)dB_s$ for all $u(x)$ and $v(t)$. However, as we know well in the theory of distributions, it is impossible to understand the pseudo-products $\psi(\dot{B})^n$ $(n \geq 2)$ in a direct way; say, for example, as the trace of the tensor products $\psi_{t_1}\dot{B}_{t_1} \otimes \dot{B}_{t_2} \otimes \cdots \otimes \dot{B}_{t_{n-1}}|_{t_1=t_2=\cdots=t}$ or, as the convolution products $\psi\dot{B} * \dot{B} * \cdots * \dot{B}$. Nevertheless the problem explained in § 1 demands us to find a good solution for this mysterious question. For this purpose we introduce the notion of the "mean derivative $\mathscr{D}_t\psi$" of a B-differentiable process ψ in which we see a natural extension of E. Nelson's mean derivative defined for continuous quasi-martingales ([6]); We set $\mathscr{D}_t\psi = D_t\left(\psi - \int^t \hat{\psi} d^\circ B\right)$. Obviously we have the relation $\mathscr{D}_t\psi = D_t\psi$ for B-differentiable processes ψ such that $\hat{\psi} = 0$. It is essential to notice that if $\mathscr{D}_t\psi$ is B-differentiable the product $(\mathscr{D}_t\psi)\dot{B}$ becomes meaningful while $(D_t\psi)\dot{B}$ does not in general.

Definition 4. For a regular process ψ such that the mean derivative $\mathscr{D}_t\psi$ exists and is B-differentiable, we understand by $(D_t\psi)\dot{B}$ the product $(\mathscr{D}_t\psi)\dot{B}$.

Remark 2. For a B-differentiable process X we have $X \cdot \dot{B} = \left(D_t \int^t Xds\right)\dot{B} = \left(\mathscr{D}_t \int^t Xds\right)\dot{B}$. The definition above is therefore consistent with that of the product $X \cdot \dot{B}$ itself.

It is interesting to see what the definition 4 means about the pseudo-products of the white noise. Let ψ be n-times B-differentiable, then from Definition 4 we see that ;

$$\psi(\dot{B})^2 = (\psi\dot{B})\dot{B} = \left(D_t\int\psi dB\right)\dot{B} = \left(\mathscr{D}_t\int\psi dB\right)\dot{B} = \frac{1}{2}\hat{\psi}\dot{B}\,,$$

hence,

$$\psi(\dot{B})^3 = (\psi(\dot{B})^2)\dot{B} = \frac{1}{2}\left(D_t\int\hat{\psi}dB\right)\dot{B} = \frac{1}{2}\left(\mathscr{D}_t\int\hat{\psi}dB\right)\dot{B} = \left(\frac{1}{2}\right)^2\hat{\psi}^{(2)}\dot{B}$$

and in general we have

(10) $$\psi(\dot{B})^m = \left(\frac{1}{2}\right)^{m-1}\hat{\psi}^{(m-1)}\dot{B} \qquad (n \geq m)$$

where $\hat{\psi}^{(j)}$ is the j-th B-derivative of ψ.

The relation (10) is just what we employed in [10] to define the pseudoproducts.

P. Lévy employed in [5] the very suggestive notation $\xi\sqrt{dt}$ (ξ; a normally distributed random variable) to represent the quantity dB_t. If we apply his idea to represent the white noise, it becomes; $\dot{B}_t = \xi\sqrt{dt}/dt = \xi/\sqrt{dt}$ which may be something like a fractional power of operator, say $\sqrt{D_t}$. Definition 4 seems to give a partial justification to his intuitive notation, as we see in the

Proposition 1. $(D_t - 2(\dot{B})^2)\int\psi dB = 0$ *for any B-differentiable process* ψ.

Proof. $D_t\int\psi dB = \psi\dot{B} = \left(\widehat{\int\psi dB}\right)\cdot\dot{B} = 2\left(\int\psi dB\right)(\dot{B})^2\,.$

With these preparations we can see the meaning of the problem (3)–(4). Especially, the initial condition $\psi(0, x) = \psi_0(x)$ in (3) must be read in such way that $\psi_0(x) \in K'(R^1)$ and that $\lim_{t\to 0}\psi_t(u) = \langle\psi_0, u\rangle$ for any $u(x)$. Because the temporal cut $\psi_t(u)$ of a B-differentiable process ψ is continuous in t, in the L^2-sense. But in what follows we always assume that the initial data $\psi_0(x)$ in (3) is of Schwartz class ($\in S'(R^1)$), the reason of which will be found in the next paragraph.

4. Let us return to the problem (3)–(4). First of all we notice that the B-shift $(\psi_0)_{kB}$ of the initial data $\psi_0(x)(\in S'(R^1))$ is a solution ([11]): The B-shift $T_{kB}(t, x)$ of a Schwartz distribution $T(x)$ is an element of $S'(\Omega)$ (and so, $\in K'(\Omega)$) such that

(11) $$\langle T_{kB}, u\cdot v\rangle = \int_0^T\langle T(\,\cdot\,), u(\cdot + kB_t)\rangle v(t)dt\,.$$

So if there holds a certain uniqueness property of solutions for the problem (3)–(4) the B-shift $(\psi_0)_{kB}$ must be the unique one. This consideration makes us to study the B-shifts of generalized random processes.

Let ψ be a regular element of $K'(\Omega)$. Then it is not difficult to see the existence of a random kernel $F(t, x, \omega)$ such that for a positive numbers m and r,

$$\langle \psi, u \cdot v \rangle = \int_0^T dt \int_{R^1} F(t, x, \omega) v(t) u^{(m)}(x) dx , \tag{12}$$

and

$$E\left[\int_K |F(t, x, \omega)|^2 \, dxdt\right] < \infty \quad \text{for any compact } K \subset [0, T] \times R^1 . \tag{13}$$

We associate to the element ψ the following bilinear form,

$$B_q(u, v) = \int_0^T \langle \psi, u(\cdot + qB_t) \rangle v(t) dt \qquad (q \in R^1) . \tag{14}$$

If the form $B_q(u, v)$ is separately continuous, then by the kernel theorem it defines a generalized random process ψ_{qB} that we shall call the B-shift of ψ. From the discussion above we confirm that the necessary and sufficient condition for this is the following

$$E\left[\int_K |F(t, x - qB_t, \omega)|^2 \, dxdt\right] < \infty \quad \text{for any compact } K \subset [0, T] \times R^1 . \tag{15}$$

We shall call such ψ a tame process whose random kernel $F(t, x, \omega)$ satisfies the condition (15) for all $q \in R^1$. The elements of $S'(R^1)$ are, for example, tame processes ([11]), while the elements of $K'(R^1)$ are not so in general. Obviously the derivatives $(\partial/\partial x)^m \psi$ of a tame process ψ are also tame processes and there holds the relation, $(\partial/\partial x)^m \psi_{qB} = [(\partial/\partial x)^m \psi]_{qB}$. On the other hand, the derivative $D_t \psi$ of a tame process may fail to be a regular process.

Remark 3. Therefore, in general case, the following formal calculus does not make sense,

$$D_t \psi_{qB} = (D_t \psi)_{qB} - q\left(\frac{\partial}{\partial x} \psi\right)_{qB} \cdot \dot{B} . \tag{16}$$

However, as we shall see later, there are the cases that the formula (16) becomes meaningful even though $D_t\psi$ fails to be a regular process.

For the next step, we prepare the

Proposition 2. *Let ψ be a twice B-differentiable tame process whose temporal cut $\psi_\cdot(u)$ satisfies the following*

$$\psi_t(u) - \psi_0(u) = \int_0^t \hat{\psi}_s(u)dB_s + X_t(u) - X_0(u) , \tag{17}$$

where X is such that $\hat{X} = 0$.

If $\hat{\psi}$ and X are tame processes and if the B-shift X_{qB} is B-differentiable with the following B-derivative

$$(X_{qB})^\wedge = -q\left(\frac{\partial}{\partial x}\right)X_{qB} . \tag{18}$$

Then the B-shift ψ_{qB} of ψ is also B-differentiable and it holds the relation

$$(\psi_{qB})^\wedge = \left(\hat{\psi} - q\frac{\partial}{\partial x}\psi\right)_{qB} . \tag{19}$$

Proof. Let $G(t, x, \omega)$ be the random kernel of $\hat{\psi}$. Then by the condition (13), which justifies the change of order of integrations $\int dB$ and $\int dx$, we find that the random kernel $\tilde{G}(t, x, \omega)$ of the element $\Phi = \int \hat{\psi}dB_s$ is represented by $\tilde{G}(t, x, \omega) = \int^t G(s, x, \omega)dB_s$. As X and ψ are supposed to be tame processes, so is the element Φ and therefore the kernel $\tilde{G}(t, x, \omega)$ satisfies the condition (15). Now from (17), we get

$$(\psi_{qB})_t(u) = C(u) + (\Phi_{qB})_t(u) + (X_{qB})_t(u) \tag{17$'$}$$

where $C(u) = \psi_0(u) - X_0(u)$.

On the other hand, by a simple application of Lemma 1–(iii), we obtain

$$\begin{aligned}(\Phi_{qB})_t(u) &= \int_{R^1} u^{(m)}(x + qB_t)\tilde{G}(t, x, \omega)dx \\ &= \int_{R^1} dx \int_0^t [qu^{(m+1)}(x + qB_s)\tilde{G}(s, x, \omega) \\ &\qquad\qquad + u^{(m)}(x + qB_s)G(s, x, \omega)]dB_s \\ &= \int_0^t dB_s \int_{R^1} [qu^{(m+1)}(x + qB_s)\tilde{G}(s, x, \omega)\end{aligned}$$

$$+ u^{(m)}(x + qB_s)G(s, x, \omega)]dB_s$$

$$= \int_0^t \left[(\hat{\psi}_{qB})_s(u) - q\left(\frac{\partial}{\partial x}\Phi_{qB}\right)_s(u) \right] dB_s \, .$$

Hence, with the relations (18) and (17)′ we get the (19). This completes the proof.

It is worthwhile to notice that the condition (18) is fulfiled when for each $u(x)$, $X_t(u)$ is absolutely continuous in t for almost all ω. Following the same argument we verify the next statement.

Proposition 3. *Let ψ be a twice B-differentiable tame process such that;*

(20) $$D_t\psi = 2(\dot{B})^2\psi + X \quad \text{where } X \text{ is a tame process} \, .$$

If the B-shift $(\hat{\psi})_{qB}$ is B-differentiable then ψ_{qB} satisfies the

(21) $$D_t\psi_{qB} = \left(\hat{\psi} - q\frac{\partial}{\partial x}\psi\right)_{qB} \cdot \dot{B} + X_{qB} \, .$$

Since we know from Proposition 2 that $\widehat{(\psi_{qB})} = (\hat{\psi} - q(\partial/\partial x)\psi)_{qB}$, the equation (21) can be rewritten in the following form;

(21)′ $$D_t\psi_{qB} = 2(\dot{B})^2\psi_{qB} + X_{qB} \, .$$

If we understand by $(\dot{\psi}\dot{B})_{qB}$ the quantity $(\hat{\psi})_{qB}\cdot\dot{B}$ (this procedure can be formalized in a concrete way), then the equation (21) justifies the formal calculus (16).

Theorem. *Let ψ be a B-differentiable tame process satisfying the equation* (20). *Moreover, if ψ satisfies the wave condition*: $\hat{\psi} + p(\partial/\partial x)\psi = 0$ *for a real number p, then ψ_{qB} solves the following Brownian particle equation,*

(21)″ $$D_t\psi_{qB} = -(p + q)\left(\frac{\partial}{\partial x}\psi_{qB}\right)\dot{B} + X_{qB} \, .$$

Proof. The wave condition implies that $(\hat{\psi})_{qB} = -p((\partial/\partial x)\psi)_{qB} = -p(\partial/\partial x)\psi_{qB}$. Since the order of two operations, $\wedge$ and $\partial/\partial x$, is changeable we see by Proposition 2 that the *B*-shift $(\hat{\psi})_{qB}$ is *B*-differentiable. Thus, the conclusion follows immediately from Proposition 3.

From this theorem comes an answer to the question of uniqueness.

Corollary 1. *The B-shift* $(\psi_0)_{kB}$ *of the initial data* $\psi_0(x)$ *is a solution of the problem* (3)–(4) *which is unique in the class of B-differentiable tame processes.*

Proof. It suffices to show that a B-differentiable tame process ψ satisfying (4) and the

$$(3)' \qquad D_t\psi = 2(\dot{B})^2\psi\ , \qquad \psi(0, x) = 0\ ,$$

is necessarily the zero element of $K'(\Omega)$.

Let ψ_{-kB} be the B-shift of such ψ. Then Theorem implies that; $D_t\psi_{-kB} = 0$, hence $\psi_{-kB} = \text{const.} = 0$ (the initial condition). From the trivial relation $(\psi_{-kB})_{kB} = \psi$ we conclude that $\psi = 0$.

As a variation of this problem we consider the following;

$$(3)'' \qquad D_t\psi = [2(\dot{B}) + U(x)]\psi\ , \qquad \psi(0, x) = \psi_0(x)$$

$$(4) \qquad \hat{\psi} + k\frac{\partial}{\partial x}\psi = 0\ .$$

For the simplicity of discussions we shall suppose that $U(x)$ is a sufficiently smooth function, bounded from above and that their derivatives are also bounded. Then it is easy to see that $X = \exp\left[\int_0^t U(x + kB_s)ds\right]\psi_0(x)$ is a tame process and so is its B-shift X_{kB}. Moreover we have the following

Corollary 2. X_{kB} *is a solution of the problem* (3)″–(4) *which is unique in the class of B-differentiable tame processes.*

Proof. Let ψ be a B-differentiable tame process solving the problem (3)″–(4)′. Then we see that $U(x)\psi$ is a tame process by the condition imposed on $U(x)$. Applying the theorem we find that; $D_t\psi_{-kB} = (U\psi)_{-kB} = U(x + kB_t)\psi_{-kB}$, hence $\psi = 0$, if $\psi_0(x) = 0$. Since it is easy to see that X_{kB} is a solution of the problem (3)″–(4), this completes the proof.

5. Encouraged by the results in the preceeding paragraph, we propose to define the functional $A_t(k)$ as an operator acting on the class of B-differentiable tame processes in such way that;

$$(22) \qquad A_t(k)T = T_{kB} = (T * \delta_{kB})(t, x)\ ,$$

where δ_{kB} is the B-shift of the Dirac delta $\delta_0(x)$, which is a fundamental solution of the Cauchy problem (3)–(4) (or, if you want, a stochastic fundamental solution of the heat equation).

Here are some examples showing that the definition is a natural one.

Example 1. By the form $A_t(k) = \exp\left[k\int_0^t (\dot{B})^2 ds\right]$ we expect to have the relation, $A_t(k_1+k_2) = A_t(k_1)A_t(k_2)$. This is true because we have, $T_{(k_1+k_2)B} = (T_{k_1B})_{k_2B}$ for any B-differentiable tame process T.

Example 2. (Interpretation of the Feynman integral in terms of the white noise)

The unique solution X_{kB} mentioned in Corollary 2 can be written in the form as follows;

$$\begin{aligned} X_{kB} &= \left[\exp\left(\int_0^t U(x+kB_s)ds\right)\psi_0(x)\right]_{kB} \\ &= \exp\left(k\int_0^t (\dot{B}_s)^2 ds\right)\left[\exp\left(\int_0^t U(x+kB_s)ds\right)\psi_0(x)\right] \\ &= \exp\left[\int_0^t U(x-k(B_t-B_s))ds\right]\exp\left(k\int_0^t (\dot{B}_s)^2 ds\right)\psi_0(x) . \end{aligned}$$

So if we admit to write the last term into the form;

$$\exp\left[\int_0^t [k(\dot{B}_s)^2 + U(x-k(B_t-B_s))]ds\right]\psi_0(x) ,$$

this procedure provides us with a good interpretation of the Feynman integral (for the case that k is real) in terms of the white noise.

References

[1] Feynman, R. P., Space-time approach to non-relativistic quantum mechanics, Rev. Modern Phys., 20-2 (1948), 367–387.
[2] Fisk, D. L., Quasi-martingales and stochastic integrals, Michigan State Univ., East Lansing, 1963.
[3] Funaki, N., The boundary value problem of the random transportation equation, Master thesis, Tokyo Univ., 1976.
[4] Kallianpur, G. and Striebel, C., A stochastic differential equation of Fisk type for estimation and nonlinear filtering problems, SIAM J. Appl. Math., 21-1 (1971), 61–72.
[5] Lévy, P., Processus stochastique et mouvement brownien, Gauthier-Villars, Paris, 1948.
[6] Nelson, E., Dynamical theory of Brownian motion, Princeton Univ. Press, 1967.

[7] Ogawa, S., Ichi-zigen ryûshi hôteishiki ron (in Japanese), Master thesis, Fac. of Engineering, Kyoto Univ., 1969.
[8] ——, On a Riemann definition of the stochastic integral, (I) and (II), Proc. Japan Acad., **46**-2 (1970), 153–161.
[9] ——, A partial differential equation with the white noise as a coefficient, Z. Wahrscheinlichkeitstheorie und Verw. Gebiete, **28** (1973), 53–71.
[10] ——, Equation de Schrödinger et équation de particule brownienne, J. Math. Kyoto Univ., **16**-1 (1976), 185–200.
[11] ——, Le bruit blanc et calcul stochastique, Proc. Japan Acad., **51**-6 (1975), 384–388.
[12] ——, Studies on the wave propagation in random media and the related subjects, Doctor thesis, Fac. of Engineering, Kyoto Univ., 1975.

Mathematical Institute
Tohoku University
Sendai 980, Japan

Proc. of Intern. Symp. SDE
Kyoto 1976, pp. 341-365

Estimation Problems for a Linear Stochastic Differential Equation in Hilbert Space

Sigeru OMATU and Takasi SOEDA

Abstract

A new approach by using functional analysis to the estimation problems for a linear stochastic distributed parameter system is proposed. It is assumed that the number of the sensor locations is finite and the error criterion is based on the unbiased and minimum variance estimations. Using the comparison theorem for the operator-valued differential equations and Itô's stochastic calculus in Hilbert spaces, we solve the abstract filtering, smoothing and prediction problems. By applying the kernel theorem due to Schwartz to these results, a class of partial differential equations for the optimal estimators is derived. Furthermore, the existence and uniqueness theorem concerning the solutions for the estimation problems is considered.

Introduction

The estimation problems, that is, the filtering, smoothing and prediction problems for a linear stochastic distributed parameter system are solved by using both the functional analysis and the comparison theorem for the operator-valued differential equations based on the unbiased and minimum variance estimation error criterion. The techniques of earlier papers on these problems include the orthogonal projection or innovation theories [2], [5]–[9], [24], [29], the maximum-likelihood approaches combined with the variational inequality or dynamic programming method [3], [4], [25], [30] and the Bayesian approach [15], [31], [32].

The approach adopted in this paper to the estimation problems was originally proposed by Balakrishnan [1] followed by Falb [9], and then applied to solve the estimation problems or the control problems for linear stochastic distributed parameter systems by Balakrishnan & Lions [3] and Omatu et al. [20]–[23]. This paper is an extension of our previous works [20]–[23] to the more general estimation problems for a linear stochastic distributed parameter system. Thus, the state of a system and the observation process are regarded as the stochastic ordinary differential equations in Hilbert spaces. Based on the unbiased and minimum variance estimation error criterion, the abstract optimal estimators are derived by using both Itô's formula in Hilbert spaces [8], [9] and the comparison theorem for the operator-valued differential equations. The new feature of this approach neither necessitates the orthogonal projection lemma [5]–[7] nor uses the innovation theory [2], [5]. Furthermore, this approach clarifies the interrelation between the innovation process and the unbiased estimate and shows that it is also possible to derive the smoothing problems from the viewpoint of the unbiased and minimum variance estimation error criterion without using the Wiener-Hopf theory. We derive the kernel representations for the optimal estimators in Hilbert spaces with an aid of the kernel theorem due to Schwartz [10]. Finally, a uniqueness property concerning the solutions for the optimal estimation problems is considered.

§ 1. Preliminaries

1.1. Evolution equations and comparison theorems

Let V and H be two separable real Hilbert spaces of real functions on D, a bounded open subset of R^r, with the boundary Γ which is an infinitely differentiable manifold of dimension $r-1$. The norms of V and H are denoted by $\|\cdot\|_V$ and $\|\cdot\|_H$ and the corresponding scalar products are denoted by $((\cdot\,,\cdot))$ and $(\cdot\,,\cdot)$, respectively. Assume that H is identified with its dual H' and

$$V \subset H \subset V'$$

where V is dense in H, H is dense in V', and the corresponding injections are continuous. Let $\langle\cdot\,,\cdot\rangle$ be the scalar product between V' and V and assume that

$$A(\cdot) \in L^\infty(0, T; \mathscr{L}(V, V')) \tag{1.1}$$

satisfying for some $\lambda \geqslant 0$, $\alpha > 0$

$$\langle -A(t)z, z\rangle + \lambda \|z\|_H^2 \geqq \alpha \|z\|_V^2 \tag{1.2}$$

for $\forall z \in V$, $\forall t \in (0, T]$. Here, $\mathscr{L}(V, V')$ denotes a space of bounded linear operators from V into V'. Note that under the preceding assumptions on $A(t)$ there exists the evolution operator $\mathscr{U}(t, s) \in \mathscr{L}(H, H)$ with the following properties [4], [14]:

(1) $\mathscr{U}(t, t_1)\mathscr{U}(t_1, s) = \mathscr{U}(t, s)$, $\mathscr{U}(s, s) = \mathscr{I}$ for $s < t_1 < t$

where $\mathscr{I}$ denotes the identity operator.

(2) $\mathscr{U}(t,s)\zeta \in L^2(s,T;V),\ d\mathscr{U}(t,s)\zeta/dt \in L^2(s,T;V')$

for $\forall\zeta \in H$ and $\forall t \in (s,T]$.

(1.3) (3) $d\mathscr{U}(t,s)\zeta/dt = A(t)\mathscr{U}(t,s)\zeta$ for $a.e.\ t \in (s,T]$, $\mathscr{U}(s,s)\zeta = \zeta$.

(1.4) (4) $\|\mathscr{U}(t,s)\|_{\mathscr{L}(H,H)} \leqq 1$ for $\forall t > s$.

Now consider the following equation for $N(t) = N^*(t) \in \mathscr{L}(H,H)$:

$$\frac{dN(t)}{dt} = A(t)N(t) + N(t)A^*(t) + g(t,N) \tag{1.5}$$

$$N(s) = N_0 \in \mathscr{L}(H,H) \tag{1.6}$$

where "$*$" denotes the adjoint operator and $g(t,N)$ is a mapping from $(s,T] \times \mathscr{L}(H,H)$ into $\mathscr{L}(H,H)$. Let us assume that $g(t,N)$ maps bounded sets into bounded sets and $g(t,N)$ is continuously Fréchet differentiable with respect to N. Then the following theorem was proved [14].

Theorem 1.1. *Under the preceding conditions for $A(t)$ and $g(t,N)$, (1.5) has a unique solution $N(t)$ for $\forall N_0 \in \mathscr{L}(H,H)$ such that $N(t) \in C(s,T;\mathscr{L}(H,H))$. The solution of (1.5) satisfies the following integral equations:*

$$N(t) = \mathscr{U}(t,s)N_0\mathscr{U}(t,s)^* + \int_s^t \mathscr{U}(t,\tau)g(\tau,N)\mathscr{U}(t,s)^*d\tau\ . \tag{1.7}$$

Note that Theorem 1.1 holds even if $A(t)$ vanishes [14]. Let $f(t,N)$ satisfies the same conditions as $g(t,N)$, and let $M(t)$ denote the solution of the equation given by exchanging $g(t,N)$ for $f(t,N)$ in (1.5) under the same initial condition as $N(t)$, that is,

$$\frac{dM(t)}{dt} = A(t)M(t) + M(t)A^*(t) + f(t,M)\ , \qquad M(s) = N_0\ . \tag{1.8}$$

Then the following theorem can be proved.

Theorem 1.2. *Assume that the following relation holds*

$$g_N(t,N)\hat{N} = \frac{1}{2}g_N(t,N)\hat{N} + \frac{1}{2}\hat{N}g_N^*(t,N)\ , \qquad \forall\hat{N} \in \mathscr{L}(H,H) \tag{1.9}$$

where $g_N(t,N)$ denotes the Fréchet derivative with respect to N. If $g(t,N) \leqq f(t,N)$ for $\forall t \in (s,T]$ and $\forall N \in \mathscr{L}(H,H)$, then

$$N(t) \leqq M(t) \qquad \textit{for } \forall t \in (s,T]$$

where the inequality $g \leqq f$ of the operators means that

$$(g\phi, \phi) \leqq (f\phi, \phi) \qquad \text{for } \forall\phi \in H\ .$$

Proof. Defining $L(t) \triangleq M(t) - N(t)$, it follows from (1.5) and (1.8) that

$$\frac{dL(t)}{dt} = A(t)L(t) + L(t)A^*(t) + f(t, M) - g(t, N)\ . \tag{1.10}$$

Applying Theorem 1.1 to (1.10), and using (1.9) and the mean value theorem for the operator-valued function [33] yields

$$L(t) = \int_\tau^t \mathscr{U}_g(t, \sigma)(f(\sigma, M) - g(\sigma, M))\mathscr{U}_g^*(t, \sigma)d\sigma \tag{1.11}$$

where

$$\frac{\partial \mathscr{U}_g(t, \sigma)}{\partial t} = \left(A(t) + \frac{1}{2}g_N(t, N + \theta L)\right)\mathscr{U}_g(t, \sigma) \qquad \text{with some } \theta \in (0, 1)$$

$$\mathscr{U}_g(\sigma, \sigma) = \mathscr{I}\ .$$

Hence, applying the assumption of the theorem to (1.11) yields

$$L(t) \geqq 0\ , \quad \text{i.e.,} \quad M(t) \geqq N(t)\ .$$

Thus, the proof of the theorem is completed. Q.E.D.

Note that Theorem 1.2 holds even if $A(t)$ vanishes.

1.2. Probability theory and stochastic processes

Let $(\Omega, \mathscr{B}(\Omega), \mu)$ be a complete probability space with Ω as a topological space, $\mathscr{B}(\Omega)$ as the Borel field generated by Ω, and μ as the Radon probability measure on Ω. Let Φ be a separable real Hilbert space and let $\langle \cdot, \cdot \rangle_\Phi$ be the scalar product between Φ' and Φ.

Definition 1.1. A function $x(\omega)$ defined on Ω with values in Φ is called measurable if $x(\omega)$ is essentially separably-valued and $x^{-1}(\mathscr{O}) \in \mathscr{B}(\Omega)$ for each open set $\mathscr{O}$ in Φ. Such a measurable function $x(\omega)$ is called a random variable with a value in Φ and the space of $x(\omega)$ is denoted by Mes $[\Omega, \mu; \Phi]$.

Definition 1.2. $x(\omega)$ is called a second order random variable if $x(\omega)$ satisfies

$$E[\|x(\omega)\|_\Phi^2] \triangleq \int_\Phi \|x(\omega)\|_\Phi^2\, d\mu(\omega) < \infty\ ,$$

and the space of $x(\omega)$ is denoted by $L^2(\Omega, \mu; \Phi)$.

Definition 1.3. $x(t, \omega)$ is called a stochastic process with values in Φ if $x(\cdot, \omega)$ is a measurable mapping from $[0, T]$ into Mes $[\Omega, \mu; \Phi]$.

Definition 1.4. The linear random functional on Φ' is defined by a linear mapping from Φ' into Mes $[\Omega, \mu; \boldsymbol{R}]$, that is, a family of real random variables $X_{\phi_*}(\omega)$ for $\forall \phi_* \in \Phi'$ such that

$$X_{\alpha_1\phi_*^1+\alpha_2\phi_*^2}(\omega) = \alpha_1 X_{\phi_*^1}(\omega) + \alpha_2 X_{\phi_*^2}(\omega) , \qquad a.e.\ \omega \in \Omega$$

for $\forall \alpha_1, \alpha_2 \in \boldsymbol{R}$ and $\forall \phi_*^1, \phi_*^2 \in \Phi'$.

If $X_{\phi_*}(\omega)$ is continuous from Φ' into $\boldsymbol{R}$, then $X_{\phi_*}(\omega)$ is given by the following relation [33]:

$$X_{\phi_*}(\omega) = \langle \phi_*, X(\omega) \rangle_\Phi , \qquad X(\omega) \in \text{Mes}\ [\Omega, \mu; \Phi] .$$

If $X_{\phi_*}(\omega)$ is integrable and the mapping from $\phi_* \in \Phi'$ into $E[X_{\phi_*}(\omega)] \in \boldsymbol{R}$ is continuous, then there exists $\bar{X} \in \Phi$ [33] such that

$$E[X_{\phi_*}(\omega)] = \langle \phi_*, \bar{X} \rangle_\Phi .$$

$\bar{X}$ is called the mean value or the expectation of $X(\omega)$ and denoted by $E[X(\omega)]$. For given X_{ϕ_*} and Y_{ϕ_*} define the covariance function $\Theta(\phi_*^1, \phi_*^2)$ for $\forall \phi_*^1, \phi_*^2 \in \Phi'$ by

$$\Theta(\phi_*^1, \phi_*^2) \triangleq E[(X_{\phi_*^1}(\omega) - E[X_{\phi_*^1}(\omega)])(Y_{\phi_*^2}(\omega) - E[Y_{\phi_*^2}(\omega)])] .$$

If the mapping from $(\phi_*^1, \phi_*^2) \in \Phi' \times \Phi'$ into $\Theta(\phi_*^1, \phi_*^2) \in \boldsymbol{R}$ is continuous, then there exists an operator $\Lambda \in \mathscr{L}(\Phi', \Phi)$ [33] such that

$$\Theta(\phi_*^1, \phi_*^2) = \langle \phi_*^1, \Lambda \phi_*^2 \rangle_\Phi . \tag{1.12}$$

Λ is called the covariance operator of $X(\omega)$ and $Y(\omega)$ and denoted by Cov $[X(\omega), Y(\omega)]$. Assume that Φ is identified with its dual Φ'. Then we have the definition of a Wiener process with values in Φ.

Definition 1.5. Assume that J_n is a finite set of nonnegative integers. $W(t, \omega)$ is called a Wiener process with values in Φ or a Φ-valued Wiener process if $W(t, \omega)$ is a random variable with values in Φ for any fixed $t \in [0, T]$ and $W_{\phi_*^j}(t_i, \omega) = (\phi_*^j, W(t_i, \omega))_\Phi$ for $\forall \phi_*^j \in \Phi$ and $\forall i, j \in J_n$ is a real Gaussian random variable such that

$$E[W(t, \omega)] = 0 , \qquad t \geqq 0 ,$$

$$E[W_{\phi_*^j}(t_i, \omega) W_{\phi_*^k}(t_n, \omega)] = \int_0^{\min(t_i, t_n)} (\phi_*^j, Q(\tau)\phi_*^k)_\Phi d\tau \tag{1.13}$$

where $(\cdot, \cdot)_\Phi$ denotes the scalar product in Φ and $Q(\cdot)$ satisfies

$$Q(\cdot) \in L^\infty(0, T; \mathscr{L}(\Phi, \Phi)) .$$

It is assumed that $Q(\tau)$ is a nonnegative, self-adjoint, and nuclear operator for a.e. $\tau \in [0, T]$. Define the notation "$x \circ y$" for $\forall x \in \Phi$ and $\forall y \in \Psi'$ [9] by

$$(x \circ y)z = x\langle y, z\rangle_\Psi , \qquad \forall z \in \Psi \tag{1.14}$$

where Ψ is a Hilbert space.

Letting $\Phi = \Psi$ yields that (1.13) can be more simply described by

$$\begin{aligned} \mathrm{Cov}\,[W(t_i, \omega), W(t_n, \omega)] &= E[W(t_i, \omega) \circ W(t_n, \omega)] \\ &= \int_0^{\min(t_i, t_n)} Q(\tau) d\tau . \end{aligned} \tag{1.15}$$

Then the following property was proved [4].

Theorem 1.3. *If $W(t, \omega)$ is a Φ-valued Wiener process, then*

$$W(t, \omega) \in C(0, T; L^2(\Omega, \mu, \Phi)) \tag{1.16}$$

with independent increments such that

$$E[(W(t_2, \omega) - W(t_1, \omega)) \circ (W(t_4, \omega) - W(t_3, \omega))] = 0$$
$$\text{for } t_1 < t_2 \leqq t_3 < t_4 .$$

It is easily seen from (1.16) that $W(t, \omega)$ is a measurable mapping from $[0, T]$ into Mes $[\Omega, \mu; \Phi]$, that is, a stochastic process with values in Φ. Let $C(\tau) \in L^\infty(0, T; \mathscr{L}(H, H))$ and let $W(t, \omega)$ be a H-valued Wiener process. Let us consider the linear stochastic system in Hilbert spaces described by

$$U(t, \omega) = \int_0^t A(\tau)U(\tau, \omega)d\tau + \int_0^t C(\tau)dW(\tau, \omega) + U_0(\omega) \tag{1.17}$$

$$U_0(\omega) \in \mathscr{H} \triangleq L^2(\Omega, \mu; H) .$$

Then the following theorem was proved [4].

Theorem 1.4. *Under the preceding conditions for $A(t)$ and $C(t)$, there exists a unique solution of* (1.17) *such that*

$$U(t, \omega) \in C(0, T; \mathscr{V}') \cap L^\infty(0, T; \mathscr{H}) \cap L^2(0, T; \mathscr{V}) \tag{1.18}$$

where

$$\mathscr{V} \triangleq L^2(\Omega, \mu; V) \subset \mathscr{H} \subset \mathscr{V}' \triangleq L^2(\Omega, \mu; V') .$$

Note that (1.17) is usually denoted by the following linear stochastic differential equation in Hilbert spaces [4], [8]:

$$dU(t, \omega) = A(t)U(t, \omega)dt + C(t)dW(t, \omega) \tag{1.19}$$

$$U(0, \omega) = U_0(\omega) . \tag{1.20}$$

Assume that $U_0(\omega)$ is independent of $W(t, \omega)$ and is a Gaussian random variable in H [4] which has the mean value and the covariance operator, respectively, given by

$$E[U_0(\omega)] = U_1 \neq 0 \tag{1.21}$$

$$\mathrm{Cov}\,[U_0(\omega), U_0(\omega)] = E[(U_0(\omega) - U_1) \circ (U_0(\omega) - U_1)] = P_0 \tag{1.22}$$

where P_0 is a nonnegative, self-adjoint and nuclear operator.

§ 2. Estimation problems in Hilbert spaces

2.1. General formulations

Let us assume that the state $U(t, \omega)$ with values in the Hilbert space V is described by (1.17). The observation $Z(t, \omega)$ of the state $U(t, \omega)$ is given by

$$dZ(t, \omega) = H(t)U(t, \omega)dt + dV(t, \omega) \tag{2.1}$$

where $V(t, \omega)$ is a Wiener process with values in a Hilbert space and is independent of both $W(t, \omega)$ and $U_0(\omega)$. Assume that the mean value and the covariance operator of $V(t, \omega)$ are, respectively, given by

$$E[V(t, \omega)] = 0$$

$$\begin{aligned} \mathrm{Cov}\,[V(t_1, \omega), V(t_2, \omega)] &= E[V(t_1, \omega) \circ V(t_2, \omega)] \\ &= \int_0^{\min(t_1, t_2)} R(\tau)d\tau \end{aligned} \tag{2.2}$$

where $R(\tau)$ is a positive, invertible, self-adjoint and nuclear operator. Since $R(t)$ is the invertible nuclear operator, the dimension of $V(t, \omega)$ is finite [33] and is assumed to be $\boldsymbol{R}^m$. If we restrict our attention to the pointwise observation, it is natural to consider the case where $H(t)$ contains the term of $\mathrm{Col}\,[\delta(x - x^1), \cdots, \delta(x - x^m)]$ for $\forall x, x^1, \cdots, x^m \in D$ where $\delta(\cdot)$ denotes Dirac's delta function defined on $D \subset \boldsymbol{R}^r$. Then denoting the Sobolev space of order $n \in \boldsymbol{R}$ defined on D by $H^n(D)$, it was proved [19] that

$$\delta(x) \in (H^{[r/2]+1}(D))'$$

where r and $[r/2]$ denote the dimension of D and the integer part of $r/2$, respectively. Hence, if

$$V \subset H^{[r/2]+1}(D)\,, \tag{2.3}$$

then the pointwise observation of $U(t, \omega) \in \mathscr{V}$ for *a.e.* $t \in (0, T]$ is possible in the sense of Lions [16] and (2.1) has meaning as a time function.

Now the estimation problems are posed as follows: Given the measurement data Z_0^t denoting a family of $Z(\tau)$ from $\tau = 0$ up to the present time t, find an estimate $\hat{U}(\tau, \omega\,|\,t)$ of the state $U(\tau, \omega)$ at only particular time τ. Specifically, for $\tau < t$ we have the smoothing problem, for $\tau = t$ the filtering problem and for $\tau > t$ the prediction problem. Here, we assume that the estimation error criterion is based on the unbiased and minimum variance estimations given by

$$E[\tilde{U}(\tau, \omega\,|\,t)] = 0 \tag{2.4}$$

$$\mathrm{Cov}\,[\tilde{U}^0(\tau, \omega\,|\,t), \tilde{U}^0(\tau, \omega\,|\,t)] \leqq \mathrm{Cov}\,[\tilde{U}(\tau, \omega\,|\,t), \tilde{U}(\tau, \omega\,|\,t)] \tag{2.5}$$

$$\tilde{U}(\tau, \omega\,|\,t) \triangleq U(\tau, \omega) - \hat{U}(\tau, \omega\,|\,t) \tag{2.6}$$

where $\tilde{U}^0(\tau, \omega\,|\,t)$ denotes the optimal estimation error function. The notation $\theta^0 \leqq \theta$ of (2.5) means that

$$(\phi, \theta^0\phi) \leqq (\phi, \theta\phi) \qquad \text{for } \forall\phi \in H\,. \tag{2.7}$$

Since $U(t, \omega)$ and $Z(t, \omega)$ are jointly Gaussian distributed [4], it is clear that the optimal minimum variance estimator is given by the linear transformation of the observed data Z_0^t. Hence, we assume that the time evolution of $\hat{U}(\tau, \omega\,|\,t)$ is given by

$$d\hat{U}(\tau, \omega\,|\,t) = K(\tau, t)\hat{U}(t, \omega\,|\,t)dt + L(\tau, t)dZ(t, \omega) \qquad \text{for } \tau \leqq t \tag{2.8}$$

and

$$d\hat{U}(\tau, \omega\,|\,t) = K(\tau, t)\hat{U}(t, \omega\,|\,t)d\tau + L_p(\tau, t)dZ(\tau, \omega) \qquad \text{for } \tau > t \tag{2.9}$$

where

$$K(\tau, \cdot) \in L^2(0, T\,; \mathscr{L}(V, V'))\,, \quad L(\tau, \cdot), L_p(\tau, \cdot) \in L^\infty(0, T\,; \mathscr{L}(R^m, H))\,.$$

Further, assume that for some $\lambda \geqslant 0$ and $\alpha > 0$

$$\langle -K(\tau, t)z, z\rangle + \lambda\|z\|_H^2 \geqq \alpha\|z\|_V^2 \qquad \text{for } \forall\tau \in (0, T], \forall z \in V\,. \tag{2.10}$$

Then it is clear from Theorem 1.4 that there exist unique solutions $\hat{U}(\tau, \omega \mid t)$ for (2.8) and (2.9). Note that setting $\tau = t$ in (2.8) yields

$$\hat{U}(t, \omega \mid t) = \mathscr{U}_K(t, 0)U_1 + \int_0^t \mathscr{U}_K(t, \sigma)L(t, \sigma)dZ(\sigma, \omega)$$

where $\mathscr{U}_K(t, \sigma)$ denotes the evolution operator of $K(t, \sigma)$. Hence, (2.8) for $\tau = t$ means that $\hat{U}(t, \omega \mid t)$ is given by the linear transformation of Z_0^t. Integrating (2.8) for $\tau < t$ from $\sigma = \tau$ to $\sigma = t$ yields

$$\hat{U}(\tau, \omega \mid t) = \hat{U}(\tau, \omega \mid \tau) + \int_\tau^t K(\tau, \sigma)\hat{U}(\sigma, \omega \mid \sigma)d\sigma + \int_\tau^t L(\tau, \sigma)dZ(\sigma, \omega) .$$

Therefore, we find that the optimal filtering estimate $\hat{U}(\tau, \omega \mid \tau)$ constitutes the initial value for the smoothing estimate $\hat{U}(\tau, \omega \mid t)$ and that $\hat{U}(\tau, \omega \mid \tau)$ is corrected by the linear transformation of the additionally available measurement data Z_τ^t. Thus, (2.8) for $\tau < t$ means that the optimal smoothing estimate $\hat{U}(\tau, \omega \mid t)$ is given by the linear transformation of Z_0^t. Taking into consideration that $\hat{U}(\tau, \omega \mid t)$ for $\tau > t$ is given by the linear transformation of Z_0^t, it is clear that $L_p(\tau, t)$ of (2.9) is equal to null operator, that is,

$$L_p(\tau, t) = 0 .$$

Therefore, we see that (2.8) and (2.9) with $L_p(\tau, t) = 0$ constitute the general forms of the optimal estimators in Hilbert spaces. From now on, we omit the generic point $\omega \in \Omega$ of the random variables since it is clear from the context.

2.2. Filtering problems

In order to derive the optimal filter, it is necessary to obtain the time evolution of $\hat{U}(t \mid t)$. It follows from (1.19), (2.1), (2.6) and (2.8) that

$$d\tilde{U}(t \mid t) = -(K(t, t) + L(t, t)H(t) - A(t))U(t)dt + K(t, t)\tilde{U}(t \mid t)dt \\ - L(t, t)dV(t) + C(t)dW(t) .$$

Since it follows from (1.19) and (1.21) that

$E[U(t)]$ $(\neq 0) \in H$ for $\forall t \in (0, T]$ and from (2.4) that $\hat{U}(t \mid t)$ is unbiased, the following relation holds

(2.11) $$K(t, t) = -L(t, t)H(t) + A(t) .$$

Then we have

(2.12) $$d\hat{U}(t \mid t) = A(t)\hat{U}(t \mid t)dt + L(t, t)d\nu(t)$$

(2.13) $$d\nu(t) \triangleq dZ(t) - H(t)\hat{U}(t\,|\,t)dt$$

(2.14) $$\begin{aligned} d\tilde{U}(t\,|\,t) &= A(t)\tilde{U}(t\,|\,t)dt - L(t,t)H(t)\tilde{U}(t\,|\,t)dt \\ &\quad - L(t,t)dV(t) + C(t)dW(t)\,. \end{aligned}$$

On the other hand, it follows from (1.12) and (1.14) that

(2.15) $$\begin{aligned} (\phi, \mathrm{Cov}\,[\tilde{U}(t\,|\,t), \tilde{U}(t\,|\,t)]\phi) &= E[(\phi, \tilde{U}(t\,|\,t))(\tilde{U}(t\,|\,t), \phi)] \\ &= (\phi, P(t\,|\,t)\phi) \qquad \text{for } \forall\phi \in H \end{aligned}$$

where

(2.16) $$P(t\,|\,t) \triangleq E[\tilde{U}(t\,|\,t) \circ \tilde{U}(t\,|\,t)]\,.$$

Accordingly, it is sufficient to determine $L(t, t)$ such that (2.15) is minimized. Applying Itô's lemma in Hilbert spaces [8], and using the follow-ing relation [8], [9]

(2.17) $$\begin{aligned} h_1 \circ (Nh_2) &= (h_1 \circ h_2)N^* \\ &\qquad \text{for } \forall h_1 \in H,\ \forall h_2 \in V \text{ and } \forall N \in \mathscr{L}(V, V')\,, \end{aligned}$$

it follows that

$$\begin{aligned} d(\tilde{U}(t\,|\,t) \circ \tilde{U}(t\,|\,t)) &= (d\tilde{U}(t\,|\,t)) \circ \tilde{U}(t\,|\,t) + \tilde{U}(t\,|\,t) \circ (d\tilde{U}(t\,|\,t)) \\ &\quad + L(t,t)R(t)L^*(t,t)dt + D_0(t)dt \end{aligned}$$

where

$$D_0(t) \triangleq C(t)Q(t)C^*(t)\,.$$

Using (2.14) and the following relation [8], [9]

$$(h_1 \circ h_2)^* = h_2 \circ h_1 \qquad \text{for } \forall h_1, h_2 \in H\,,$$

and taking the expectation of each side yields

(2.18) $$\begin{aligned} \frac{dP(t\,|\,t)}{dt} &= K(t,t)P(t\,|\,t) + P(t\,|\,t)K^*(t,t) \\ &\quad + L(t,t)R(t)L^*(t,t) + D_0(t)\,. \end{aligned}$$

Substituting (2.11) into (2.18) yields

(2.19) $$\begin{aligned} \frac{dP(t\,|\,t)}{dt} &= A(t)P(t\,|\,t) + P(t\,|\,t)A^*(t) \\ &\quad + D_0(t) - P(t\,|\,t)D(t)P(t\,|\,t) + \Sigma_\Delta(t) \end{aligned}$$

where

$$D(t) \triangleq H^*(t)R^{-1}(t)H(t)$$

(2.20) $$\Sigma_\Delta(t) \triangleq T_\Delta(t)R^{-1}(t)T_\Delta^*(t)$$

$$T_\Delta(t) \triangleq L(t, t)R(t) - P(t, t)H^*(t) .$$

Since $R(t)$ is positive, applying Theorem 1.2 to (2.19) yields that $T_\Delta(t)$ is null operator if and only if $L(t, t)$ minimizes the scalar product (2.15). Hence, it follows from (2.20) that

(2.21) $$L(t, t) = P(t\,|\,t)H^*(t)R^{-1}(t) .$$

Therefore, the minimum error covariance operator $P(t\,|\,t)$ satisfies

(2.22) $$\frac{dP(t\,|\,t)}{dt} = A(t)P(t\,|\,t) + P(t\,|\,t)A^*(t) + D_0(t) - P(t\,|\,t)D(t)P(t\,|\,t) .$$

The initial condition of (2.22) is given from (1.22) and (2.16) by

(2.23) $$P(0\,|\,0) = P_0 .$$

The optimal filtering estimator is given from (2.12) and (2.21) by

(2.24) $$d\hat{U}(t\,|\,t) = A(t)\hat{U}(t\,|\,t)dt + P(t\,|\,t)H^*(t)R^{-1}(t)d\nu(t) .$$

The initial value of $\hat{U}(t\,|\,t)$ is obviously given from (1.21) and (2.4) by

(2.25) $$\hat{U}(0\,|\,0) = U_1 .$$

2.3. Smoothing problems

From (1.19), (2.1), (2.6) and (2.8), we have the time evolution of $\tilde{U}(\tau\,|\,t)$ given by

$$d\tilde{U}(\tau\,|\,t) = -(K(\tau, t) + L(\tau, t)H(t))U(t)dt + K(\tau, t)\tilde{U}(t\,|\,t)dt - L(\tau, t)dV(t) .$$

Since $\tilde{U}(\tau\,|\,t)$ and $\tilde{U}(t\,|\,t)$ are unbiased estimators, we have

(2.26) $$K(\tau, t) = -L(\tau, t)H(t) .$$

Then we have

(2.27) $$d\hat{U}(\tau\,|\,t) = L(\tau, t)(dZ(t) - H(t)\hat{U}(t\,|\,t)dt) = L(\tau, t)d\nu(t)$$

and

$$d\tilde{U}(\tau \mid t) = K(\tau, t)\tilde{U}(t \mid t)dt - L(\tau, t)dV(t) . \tag{2.28}$$

On the other hand, it follows from (1.12) and (1.14) that

$$(\phi, \mathrm{Cov}\,[\tilde{U}(\tau \mid t), \tilde{U}(\tau \mid t)]\phi) = (\phi, P(\tau \mid t)\phi) \qquad \text{for } \forall\phi \in H \tag{2.29}$$

where

$$P(\tau \mid t) \triangleq E[\tilde{U}(\tau \mid t) \circ \tilde{U}(\tau \mid t)] . \tag{2.30}$$

Hence, we must determine $L(\tau, t)$ such that (2.29) is minimized. Applying Itô's lemma in Hilbert spaces [8], we have

$$\frac{dP(\tau \mid t)}{dt} = K(\tau, t)P^*(\tau, t) + P(\tau, t)K^*(\tau, t) + L(\tau, t)R(t)L^*(\tau, t) \tag{2.31}$$

where

$$P(\tau, t) \triangleq E[\tilde{U}(\tau \mid t) \circ \tilde{U}(t \mid t)] . \tag{2.32}$$

Substituting (2.26) into (2.31) yields

$$\frac{dP(\tau \mid t)}{dt} = -P(\tau, t)D(t)P^*(\tau, t) + \Sigma_p(t) \tag{2.33}$$

where

$$\begin{aligned} \Sigma_p(t) &\triangleq T(t)R^{-1}(t)T^*(t) \\ T(t) &\triangleq L(\tau, t)R(t) - P(\tau, t)H^*(t) . \end{aligned} \tag{2.34}$$

Applying Theorem 1.2 to (2.33) yields that $T(t)$ is null operator if and only if $L(\tau, t)$ minimizes the scalar product (2.29). Hence, it follows from (2.34) that

$$L(\tau, t) = P(\tau, t)H^*(t)R^{-1}(t) . \tag{2.35}$$

Therefore, the minimum estimation error covariance operator satisfies

$$\frac{dP(\tau \mid t)}{dt} = -P(\tau, t)D(t)P^*(\tau, t) . \tag{2.36}$$

Applying Theorem 1.1 to (2.36) yields

(2.37) $$P(\tau\,|\,t) = P(\tau\,|\,\tau) - \int_\tau^t P(\tau, \sigma)D(\sigma)P^*(\tau, \sigma)d\sigma \ .$$

We have from (2.27) and (2.35)

(2.38) $$\tilde{U}(\tau\,|\,t) = \hat{U}(\tau\,|\,\tau) + \int_\tau^t P(\tau, \sigma)H^*(\sigma)R^{-1}(\sigma)d\nu(\sigma) \ .$$

It remains to derive the time evolution of $P(\tau, \sigma)$. Using Itô's lemma in Hilbert spaces [8], we have

(2.39) $$\begin{aligned} d(\hat{U}(\tau\,|\,\sigma)\circ\tilde{U}(\sigma\,|\,\sigma)) &= (d\tilde{U}(\tau\,|\,\sigma))\circ\tilde{U}(\sigma\,|\,\sigma) + \tilde{U}(\tau\,|\,\sigma)\circ(d\tilde{U}(\sigma\,|\,\sigma)) \\ &\quad + L(\tau, \sigma)R(\sigma)L^*(\sigma, \sigma)d\sigma \ . \end{aligned}$$

Substituting (2.14) and (2.28), and taking the expectation of each side of (2.39) yields

$$\begin{aligned} \frac{dP(\tau, \sigma)}{d\sigma} &= K(\tau, \sigma)P(\sigma\,|\,\sigma) + P(\tau, \sigma)(A(\sigma) - L(\sigma, \sigma)H(\sigma))^* \\ &\quad + L(\tau, \sigma)R(\sigma)L^*(\sigma, \sigma) \ . \end{aligned}$$

It follows from (2.26) that

$$\begin{aligned} K(\tau, \sigma)P(\sigma\,|\,\sigma) &= -L(\tau, \sigma)H(\sigma)P(\sigma\,|\,\sigma) \\ &= -L(\tau, \sigma)R(\sigma)(P(\sigma\,|\,\sigma)H^*(\sigma)R^{-1}(\sigma))^* \\ &= -L(\tau, \sigma)R(\sigma)L^*(\sigma, \sigma) \ . \end{aligned}$$

Hence, we have

(2.40) $$\frac{dP(\tau, \sigma)}{d\sigma} = P(\tau, \sigma)(A(\sigma) - L(\sigma, \sigma)H(\sigma))^*$$

(2.41) $$P(\tau, \tau) = P(\tau\,|\,\tau) \ .$$

Letting the evolution operator of $A(\sigma) - L(\sigma, \sigma)H(\sigma)$ be $\mathscr{U}_A(\sigma, \tau)$ yields

(2.42) $$\begin{aligned} \frac{d\mathscr{U}_A(\sigma, \tau)}{d\sigma} &= (A(\sigma) - L(\sigma, \sigma)H(\sigma))\mathscr{U}_A(\sigma, \tau) \\ \mathscr{U}_A(\tau, \tau) &= \mathscr{I} \ . \end{aligned}$$

Then we have from (2.37), (2.38), (2.40) and (2.42)

(2.43) $$P(\tau, \sigma) = P(\tau\,|\,\tau)\mathscr{U}_A^*(\sigma, \tau)$$

$$P(\tau \mid t) = P(\tau \mid \tau) - P(\tau \mid \tau) \int_\tau^t \mathscr{U}_A^*(\sigma, \tau) D(\sigma) \mathscr{U}_A(\sigma, \tau) d\sigma P(\tau \mid \tau) \tag{2.44}$$

$$\hat{U}(\tau \mid t) = \hat{U}(\tau \mid \tau) + P(\tau \mid \tau)\lambda(\tau, t) \tag{2.45}$$

$$\lambda(\tau, t) \triangleq \int_\tau^t \mathscr{U}_A^*(\sigma, \tau) H^*(\sigma) R^{-1}(\sigma) d\nu(\sigma) . \tag{2.46}$$

Based on these expressions, the various types smoothing estimators can be derived.

i) Fixed-interval smoothing estimator: $\tau = t$, $t = T$ fixed.
Substituting $\tau = t$ and $t = T$ fixed into (2.44) and (2.45) yields

$$d\hat{U}(t \mid T) = A(t)\hat{U}(t \mid T)dt + D_0(t)P^{-1}(t \mid t)(\hat{U}(t \mid T) - \hat{U}(t \mid t))dt \tag{2.47}$$

$$\begin{aligned} \frac{dP(t \mid t)}{dt} &= (A(t) + D_0(t)P^{-1}(t \mid t))P(t \mid T) \\ &\quad + P(t \mid T)(A(t) + D_0(t)P^{-1}(t \mid t))^* - D_0(t) . \end{aligned} \tag{2.48}$$

ii) Fixed-point smoothing estimator: $\tau = t_1$, $t \in (t_1, T]$.
Substituing $\tau = t_1$ into (2.43)–(2.45) yields

$$d\hat{U}(t_1 \mid t) = P(t_1, t)H^*(t)R^{-1}(t)d\nu(t) \tag{2.49}$$

$$\begin{aligned} \frac{dP(t_1, t)}{dt} &= P(t_1, t)(A(t) - L(t, t)H(t))^* \\ P(t_1, t_1) &= P(t_1 \mid t_1) \end{aligned} \tag{2.50}$$

$$\frac{dP(t_1 \mid t)}{dt} = -P(t_1, t)D(t)P^*(t_1, t) . \tag{2.51}$$

iii) Fixed-lag smoothing estimator: $\tau = t$, $t = t + \Delta$, $\Delta > 0$.
Substituting $\tau = t$ and $t = t + \Delta$ into (2.44) and (2.45) yields

$$\begin{aligned} d\hat{U}(t \mid t + \Delta) &= A(t)\hat{U}(t \mid t + \Delta)dt \\ &\quad + J(t + \Delta)L(t + \Delta, t + \Delta)d\nu(t + \Delta) \\ &\quad + D_0(t)P^{-1}(t \mid t)[\hat{U}(t \mid t + \Delta) - \hat{U}(t \mid t)]dt \end{aligned} \tag{2.52}$$

$$\begin{aligned} \frac{dJ(t + \Delta)}{dt} &= [A(t) + D_0(t)P^{-1}(t \mid t)]J(t + \Delta) \\ &\quad - J(t + \Delta)[A(t + \Delta) \\ &\qquad + D_0(t + \Delta)P^{-1}(t + \Delta \mid t + \Delta)] \end{aligned} \tag{2.53}$$

$$J(\Delta) = P(0, \Delta)P^{-1}(\Delta|\Delta)$$

$$\begin{aligned}\frac{dP(t|t+\Delta)}{dt} &= [A(t) + D_0(t)P^{-1}(t|t)]P(t|t+\Delta) \\ &\quad + P(t|t+\Delta)[A(t) + D_0(t)P^{-1}(t|t)]^* \\ &\quad - J(t+\Delta)L(t+\Delta, t+\Delta)R(t+\Delta) \\ &\quad L^*(t+\Delta, t+\Delta)J^*(t+\Delta) - D_0(t) .\end{aligned} \tag{2.54}$$

2.4. Prediction problems

It follows from (2.6) and (2.9) that

$$\begin{aligned}d\tilde{U}(\tau|t) &= (A(\tau)\mathscr{U}(\tau, t) - K(\tau, t))U(t)d\tau \\ &\quad + K(\tau, t)\tilde{U}(t|t)d\tau + C(\tau)dW(\tau) \\ &\quad + A(\tau)\int_t^\tau \mathscr{U}(\tau, \sigma)C(\sigma)dW(\sigma)d\tau .\end{aligned}$$

Hence, we have from (2.4)

$$K(\tau, t) = A(\tau)\mathscr{U}(\tau, t) . \tag{2.55}$$

Then it follows from (2.9) that

$$d\hat{U}(\tau|t) = A(\tau)\mathscr{U}(\tau, t)\hat{U}(t|t)d\tau \tag{2.56}$$

$$\begin{aligned}d\tilde{U}(\tau|t) &= K(\tau, t)\tilde{U}(t|t)d\tau + C(\tau)dW(\tau) \\ &\quad + A(\tau)\int_t^\tau \mathscr{U}(\tau, \sigma)C(\sigma)dW(\sigma)d\tau .\end{aligned} \tag{2.57}$$

Substituting (1.3) into (2.56), we have

$$d\hat{U}(\tau|t) = d\mathscr{U}(\tau, t)\hat{U}(t|t) .$$

Integrating each side from $\tau = t$ to $\tau = \sigma$ yields

$$\hat{U}(\sigma|t) = \mathscr{U}(\sigma, t)\hat{U}(t|t) \qquad \text{for } \sigma > t .$$

Hence, we have from (2.56)

$$d\hat{U}(\tau|t) = A(\tau)\hat{U}(\tau|t)d\tau \qquad \text{for } \tau > t . \tag{2.58}$$

It follows from (2.57) and Itô's lemma in Hilbert spaces [8] that

$$d(\tilde{U}(\tau|t)\circ\tilde{U}(\tau|t)) = K(\tau,t)\tilde{U}(t|t)\circ\tilde{U}(\tau|t)d\tau + \tilde{U}(\tau|t)\circ\tilde{U}(t|t)K^*(\tau,t)d\tau + D_0(\tau)d\tau\ . \tag{2.59}$$

Taking the expectation of each side of (2.59), and substituting (2.58) into (2.59) yields

$$\frac{dP(\tau|t)}{d\tau} = A(\tau)P(\tau|t) + P(\tau|t)A^*(\tau) + D_0(\tau)\ . \tag{2.60}$$

The initial condition of $P(\tau|t)$ is obviously given by $P(t|t)$.

If $V = H = \boldsymbol{R}^n$, then the underlying spaces are finite dimensional and the oprator "$h_1\circ h_2$" corresponds to the matrix $h_1h_2^T$ where h_2^T denotes the transpose of h_2. The results obtained here can be reduced to those of lumped parameter systems [12], [13] [17], [18], [26]–[28].

§ 3. Estimation problems for a distributed parameter system

Let us now apply the results in the Hilbert spaces to the estimation problems for a distributed parameter system. Assume that $V = H^1(D)$ and $H = L^2(D)$. Let M be $[0,t]\times D$ and assume that

$$\langle -A(t)z_1, z_2\rangle \triangleq \sum_{i,j=1}^{r}\int_D a_{ij}(t,x)\frac{\partial z_1(x)}{\partial x_i}\frac{\partial z_2(x)}{\partial x_j}dx \tag{3.1}$$

for $\forall z_1, z_2 \in V$, where $a_{ij}(t,x)\in L^\infty(M)$ satisfies for some $\beta > 0$

$$\sum_{i,j=1}^{r} a_{ij}(t,x)y_iy_j \geqq \beta(y_1^2+\cdots+y_r^2)\ , \qquad \forall y_i\in\boldsymbol{R},\ \forall(t,x)\in M\ .$$

In this case, it is easily verified from (3.1) that (1.1) is satisfied since $V = H^1(D)$. Then (1.19) is equivalent to the following equation:

$$d(U(t),z) = \langle A(t)U(t),z\rangle\ \ dt + (C(t)dW(t),z) \quad \text{for } \forall z\in V\subset H\ . \tag{3.2}$$

Let $\mathscr{D}(D)$ be a space of infinitely differentiable functions with compact support in D. Since $\forall\phi\in\mathscr{D}(D)\subset V$, it follows that

$$\int_D dU(t,x)\phi(x)dx = \int_D (A_xU(t,x))\phi(x)dxdt + \int_D C(t,x)dW(t,x)\phi(x)dx \tag{3.3}$$

where

$$A_x(\cdot) = \sum_{i,j=1}^{r} \frac{\partial}{\partial x_i}\left(a_{ij}(t,x)\frac{\partial(\cdot)}{\partial x_j}\right).$$

Therefore, it holds that

(3.4) $dU(t,x) = A_x U(t,x)dt + C(t,x)dW(t,x)$ for a.e. $(t,x) \in M$

(3.5) $$U(0,x) = U_0(x).$$

Using Green's formula [16], [19] for $\forall z \in V$ yields

(3.6) $$\langle -A(t)U(t,x), z\rangle = \int_\Gamma \frac{\partial U(t,\xi)}{\partial \boldsymbol{n}} z(\xi)d\xi + \int_D (A_x U(t,x))z(x)dx$$

where $\boldsymbol{n}$ denotes the normal vector to the boundary Γ and

$$\partial(\cdot)/\partial \boldsymbol{n} \triangleq \sum_{i,j=1}^{r} a_{ij}(t,\xi)\cos(\boldsymbol{n}, x_i)\partial(\cdot)/\partial x_j .$$

Then it follows from (3.3), (3.4) and (3.6) that

(3.7) $\partial U(t,\xi)/\partial \boldsymbol{n} = 0$ for $\forall \xi \in \Gamma$.

Hence, the general formulations of (1.19) and (1.20) include the distributed parameter system described by (3.4), (3.5) and (3.7). Letting $H(t)(\cdot) = G(t)\int_D \text{Col}\,[\delta(x - x^1), \cdots, \delta(x - x^m)](\cdot)dx$ yields

(3.8) $$dZ(t) = G(t)U_m(t)dt + dV(t)$$
$$U_m(t) \triangleq \text{Col}\,[U(t,x^1), \cdots, U(t,x^m)]$$

where $G(t)$ is an $m \times m$-matrix.

In order for the pointwise observation to be possible, it is necessary from (2.3) that the dimension r of the spatial domain D is equal to one. Note that if $V = H^2(D)$ which is true for the case that $a_{ij}(t,x)$ and $C(t,x)$ are sufficiently smooth on M [4], [16], then r can take the integers such that $1 \leqq r \leqq 3$. Because from (2.3) it is necessary for the pointwise observation to be possible that

$$[r/2] + 1 \leqq 2, \quad \text{that is,} \quad r = 1, 2, 3.$$

To find the optimal estimator described by the stochastic partial differential equations, Schwartz's kernel theorem [10] is reformulated to obtain a more suitable form for the estimation problems in the Hilbert spaces. Hence, the following lemma will be proved.

Lemma 3.1. *The operator $h_1(t) \circ h_2(t)$ possesses the kernel representation given by*

$$(3.9) \qquad (h_1(t) \circ h_2(t))h_3(t) = \int_D H(t, x, y)h_3(t, y)dy$$
$$H(t, x, y) \triangleq h_1(t, x)h_2(t, y)$$

for $\forall h_1(t), h_2(t), h_3(t) \in H$ *and a fixed* $t \in (0, T]$.

Proof. From the definition (1.14) of the operator "$\circ$" we have

$$(h_1(t) \circ h_2(t))h_3(t) = h_1(t)(h_2(t), h_3(t)) = h_1(t, x)\int_D h_2(t, y)h_3(t, y)dy$$
$$= \int_D H(t, x, y)h_3(t, y)dy \; .$$

Hence, the proof of the lemma is completed. Q.E.D.

Lemma 3.2. *The operator $P(\tau | t)A^*(t)$ possesses the following kernel representation*:

$$(3.10) \qquad P(\tau | t)A^*(t)\phi(t) = \int_D (A_y P(\tau, x, y | t))\phi(t, y)dy$$
$$P(\tau, x, y | t) = E[\tilde{U}(\tau, x | t)\tilde{U}(\tau, y | t)]$$

for $\forall \phi(t) \in \mathscr{D}(D)$ *and a fixed* $t, \tau \in (0, T]$ *where*

$$\tilde{U}(\tau, x | t) \triangleq U(\tau, x) - \hat{U}(\tau, x | t) \; .$$

Proof. From the definition (2.30) of $P(\tau | t)$ and the property of (2.17) for the operator "$\circ$", it follows that

$$(P(\tau | t)A^*(t))\phi(t) = E[\tilde{U}(\tau, x | t)\langle A(t)\tilde{U}(\tau | t), \phi(t)\rangle]$$
$$= \int_D E[\tilde{U}(\tau, x | t)(A_y\tilde{U}(\tau, y | t))]\phi(t, y)dy$$
$$= \int_D (A_y P(\tau, x, y | t))\phi(t, y)dy \; .$$

Hence, the proof of the lemma is completed. Q.E.D.

Then applying (3.9) and (3.10) to (2.22), (2.36), (2.40) and (2.60), we have

$$(3.11) \qquad \frac{\partial P(t, x, y | t)}{\partial t} = A_x P(t, x, y | t) + A_y P(t, x, y | t) + D_0(t, x, y) - P_m(t, x | t)D(t)P_m^T(t, y | t)$$

$$\frac{\partial P(\tau, x, y, t)}{\partial t} = A_y P(\tau, x, y, t) - P_m(\tau, x, t) D(t) P_m^T(t, y \mid t) \tag{3.12}$$

for $\tau < t$

$$\frac{\partial P(\tau, x, y \mid t)}{\partial t} = -P_m(\tau, x, t) D(t) P_m^T(\tau, y, t) \qquad \text{for } \tau < t$$

$$\frac{\partial P(\tau, x, y \mid t)}{\partial t} = A_x P(\tau, x, y \mid t) + A_y P(\tau, x, y \mid t) + D_0(t, x, y) \qquad \text{for } \tau > t \tag{3.13}$$

where

$$P_m(t, x \mid t) \triangleq (P(t, x, x^1 \mid t), \cdots, P(t, x, x^m \mid t))$$
$$P_m(\tau, x, t) \triangleq (P(\tau, x, x^1, t), \cdots, P(\tau, x, x^m, t))$$
$$D_0(t, x, y) \triangleq C(t, x) Q(t, x, y) C(t, y) .$$

The boundary conditions of (3.11)–(3.13) are

$$\partial[\cdot]/\partial \boldsymbol{n} = 0 ,$$
$$[\cdot] = P(t, x, \xi \mid t) , \quad P(\tau, x, \xi, t) \quad \text{and} \quad P(\tau, x, \xi \mid t) \quad \text{for } \forall \xi \in \Gamma .$$

We shall consider the existence and uniqueness theorem concerning the solutions of the equations for the estimation error covariance functions and the optimal estimators. For the filtering problem, it was proved [4], [6] that there exists a bounded unique solution $P(t, x, y \mid t)$ of (3.11). Since it is clear from (3.13) that there exists a unique solution of the prediction problem, we shall prove the existence and uniqueness theorem of the solution only for the smoothing problem. It was proved that the evolution operator $\mathscr{U}(t, \tau)$ has the following kernel representation [14], [19]:

$$\mathscr{U}(t, \tau)\phi(\tau) = \int_D \Sigma(t, x, \tau, y)\phi(\tau, y) dy \qquad \text{for } \forall \phi(\tau) \in \mathscr{D}(D) .$$

Since $\mathscr{U}(t, \tau)$ is the evolution operator of $A(t)$, it follows from [11] that

$$\frac{\partial \Sigma(t, x, \tau, y)}{\partial t} = A_x \Sigma(t, x, \tau, y) \tag{3.14}$$

$$\partial \Sigma(t, \xi, \tau, y)/\partial \boldsymbol{n} = 0 \tag{3.15}$$

$$\Sigma(\tau, x, \tau, y) = \delta(x - y) \tag{3.16}$$

$$\int_D |\Sigma(t, x, \tau, y)|\, dy \leqq 1 \qquad \text{for } \forall t \in (\tau, T]\,. \tag{3.17}$$

Then the following theorem can be proved.

Theorem 3.3. *There exists the unique solution of* (3.12) *and the following relation holds*:

$$\begin{aligned} P(\tau, x, y, t) &= \Psi(t, x, y) \\ &\quad - \int_\tau^t \int_D P_m(\tau, x, \sigma) D(\sigma) P_m^T(\sigma, \xi \,|\, \sigma) \Sigma(t, y, \sigma, \xi)\, d\xi\, d\sigma \end{aligned} \tag{3.18}$$

where

$$\Psi(t, x, y) = \int_D P(\tau, x, \tau, \xi) \Sigma(t, y, \tau, \xi)\, d\xi\,.$$

Proof. Define $P^i(\tau, x, y, t)$, $i = 1, 2, \cdots$, by induction as

$$\frac{\partial P^{i+1}(\tau, x, y, t)}{\partial t} = A_y P^{i+1}(\tau, x, y, t) - P_m^i(\tau, x, t) D(t) T_m^T(t, y \,|\, t) \tag{3.19}$$

where

$$P^1(\tau, x, y, t) = \Psi(t, x, y)$$
$$P^i(\tau, x, y, \tau) = P(\tau, x, y, \tau)\,, \qquad i = 1, 2, \cdots\,.$$

Using the Green function $\Sigma(t, x, \sigma, y)$, (3.19) can be written as follows [11]:

$$\begin{aligned} P^{i+1}(\tau, x, y, t) &= \Psi(t, x, y) \\ &\quad - \int_\tau^t \int_D P_m^i(\tau, x, \sigma) D(\sigma) P_m^T(\tau, z \,|\, \sigma) \Sigma(t, y, \sigma, z)\, dz\, d\sigma\,. \end{aligned} \tag{3.20}$$

Letting the function $M_{i+1}(t)$ be

$$M_{i+1}(t) \triangleq \sup_{\substack{\sigma \leqq t \\ x, y}} |P^{i+1}(\tau, x, y, \sigma) - P^i(\tau, x, y, \sigma)|$$

where the supremum is taken for all $\sigma \in (\tau, t]$ and $(x, y) \in D \times D$, then it follows from (3.17), (3.20) and the boundedness of $P(t, x, y \,|\, t)$ that

$$M_{i+1}(t) \leqq K \int_\tau^t M_i(\sigma)\, d\sigma$$

where

$$K \triangleq \sup_{\sigma,x,y} |e^T D(\sigma) e P(\sigma, x, y | \sigma)| , \qquad e \triangleq \text{Col}\,[1, \cdots, 1] .$$

Hence, it follows that

$$M_i(t) \leqq \frac{K^{i-1}(t-\tau)^i}{i!} M , \qquad M \triangleq \sup_{\sigma,x,y} |e^T D(\sigma) e \Psi(\sigma, x, y)| ,$$
$$i = 1, 2, \cdots .$$

Then $M_i(t)$ converges uniformly to zero as i goes to infinity, that is,

$$\text{(3.21)} \qquad \lim_{i \to \infty} P^i(\tau, x, y, t) = P(\tau, x, y, t)$$

exists. Since (3.21) holds uniformly, it is possible to interchange the limit operation and the integration. Hence, it is concluded that $P(\tau, x, y, t)$ satisfies (3.18). Since $P(\tau, x, y, t)$ given by (3.18) satisfies the partial differential equation of (3.12), it is proved that there exists a solution of (3.12). In order to show that $P(\tau, x, y, t)$ satisfying (3.12) is represented by (3.18), we assume that $P(\tau, x, \sigma, \xi)$ satisfies (3.12). Let us define I_σ by

$$I_\sigma \triangleq \int_D \Sigma(t, y, \sigma, \xi) P(\tau, x, \sigma, \xi) d\xi .$$

Differentiating I_σ with respect to σ yields

$$\frac{\partial I_\sigma}{\partial \sigma} = \int_D \left(\frac{\partial \Sigma}{\partial \sigma} P + \Sigma \frac{\partial P}{\partial \sigma} \right) d\xi .$$

On the other hand, from (3.12) and (3.14) it follows that

$$\frac{\partial I_\sigma}{\partial \sigma} = \int_D [-(A_\xi \Sigma) P + \Sigma (A_\xi P) - \Sigma P_m(\tau, x | \sigma) D(\sigma) P_m(\sigma, \xi | \sigma))] d\xi .$$

Using (3.15), the boundary condition of $P(\tau, x, \sigma, \xi)$, and Green's formula yields

$$\frac{\partial I_\sigma}{\partial \sigma} = -\int_D \Sigma(t, y, \sigma, \xi) P_m(\tau, x | \sigma) D(\sigma) P_m(\sigma, \xi | \sigma) d\xi .$$

Integrating with respect to σ from $\sigma = \tau$ to $\sigma = t$ yields

$$I_t = I_\tau - \int_\tau^t \int_D \Sigma(t, y, \sigma, \xi) P_m(\tau, x | \sigma) D(\sigma) P_m(\sigma, \xi | \sigma) d\xi d\sigma .$$

Using (3.16), we have

$$I_t = P(\tau, x, t, y) \ , \quad I_\tau = \Psi(t, x, y) \ .$$

Hence, it is proved that $P(\tau, x, y, t)$ satisfying (3.12) is represented by (3.16). Therefore, we can conclude that (3.12) is equivalent to (3.18). We must prove the uniqueness property of (3.16). If we assume that $P_1(\tau, x, t, y)$ and $P_2(\tau, x, t, y)$ are two solutions of (3.16), then we have from (3.18)

$$\tilde{D}(\tau, x, t, y) = -\int_\tau^t \int_D \Sigma(t, y, \sigma, \xi)\tilde{D}_m(\tau, x \mid \sigma)D(\sigma)P_m(\sigma, \xi \mid \sigma)d\xi d\sigma \tag{3.22}$$

where

$$\tilde{D}(\tau, x, t, y) = P_1(\tau, x, t, y) - P_2(\tau, x, t, y) \ ,$$
$$\tilde{D}_m(\tau, x \mid \sigma) = P_{1m}(\tau, x \mid \sigma) - P_{2m}(\tau, x \mid \sigma) \ .$$

If we let $n(t)$ be

$$n(t) = \sup_{\substack{\sigma \leqq t \\ x, y}} |\tilde{D}(\tau, x, \sigma, y)| \ ,$$

then we have from (3.17) and (3.22)

$$0 \leqq n(t) \leqq K \int_\tau^t n(\sigma)d\sigma \leqq Kn(t)(t - \tau) \ .$$

Hence, we have for $t \in (\tau, \tau + 1/K)$

$$n(t) = 0 \ .$$

Repeating the same procedure by taking $s + 1/K$ as the initial time and so on, we have

$$n(t) = 0 \qquad \text{for } \forall t \in (\tau, T] \ .$$

Therefore, it is proved that the solution of (3.18) is unique.

Hence, the proof of the theorem is completed. Q.E.D.

Since $K(\tau, t, x) = -P_m(\tau, x, t)G^T(t)R^{-1}(t)H(t) \in L^2(\tau, T; \mathscr{L}(V, V'))$, $L(\tau, t, x) = -P_m(\tau, x, t)G^T(t)R^{-1}(t) \in L^\infty(\tau, T; \mathscr{L}(R^m, H))$, and $K(\tau, t, x)$ satisfies (2.10), the next theorem concerning the existence and uniqueness of the solution for smoothing problems follows from Theorem 1.4.

Theorem 3.4. *There exists the unique solution $\hat{U}(\tau, x \mid t)$ of the smoothing estimator* (2.8) *such that*

$$\hat{U}(\tau, x \mid t) \in C(\tau, T; \mathscr{V}') \cap L^\infty(\tau, T; \mathscr{H}) \cap L^2(\tau, T; \mathscr{V}) \ .$$

Theorem 3.4 means that the unique solution of (2.8) exists in the sense of the distribution. It is in general difficult to physically interpret the solution in the sense of the distribution, which may limit the usefulness of Theorem 3.4. However, it is well-known that for the deterministic system the conditions under which the distribution is identified with the ordinary function are given by Nikodym's theorem [19]. Hence, if Nikodym's theorem can be extended to a stochastic system, Theorem 3.4 will become more useful.

§ 4. Conclusions

We derived the optimal estimators by using functional analysis. The fundamental approach to solve the estimation problems is the Wiener-Hopf theory which is originated by Kalman [14], and it enables us to sove the estimation problems except for the smoothing problem. Although we have shown already that it is possible to derive the fixed-point smoothing estimator by using the Wiener-Hopf theory [23], it is still difficult to derive the other types smoothing estimators except for the fixed-point smoothing. The powerful approach to solve the estimation problems is the innovation method which was originated by Kailath [12]. Although Curtain [5], [6] solved the smoothing problems by this approach, the approach also necessitates the Wiener-Hopf theory which is based on the minimum variance estimation problem considered in this paper.

On the other hand, the approach proposed in this paper shows that the three kinds of smoothing estimators are derived by the same procedures as well as the other estimators based on the unbiased and minimum variance estimation error criterion. Furthermore, the approach shows that the innovation process is closely related to the unbiased estimation as shown in (2.12) and (2.27). Therefore, it seems that the approach adopted here is the unified method to solve the estimation problems.

References

[1] Balakrishnan, A. V., Optimal control problems in Banach spaces, SIAM J. Control, 3-8 (1965), 152–180.

[2] ——, Stochastic control, A function space approach, SIAM J. Control, **10**-2 (1972), 285–297.

[3] Balakrishnan, A. V. and Lions, J. L., State estimation for infinite-dimensional systems, Internat. J. Comput. System Sci., 1-4 (1967), 391–403.

[4] Bensoussan, A., Filtrage optimal des systèmes linéaires, Paris, Dunod, 1971.

[5] Curtain, R. F., Estimation theory for abstract evolution equations excited

by general white noise processes, SIAM J. Control and Optimization, 14-6 (1976), 1124–1150.

[6] ——, Infinite dimensional estimation theory for linear systems, Rep. No. 38, Control Theory Centre, Univ. of Warwick, England, (1975), 1–51.

[7] ——, A survey of infinite dimensional filtering, SIAM Rev., 17-3 (1975), 395–411.

[8] Curtain, R. F. and Falb, P. L., Itô's lemma in infinite dimensions, J. Math. Anal. Appl., 31 (1970), 434–448.

[9] Falb, P. L., Infinite-dimensional filtering, The Kalman-Bucy filter in Hilbert space, Information and Control, 11 (1967), 102–137.

[10] Gel'fand, I. M. and Vilenkin, N. Ya., Generalized functions, Vol. 4, New York and London, Academic Press, 1964.

[11] Itô, S., Fundamental solutions of parabolic differential equations and boundary value problems, Japan. J. Math., 27 (1957), 55–102.

[12] Kailath, T. and Frost, P., An innovation approach to least-square estimation Part II: Linear smoothing in additive white noise, IEEE Trans. Automatic Control, AC-13-6 (1968), 655–660.

[13] Kalman, R. E., New approach to linear filtering and prediction problems, Trans. ASME, J. Basic Eng., 82 (1960), 34–45.

[14] Kato, T., Abstract evolution equation of parabolic type in Banach space and Hilbert space, Nagoya Math. J., 19 (1961), 93–125.

[15] Kushner, H. J., Filtering for linear distributed parameter systems, SIAM J. Control, 8-3 (1970), 346–359.

[16] Lions, J. L., Contrôle optimal de systèmes gouvernés par des équations aux dérivées partielles, Paris, Dunod, 1968.

[17] Meditch, J. S., Stochastic optimal linear estimation and control, New York, McGraw-Hill, 1969.

[18] ——, Least-square filtering and smoothing for distributed parameter systems, Automatica, 7 (1971), 315–322.

[19] Mizohata, S., The theory of partial differential equations, London, Cambridge Univ. Press, 1973.

[20] Omatu, S., Sibata, H. and Hata, S., Optimal boundary control for a linear stochastic distributed parameter system using functional analysis, Information and Control, 24-3 (1974), 264–278.

[21] Omatu, S. and Soeda, T., Optimal sensor location in a linear distributed parameter system, Preprint of IFAC Environmental Systems Planning, Design and Control Symposium, Y. Sawaragi ed., (1977), 233–240.

[22] Omatu, S., Tomita, Y. and Soeda, T., Linear fixed-point smoothing by using functional analysis, IEEE Trans. Automatic Control, AC-22-1 (1977), 9–18.

[23] ——, Fixed-point smoothing in Hilbert spaces, to appear in Information and Control, 34-4, (1977), 324–338.

[24] Sakawa, Y., Optimal filtering in linear distributed-parameter systems, Internat. J. Control, 16-1 (1972), 115–127.

[25] Thau, F. E., On optimal filtering for a class of linear distributed-parameter systems, Proc. 1968 JACC, 610–618.

[26] Tomita, Y., Omatu, S. and Soeda, T., An application of the information theory to the estimation problems, Information and Control, 32-2 (1976),

101–111.

[27] ——, An application of the information theory to the fixed-point smoothing problems, Internat. J. Control, **23**-4 (1976), 525–534.

[28] ——, An application of the information theory to the filtering problems, Information Sci., **11**-1 (1976), 13–27.

[29] Tzafestas, S. G. and Nightingale, J. M., Optimal filtering, smoothing and prediction in linear distributed-parameter systems, Proc. IEEE, **115**-8 (1968), 1207–1212.

[30] ——, Maximum-likelihood approach to the optimal filtering of distributed-parameter systems, Proc. IEEE, **116**-6 (1969), 1085–1093.

[31] Tzafestas, S. G., On optimal distributed-parameter filtering and fixed-interval smoothing for colored noise, IEEE trans. Automatic Control, AC-**17**-4 (1972), 448–458.

[32] ——, Bayesian approach to distributed-parameter filtering and smoothing, Internat. J. Control, **15**-2 (1972), 273–295.

[33] Yosida, K., Functional Analysis, New York, Springer-Verlag, 1968.

Department of Information Science
and Systems Engineering
Faculty of Engineering
University of Tokushima
Tokushima 770, Japan

Proc. of Intern. Symp. SDE
Kyoto 1976, pp. 367-383

Convergence of a Class of Markov Chains to Multi-dimensional Degenerate Diffusion Processes

Ken-iti SATO

§1. Introduction

We study a class of Markov chains, which includes multi-allele models of gene frequencies involving mutation, migration and a general type of selection, and prove that the Markov chain converges to a multi-dimensional degenerate diffusion process, as the population size tends to infinity and the time per generation tends to zero. Our Markov chain is a generalization of the induced Markov chain of Karlin and McGregor [6].

For each positive integer N, let $Z^{(N)}(n)$ be a time-homogeneous Markov chain on $\boldsymbol{Z}_+^d$, the set of d-dimensional non-negative lattice points, with time parameter $n = 0, 1, \cdots$. For $j = (j_1, \cdots, j_d) \in \boldsymbol{Z}_+^d$ we use the notation $|j| = \sum_{p=1}^d j_p$. Let $Q_{jk}^{(N)}$ be its one-step transition probability from j to k. We assume that, for every sufficiently large N, $Q_{jk}^{(N)}$ is the coefficient of $s_1^{k_1} \cdots s_d^{k_d}$ in the power series expansion of

$$(1.1)\qquad g(s_1, \cdots, s_d) \prod_{p=1}^d \left\{ f\left(\sum_{q=1}^d (\delta_{pq} + N^{-1}\alpha_{pq}) s_q \right) \right\}^{j_p(1+N^{-1}\gamma_p(|j|^{-1}j))}$$

for every $j = (j_1, \cdots, j_d) \in \boldsymbol{Z}_+^d$ such that $|j| = N$ and for every $k \in \boldsymbol{Z}_+^d$. Here f is the generating function of a probability distribution on $\boldsymbol{Z}_+^1$, that is,

$$(1.2)\qquad f(w) = \sum_{n=0}^\infty c_n w^n \ , \quad c_n \geq 0 \ , \quad \sum_{n=0}^\infty c_n = 1 \ ;$$

γ_p, $p = 1, \cdots, d$, are real-valued functions defined on the set of $x = (x_1, \cdots, x_d) \in \boldsymbol{R}^d$ such that $x_1 \geq 0, \cdots, x_d \geq 0$, $\sum_{p=1}^d x_p = 1$; $\alpha_{pq} \geq 0$ for $p \neq q$, $\alpha_{pp} \leq 0$, $\sum_{q=1}^p \alpha_{pq} = 0$; g is the generating function of a probability distribution on $\boldsymbol{Z}_+^d$; and δ_{pq} is Kronecker's delta. (1.1) can be interpreted as follows. Suppose that $Z^{(N)}(n) = j = (j_1, \cdots, j_d)$, which means that there are j_p particles (or genes) of type A_p ($p = 1, \cdots, d$) in the n-th generation. Particles of the $(n + 1)$-st generation come from two sources: inside and outside. The formation of the contribution from inside to the new generation proceeds in two steps. First, the particles reproduce their children particles independently in such a way that the distribution of the number of particles reproduced by a particle of type A_p has the generating function $f(w)^{1+N^{-1}\gamma_p(|j|-1j)}$. In this step, the types of the children particles are the same as their parents. As the second step, the new particles independently have a chance to change their types, the probability of change from A_p into A_q being $N^{-1}\alpha_{pq}$ for $p \neq q$. In this sense, $N^{-1}\gamma_p$ represents selection force and $N^{-1}\alpha_{pq}$ represents mutation pressure. Contribution of the outside source is immigration independent of the state of the n-th generation. Let b_k, $k = (k_1, \cdots, k_d)$, be the probability of immigration of k_p particles of type A_p $(p = 1, \cdots, d)$. Then

$$(1.3) \qquad g(s_1, \cdots, s_d) = \sum_{k \in Z_+^d} b_k s_1^{k_1} \cdots s_d^{k_d} .$$

Let $J^{(N)}$ be the set of $j \in Z_+^d$ such that $|j| = N$. The sequence of Markov chains that we are interested in is not $Z^{(N)}(n)$ itself, but the Markov chains $X^{(N)}(n)$ on $J^{(N)}$ with one-step transition probability $P_{jk}^{(N)}$ defined by

$$(1.4) \quad P_{jk}^{(N)} = P(Z^{(N)}(n+1) = k \mid Z^{(N)}(n) = j,\ Z^{(N)}(n+1) \in J^{(N)})$$

for $j, k \in J^{(N)}$. Note that we have imposed the condition that the total number of particles in each generation is fixed at N. Let $Y^{(N)}(t)$ and $\tilde{Y}^{(N)}(t)$ be stochastic processes with continuous time parameter defined by

$$(1.5) \quad Y^{(N)}(N^{-1}n) = \tilde{Y}^{(N)}(N^{-1}n) = (N^{-1}X_1^{(N)}(n), \cdots, N^{-1}X_{d-1}^{(N)}(n)) ,$$

$$(1.6) \quad Y^{(N)}(t) = (n+1-Nt)Y^{(N)}(N^{-1}n) + (Nt-n)Y^{(N)}(N^{-1}(n+1)) \quad \text{for } N^{-1}n \le t \le N^{-1}(n+1) ,$$

$$(1.7) \qquad \tilde{Y}^{(N)}(t) = \tilde{Y}^{(N)}(N^{-1}[Nt])$$

where $[Nt]$ is the greatest integer not exceeding Nt. Let K be the set of points $x = (x_1, \cdots, x_{d-1}) \in R^{d-1}$ such that $x_1, \cdots, x_{d-1}$ and $1 - \sum_{p=1}^{d-1} x_p$ are non-negative. Then $Y^{(N)}(\cdot)$ and $\tilde{Y}^{(N)}(\cdot)$ are random elements in $C = C([0, \infty), K)$ and $D = D([0, \infty), K)$, respectively. Hence they induce probability measures $P^{(N)}$ and $\tilde{P}^{(N)}$ on C and D, respectively.

Let $Y^{(N)}(0) = \tilde{Y}^{(N)}(0) = x^{(N)}$, which is non-random. Assume that $x^{(N)} \to x$, as $N \to \infty$. Our main result is that, under some regularity conditions, $P^{(N)}$ and $\tilde{P}^{(N)}$ weakly converge, as $N \to \infty$, to the probability measure P_x induced by a $(d-1)$-dimensional diffusion process on K starting at x with diffusion coefficients

$$a_{pq}(x) = \sigma^2 x_p(\delta_{pq} - x_q)\,, \quad p, q = 1, \cdots, d-1\,, \tag{1.8}$$

and drift coefficients

$$b_p(x) = x_p\Big\{\gamma_p(x) - \sum_{q=1}^{d} x_q\gamma_q(x)\Big\} + \sum_{q=1}^{d} x_q\alpha_{qp} + \mu_p - x_p\sum_{q=1}^{d}\mu_q\,, \qquad p = 1, \cdots, d-1\,, \tag{1.9}$$

where

$$x_d = 1 - \sum_{p=1}^{d-1} x_p \tag{1.10}$$

and

$$\gamma_p(x_1, \cdots, x_{d-1}) = \gamma_p(x_1, \cdots, x_{d-1}, x_d)\,. \tag{1.11}$$

Here σ^2 is a positive constant determined by f; μ_p, $p = 1, \cdots, d$, are non-negative constants determined by g and f. P_x is a probability measure on $\boldsymbol{C}$, but it is extended to $\boldsymbol{D}$ by assigning 0 to the complement of $\boldsymbol{C}$. The result in $\boldsymbol{C}$ generalizes the papers [11] and [12]. The case of no selection was considered in [11]; the case of no mutation and no migration was treated in [12].

The limiting diffusion corresponding to (1.8), (1.9) is uniquely defined as the solution of a suitable stochastic differential equation or, equivalently, of the martingale problem. This is shown in [11] and [12] in special cases and proved by Ethier [4] in general. It never gets out of the set $\boldsymbol{K}$. No boundary condition is necessary; the boundary behavior is implicitly prescribed in a unique way by the stochastic differential equation (or the martingale problem), although a part of the boundary may be regular.

In case $d = 2$, the limiting diffusion has the Kolmogorov backward equation

$$\frac{\partial u}{\partial t} = \frac{\sigma^2}{2} x_1(1 - x_1)\frac{\partial^2 u}{\partial x_1^2} + \{x_1(1 - x_1)(\gamma_1(x_1) - \gamma_2(x_1)) - (\alpha_{12} + \mu_2)x_1 + (\alpha_{21} + \mu_1)(1 - x_1)\}\frac{\partial u}{\partial x_1}\,, \qquad 0 \le x_1 \le 1\,. \tag{1.12}$$

This equation was derived by Fisher in a special case. With some choices of γ_1 and γ_2, it was studied by Feller, Kimura, and others as diffusion approximation to the Wright models in population genetics (see Crow and

Kimura [2] for bibliography). Our result shows that natural generalization of (1.12) to higher dimensions is the equation

$$(1.13)\qquad \frac{\partial u}{\partial t} = \frac{1}{2}\sum_{p,q=1}^{d-1} a_{pq}(x)\frac{\partial^2 u}{\partial x_p \partial x_p} + \sum_{p=1}^{d-1} b_p(x)\frac{\partial u}{\partial x_p}$$

on K with a_{pq} and b_p of (1.8) and (1.9). Zygotic selection corresponds to the choice

$$(1.14)\qquad \gamma_p(x) = \sum_{q=1}^{d} \lambda_{pq} x_q\,, \qquad p = 1, \cdots, d\,,$$

with constants λ_{pq} such that $\lambda_{pq} = \lambda_{qp}$. Gametic selection corresponds to the case where $\gamma_p(x)$, $p = 1, \cdots, d$, are constant functions. In the zygotic selection case, Ethier [3] derived the same equation as (1.13) and proved, among others, convergence in D of the multi-allele Wright model.

If f is Poisson with mean $\lambda > 0$ and g is the direct product of Poisson distributions with means $\mu_p > 0$, that is,

$$f(w) = e^{-\lambda+\lambda_w}\,, \qquad g(s_1, \cdots, s_d) = \prod_{p=1}^{d} e^{-\mu_p+\mu_p s_p}\,,$$

then we have

$$(1.15)\qquad P_{jk}^{(N)} = N!\prod_{p=1}^{d}(k_p!)^{-1}u_p^{k_p}$$

with

$$(1.16)\qquad u_p = c\left\{\mu_p + \lambda\sum_{q=1}^{d}(\delta_{qp} + N^{-1}\alpha_{qp})j_q(1 + N^{-1}\gamma_q(|j|^{-1}j))\right\}$$

where c is a normalizing constant. If $\gamma_p(x)$ are of the form (1.14), then (1.15) is the transition probability of multi-allele Wright model with zygotic selection, mutation, and migration. This is the model studied by Ethier [3]. In case f is binomial distribution, application to Kimura's model of polysomic inheritance is found in Karlin and McGregor [6].

Results related to this paper on convergence in D are found in Guess [5] and Kushner [7]. Norman [9] should also be mentioned in this connection.

As is well known, weak convergence of probability measures on C is strictly stronger than convergence of finite-dimensional distributions. The same is true for D when the limit measure is concentrated on C. It would be possible to apply our results to convergence of functionals such as fixation probabilities and fixation time distributions. Guess [5] already treated them in some cases.

§ 2. Assumptions and results

We give assumptions on regularity of the distributions generated by f and g and define the constants σ^2, μ_p in (1.8), (1.9). In (1.1) we take $s_1, \cdots, s_d$ to be variables in $[0, 1]$. We have implicitly assumed that if N is large and $|j| = N$, then (1.1) is the generating function of a probability distribution on Z^d_+.

Assumption 2.1. If N is large, then, for each $j \in J^{(N)}$ and p, $f(w)^{j_p(1+N^{-1}\gamma_p(N^{-1}j))}$ $(0 \leq w \leq 1)$ is the generating function of a probability distribution on Z^1_+. c_0 is positive. The maximum span of $\{c_n\}$ is 1 (that is, there is no pair of $\gamma > 1$ and δ such that $\sum_n c_{n\gamma+\delta} = 1$). Let $a = f'(1) = \sum_{n=0}^{\infty} nc_n$ (mean), $M(w) = f(e^w)$ (moment generating function), $F(w) = M(w)e^{-w}$, $b = \sup\{w : M(w) < \infty\}$. Then, one of the following holds:

(a) $1 < a \leq +\infty$;
(b) $a = 1$ *and* $b > 0$;
(c) $a < 1$ *and* $\lim_{w \uparrow b} F'(w) > 0$.

Let $K(w) = \log M(w)$ for $w < b$. The following can be proved from this assumption (see [10]): 1° $b > 0$ in Case (c). 2° If $a < 1$, $b > 0$ and $\lim_{w \uparrow b} M(w) = \infty$, then (c) holds. 3° There exists unique $\beta \in (-\infty, b)$ such that $F'(\beta) = 0$. 4° β is negative, zero, positive in Cases (a), (b), (c), respectively. 5° $K'(\beta) = 1$ and $K''(\beta) > 0$. We define $\sigma^2 = K''(\beta)$. Let us define an associated distribution $\{\hat{c}_n : n \in Z^1_+\}$ of $\{c_n\}$ by $\hat{c}_n = c_n e^{n\beta}/M(\beta)$. 6° The distribution $\{\hat{c}_n\}$ has mean 1 and variance σ^2.

Assumption 2.2. $g(e^{\beta+\varepsilon}, \cdots, e^{\beta+\varepsilon}) < \infty$ *for some* $\varepsilon > 0$.

In Case (a), this assumption is automatically satisfied. Let $\{\hat{b}_k : k \in Z^d_+\}$ be an associated distribution of $\{b_k\}$ defined by $\hat{b}_k = b_k e^{|k|\beta}/g(e^\beta, \cdots, e^\beta)$. We define $(\mu_1, \cdots, \mu_d)$ as the mean vector of the distribution $\{\hat{b}_k\}$, that is, $\mu_p = \sum_{k \in Z^d_+} k_p \hat{b}_k$ for $p = 1, \cdots, d$. We write $|\mu| = \sum_{p=1}^d \mu_p$.

Further we make the following

Assumption 2.3. $\gamma_p(x)$, $p = 1, \cdots, d$, *are functions of class* C^5.

Under the above three assumptions, we will prove the following results.

Lemma 2.1. *If N is large enough, then*

(2.1) $P(Z^{(N)}(n+1) \in \boldsymbol{J}^{(N)} \mid Z^{(N)}(n) = j) > 0$ *for all* $j \in \boldsymbol{J}^{(N)}$.

Hence, for sufficiently large N, we can define $P_{jk}^{(N)}$ for $j, k \in \boldsymbol{J}^{(N)}$ by (1.4). We define Markov chains $X^{(N)}(n)$ and continuous time parameter processes $Y^{(N)}(t)$ and $\tilde{Y}^{(N)}(t)$ as in § 1.

Let $\boldsymbol{C}$ be the space of continuous paths $\omega: [0, \infty) \to \boldsymbol{K}$ with the topology of uniform convergence on compact sets of $[0, \infty)$. Let $\boldsymbol{D}$ be the space of paths $\omega: [0, \infty) \to \boldsymbol{K}$ right continuous with left limits. The topology of $\boldsymbol{D}$ is defined as in [8]. Namely, $\omega_n \to \omega$ in $\boldsymbol{D}$ if and only if there is a sequence of continuous strictly increasing functions λ_n of $[0, \infty)$ onto $[0, \infty)$ such that $\sup_{t<\infty} |\lambda_n(t) - t| \to 0$ and, for every $t_0 < \infty$, $\sup_{t \le t_0} |\omega_n(\lambda_n(t)) - \omega(t)| \to 0$. These are complete separable metric topologies of $\boldsymbol{C}$ and $\boldsymbol{D}$. Let $\mathcal{M}$ and $\tilde{\mathcal{M}}$ be the topological σ-algebras of $\boldsymbol{C}$ and $\boldsymbol{D}$, respectively. Let $\mathcal{M}_t$ be the σ-algebra on $\boldsymbol{C}$ generated by $x(s, \omega) = \omega(s)$, $s \le t$.

For $a = (a_{pq})_{p,q=1,\cdots,d-1}$ and $b = (b_p)_{p=1,\cdots,d-1}$ of (1.8) and (1.9), the martingale problem $(\boldsymbol{K}, a, b, x)$ consists of finding a probability measure P_x on $(\boldsymbol{C}, \mathcal{M})$ such that $P_x(x(0) = x) = 1$ and, for every $\theta \in \boldsymbol{R}^{d-1}$, $(M_\theta(t), \mathcal{M}_t, P_x : 0 \le t < \infty)$ is a martingale, where

$$M_\theta(t) = \exp\left\{\langle\theta, x(t) - x(0)\rangle - \int_0^t \langle\theta, x(u)\rangle du - \frac{1}{2}\int_0^t \langle\theta, a(x(u))\theta\rangle du\right\} .$$

For each $x \in \boldsymbol{K}$, there exists one and only one solution P_x of the martingale problem $(\boldsymbol{K}, a, b, x)$. This is a consequence of Ethier [4]. We extend the measure P_x to a measure $\tilde{P}_x$ on $\boldsymbol{D}$ by letting $\tilde{P}_x(\boldsymbol{D} - \boldsymbol{C}) = 0$. Note that $\boldsymbol{C} \in \tilde{\mathcal{M}}$ and that $\mathcal{M}$ coincides with the restriction of $\tilde{\mathcal{M}}$ to $\boldsymbol{C}$. Now we give our results for $P^{(N)}$ and $\tilde{P}^{(N)}$ defined in § 1.

Theorem 2.1. *If $x^{(N)} \to x$ as $N \to \infty$, then $P^{(N)}$ weakly converges to P_x on $\boldsymbol{C}$ as $N \to \infty$.*

Theorem 2.2. *If $x^{(N)} \to x$ as $N \to \infty$, then $\tilde{P}^{(N)}$ weakly converges to $\tilde{P}_x$ on $\boldsymbol{D}$ as $N \to \infty$.*

§ 3. Proof of convergence on C

Lemma 2.1 and Theorem 2.1 are proved by the same method as the theorems in [11] and [12], although computations are more complicated.

Since Lemma 2.1 will be shown in the course of the proof of Theorem 2.1, we assume that N is large enough and we can define $P_{jk}^{(N)}$. Let $\boldsymbol{J}_0^{(N)}$ be the set of points $j = (j_1, \cdots, j_{d-1}) \in \boldsymbol{Z}_+^{d-1}$ such that $\sum_{p=1}^{d-1} j_p \leq N$ and let $\boldsymbol{K}^{(N)} = \{N^{-1}j : j \in \boldsymbol{J}_0^{(N)}\}$. Sometimes we identify j in $\boldsymbol{J}_0^{(N)}$ with $(j_1, \cdots, j_{d-1}, N - \sum_{p=1}^{d-1} j_p)$ in $\boldsymbol{J}^{(N)}$. For $j \in \boldsymbol{J}_0^{(N)}$, let

$$b_p^{(N)}(N^{-1}j) = N \sum_{k \in \boldsymbol{J}^{(N)}} (N^{-1}k_p - N^{-1}j_p) P_{jk}^{(N)} \tag{3.1}$$

$$a_{pq}^{(N)}(N^{-1}j) = N \sum_{k \in \boldsymbol{J}^{(N)}} (N^{-1}k_p - N^{-1}j_p)(N^{-1}k_q - N^{-1}j_q) P_{jk}^{(N)} \tag{3.2}$$

$$e_p^{(N)}(N^{-1}j) = N \sum_{k \in \boldsymbol{J}^{(N)}} (N^{-1}k_p - N^{-1}j_p)^4 P_{jk}^{(N)} . \tag{3.3}$$

By Theorem 3.1 of [11], it suffices to show that

$$\lim_{N\to\infty} \sup_{j \in \boldsymbol{J}_0^{(N)}} |b_p^{(N)}(N^{-1}j) - b_p(N^{-1}j)| = 0 , \qquad p = 1, \cdots, d-1 , \tag{3.4}$$

$$\lim_{N\to\infty} \sup_{j \in \boldsymbol{J}_0^{(N)}} |a_{pq}^{(N)}(N^{-1}j) - a_{pq}(N^{-1}j)| = 0 , \qquad p, q = 1, \cdots, d-1 , \tag{3.5}$$

$$\lim_{N\to\infty} \sup_{j \in \boldsymbol{J}_0^{(N)}} N \sum_{N^{-1}k \in U_\varepsilon^c(N^{-1}j)} P_{jk}^{(N)} = 0 , \qquad \varepsilon > 0 , \tag{3.6}$$

where $U_\varepsilon^c(N^{-1}j)$ is the set of $N^{-1}k$ in $\boldsymbol{K}^{(N)}$ such that $|N^{-1}k - N^{-1}j| \geq \varepsilon$. (3.6) follows from

$$\lim_{N\to\infty} \sup_{j \in \boldsymbol{J}_0^{(N)}} e_p^{(N)}(N^{-1}j) = 0 , \qquad p = 1, \cdots, d-1 . \tag{3.7}$$

We will check (3.4), (3.5), and (3.7).

Given N and $j \in \boldsymbol{J}^{(N)}$, let

$$\begin{aligned}
x &= (x_1, \cdots, x_d) = |j|^{-1}j = N^{-1}j = (N^{-1}j_1, \cdots, N^{-1}j_d) , \\
\gamma_p &= \gamma_p(x) = \gamma_p(N^{-1}j) , \\
y &= \sum_{p=1}^{d} x_p \gamma_p = N^{-1} \sum_{p=1}^{d} j_p \gamma_p(N^{-1}j) , \\
u_p &= \sum_{l=1}^{d} x_l \alpha_{lp} = N^{-1} \sum_{l=1}^{d} j_l \alpha_{lp} , \\
j_p^* &= j_p + x_p \gamma_p .
\end{aligned}$$

Let

$$\alpha_{pq}^{(N)} = \delta_{pq} + N^{-1}\alpha_{pq} .$$

Let

$$\Phi(w, s_1, \cdots, s_d) = \prod_{p=1}^{d} f(ws_p)^{j_p^*} ,$$

$$\Psi(w, s_1, \cdots, s_d) = \Phi\Big(w, \sum_{l=1}^{d} \alpha_{1l}^{(N)} s_l, \cdots, \sum_{l=1}^{d} \alpha_{dl}^{(N)} s_l\Big) ,$$

$$\Theta(w, s_1, \cdots, s_d) = \Psi(w, s_1, \cdots, s_d) g(ws_1, \cdots, ws_d) .$$

Then

$$Q_{jk}^{(N)} = \text{coefficient of } w^N s_1^{k_1} \cdots s_d^{k_d} \text{ in } \Theta(w, s_1, \cdots, s_d) , \qquad k \in \boldsymbol{J}^{(N)} . \tag{3.8}$$

As in [12], we use the following functions:

$$\begin{aligned}
h_1(w) &= f(w)^{N+y} , \\
h_2(w) &= f(w)^{N+y-1} f'(w) w , \\
h_3(w) &= f(w)^{N+y-2} f'(w)^2 w^2 , \\
h_4(w) &= f(w)^{N+y-1} f''(w) w^2 , \\
h_5(w) &= f(w)^{N+y-3} f'(w)^3 w^3 , \\
h_6(w) &= f(w)^{N+y-2} f'(w) f''(w) w^3 , \\
h_7(w) &= f(w)^{N+y-1} f'''(w) w^3 , \\
h_8(w) &= f(w)^{N+y-4} f'(w)^4 w^4 , \\
h_9(w) &= f(w)^{N+y-3} f'(w)^2 f''(w) w^4 , \\
h_{10}(w) &= f(w)^{N+y-2} f''(w)^2 w^4 , \\
h_{11}(w) &= f(w)^{N+y-2} f'(w) f'''(w) w^4 , \\
h_{12}(w) &= f(w)^{N+y-1} f''''(w) w^4 .
\end{aligned}$$

These are power series of w. Let

$$A_\nu(N) = \text{coefficient of } w^N \text{ in } h_\nu(w) g(w, \cdots, w) ,$$

$$A_{\nu p_1 \cdots p_m}(N) = \text{coefficient of } w^N \text{ in } h_\nu(w) D_{p_1 \cdots p_m} g(w, \cdots, w) w^m ,$$

where

$$D_{p_1 \cdots p_m} = \frac{\partial^m}{\partial s_{p_1} \cdots \partial s_{p_m}} .$$

Since

$$P(Z^{(N)}(n+1) \in \boldsymbol{J}^{(N)} \mid Z^{(N)}(n) = j) = A_1(N) , \tag{3.9}$$

we have

(3.10) $$P_{jk}^{(N)} = A_1(N)^{-1}Q_{jk}^{(N)}\ .$$

Let

$$G(s_1, \cdots, s_d) = \sum_{k \in J(N)} P_{jk}^{(N)} s_1^{k_1} \cdots s_d^{k_d}.$$

By (3.8) and (3.9),

$$G(s_1, \cdots, s_d) = A_1(N)^{-1}(\text{coefficient of } w^N \text{ in } \Theta(w, s_1, \cdots, s_d))\ .$$

Let

$$C_{p_1 \cdots p_m} = D_{p_1 \cdots p_m} G(1, \cdots, 1)\ .$$

we have

(3.11) $$C_{p_1 \cdots p_m} = A_1(N)^{-1}(\text{coefficient of } w^N \text{ in } D_{p_1 \cdots p_m}\Theta(w, 1, \cdots, 1))\ .$$

We need to estimate C_p, C_{pq}, C_{ppp}, and C_{pppp} for large N. Their estimation is reduced to the estimation of $A_\nu(N)$ and $A_{\nu p_1 \cdots p_m}(N)$.

It follows from Assumption 2.1 that $h_\nu(e^w)$, $\nu = 1, \cdots, 12$, have analytic extensions $h_\nu(e^z)$ for complex z with $\operatorname{Re} z < b$. By Assumption 2.2, $g(e^w, \cdots, e^w)$ has an analytic extension $g(e^z, \cdots, e^z)$ for complex z with $\operatorname{Re} z < \beta + \varepsilon$. It is easy to see that

(3.12) $$A_\nu(N) = (2\pi i)^{-1} \int_{\beta - i\pi}^{\beta + i\pi} h_\nu(e^z) g(e^z, \cdots, e^z) e^{-Nz} dz\ ,$$

(3.13) $$A_{\nu p_1 \cdots p_m}(N) = (2\pi i)^{-1} \int_{\beta - i\pi}^{\beta + i\pi} h_\nu(e^z) D_{p_1 \cdots p_m} g(e^z, \cdots, e^z) e^{mz - Nz} dz\ ,$$

where the integrals are along the line segment from $\beta - i\pi$ to $\beta + i\pi$. Lemma 3.1 of [12] applies to estimation of the right-hand sides of (3.12), (3.13).

Lemma 3.1.

(3.14) $$C_p = j_p + x_p(\gamma_p - y - |\mu|) + u_p + \mu_p + O(N^{-1})$$

uuiformly in $j \in J^{(N)}$,

(3.15) $$b_p^{(N)}(x) = x_p(\gamma_p - y - |\mu|) + u_p + \mu_p + O(N^{-1})$$

uniformly in $j \in J^{(N)}$,

as $N \to \infty$.

What we mean by (3.14) is

$$\limsup_{N\to\infty} \sup_{j\in J(N)} N|C_p - \{j_p + x_p(\gamma_p - y - |\mu|) + u_p + \mu_p\}| < \infty .$$

(3.15) and similar expressions are understood in the same way.

Proof. We have

$$D_p\Theta(w, s_1, \cdots, s_d) = D_p\Psi(w, s_1, \cdots, s_d)g(ws_1, \cdots, ws_d) + \Psi(w, s_1, \cdots, s_d)D_pg(ws_1, \cdots, ws_d)w ,$$

$$D_p\Psi(w, s_1, \cdots, s_d) = \sum_l D_l\Phi\left(w, \sum_m \alpha_{1m}^{(N)}s_m, \cdots, \sum_m \alpha_{dm}^{(N)}s_m\right)\alpha_{lp}^{(N)} ,$$

$$D_l\Phi(w, s_1, \cdots, s_d) = f(ws_1)^{j_1^*}\cdots\{j_l^*f(ws_l)^{j_l^*-1}f'(ws_l)w\}\cdots f(ws_d)^{j_d^*} .$$

Hence $D_l\Phi(w, 1, \cdots, 1) = j_l^*h_2(w)$ and

$$D_p\Psi(w, 1, \cdots, 1) = \left(j_p^* + u_p + N^{-1}\sum_l x_l\gamma_l\alpha_{lp}\right)h_2(w) . \tag{3.16}$$

Noting that

$$\Psi(w, 1, \cdots, 1) = \Phi(w, 1, \cdots, 1) = h_1(w) , \tag{3.17}$$

we have

$$D_p\Theta(w, 1, \cdots, 1) = \left(j_p^* + u_p + N^{-1}\sum_l x_l\gamma_l\alpha_{lp}\right)h_2(w)g(w, \cdots, w) + h_1(w)D_pg(w, \cdots, w)w .$$

Therefore we have, by (3.11),

$$C_p = A_1(N)^{-1}\left\{\left(j_p + x_p\gamma_p + u_p + N^{-1}\sum_l x_l\gamma_l\alpha_{lp}\right)A_2(N) + A_{1p}(N)\right\} . \tag{3.18}$$

Let $L(s_1, \cdots, s_d) = g(e^{s_1}, \cdots, e^{s_d})$. Let $M(z)$, $K(z)$, and $L(z_1, \cdots, z_d)$ be analytic extensions of $M(w)$, $K(w)$, and $L(s_1, \cdots, s_d)$. Let

$$K(z) = \sum_{n=0}^{\infty} \kappa_n(z-\beta)^n , \qquad L(z, \cdots, z) = \sum_{n=0}^{\infty} \rho_n(z-\beta)^n$$

in a neighborhood of β. We see that $\kappa_0 = K(\beta)$, $\kappa_1 = 1$, $\kappa_2 = 2^{-1}\sigma^2$, and

$$\rho_0 = g(e^\beta, \cdots, e^\beta) = \sum_k b_k e^{|k|\beta} ,$$

$$\rho_1 = \rho_0 \sum_k |k|\, b_k e^{|k|\beta} = \rho_0 \sum_k |k|\, \hat{b}_k = \rho_0 |\mu| .$$

From now on, the integral sign denotes the line integral along the segment from $\beta - i\pi$ to $\beta + i\pi$. By (3.12) and (3.13),

$$A_1(N) = (2\pi i)^{-1} \int M(z)^{N+y} L(z, \cdots, z) e^{-Nz} dz ,$$

$$A_2(N) = (2\pi i)^{-1} \int M(z)^{N+y-1} M'(z) L(z, \cdots, z) e^{-Nz} dz ,$$

$$A_{1p}(N) = (2\pi i)^{-1} \int M(z)^{N+y} D_p L(z, \cdots, z) e^{-Nz} dz .$$

Using Lemma 3.1 of [12], we get

(3.19) $$A_1(N) = \Delta_N e^{yK(\beta)}(\rho_0 + N^{-1}a_1 + O(N^{-2})) ,$$

(3.20) $$A_2(N) = \Delta_N e^{yK(\beta)}(\rho_0 + N^{-1}a_2 + O(N^{-2})) ,$$

(3.21) $$A_{1p}(N) = \Delta_N e^{yK(\beta)}(\rho_0 \mu_p + O(N^{-1}))$$

uniformly in y, where

$$\begin{aligned} \Delta_N &= \sigma^{-1}(2\pi N)^{-1/2} e^{N(K(\beta)-\beta)} , \\ a_1 &= 2^{-1}\rho_0(6\sigma^{-4}\kappa_4 - 15\sigma^{-6}\kappa_3^2 - y - \sigma^{-2}y^2 + 6\sigma^{-4}y\kappa_3) \\ &\quad + \rho_1(3\sigma^{-4}\kappa_3 - \sigma^{-2}y) - \rho_2\sigma^{-2} , \\ a_2 &= a_1 - \rho_0 y - \rho_1 . \end{aligned}$$

In getting a_2, we have used that

$$\begin{aligned} M(z)^{-1}M'(z)L(z, \cdots, z) = \rho_0 &+ (\rho_0\sigma^2 + \rho_1)(z - \beta) \\ &+ (3\rho_0\kappa_3 + \rho_1\sigma^2 + \rho_2)(z - \beta)^2 + \cdots . \end{aligned}$$

Now (3.18)–(3.21) together yield (3.14). (3.15) follows from (3.14) since $b_p^{(N)}(x) = C_p - j_p$. This proves the lemma.

Incidentally, we have proved Lemma 2.1 by (3.9) and (3.19).

Lemma 3.2.

(3.22) $$\begin{aligned} C_{pq} = j_p j_q - N^{-1} j_p j_q(\sigma^2 + 2y + 2|\mu|) &+ j_p(x_q\gamma_q + u_q + \mu_q) \\ + j_q(x_p\gamma_p + u_p + \mu_p) &+ \delta_{pq} j_p(\sigma^2 - 1) + O(1) \end{aligned}$$

uniformly in $j \in \boldsymbol{J}^{(N)}$,

(3.23) $$a_{pq}^{(N)}(x) = \sigma^2 x_p(\delta_{pq} - x) + O(N^{-1}) \qquad \textit{uniformly in } j \in \boldsymbol{J}^{(N)} .$$

Proof. By (3.11), we have to estimate coefficients in $D_{pq}\Theta(w, 1, \cdots, 1)$. We have

$$D_{pq}\Theta(w, 1, \cdots, 1) = D_{pq}\Psi(w, 1, \cdots, 1)g(w, \cdots, w) + D_p\Psi(w, 1, \cdots, 1)D_q g(w, \cdots, w)w + D_q\Psi(w, 1, \cdots, 1)D_p g(w, \cdots, w)w + \Psi(w, 1, \cdots, 1)D_{pq}g(w, \cdots, w)w^2 .$$

Calculating $D_{pq}\Psi(w, s_1, \cdots, s_d)$ and letting $s_1 = \cdots = s_d = 1$, we get

$$(3.24) \qquad D_{pq}\Psi(w, 1, \cdots, 1) = \Big\{\sum_{l,m} j_l^* j_m^* \alpha_{lp}^{(N)} \alpha_{mq}^{(N)} - \sum_l j_l^* \alpha_{lp}^{(N)} \alpha_{lq}^{(N)}\Big\} h_3(w) + \sum_l j_l^* \alpha_{lp}^{(N)} \alpha_{lq}^{(N)} h_4(w) .$$

Noting that

$$\sum_l j_l^* \alpha_{lp}^{(N)} = j_p + x_p\gamma_p + u_p + O(N^{-1}) , \qquad \sum_l j_l^* \alpha_{lp}^{(N)} \alpha_{lq}^{(N)} = \delta_{pq} j_p + O(1)$$

and using (3.16), (3.17), (3.24), we see that

$$C_{pq} = A_1(N)^{-1}\{(j_p j_q + j_p x_q \gamma_q + j_p u_q + j_q x_p \gamma_p + j_q u_p)A_3(N) + j_p A_{2q}(N) + j_q A_{2p}(N) + \delta_{pq}(-j_p A_3(N) + j_p A_4(N))\} + O(1) .$$

Now

$$A_3(N) = (2\pi i)^{-1} \int M(z)^{N+y-2} M'(z)^2 L(z, \cdots, z) e^{-Nz} dz ,$$

$$A_4(N) = (2\pi i)^{-1} \int M(z)^{N+y-1} (M''(z) - M'(z)) L(z, \cdots, z) e^{-Nz} dz ,$$

$$A_{2p}(N) = (2\pi i)^{-1} \int M(z)^{N+y-1} M'(z) D_p L(z, \cdots, z) e^{-Nz} dz ,$$

and expansions around β are

$$M(z)^{-2} M'(z)^2 L(z, \cdots, z) = \rho_0 + (2\sigma^2\rho_0 + \rho_1)(z - \beta) + (6\kappa_3\rho_0 + \sigma^4\rho_0 + 2\sigma^2\rho_1 + \rho_2)(z - \beta)^2 + \cdots ,$$

$$M(z)^{-1}(M''(z) - M'(z))L(z, \cdots, z) = \rho_0\sigma^2 + \cdots ,$$

$$M(z)^{-1} M'(z) D_p L(z, \cdots, z) = \rho_0 \mu_p + \cdots .$$

It follows that

$$A_3(N) = \Delta_N e^{yK(\beta)}(\rho_0 + N^{-1}a_3 + O(N^{-2})) ,$$
$$a_3 = a_1 - \rho_0(\sigma^2 + 2y) - 2\rho_1 ,$$
$$A_4(N) = \Delta_N e^{yK(\beta)}(\rho_0\sigma^2 + O(N^{-1})) ,$$
$$A_{2p}(N) = \Delta_N e^{yK(\beta)}(\rho_0\mu_p + O(N^{-1})) .$$

Combining these estimates, we can conclude (3.22). Since

$$a_{pq}^{(N)}(x) = N^{-1}(C_{pq} - j_pC_q - j_qC_p + j_pj_q + \delta_{pq}C_p) ,$$

(3.23) follows from (3.22).

Lemma 3.3.

$$(3.25)\quad C_{ppp} = j_p^3 + 3j_p^2(\sigma^2 - 1 + u_p + \mu_p + x_p\gamma_p - x_p(\sigma^2 + y + |\mu|)) + O(N) \qquad \textit{uniformly in } j \in \boldsymbol{J}^{(N)} .$$

Proof. Fix p and let

$$U_{mn} = A_1(N)^{-1}\Big(\text{coefficient of } w^N \text{ in } D_{\underbrace{p \dots p}_{m}}\Psi(w, 1, \cdots, 1)D_{\underbrace{p \dots p}_{n}}g(w, \cdots, w)w^n\Big) .$$

By (3.11),

$$C_{ppp} = U_{30} + 3U_{21} + 3U_{12} + U_{03} .$$

Calculating $D_{ppp}\Psi(w, s_1, \cdots, s_d)$ as in [11], we get

$$U_{30} = A_1(N)^{-1}\{(S_1 - 3S_2 + 2S_3)A_5(N) + (3S_2 - 3S_3)A_6(N) + S_3A_7(N)\}$$

with

$$S_1 = \Big(\sum_l j_l^*\alpha_{lp}^{(N)}\Big)^3 = j_p^3 + 3j_p^2(u_p + x_p\gamma_p) + O(N) ,$$
$$S_2 = \sum_{l,m} j_l^*j_m^*(\alpha_{lp}^{(N)})^2\alpha_{mp}^{(N)} = j_p^2 + O(N) ,$$
$$S_3 = \sum_l j_l^*(\alpha_{lp}^{(N)})^3 = O(N) .$$

By the observation in the proof of Lemmas 3.1 and 3.2,

$$U_{21} = j_p^2A_1(N)^{-1}A_{3p}(N) + O(N) , \quad U_{12} = O(N) , \quad U_{03} = O(1) .$$

It follows that

$$C_{ppp} = A_1(N)^{-1}\{(j_p^3 + 3j_p^2(u_p + x_p\gamma_p - 1))A_5(N) + 3j_p^2 A_6(N) + 3j_p^2 A_{3p}(N)\} + O(N) .$$

We have

$$A_5(N) = (2\pi i)^{-1}\int M(z)^{N+y-3}M'(z)^3 L(z, \cdots, z)e^{-Nz}dz$$
$$= \Delta_N e^{yK(\beta)}(\rho_0 + N^{-1}a_5 + O(N^{-2})) ,$$
$$a_5 = a_1 - 3\rho_0(y + \sigma^2) - 3\rho_1 ,$$

using

$$M^{-3}(z)M'(z)^3 L(z, \cdots, z) = \rho_0 + (3\sigma^2\rho_0 + \rho_1)(z - \beta) + (3\sigma^4\rho_0 + 9\kappa_3\rho_0 + 3\sigma^2\rho_1 + \rho_2)(z - \beta)^2 + \cdots ,$$

and

$$A_6(N) = (2\pi i)^{-1}\int M(z)^{N+y-2}M'(z)(M''(z) - M'(z))L(z, \cdots, z)e^{-Nz}dz$$
$$= \Delta_N e^{yK(\beta)}(\sigma^2\rho_0 + O(N^{-1})) ,$$
$$A_{3p}(N) = (2\pi i)^{-1}\int M(z)^{N+y-2}M'(z)^2 D_p L(z, \cdots, z)e^{-Nz}dz$$
$$= \Delta_N e^{yK(\beta)}(\rho_0\mu_p + O(N^{-1})) .$$

These estimates together prove Lemma 3.3.

Lemma 3.4.

$$C_{pppp} = j_p^4 + 2j_p^3(3\sigma^2 - 3 + 2u_p + 2\mu_p + 2x_p\gamma_p - x_p(3\sigma^2 + 2y + 2|\mu|)) + O(N^2) \tag{3.26}$$

uniformly in $j \in J^{(N)}$.

Proof. By (3.11),

$$C_{pppp} = U_{40} + 4U_{31} + 6U_{22} + 4U_{13} + U_{04} .$$

Calculation of $D_{pppp}\Psi(w, s_1, \cdots, s_d)$ similar to that in [11] yields

$$U_{40} = A_1(N)^{-1}\Big\{\Big(\sum_l j_l^*\alpha_{lp}^{(N)}\Big)^4 A_8(N) + 6\Big(\sum_l j_l^*(\alpha_{lp}^{(N)})^2\Big)\Big(\sum_l j_l^*\alpha_{lp}^{(N)}\Big)^2(A_9(N) - A_8(N))\Big\} + O(N^2) .$$

Hence

$$U_{40} = A_1(N)^{-1}\{(j_p^4 + 4j_p^3(x_p\gamma_p + u_p) - 6j_p^3)A_8(N) + 6j_p^3A_9(N)\} + O(N^2) .$$

Calculation of U_{31} is almost the same as that of U_{30} and we get

$$U_{31} = j_p^3A_1(N)^{-1}A_{5p}(N) + O(N^2) .$$

Further we have $U_{22} = O(N^2)$, $U_{13} = O(N)$, $U_{04} = O(1)$. Hence,

$$C_{pppp} = A_1(N)^{-1}\{(j_p^4 + 2j_p^3(2x_p\gamma_p + 2u_p - 3))A_8(N) + 6j_p^3A_9(N) + 4j_p^3A_{5p}(N)\} + O(N^2) .$$

This time we need

$$A_8(N) = (2\pi i)^{-1}\int M(z)^{N+y-4}M'(z)^4L(z, \cdots, z)e^{-Nz}dz = \Delta_N e^{yK(\beta)}(\rho_0 + N^{-1}a_8 + O(N^{-2})) ,$$
$$a_8 = a_1 - 2\rho_0(3\sigma^2 + 2y) - 4\rho_1 ,$$

which follows from

$$M(z)^{-4}M'(z)^4L(z, \cdots, z) = \rho_0 + (4\sigma^2\rho_0 + \rho_1)(z - \beta) + (12\kappa_3\rho_0 + 6\sigma^4\rho_0 + 4\sigma^2\rho_1 + \rho_2)(z - \beta)^2 + \cdots ,$$

and

$$A_9(N) = (2\pi i)^{-1}\int M(z)^{N+y-3}M'(z)^2(M''(z) - M'(z))L(z, \cdots, z)e^{-Nz}dz = \Delta_N e^{yK(\beta)}(\sigma^2\rho_0 + O(N^{-1})) ,$$

$$A_{5p}(N) = (2\pi i)^{-1}\int M(z)^{N+y-3}M'(z)^3D_pL(z, \cdots, z)e^{-Nz}dz = \Delta_N e^{yK(\beta)}(\rho_0\mu_p + O(N^{-1})) .$$

Now (3.26) follows and the proof is complete.

Let us finish the proof of Theorem 2.1. We have defined $e_p^{(N)}(x)$ by (3.3). It is expressed by

$$e_p^{(N)}(x) = N^{-3}\{C_{pppp} + (-4j_p + 6)C_{ppp} + (6j_p^2 - 12j_p + 7)C_{pp} + (-4j_p^3 + 6j_p^2 - 4j_p + 1)C_p + j_p^4\}$$

as in [11]. Use Lemmas 3.1–3.4. By a careful algebra we see that all terms except terms of $O(N^{-1})$ cancel out. We thus have

$$(3.27) \qquad e_p^{(N)}(x) = O(N^{-1}) \qquad \text{uniformly in } j \in \boldsymbol{J}^{(N)}\,.$$

(3.15) and (3.23) in Lemmas 3.1 and 3.2 show that

$$(3.28) \qquad b_p^{(N)}(x) = b_p(x) + O(N^{-1}) \qquad \text{uniformly in } j \in \boldsymbol{J}^{(N)}\,,$$

$$(3.29) \qquad a_{pq}^{(N)}(x) = a_{pq}(x) + O(N^{-1}) \qquad \text{uniformly in } j \in \boldsymbol{J}^{(N)}\,.$$

Hence we have obtained (3.4), (3.5), (3.7) together with estimate of speed of convergence. This is more than needed for the proof of Theorem 2.1.

§ 4. Proof of convergence on D

We can prove Theorem 2.2 along the same line as the proof of Theorem 2.1, because we can prove in $\boldsymbol{D}$ an invariance principle analogous to Theorem 3.1 of [11]. But Theorem 2.2 is a consequence of Theorem 2.1, which is shown below.

For each N let φ_N be a mapping of $\boldsymbol{C}$ into $\boldsymbol{D}$ defined by $\varphi_N(\omega)(t) = \omega(N^{-1}[Nt])$ for $\omega \in \boldsymbol{C}$. Then $\tilde{Y}^{(N)}(t) = \varphi_N(Y^{(N)}(\cdot))(t)$ and hence $\tilde{P}^{(N)} = P^{(N)}\varphi_N^{-1}$. Let φ be the identity mapping of $\boldsymbol{C}$ into $\boldsymbol{D}$. We have $\tilde{P}_x = P_x\varphi^{-1}$. If $\omega_N \to \omega$ in $\boldsymbol{C}$, then $\varphi_N(\omega_N) \to \varphi(\omega)$, because

$$\begin{aligned} &|\varphi_N(\omega_N)(t) - \varphi(\omega)(t)| \\ &\qquad \le |\omega_N(N^{-1}[Nt]) - \omega(N^{-1}[Nt])| + |\omega(N^{-1}[Nt]) - \omega(t)|\,. \end{aligned}$$

Theorefore, by Theorem 5.5 of Billingsley [1], $P^{(N)} \to P$ implies $\tilde{P}^{(N)} \to \tilde{P}_x$.

Finally we remark that the same method can be applied to the proof that $\tilde{P}^{(N)} \to \tilde{P}_x$ implies $P^{(N)} \to P_x$.

References

[1] Billingsley, P., Convergence of probability measures, John Wiley, New York, 1968.

[2] Crow, J. F. and Kimura, M., An introduction to population genetics theory, Harper and Row, New York, 1970.

[3] Ethier, S. N., An error estimate for the diffusion approximation in population genetics, Ph.D. Thesis, University of Wisconsin, 1975.

[4] ——, A class of degenerate diffusion processes occurring in population genetics, Comm. Pure Appl. Math., 29 (1976), 483–493.

[5] Guess, H. A., On the weak convergence of Wright-Fisher models, Stochastic Processes Appl., 1 (1973), 287–306.

[6] Karlin, S. and McGregor, J., Direct product branching processes and related Markov chains, Proc. Nat. Acad. Sci. U.S.A., **51** (1964), 598–602.

[7] Kushner, H. J., On the weak convergence of interpolated Markov chains to a diffusion, Ann. Probability, **2** (1974), 40–50.

[8] Lindvall, T., Weak convergence of probability measures and random functions in the function space $D[0, \infty)$, J. Appl. Probability, **10** (1973), 109–121.

[9] Norman, M. F., Diffusion approximation of non-Markovian processes, Ann. Probability, **3** (1975), 358–364.

[10] Sato, K., Asymptotic properties of eigenvalues of a class of Markov chains induced by direct product branching processes, J. Math. Soc. Japan, **28** (1976), 192–211.

[11] ——, Diffusion processes and a class of Markov chains related to population genetics, Osaka J. Math., **13** (1976), 631–659.

[12] ——, A class of Markov chains related to selection in population genetics, J. Math. Soc. Japan, **28** (1976), 621–636.

College of Liberal Arts
Kanazawa University
Kanazawa 920, Japan

Proc. of Intern. Symp. SDE
Kyoto 1976, pp. 385–395

Construction of a Solution of Linear Stochastic Evolution Equations on a Hilbert Space

Akinobu SHIMIZU

Let us consider the stochastic evolution equation

$$(1) \qquad du_t = -Au_t dt + Bu_t dB_t ,$$

taking values in a real separable Hilbert space $\mathscr{H}$. Here, $-A$ is the infinitesimal generator of a semi-group on $\mathscr{H}$, B is a linear mapping of a subspace of $\mathscr{H}$ into the space of Hilbert-Schmidt operators on H, and $\{B_t\}$ is a cylindrical Brownian motion on $\mathscr{H}$. The aim of this article is to obtain an explicit expression of the solution of (1) under suitable conditions of A and B.

For our purpose, it is convenient to represent the mapping B in the form

$$(2) \qquad (Bu)v = \sum_{k=1}^{\infty} (v, e_k)B_k u , \qquad \forall u \in D(B) , \quad \forall v \in \mathscr{H} ,$$

where $\{e_k\}$ is an orthonormal system of $\mathscr{H}$, B_k are linear operators on $\mathscr{H}$, and $\mathscr{D}(B)$ equals the space

$$\left\{ u \in H ; u \in D(B_k),\ k = 1, 2, \cdots, \text{ and } \sum_{k=1}^{\infty} \|B_k u\|^2 < \infty \right\} .$$

First, we assume in § 1 that the operators B_k are bounded self-adjoint operators with common resolution of the identity, and that e^{-tA} and B_k are commutative. The existence and uniqueness of the solution is known in this case. Under some additional conditions, we shall obtain an explicit expression of the solution of the equation (1) (Theorem 1). Next, we assume in § 2 that the operator A equals zero, and that the operators B_k are unbounded self-adjoint operators with common resolution of the identity. We shall establish the existence and uniqueness of the solution of the equation (1). In this case, we can prove the existence, getting an explicit solution of the equation (1) (Theorem 3).

§ 1. The case that the operators B_k are bounded self-adjoint

Let $\{T_t\}_{t\geq 0}$ be a strongly continuous semigroup on $\mathscr{H}$ with infinitesimal generator $-A$, and $\|T_t\| \leqq e^{\alpha t}$, $\alpha > 0$. In this section, we regard the equation (1) as the following equation,

$$(3) \qquad u_t = T_t u_0 + \sum_{k=1}^{\infty} \int_0^t T_{t-s} B_k u_s dB_s(e_k) .$$

Here, $B_t(e_k)$, $k = 1, 2, \cdots$, are independent one-dimensional Brownian motions, which is deduced from the definition of a cylindrical Brownian motion [9].

We assume that the operators B_k are represented in the form

$$(4) \qquad B_k = \int_0^{\infty} f_k(\lambda) dE_\lambda ,$$

where $\{E_\lambda\}$ is a resolution of the identity, that the equality

$$(5) \qquad T_t E_\lambda = E_\lambda T_t , \qquad \text{for any } t \geqq 0 , \lambda \geqq 0 ,$$

holds, and that there exists a sequence $\{a_k\} \in l_2$ such that

$$(6) \qquad \sum_{k=1}^{\infty} \frac{|f_k(\lambda)|^2}{a_k^2} \text{ is bounded in } \lambda \in [0, +\infty) .$$

Now, we will try to solve the equation (3) by the successive approximation. We set

$$u_t^0 = T_t u_0 ,$$

$$u_t^n = T_t u_0 + \int_0^t T_{t-s} B u_s^{n-1} dB_s ,$$

$$v_t^n = u_t^n - u_t^{n-1} , \qquad n \geqq 1 ,$$

and

$$v_t^0 = u_t^0 .$$

The sum $\sum_{n=0}^{\infty} v_t^n$ is the solution of the equation (3), which should be expressed explicitly. By the definition of the stochastic integral and (5), we can easily verify the equality,

$$(7) \qquad v_t^n = T_t \sum_{k_1, \cdots, k_n = 1}^{\infty} \int_0^t \int_0^{s_1} \cdots \int_0^{s_{n-1}} B_{k_1} B_{k_2} \cdots B_{k_2} \cdots B_{k_n} u_0 dB(e_{k_1}) \cdots dB(e_{k_n}) .$$

We can prove without difficulty

Lemma 1. *The equality*

$$\sum_{k_1,\cdots,k_n=1}^{\infty}\int_0^t\int_0^{s_1}\cdots\int_0^{s_{n-1}} B_{k_1}B_{k_2}\cdots B_{k_n}u_0 dB(e_{k_1})\cdots dB(e_{k_n}) = \int_0^\infty \|y_\lambda\|^n H_n\left(B_t\left(\frac{y_\lambda}{\|y_\lambda\|}\right);t\right)dE_\lambda u_0 , \tag{8}$$

holds, where $H_n(z;t)$ is the Hermite polynomial of degree n with parameter t, and

$$y_\lambda = \sum_{k=1}^{\infty} f_k(\lambda)e_k . \tag{9}$$

We set

$$S_N^t(\lambda) = 1 + \sum_{k=1}^{N}\|y_\lambda\|^k H_k\left(B_t\left(\frac{y_\lambda}{\|y_\lambda\|}\right);t\right) \tag{10}$$

$$S_\infty^t(\lambda) = 1 + \sum_{k=1}^{\infty}\|y_\lambda\|^k H_k\left(B_t\left(\frac{y_\lambda}{\|y_\lambda\|}\right);t\right) \tag{11}$$

Since $B_t(y_\lambda/\|y_\lambda\|)$ is a one-dimensional Brownian motion for each λ, we get

$$\begin{aligned} S_\infty^t(\lambda) &= \exp\left[\|y_\lambda\|\,B_t\left(\frac{y_\lambda}{\|y_\lambda\|}\right) - \frac{1}{2}t\,\|y_\lambda\|^2\right] \\ &= \exp\left[B_t(y_\lambda) - \frac{1}{2}t\,\|y_\lambda\|^2\right] . \end{aligned} \tag{12}$$

By (7), (8) and (10), we have

$$\sum_{n=0}^{N} v_t^n = T_t\int_0^\infty S_N^t(\lambda)dE_\lambda u_0 . \tag{13}$$

Lemma 2. *Under the assumption* (6), *the operator* $\int_0^\infty S_\infty^t(\lambda)dE_\lambda$ *is a bounded operator on H with probability* 1.

Proof. It is sufficient to show that $B_t(y_\lambda)$ is bounded in $(\lambda, t)\in[0,\infty)\times[0,T]$ for any finite T. Since we have $B_t(y_\lambda) = \sum_{k=1}^\infty f_k(\lambda)B_t(e_k)$, we get

$$(14) \qquad B_t(y_\lambda) \leqq \sqrt{\sum_{k=1}^{\infty} \frac{|f_k(\lambda)|^2}{a_k^2}} \sqrt{\sum_{k=1}^{\infty} a_k^2 B_t(e_k)^2} \,.$$

By (6), the continuity of paths and (14), we get the conclusion.

Now, we will state our theorem.

Theorem 1. *Under the assumption* (4), (5) *and* (6), *the solution of the equation* (3) *is expressed in the form,*

$$(15) \qquad u_t = T_t \int_0^{\infty} \exp\left\{B_t(y_\lambda) - \frac{1}{2} t \,\|y_\lambda\|^2\right\} dE_\lambda u_0 \,,$$

where y_λ is given by (9).

Proof. It is sufficient to prove that

$$(16) \qquad E\left[\left\|T_t \int_0^{\infty} S_N^t(\lambda) dE_\lambda u_0 - T_t \int_0^{\infty} S_\infty^t(\lambda) dE_\lambda u_0\right\|^2\right] \to 0 \,,\; N \to +\infty \,.$$

Noting that the inequalities

$$\begin{aligned}
&\left\|T_t \int_0^{\infty} S_N^t(\lambda) dE_\lambda u_0 - T_t \int_0^{\infty} S_\infty^t(\lambda) dE_\lambda u_0\right\|^2 \\
&\leqq \text{const.} \left\|\int_0^{\infty} S_N^t(\lambda) dE_\lambda u_0 - \int_0^{\infty} S_\infty^t(\lambda) dE_\lambda u_0\right\|^2 \\
&\leqq \text{const.} \int_0^{\infty} \{S_N^t(\lambda) - S_\infty^t(\lambda)\}^2 d(E_\lambda u_0, u_0) \,,
\end{aligned}$$

and that

$$E[\{S_N^t(\lambda) - S_\infty^t(\lambda)\}^2] \leqq \sum_{k=N+1}^{\infty} \frac{(Mt)^k}{k!}$$

for some constant M, we get the statement (16). Thus the proof is complete.

Remark 1. The solution u_t is continuous in t with probability 1. In fact, the process $\int_0^{\infty} S_N^t(\lambda) dE^\lambda u_0$ is a continuous martingale taking values in $\mathscr{H}$ ([6], [9]), which converges to $\int_0^{\infty} S_\infty^t(\lambda) dE_\lambda u_0$ in norm $E[\|\cdot\|^2]$. Hence, the process $\int_0^{\infty} S_\infty^t(\lambda) dE_\lambda u_0$ is a continuous martingale. By Theorem 1, we can see that u_t is strongly continuous.

Remark 2. Assume that the operator A and the semi-group T_t are are given by $A = \int_0^\infty \lambda dE_\lambda$, $T_t = \int_0^\infty e^{-\lambda t} dE_\lambda$.

Then, $u_0 \in \mathscr{D}(A)$ implies $u_t \in \mathscr{D}(A)$. Indeed, we have

$$\int_0^\infty \lambda^2 d(E_\lambda u_t, u_t) = \int_0^\infty \lambda^2 S_\infty^t(\lambda)^2 e^{-2\lambda t} d(E_\lambda u_0, u_0)$$
$$\leqq \text{const.} \int_0^\infty \lambda^2 d(E_\lambda u_0, u_0) < +\infty$$

because $S_\infty^t(\lambda)$ is bounded under the assumption (6).

Remark 3. Making use of Itô's formula ([9], Théorèm III.1.3.), we can verify that the process u_t given by (15) satisfies the equation (1) under the same assumption as in Remark 2.

We will give examples of a semi-group and a resolution of the identity satisfying the assumption (5).

Example 1. We set $\mathscr{H} = L^2(D)$, where D is a domain of R^d. Let $\{E_\lambda\}$ be any resolution of the identity on $\mathscr{H}$, and set $T_t = \int_0^\infty e^{-(\lambda+\alpha)t} dE_\lambda$, with constant α. Clearly, T_t and E_λ are commutative.

Example 2. We set $\mathscr{H} = L^2(R^d)$ and denote by $\{E_\lambda\}$ the resolution resolution of the identity of the operator- Δ. Set $T_t = \exp\{t(\sum_{i,j} a_{ij}(\partial^2 /\partial x_i \partial x_j) + \sum_i b_i(\partial/\partial x_i) + c)$, where a_{ij}, b_i and c are constants, and the matrix (a_{ij}) is positive definite. We can easily verify that $T_t E_\lambda = E_\lambda T_t$, $t \geqq 0$, $t \geqq 0$.

§ 2. The case that the operators B_k are unbounded self-adjoint

From now on, we assume that the operator A equals zero, and that the operators B_k are represented in the form (4). That is, the equation (1) is reduced to the equation,

(17) $$du_t = Bu_t dB_t = \sum_{k=1}^\infty B_k u_t dB_t(e_k) .$$

In order to prove the uniqueness of solutions of the equation (17), we assume the following condition,

(18) $\sum_{k=1}^\infty |f_k(\lambda)|^2$ is bounded on each compact interval of λ .

Let a process u_t be a solution of the equation (17). We set

$$E_{[n,n+1]} = E_{n+1} - E_{n-0} ,$$
$$u_t^n = E_{[n,n+1]} u ,$$
$$B_k^{[n,n+1]} = B_k E_{[n,n+1]} .$$

Then the process u_t^n satisfies the equation,

(19) $$u_t^n = u_0^n + \sum_{k=1}^{\infty} \int_0^t B_k^{[n,n+1]} u_s^n dB_s(e_k) , \qquad u_0^n = E_{[n,n+1]} u_0 .$$

This equation (19) has a unique solution, because of (18). Hence, the equation (17) has at most a solution. Indeed, let u_t and v_t be solutions of the equation (17), then $E_{[n,n+1]} u_t$ and $E_{[n,n+1]} v_t$ satisfy the equation (19). Hence, we get

(20) $$E_{[n,n+1]} u_t = E_{[n,n+1]} v_t , \qquad \text{for any integer } n \geqq 0 .$$

Summing the both sides of (20), we have

$$u_t = v_t , \qquad \text{with probability 1 .}$$

Hence we get

Theorem 2. *The uniqueness of solutions of the equation* (17) *holds under the conditions* (4) *and* (18).

Next, we will construct a solution of the equation (17). We have to assume that there exists a sequence $\{a_k\} \in l_2$ for any n such that

(21) $$\sum_{k=1}^{\infty} \frac{|f_k(\lambda)|^2}{a_k^2} \text{ is bounded on } [0, n] .$$

The condition (21) implies (18).

We set

$$B_k^{[0,n]} = \int_0^n f_k(\lambda) dE_\lambda = B_k E_{[0,n]} .$$

We have shown in § 1 that the solution of the equation

(22) $$v_t^n = v_0^n + \sum_{k=1}^{\infty} \int_0^t B_k^{[0,n]} v_s^n dB_s(e_k) , \qquad v_0^n = E_{[0,n]} v_0 , \quad v_0 \in \mathscr{H} ,$$

is expressed in the form,

$$v_t^n = \int_0^n \exp\left\{B_t(y_\lambda) - \frac{t}{2}\sum_{k=1}^{\infty}|f_k(\lambda)|^2\right\} dE_\lambda v_0 ,$$

where $y_\lambda = \sum_{k=1}^{\infty} f_k(\lambda) e_k$, under the assumption (21).

We shall show that the sequence v_t^n converges to a process u_t, which is a solution of the stochastic evolution equation (17). We assume that the initial vector v_0 is given by

$$(23) \qquad v_0 = \int_0^\infty \exp\left\{-\varepsilon\left(\sum_{k=1}^{\infty}|f_k(\lambda)|^2\right)^{1+\delta}\right\} dE_\lambda g , \qquad g \in \mathscr{H} ,$$

for some positive ε and δ. We put

$$(24) \qquad u_t = \int_0^\infty \exp\left\{B_t(y_\lambda) - \frac{t}{2}\sum_{k=1}^{\infty}|f_k(\lambda)|^2 - \varepsilon\left(\sum_{k=1}^{\infty}|f_k(\lambda)|^2\right)^{1+\delta}\right\} dE_\lambda g .$$

Then, we have

$$\begin{aligned} E\,\|v_t^n - u_t\|^2 &= \int_n^\infty E\left[\exp\left\{2B_t(y_\lambda) - t\sum_{k=1}^{\infty}|f_k(\lambda)|^2 - 2\varepsilon\left(\sum_{k=1}^{\infty}|f_k(\lambda)|^2\right)^{1+\delta}\right\}\right] d(E_\lambda g, g) \\ &= \int_n^\infty \exp\left\{t\sum_{k=1}^{\infty}|f_k(\lambda)|^2 - 2\varepsilon\left(\sum_{k=1}^{\infty}|f_k(\lambda)|^2\right)^{1+\delta}\right\} d(E_\lambda g, g) . \end{aligned}$$

Since $\exp\{t\sum_{k=1}^{\infty}|f_k(\lambda)|^2 - 2\varepsilon(\sum_{k=1}^{\infty}|f_k(\lambda)|^2)^{1+\delta}\}$ is bounded in $(t, \lambda) \in [0, T] \times [0, \infty)$, we get $E\,\|v_t^n - u_t\|^2 \to 0$. On the other hand, we can show that

$$\sum_{k=1}^{\infty}\int_0^t B_k^{[0,n]} v_s^n dB_s(e_k) \to \sum_{k=1}^{\infty}\int_0^t B_k u_s dB_s(e_k) , \quad \text{as } n \to +\infty .$$

Indeed, we have

$$\begin{aligned} E&\left\|\sum_{k=1}^{\infty}\int_0^t B_k^{[0,n]} v_s^n dB_s(e_k) - \sum_{k=1}^{\infty}\int_0^t B_k u_s dB_s(e_k)\right\|^2 \\ &= \int_0^t \sum_{k=1}^{\infty} E[\|(B_k^{[0,n]} - B_k)u_s\|^2] ds \\ &= \int_0^t ds \int_n^\infty \sum_{k=1}^{\infty} f_k(\lambda)^2 E\left[\exp\left\{2B_s(y_\lambda) - s\sum_{k=1}^{\infty}|f_k(\lambda)|^2 - 2\varepsilon\left(\sum_{k=1}^{\infty}|f_k(\lambda)|^2\right)^{1+\delta}\right\}\right] d(E_\lambda g, g) \end{aligned}$$

$$= \int_0^t ds \int_n^\infty \sum_{k=1}^\infty |f_k(\lambda)|^2 \exp\left\{s \sum_{k=1}^\infty |f_k(\lambda)|^2 - 2\varepsilon\left(\sum_{k=1}^\infty |f_k(\lambda)|^2\right)^{1+\delta}\right\} \times d(E_\lambda g, g)$$

$$\leqq \int_0^t ds \int_n^\infty c \exp\left\{(s+s_0) \sum_{k=1}^\infty |f_k(\lambda)|^2 - 2\varepsilon\left(\sum_{k=1}^\infty |f_k(\lambda)|^2\right)^{1+\delta}\right\} \times d(E_\lambda g, g)$$

$$\to 0 .$$

Thus, we have

Theorem 3. *The equation* (17) *has a solution, which is expressed in the form* (24), *under the assumptions* (4), (21) *and* (23).

We will give two examples.

Example 3. We consider the equation

$$du(t, x, \omega) = \Delta u(t, x, \omega) dB_t , \qquad x \in D \subset R^d , \tag{25}$$

with boundary condition

$$u(t, x, \omega) = 0 , \qquad \text{if } x \in \partial D , \tag{26}$$

where Δ is the Laplacian, D is a bounded domain with smooth boundary ∂D, and $\{B_t\}$ is a one-dimensional Brownian motion. This case is reduced to the equation (17), setting that the operator B_1 be the self-adjoint operator in $L^2(D)$ which is determined by the Laplace operator and the boundary condition (26), and that the operators B_k ($k \geqq 2$) be zero. It is clear that the condition (4) and (21) are satisfied. Let λ_j, ϕ_j, ($j=1, 2, \cdots$) be the eigenvalues and the eigenfunctions of the operator $-B_1$ respectively. Then, the condition (23) means that the initial function v_0 is represented in the form,

$$u_0 = \sum_{j=1}^\infty b_j e^{-\lambda_j^{2+\delta}} \phi_j , \qquad \sum_{j=1}^\infty b_j^2 < \infty , \quad \delta > 0 .$$

Example 4. We consider the equation

$$du(t, x, \omega) = \phi(x) u(t, x, \omega) dB_t(x, \omega) , \qquad x \in D , \tag{27}$$

where D is a domain of R^1, where B_t is a $L^2(D)$-valued Brownian motion, and where $\phi(x)$ is an unbounded function. This Brownian motion $\{B_t\}$ is represented as follows;

$$B_t(x, \omega) = \sum_{k=1}^\infty c_k \phi_k(x) \xi_t^k(\omega) ,$$

where $\sum_{k=1}^{\infty} c_k^2 < +\infty$, $\{\phi_k(x)\}$ is a complete orthonormal system of $L^2(D)$, and ξ_t^k are independent one-dimensional Brownian motions. Therefore, the equation (27) is expressed in the form,

$$du(t, x, \omega) = \sum_{k=1}^{\infty} c_k \phi(x) \phi_k(x) u(t, x, \omega) d\xi_t^k(\omega) . \tag{28}$$

Setting that

$$B_k u = c_k \phi(x) \phi_k(x) u ,$$

the equation (28) is reduced to the equation (17). Assume that the eigenfunction $\phi_k(x)$ are uniformly bounded, and that the sequence $\{c_k\}$ is represented in the form,

$$c_k = \alpha_k \beta_k , \quad \{\alpha_k\} \in l_2 , \quad \{\beta_k\} \in l_2 . \tag{29}$$

For simplicity, we consider only the case that $D = R^1$. In this example, the equalities

$$E_\lambda = \text{the projection to the space } L^2(-\infty, \lambda) ,$$
$$f_k(\lambda) = c_k \phi(\lambda) \phi_k(\lambda) ,$$

hold, so that

$$\sum_k^{\infty} \frac{|f_k(\lambda)|^2}{\alpha_k^2} \leq \text{const.}\ \phi(\lambda)^2 \sum_k^{\infty} \beta_k^2 .$$

Hence, the assumption (21) is satisfied if $\phi(x)$ is locally bounded. The condition (23) means that the initial function u_0 is expressed as follows;

$$u_0 = e^{-K\phi(x)^{2+\delta}} g(x) , \quad g(x) \in L^2(R^1) , \quad K, \delta > 0 .$$

Next, we will show the non-existence of a solution satisfying $E\,\|u(t, x, \omega)\|^2 < \infty$, when the initial function u_0 does not satisfy (23). Under the condition (29), the inequality

$$\left[\sum_{k=i}^{j} |c_k \phi_k(x) \xi_t^k(\omega)|\right]^2 \leq \text{const.} \sqrt{\sum_k \alpha_k^2} \times \sqrt{\sum_k \beta_k^2 \xi_t^k(\omega)^2}$$

holds, therefore $B_t(x, \omega)$ is (t, x)- continuous with probability 1, by t-continuity of the second term of the right-hand side. Setting

$$Q(x) = \sum_k c_k^2 \phi_k(x)^2 ,$$

we can verify that the process $[B_t(x, \omega)]/Q(x)$ is a one-dimensional

Brownian motion for each x. Hence, we can solve the equation (27) for each x, and we get

$$u(t, x, \omega) = \exp\left\{\phi(x)B_t(x, \omega) - \frac{t}{2}\phi(x)^2 Q(x)\right\} g(x) ,$$

so that

$$E[u(t, x, \omega)^2] = \exp\{t\phi(x)^2 Q(x)\} g(x)^2 .$$

Let $\phi_0(x)$ be a function satisfying the conditions;

$$\exp\left\{-\frac{1}{2}\phi_0(x)^2\right\} \in L^2(R^1) , \quad \text{and} \quad \frac{\phi_0(x)}{\sqrt{Q(x)}}$$

is unbounded.
We set

$$g(x) = \exp\left\{-\frac{1}{2}\phi_0(x)^2\right\} , \qquad \phi(x) = \frac{\phi_0(x)}{\sqrt{Q(x)}} .$$

Then, we get

$$\begin{aligned} E[\| u(t, x, \omega)\|^2] &= \int_{-\infty}^{+\infty} \exp\{t\phi(x)^2 Q(x)\} g(x)^2 dx \\ &= \int_{-\infty}^{+\infty} \exp\{(t-1)\phi_0(x)^2\} dx = +\infty , \qquad \text{if } t = 1 . \end{aligned}$$

Finally, we will give a remark on the case that the operator A does not equal zero in the equation (1). We assume the conditions (4), (5) and (18). Then, the uniqueness of solutions of the equation (3) holds. Next, assume the conditions (4), (5) and (21). Then, we can construct a solution u_t of the equation (3), if the initial function u_0 satisfies (23), and it is expressed in the form,

$$(30) \qquad u_t = T_t \int_0^\infty \exp\left\{B_t(y_\lambda) - \frac{t}{2}\sum_{k=1}^\infty |f_k(\lambda)|^2 - \varepsilon\left(\sum_{k=1}^\infty |f_k(\lambda)|^2\right)^{1+\delta}\right\} dE_\lambda g .$$

References

[1] Balakrishnan, A. V., Stochastic bilinear partial differential equations, the U.S.-Italy Conference on variable structure systems, Oregon, May (1974).

[2] Dawson, D. A., Stochastic evolution equations and related measure processes, J. Multivariate Anal., **5**-1 (1975), 1–52.

[3] Fleming, W. H., Distributed parameter stochastic systems in population biology, Lecture Notes in Economics and Mathematical Systems, Vol. 107 (1975).

[4] Hida, T., Brownian motion (in Japanese), Iwanami Co. Pub. (1975).

[5] ——, Analysis of Brownian functionals, Carleton Univ. lecture notes (1975).

[6] Kunita, H., Stochastic integrals based on martingales taking values in Hilbert space, Nagoya Math. J., **38** (1970), 41–52.

[7] Б. Н. Розовский, Стохастичские дифференциальные уравнения в Бесконечномерных пространствах и проблемы фильтрции, Труды школы-семинара по теории случайных процессов, Вильнюс (1975), 147–194.

[8] Shimizu, A., Construction of a solution of a certain evolution equation, Nagoya Math. J., **66** (1977), 23–36.

[9] Yor, M., Existence et unicité de diffusions à valeurs dans un espace de Hilbert, Ann. Inst. H. Poincaré, **X**-1 (1974), 55–88.

Department of Mathematics
Nagoya Institute of Technology
Nagoya 466, Japan

Proc. of Intern. Symp. SDE
Kyoto 1976, pp. 397–408

On the Optimal Control for Distributed Parameter Systems with White Noise Coefficients

Yoshifumi SUNAHARA, Shin'ichi AIHARA
Muneshi KOYAMA and Fumio KOJIMA

Abstract

The purpose of this paper is to find the optimal feedback control strategy for linear distributed parameter systems with random coefficients, whose statistics are given by the white Gaussian process.

First, parameter statistics contained in the system are specified by the notion of random eigenvalue problems. Secondly, based on the concept of functional analysis, the optimal control for the quadratic performance is obtained by using the Dynamic Programming approach.

For the purpose of supporting the theoretical aspects developed here, results of simulation studies are also demonstrated.

§ 1. Introduction

Based on the fact that measurements of physical properties of the system considered inherently exhibit various kinds of uncertainties, many physical systems are modeled by a class of partial differential equations with randomly varying coefficients. We may find many physical examples of distributed parameter systems with random coefficients in engineering, biological and environmental sciences.

Since coefficients contained in a differential operator are random, the problem to find stochastic properties of a solution process to the partial differential equation from the known statistics of random coefficients is included in the mathematical literature of random eigenvalue problems. This represents a mathematically formidable subject and consequently various methods of approximation have been developed by Keller [1]. Recently, a method of finding the sub-optimal control was developed by the author for a class of distributed parameter systems with stochastic coefficients modeled by a random constant [2]. One of our principal motivation is, in this paper, to consider a more practical situation in which stochastic coefficients are modeled by the white Gaussian process.

In Section 2, the mathematical model of the system considered here is given in a form of partial differential equation whose differential operator involves stochastic parameters. Two basic assumptions of the eigenvalues and eigenvectors are made, reflecting stochastic properties of system parameters. In Section 3, a differential rule on Hilbert space is derived which plays an important role to determine the optimal control. In Section 4, the optimal control law is derived within the framework of Dynamic Programming under the criterion that a quadratic pay-off functional becomes minimal. Section 5 is devoted to show an example for the purpose of interpreting the general theory.

§ 2. Mathematical model and basic assumptions

We denote by G a bounded domain in the n-dimensional Euclidean space with smooth boundary ∂G. Let x be an n-dimensional spacial variable. We shall consider a dynamical system described by the following stochastic partial differential equation,

$$\text{(2.1a)} \qquad \frac{\partial u(t,x)}{\partial t} + \mathscr{A}(t,\omega;D_x)u(t,x) = B(t,x)f(t,x)\,, \qquad (t,x) \in T \times G$$

together with the initial and boundary conditions,

$$\text{(2.1b)} \qquad u(t_0, x) = u_0(x) \qquad \text{for } x \in G$$

$$\text{(2.1c)} \qquad \mathscr{B}_j(t,x)u(t,x) = 0 \qquad \text{for } (t,x) \in T \times \partial G,\ T =]t_0, t_f[\ ,$$

where u is the scalar stochastic state variable and $\mathscr{A}(t,\omega;D_x)$ is the partial differential operator with stochastic coefficients such that

$$\text{(2.1d)} \qquad \mathscr{A}(t,\omega;D_x) = \sum_{|p|\le n} a_p(t,\omega)D_x^p\,,$$

and furthermore $\mathscr{B}_j$ $(j = 1, 2, \cdots, n/2)$ is a deterministic bundary operator. In Eq. (2.1d), we express the generic point on the sample space Ω by ω and use the notation $D_x{}^p = \partial^{|p|}/\partial x_1^{p_1}\partial x_2^{p_2} \cdots \partial x_n^{p_n}$.

Let $\mathscr{H}$ be the Hilbert space $L^2(G)$, let $\mathscr{H}^{n/2}(G)$ be the Sobolev space of order $n/2$ on G and let $\mathscr{H}_0^{n/2}(G) = \{u \in \mathscr{H}\,;\, D_x^\alpha u \in \mathscr{H}$ for $|\alpha| \le n/2$, $\mathscr{B}_j(t,x)u = 0$, for $j = 1, 2, \cdots, n/2\}$. Guided by the well-known concept of Sobolev spaces [3], we shall denote $\mathscr{V}$ by

$$\mathscr{V} = \mathscr{H}_0^{n/2}(G) \cap \mathscr{H}^n(G)$$

and then the dual of $\mathscr{V}$ is

$$\mathscr{V}' = \mathscr{H}^{-n/2}(G)\,.$$

Let $(\Omega, \sigma(\Omega), m)$ be the probability space where $\sigma(\Omega)$ is the minimal

σ-algebra of Borel sets of Ω and m is the stochastic measure defined on Ω.

Based on the concept of random eigenvalue problem, the following assumptions are made on $\mathscr{A}(t, \omega; D_x)$;

(A–1) The Hilbert space $\mathscr{V}$ is separable. Then, the orthonormal basis of $\mathscr{H}$ may be made up with elements of $\mathscr{V}$. We denote those by $\phi_1, \phi_2, \cdots$.

(A–2) There exists a sequence $\{\lambda_i(t, \omega), \phi_i; i = 1, 2, \cdots\}$ of stochastic eigenvalues and deterministic eigenvectors such that, for $\forall \psi \in \mathscr{H}$,

$$(\mathscr{A}(t, \omega; D_x)\phi_i, \psi) = (\lambda_i(t, \omega)\phi_i, \psi) , \tag{2.2}$$

where $(\cdot, \cdot)$ express the inner product on $\mathscr{H}$.

(A–3) The stochastic eigenvalue $\lambda_i(t, \omega)$ is expressed by

$$\lambda_i(t, \omega) = \tilde{\lambda}_i(t) + \mu_i(t)\eta(t, \omega) , \tag{2.3}$$

where both $\tilde{\lambda}_i(t)$ and $\mu_i(t)$ are deterministic functions and $\eta(t, \omega)$ is a white Gaussian process with zero mean and unit variance. Furthermore, the following operators $\tilde{A}(t)$ and $G(t)$ exist such that

$$(\tilde{A}(t)\phi_i, \psi) = (\tilde{\lambda}_i(t)\phi_i, \psi) \tag{2.4}$$

and

$$(G(t)\phi_i, \psi) = (\mu_i(t)\phi_i, \psi) . \tag{2.5}$$

(A–4) The operators defined by Eqs. (2.4) and (2.5) satisfy the following coercivity condition: there exists a positive constant α such that for $\forall u \in \mathscr{V}$

$$2\langle \tilde{A}(t)u, u\rangle - (G(t)u, G(t)u) \geq \alpha \| u \|_{\mathscr{V}}^2 , \tag{2.6}$$

where $\langle \cdot, \cdot \rangle$ denotes the duality between $\mathscr{V}$ and $\mathscr{V}'$, and where $\| \cdot \|_{\mathscr{V}}$ and $| \cdot |$ denote the norm in $\mathscr{V}$ and $\mathscr{H}$ respectively.

From Eq. (2.6) it is easy to find $\tilde{A}(t) \in \mathscr{L}(\mathscr{V}; \mathscr{V}')$ and $G^*(t)G(t) \in \mathscr{L}(\mathscr{V}; \mathscr{V}')$ where $\mathscr{L}$ is the set of all linear mappings from $\mathscr{V}$ to $\mathscr{V}'$.

Since the partial differential operator $\mathscr{A}(t, \omega; D_x)$ in Eq. (2.1) contains the white Gaussian noise, Eq. (2.1) has no mathematically precise meaning. Consequently, with the assumptions (A–1) to (A–4), we shall assume that Eq. (2.1) is rewritten by the following stochastic equation on the spaces $\mathscr{V}, \mathscr{H}, \mathscr{V}'$;

$$u(t) + \int_{t_0}^{t} \tilde{A}(s)u(s)ds + \int_{t_0}^{t} G(s)u(s)dw(s) = \int_{t_0}^{t} B(s)f(s)ds + u_0 , \tag{2.7}$$

where the third term of the left hand side of Eq. (2.7) is a stochastic integral such that

(2.8) $$\int_{t_0}^{t} G(s)u(s)dw(s) \triangleq \int_{t_0}^{t} \sum_{i=1}^{\infty} \mu_i(s)(u(s), \phi_i)\phi_i dw(s)$$

and

(2.9) $$E\left\{\left|\int_{t_0}^{t} G(s)u(s)dw(s)\right|^2\right\} = E\left\{\int_{t_0}^{t} \langle G^*(s)G(s)u(s), u(s)\rangle ds\right\} ,$$

where $u \in L^2(T; L^2(\Omega, m; \mathscr{V}))$ and $G(t)u(t)$ is $\mathscr{F}_t$ measurable and where $\mathscr{F}_t$ is an increasing family of σ-algebras.

§ 3. Stochastic differential equation and differential rule for quadratic functional

We will discuss Eq. (2.7) in the special case $f(t) = 0$.

Theorem 1. *With the coericivity condition (2.6), there exists a unique stochastic process $u(t)$ such that, for any $t \in T$,*

(3.1) $$u \in L^2(T; L^2(\Omega, m; \mathscr{V})) \cap C(T; L^2(\Omega, m; \mathscr{H})) ,$$

(3.2) $$u(t) + \int_{t_0}^{t} \tilde{A}(s)u(s)ds + \int_{t_0}^{t} G(s)u(s)dw(s) = u_0$$

and we have the energy inequality

(3.3) $$E\{|u(t)|^2\} + \alpha E\left\{\int_{t0}^{t} \|u(s)\|_{\mathscr{V}}^2 ds\right\} \leq E\{|u_0|^2\} .$$

Theorem 2. *Define the quadratic functional*

(3.4) $$V(t, u) \triangleq (u(t), \Pi(t)u(t)) ,$$

where $\Pi(t)$ is a deterministic symmetric bounded operator.

We have the following stochastic differential,

(3.5) $$\begin{aligned} dV(t,\omega) = (u(t), [\dot{\Pi}(t) - \tilde{A}^*(t)\Pi(t) - \Pi(t)\tilde{A}(t) \\ + G^*(t)\Pi(t)G(t)]u(t))dt + 2(G(t)u(t), \Pi(t)u(t))dw(t) . \end{aligned}$$

The proofs of Theorems 1 and 2 will respectively be demonstrated in Appendixies 1 and 2.

§ 4. Derivation of optimal control

First of all, in this section, we consider a system whose evolution is governed by Eq. (3.2) with the control term $B(t)f(t)$. We assume $B(t) \in$

$\mathscr{L}(\mathscr{H};\mathscr{H})$ and $f \in \mathscr{W}$, where $\mathscr{W}$ denotes a class of admissible control,

$$\text{(4.1)} \quad \mathscr{W} = \{f; f \in L^2(T; L^2(\Omega, m; \mathscr{H})),\ f(t) \text{ is } \mathscr{F}(t,\omega) \text{ measurable and } f(t) \in \text{convex closed subset of } \mathscr{H}\}\,,$$

where $\mathscr{F}(t,\omega)$ is the σ-algebra generated by the random variable $u(s,\omega)$ for $t_0 \leq s \leq t$, i.e.,

$$\text{(4.2)} \quad \mathscr{F}(t,\omega) = \sigma\{u(s,\omega); t_0 \leq s \leq t\}\,.$$

It is not easy to extend Theorem 1 and Theorem 2 to the case where $f \in \mathscr{W}$, but assuming that this extension is possible we will proceed further. Let $u(t)$ be the solution of

$$\text{(4.3)} \quad u(t) + \int_{t_0}^{t} \tilde{A}(s)u(s)ds + \int_{t_0}^{t} G(s)u(s)dw(s) = \int_{t_0}^{t} B(s)f(s)ds + u_0$$

and consider the following quadratic pay-off functional,

$$\text{(4.4)} \quad L(t,u,f) = (P(t)u(t), u(t)) + (Q(t)f(t), f(t))\,,$$

where $P(t)$ and $Q(t)$ are bounded positive definite and self-adjoint operators, respectively. Our problem is to find the feedback optimal control $f^\circ(t)$ in such a way that the functional,

$$\text{(4.5)} \quad J(f) = E\left\{\int_{t_0}^{t_f} L(t,u,f)dt\right\}\,,$$

becomes minimal with respect to $f \in \mathscr{W}$.

Define

$$\text{(4.6)} \quad V(t,\omega) \triangleq \min_{f\in\mathscr{W}} E\left\{\int_{t}^{t_f} L(\tau,u,f)d\tau \,|\, u(t) = \overline{u(t)}\right\}\,.$$

Bearing in mind the feedback optimal control, the minimal cost functional $V(t,\omega)$ is formally defined,

$$\text{(4.7)} \quad V(t,\omega) = (\bar{u}, \Pi(t)\bar{u})\,.$$

From (4.7), by applying the principle of optimality [4] and by Theorem 2, the following basic equation is derived;

$$\text{(4.8)} \quad \begin{aligned} \min_{f\in\mathscr{W}} [&(P(t)\bar{u},\bar{u}) + (Q(t)f,f) + (\bar{u}, \dot{\Pi}(t)\bar{u}) \\ &+ (\bar{u}, [G^*(t)\Pi(t)G(t) - \tilde{A}^*(t)\Pi(t) - \Pi(t)\tilde{A}(t)]\bar{u}) \\ &+ (\bar{u}, \Pi(t)B(t)f) + (f, B^*(t)\Pi(t)\bar{u})] = 0\,. \end{aligned}$$

From (4.8), it follows that

$$
\begin{aligned}
(4.9)\qquad &\min_{f\in\mathscr{W}} [(Q(t)f + B^*(t)\Pi(t)\bar{u}, Q(t)^{-1}(Q(t)f + B^*(t)\Pi(t)\bar{u}))\\
&+ (\bar{u}, (\dot{\Pi}(t) - \Pi(t)B(t)Q(t)^{-1}B^*(t)\Pi(t) + P(t))\bar{u})\\
&+ (\bar{u}, (G^*(t)\Pi(t)G(t) - \tilde{A}^*(t)\Pi(t) - \Pi(t)\tilde{A}(t))\bar{u})] = 0 .
\end{aligned}
$$

Then, noting that the first term of left hand side of (4.9) is a quadratic form, the optimal control $f^\circ(t)$ is given by

$$(4.10)\qquad f^\circ(t) = -Q(t)^{-1}B^*(t)\Pi(t)\bar{u} ,$$

where $\Pi(t)$ satisfies

$$
\begin{aligned}
(4.11a)\qquad &\dot{\Pi}(t) - \tilde{A}^*(t)\Pi(t) - \Pi(t)\tilde{A}(t) + G^*(t)\Pi(t)G(t)\\
&- \Pi(t)B(t)Q(t)^{-1}B^*(t)\Pi(t) + P(t) = 0
\end{aligned}
$$

with the terminal condition,

$$(4.11b)\qquad \Pi(t_f) = 0 .$$

§ 5. An illustrative example

For the purpose of better understanding, we shall consider a somewhat artificial but important class of systems.

Example. Consider the 4th order parabolic systems described by

$$
\begin{aligned}
(5.1a)\qquad &\frac{\partial u(t, x)}{\partial t} + a_2(t)\frac{\partial^4 u(t, x)}{\partial x^4} + (a_1(t) + g(t)\eta(t, \omega))\frac{\partial^2 u(t, x)}{\partial x^2}\\
&= B(t)f(t, x) , \qquad \text{for } (t, x) \in]t_0, t_f[\times]0, 1[
\end{aligned}
$$

with the initial condition

$$(5.1b)\qquad u(t_0, x) = u_0(x) , \qquad \text{for } x \in]0, 1[$$

and the boundary conditions

$$
\begin{aligned}
(5.1c)\qquad &u(t, x)|_{x=0} = 0 \qquad u(t, x)|_{x=1} = 0\\
&\left.\frac{\partial^2 u(t, x)}{\partial x^2}\right|_{x=0} = 0 \qquad \left.\frac{\partial^2 u(t, x)}{\partial x^2}\right|_{x=1} = 0 ,
\end{aligned}
$$

where $a_1(t)$, $a_2(t)$, $g(t)$ and $B(t)$ are scalar functions, respectively. From (2.6) of (A–4), it is assumed that

(5.2) $$a_2(t) - \frac{1}{2}g^2(t) > 0\,, \quad \text{and } a_1(t) < 0 \quad \text{for } \forall t \in [t_0, t_f]\,.$$

In this example, the deterministic eigenfunctions can be selected as

(5.3) $$\phi_i = \sqrt{2}\,\sin i\pi x\,.$$

Hence, from (2.3) in (A–3), the stochastic eigenvalue becomes

(5.4) $$\lambda_i(t, \omega) = (i\pi)^4 a_2(t) - (i\pi)^2(a_1(t) + g(t)\eta(t, \omega))\,.$$

From (2.5), we have

(5.5) $$\tilde{\lambda}_i(t) = (i\pi)^4 a_2(t) - (i\pi)^2 a_1(t)$$

(5.6) $$\mu_i(t) = -(i\pi)^2 g(t)\,.$$

Bearing in mind (2.4), (2.5), the partial differential operators $\tilde{A}(t)$ and $G(t)$ are respectively expressed by

(5.7) $$\tilde{A}(t) = a_2(t)\frac{\partial^4(\cdot)}{\partial x^4} + a_1(t)\frac{\partial^2(\cdot)}{\partial x^2}$$

and

(5.8) $$G(t) = g(t)\frac{\partial^2(\cdot)}{\partial x^2}\,.$$

Consider the following performance criteria

(5.9) $$J(f) = E\left\{\int_{t_0}^{t_f} [(u(t), u(t)) + (f(t), f(t))]dt\right\}\,.$$

From (4.10) and (4.11), the optimal control $f^\circ(t)$ is obtained by

(5.10) $$f^\circ(t, x) = -B(t)\int_0^1 \Pi(t, x, y)\bar{u}(t, y)dy\,,$$

where

$$\frac{\partial \Pi(t, x, y)}{\partial t} = \left[a_2(t)\left(\frac{\partial^4}{\partial x^4} + \frac{\partial^4}{\partial y^4}\right) + a_1(t)\left(\frac{\partial^2}{\partial x^2} + \frac{\partial^2}{\partial y^2}\right)\right]\Pi(t, x, y)$$

(5.11) $$- \frac{1}{2}g^2(t)\left[\frac{\partial^4}{\partial x^2 \partial y^2} + \frac{\partial^4}{\partial y^2 \partial x^2}\right]\Pi(t, x, y) - \delta(x - y)$$

$$+ \int_0^1\int_0^1 \Pi(t, x, z_1)B(t)\delta(z_1 - z_2)B(t)\Pi(t, z_2, y)dz_1 dz_2$$

with the terminal condition $\Pi(t_f, x, y) = 0$ and boundary conditions

$$\Pi(t, x, y)|_{x=0,\, x=1} = 0 \qquad \Pi(t, x, y)|_{y=0,\, y=1} = 0$$

(5.12)
$$\left.\frac{\partial^2 \Pi(t, x, y)}{\partial x^2}\right|_{x=0,\, x=1} = 0 \qquad \left.\frac{\partial^2 \Pi(t, x, y)}{\partial y^2}\right|_{y=0,\, y=1} = 0 .$$

In this example, Eq. (5.1) was simulated on a digital computer and the optimal control $f^\circ(t, x)$ was determined by (5.10) with the solution to (5.11). A wide variety of sample runs was simulated. The results presented below are representative of simulation experiments. In all experiments, the values of a_1, a_2, g and B were respectively set as $a_1 = -0.1$, $a_2 = 0.002$, $g = 0.015$ and $B = 0.5$. The initial condition (5.1b) was given by $u_0(x) = \sin^2 \pi x$. Throughout the experiments, the partitioned time interval and spacial variable were $\Delta t = 2.5 \times 10^{-4}$ and $\Delta x = 0.05$, respectively.

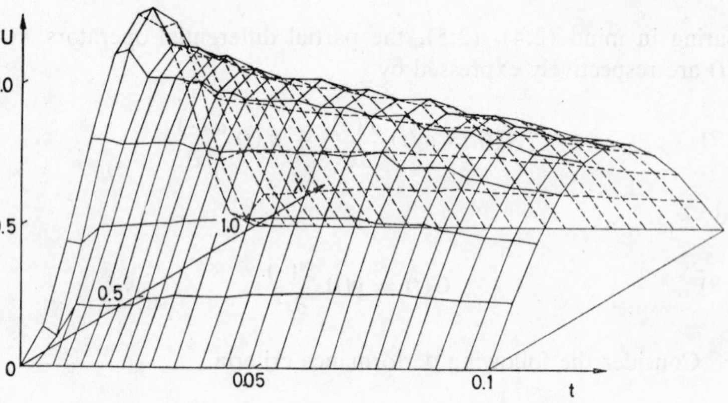

Fig. 1. Sample run of the system without control in Example

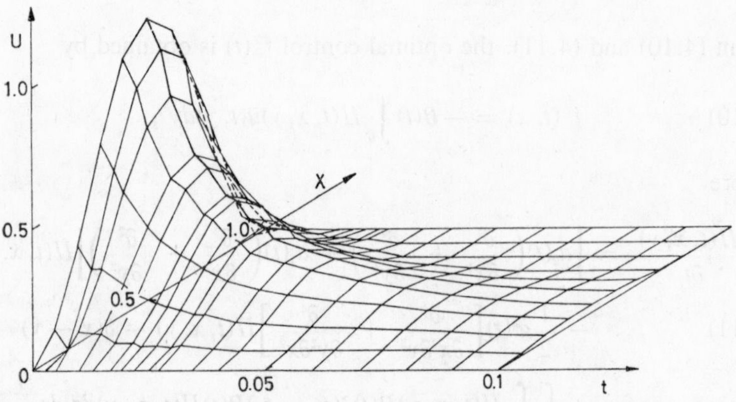

Fig. 2. Sample run of the controlled system in Example

Figure 1 shows a representative of sample runs of the system without control. A sample run of the system driven by the optimal control signal $f^{\circ}(t, x)$ is shown in Fig. 2.

§ 6. Conclusions

In this paper, first the mathematical model for distributed parameter systems with random coefficients has been established by introducing the concept of stochastic eigenvalue problem. The random coefficient is modeled by white Gaussian process. In order to derive the optimal control, the differential rules were given by using the eigenfunction expansion method.

By using the Dynamic Programming approach, the optimal control for the quadratic performance was obtained.

References

[1] Keller, J. B., Stochastic equation and wave propagation in random media, Proc. Symp. Appl. Math., **16** (1960), 145–170.

[2] Sunahara, Y. and Hamatsuka, T., Optimal control for distributed parameter systems with random coefficients, Proc. the 6th Boston IFAC Congress, Boston, Aug., 1975.

[3] Lions, J. L., Contrôle optimal des systèmes gouverniés par des équations aux dérivées partielles, Dunod, Paris, 1968.

[4] Bellman, R., Adaptive control process, A guided tour, Princeton, N.Y., Princeton University Press, 1961.

[5] Itô, K., On a formula concerning stochastic differential, Nagoya Math. J., **3** (1950), 55–65.

Appendix 1. Proof of Theorem 1

Let $\mathscr{V}_m = [\phi_1, \phi_2, \cdots, \phi_m]$ and $\tilde{A}_m(t) \in \mathscr{L}(\mathscr{V}_m; \mathscr{V}'_m)$ be defined by

$$\tilde{A}_m(t) \triangleq \sum_{i=1}^{m} < \tilde{A}(t)(\cdot), \phi_i > \phi_i \,. \tag{A–1}$$

Furthermore, define

$$G_m(t) = \sum_{i=1}^{m} (G(t)(\cdot), \phi_i)\phi_i \tag{A–2}$$

and

$$u_{0m} = \sum_{i=1}^{m} (u_0, \phi_i)\phi_i \, . \tag{A–3}$$

We approximate the stochastic equation (3.2) by [5]

$$u_m(t) + \int_{t_0}^{t} A_m(s)u_m(s)ds + \int_{t_0}^{t} G_m(s)u_m(s)dw(s) = u_{0m} \, . \tag{A–4}$$

By using the well-known Itô's differential rule, it follows that

$$\begin{aligned} \frac{1}{2}|u_m(t)|^2 = \frac{1}{2}|u_{0m}|^2 &- \int_{t_0}^{t} (u_m(s), \tilde{A}_m(s)u_m(s))ds \\ &+ \int_{t_0}^{t} (u_m(s), G_m(s)u_m(s))dw(s) \\ &+ \frac{1}{2}\int_{t_0}^{t} (G_m(s)u_m(s), G_m(s)u_m(s))ds \, . \end{aligned} \tag{A–5}$$

From Eq. (A–1), we have

$$\begin{aligned} \int_{t_0}^{t} (u_m(s), \tilde{A}_m(s)u_m(s))ds &= \int_{t_0}^{t} \left(u_m(s), \sum_{i=1}^{m} \langle \tilde{A}(s)u_m(s), \phi_i \rangle \phi_i\right) ds \\ &= \int_{t_0}^{t} \langle \tilde{A}(s)u_m(s), u_m(s) \rangle ds \, . \end{aligned} \tag{A–6}$$

From Eq. (A–2), it follows that

$$\begin{aligned} \int_{t_0}^{t} (G_m(s)u_m(s), G_m(s)u_m(s))ds &= \int_{t_0}^{t} \sum_{i=1}^{m} (G(s)u_m(s), \phi_i)^2 ds \\ &\le \int_{t_0}^{t} (G(s)u_m(s), G(s)u_m(s))ds \, . \end{aligned} \tag{A–7}$$

By using Eqs. (A–6) and (A–7), the following inequality holds:

$$\begin{aligned} \frac{1}{2}E\{|u_m(t)|^2\} + E\Big\{\int_{t_0}^{t} \Big[&\langle \tilde{A}(s)u_m(s), u_m(s) \rangle \\ &- \frac{1}{2}(G(s)u_m(s), G(s)u_m(s)) \Big] ds \Big\} \le \frac{1}{2}E\{|u_{0m}|^2\} \, . \end{aligned} \tag{A–8}$$

On the other hand, from the coercivity condition, it follows that

$$E\{|u_m(t)|^2\} + \alpha E\Big\{\int_{t_0}^{t} \| u_m(s) \|_{\mathscr{V}}^2 \, ds\Big\} \le E\{|u_{0m}|^2\} \, . \tag{A–9}$$

Then, for $\forall t \in T$, we have

(A–10) $$E\{|u_m(t)|^2\} \leq C_1$$

and

(A–11) $$E\left\{\int_{t_0}^{t} \|u_m(s)\|_{\mathscr{V}}^2\, ds\right\} \leq C_2 .$$

Consequently, we extract a subsequence $u_m \to u$ in $L^2(\Omega, m; L^2(T; \mathscr{V}))$. Let ϕ_i be an arbitrary but fixed element of the basis. For $m \geq i$, we have

(A–12) $$\begin{aligned}(u_m(t), \phi_i) - (u_{0m}, \phi_i) + \int_{t_0}^{t} \langle \tilde{A}(s)u_m(s), \phi_i\rangle ds \\ + \int_{t_0}^{t} (G(s)u_m(s), \phi_i)dw(s) = 0 .\end{aligned}$$

Passing to the limit, it follows that

(A–13) $$\begin{aligned}(u(t), \phi_i) - (u_0, \phi_i) + \int_{t_0}^{t} \langle \tilde{A}(s)u(s), \phi_i\rangle ds \\ + \int_{t_0}^{t} (G(s)u(s), \phi_i)dw(s) = 0 .\end{aligned}$$

From Eq. (A–13), it is easy to show that there is a solution to Eq. (3.2) which belongs to the class

$$L^2(T; L^2(\Omega, m; \mathscr{V})) \cap L^\infty(T; L^2(\Omega, m; \mathscr{H})) .$$

Furthermore from the weak convergence of $u_m \to u$, we obtain the energy inequality,

(A–14) $$E\{|u(t)|^2\} + \alpha E\left\{\int_{t_0}^{t} \|u(s)\|_{\mathscr{V}}^2\, ds\right\} \leq E\{|u_0|^2\} .$$

From Eq. (A–14), the remainder half of the proof can easily be completed.

Appendix 2. Proof of Theorem 2

By using the same procedure as demonstrated in Appendix 1 we can approximate the quadratic functional by

(B–1) $$V(t, u_m(t)) = (u_m(t), \Pi(t)u_m(t)) .$$

From Eq. (A–4), we have

(B–2) $$\begin{aligned}dV(t, u_m(t)) = (u_m(t), \dot{\Pi}(t)u_m(t))dt - 2\langle \tilde{A}_m(t)u_m(t), \Pi(t)u_m(t)\rangle dt \\ + (G_m(t)u_m(t), \Pi(t)G_m(t)u_m(t))dt \\ + 2(G_m(t)u_m(t), \Pi(t)u_m(t))dw(t) .\end{aligned}$$

Bearing in mind Eqs. (A–10) and (A–11), and passing to the limit, it follows that

$$V(t, u(t)) - V(t_0, u_0) = \int_{t_0}^{t} (u(s), [\dot{\Pi}(s) - \tilde{A}^*(s)\Pi(s) - \Pi(s)\tilde{A}(s) + G^*(s)\Pi(s)G(s)]u(s))ds + 2\int_{t_0}^{t} (G(s)u(s), \Pi(s)u(s))dw(s) . \tag{B–3}$$

KYOTO INSTITUTE OF TECHNOLOGY
MATSUGASAKI KYOTO 606, JAPAN

Proc. of Intern. Symp. SDE
Kyoto 1976, pp. 409–425

On the Uniqueness of Markov Process Associated with the Boltzmann Equation of Maxwellian Molecules

Hiroshi TANAKA

§1. Introduction

The spatially homogeneous Boltzmann equation of Maxwellian molecules takes the following form:

$$(1.1) \qquad \frac{\partial u(t,x)}{\partial t} = \int (u'u'_1 - uu_1)Q(d\theta)d\varepsilon dx_1 \,,$$

where $Q(d\theta) = Q_M(\theta)\sin\theta d\theta$ with a positive decreasing function $Q_M(\theta)$ such that $Q_M(\theta) \sim \text{const.} \times \theta^{-5/2}$, $\theta \downarrow 0$, and the integration is carried out over $(0,\pi) \times (0,2\pi) \times R^3$. In this paper we consider the following weak version of (1.1):

$$(1.2) \qquad \frac{d}{dt}\langle u, \varphi\rangle = \langle u \otimes u, K\varphi\rangle \,, \qquad \varphi \in C_0^\infty(R^3) \,,$$

where $C_0^\infty(R^3)$ is the space of real valued C^∞-functions on R^3 with compact supports and $(K\varphi)(x, x_1) = \int \{\varphi(x') - \varphi(x)\}Q(d\theta)d\varepsilon$; x' is on the sphere S_{x,x_1} with center $(x+x_1)/2$ and diameter $|x - x_1|$; θ and ε are the colatitude and the longitude of x', respectively, with respect to a spherical coordinate system on S_{x,x_1} having $x - x_1$ for the polar axis, and finally $u = u(t,\cdot)$ is a probability measure solution to be found.

McKean [3] introduced a class of Markov processes associated with certain nonlinear parabolic equations. The existence problem for (1.1) or (1.2) is almost equivalent to the problem of constructing an associated Markov process of the type introduced by McKean, but it seems that no rigorous results on the existence problem for (1.1) or (1.2), from (non-probabilistic) analysis, have been obtained. The difficulty lies in that $\int_0^\pi Q(d\theta)$ diverges. In [4] I have constructed a Markov process associated with (1.2) by solving the stochastic differential equation:

$$(1.3a) \qquad X_t = X_0 + \int_{(0,t]\times S} a(X_s, Y_s(\alpha), \theta, \varepsilon)\hat{p}(dsd\theta d\varepsilon d\alpha) \,,$$

where $S = (0, \pi) \times (0, 2\pi) \times (0, 1)$, $a(x, x_1, \theta, \varepsilon) = x' - x$ and

(i) $\hat{p}(dsd\theta d\varepsilon d\alpha)$ is a Poisson random measure on $(0, \infty) \times S$ associated with $dsQ(d\theta)d\varepsilon d\alpha$,

(ii) $\{Y_t(\alpha),\ t \geq 0\}$ is a stochastic process defined on the probability space $\{(0, 1), d\alpha\}$, and is equivalent in law to the process $\{X_t, t \geq 0\}$ to be found.

An alternative form of (1.3a) is

$$X_t = X_0 + \sum_{s \leq t} a(s, X_s, \hat{Z}_s)\ , \tag{1.3b}$$

where $a(s, x, \sigma) = a(x, Y_s(\alpha), \theta, \varepsilon)$ for $\sigma = (\theta, \varepsilon, \alpha)$ and $\{\hat{Z}_t, t > 0\}$ is the Poisson point process on S corresponding to the Poisson random measure $\hat{p}$, i.e., defined by the relation $\hat{p}(A) = \sum \chi_A(s, \hat{Z}_s)$ for $A \in \mathscr{B}(R_+ \times S)$. It has been also proved in [4] that the uniqueness in the law sense for solutions of (1.3) holds, however this uniqueness does not mean the uniqueness of the associated Markov process.

The purpose of this paper is to fill this gap by showing that *path functions of any Markov process associated with* (1.2) *can be represented as solutions of* (1.3) *after a suitable extension of basic probability space.* Our task is to find a Poisson point process $\{\hat{Z}_t,\ t > 0\}$ on S with characteristic measure $Q(d\theta)d\varepsilon d\alpha$ such that (1.3b) holds. For this we have to find the compensator of the point process $\{Z_t,\ t > 0\}$ on $R^3 - \{0\}$ defined by $Z_t = X_t - X_{t-}$; this will be done in § 3 and it will follow that $X_t = X_0 + \sum_{s \leq t} Z_s$. Once the compensator of $\{Z_t\}$ is found, a Poisson point process $\{\hat{Z}_t\}$ will be obtained by an application of general results on point processes due to Grigelionis [1] and Karoui-Lepeltier [2]; we will give a construction of $\{\hat{Z}_t\}$ in a simplified form adapted to the present situation.

§ 2. Markov process associated with (1.2)

We begin with the definition of transition function associated with (1.2). Denote by $\mathscr{P}$ the family of probability distributions f on R^3 satisfying $\int_{R^3} |x| f(dx) < \infty$, and for $f \in \mathscr{P}$ we put

$$(K_f\varphi)(x) = \int_{R^3} (K\varphi)(x, x_1) f(dx_1)\ .$$

Definition. $\{e_f(t, x, \cdot) : f \in \mathscr{P},\ t \geq 0,\ x \in R^3\}$ is called a *transition function associated with* (1.2), if the following five conditions are satisfied.

(e.1) For fixed $f \in \mathscr{P}$, $t \geq 0$ and $x \in R^3$, $e_f(t, x, \cdot)$ is a probability measure on R^3.

(e.2) For fixed $A \in \mathscr{B}(R^3)$, $e_f(t, x, A)$ is jointly measurable in $(f, t, x) \in \mathscr{P} \times R_+ \times R^3$.

(e.3) For each $t \geq 0$ and $f \in \mathscr{P}$, there exists a constant c depending only upon t and f such that

$$\int_{R^3} |y|\, e_f(s, x, dy) \leq c(1 + |x|)\,, \qquad 0 \leq s \leq t\,, \quad x \in R^3\,.$$

(e.4) If we set $u(t, \cdot) = \int_{R^3} f(dx) e_f(t, x, \cdot)$, then

$$\langle e_f(t, x, \cdot), \varphi \rangle = \varphi(x) + \int_0^t \langle e_f(s, x, \cdot), K_{u(s)}\varphi \rangle ds\,, \qquad \varphi \in C_0^\infty(R^3)\,.$$

(e.5) (Kolmogorov-Chapman equation)

$$e_f(t, x, \cdot) = \int_{R^3} e_f(s, x, dy) e_{u(s)}(t - s, y, \cdot)\,, \qquad 0 \leq s \leq t\,,$$

where u is the same as in (e.4).

Given a transition function $\{e_f(t, x, \cdot)\}$ associated with (1.2), we can construct a Markov process on R^3. To be precise, let Ω be the space of R^3-valued functions on R_+, and denote the value $\omega(t)$ of $\omega(\in \Omega)$ at t by $X_t(\omega)$ or X_t. We put $\mathscr{B} = \sigma\{X_t : t < \infty\}$ and $\mathscr{B}_t = \sigma\{X_s : s \leq t\}$, where $\sigma\{\text{—}\}$ denotes the smallest σ-field on Ω that makes $\{\text{—}\}$ measurable. Then there exists a unique family $\{P_f^x : x \in R^3, f \in \mathscr{P}\}$ of probability measures on $(\Omega, \mathscr{B})$ with the following properties (i) and (ii).

(i) For fixed $A \in \mathscr{B}$, $P_f^x(A)$ is jointly measurable in $(x, f) \in R^3 \times \mathscr{P}$.

(ii) For $0 < t_1 < \cdots < t_n$ and $A_k \in \mathscr{B}(R^3)$, $0 \leq k \leq n$,

$$\begin{aligned} P_f^x\{X_0 \in A_0, X_{t_1} \in A_1, \cdots, X_{t_n} \in A_n\} \\ = \delta_x(A_0) \int_{A_1} e_f(t_1, x, dx_1) \int_{A_2} e_{u(t_1)}(t_2 - t_1, x_1, dx_2) \\ \cdots \int_{A_n} e_{u(t_{n-1})}(t_n - t_{n-1}, x_{n-1}, dx_n)\,. \end{aligned}$$

We put $P_f(\cdot) = \int_{R^3} f(dx) P_f^x(\cdot)$, and denote by E_f (or E_f^x) the expectation with respect to P_f (or P_f^x). Then the following Markov property is proved by a routine method. For a nonegative $\mathscr{B}$-measurable function Φ on Ω we have

(2.1a) $$E_f^x\{\Phi(\Theta_t\omega)\,|\,\mathscr{B}_t\} = E_{u(t)}^{X_t}\{\Phi\}\ , \qquad P_f^x\text{-a.s.}\ ,$$

(2.1b) $$E_f\{\Phi(\Theta_t\omega)\,|\,\mathscr{B}_t\} = E_{u(t)}^{X_t}\{\Phi\}\ , \qquad P_f\text{-a.s.}\ ,$$

where $\Theta_t : \Omega \to \Omega$ is the shift operator defined by $X_s(\Theta_t\omega) = X_{t+s}(\omega)$ for $0 \leq s < \infty$.

Lemma 1. *If $\varphi : R^3 \to R$ is Lipschitz continuous, then*

(i) *there exists a constant c_1 depending only upon the Lipschitz constant of φ such that $|(K\varphi)(x, x_1)| \leq c_1|x - x_1|$,*

(ii) *for each $t \geq 0$ and $f \in \mathscr{P}$, there exists a constant c_2 depending only upon t, f and the Lipschitz constant of φ such that*

$$|(K_{u(s)}\varphi)(x)| \leq c_2(1 + |x|)\ , \qquad 0 \leq \forall s \leq t\ ,$$

where $u(s) = \int_{R^3} f(dx)e_f(s, x, \cdot)\ .$

Proof. From $|x' - x| \leq \theta|x - x_1|/2$, we have $|\varphi(x') - \varphi(x)| \leq \text{const.}\,|x - x_1|\theta$ and hence

$$|(K\varphi)(x, x_1)| \leq \text{const.}\,|x - x_1| \int_{(0,\pi)\times(0,2\pi)} \theta Q(d\theta)d\varepsilon = c_1|x - x_1|\ ,$$

proving (i). (ii) follows from (i) and (e.3).

The following lemma is an immediate consequence of the above lemma and (e.4),

Lemma 2. *If $\varphi : R^3 \to R$ is Lipschitz continuous, then*

$$\langle e_f(t, x, \cdot), \varphi\rangle = \varphi(x) + \int_0^t \langle e_f(s, x, \cdot), K_{u(s)}\varphi\rangle ds\ .$$

Lemma 3. $E_f^x\{|X_t - X_s|\} \leq \text{const.}\,(1 + |x|)\,|t - s|$ *for* $0 \leq s,\ t \leq T$, *where* const. *may depend upon f and T but not upon x.*

Proof. For $0 < s < t$ we have

$$E_f^x\{|x_t - x_s|\} = \int_{R^3} e_f(s, x, dy) \int_{R^3} e_{u(s)}(t - s, y, dz)\,|z - y|\ .$$

We now put $\varphi(z) = |z - y|$ for fixed y and apply Lemma 2:

$$\int_{R^3} e_{u(s)}(t - s, y, dz)\,|z - y| = \varphi(y) + \int_0^{t-s} \langle e_{u(s)}(\tau, y, \cdot), K_{u(s+\tau)}\varphi\rangle d\tau$$

$$\leq \int_0^{t-s} \langle e_{u(s)}(\tau, y, \cdot), |K_{u(s+\tau)}\varphi|\rangle d\tau$$

$$\leq c_2 \int_0^{t-s} d\tau \int_{R^3} e_{u(s)}(\tau, y, dz)(1 + |z|) .$$

Therefore we have

$$\begin{aligned} E_f^x\{|X_t - X_s|\} &\leq c_2 \int_0^{t-s} d\tau \int_{R^3} e_f(s, x, dy) \int_{R^3} e_{u(s)}(\tau, y, dz)(1 + |z|) \\ &= c_2 \int_0^{t-s} d\tau \int_{R^3} e_f(s + \tau, x, dz)(1 + |z|) \\ &\leq \text{const. } (1 + |x|)(t - s) . \end{aligned}$$

Theorem 1. *The stochastic process $\{X_t, P_f^x\}$ has a modification which is right continuous and of bounded variation on each finite t-interval.*

Proof. Let Y_t be any component of X_t. Let $Q_+ = \{r_k\}_{k \geq 1}$ be the set of nonnegative rational numbers, and for any partition of $[0, t]$:

$$\Delta : 0 = t_0 < t_1 < \cdots < t_n = t , \quad t_j \in Q_+ \quad (0 \leq j \leq n) ,$$

we put

$$V_t^\Delta = \sum_{j=1}^n |Y_{t_j} - Y_{t_{j-1}}| ,$$

$$U_t^\Delta = V_t^\Delta - (Y_t - Y_0) = 2 \sum_{k=1}^n (Y_{t_j} - Y_{t_{j-1}})^- ,$$

and then

$$V_t = \sup_\Delta V_t^\Delta , \quad U_t = \sup_\Delta U_t^\Delta , \quad t \in Q_+ ,$$

where the supremum is taken over all such partitions Δ. For each $t \in Q_+$ and $k \geq 1$ we write $\{r_1, \cdots, r_k\} \cap [0, t] = \{\tau_j\}_{1 \leq j \leq n}$ with $0 = \tau_0 < \cdots < \tau_n = t$ and then denote by Δ_k the partition of $[0, t]$ with partitioning points $\{\tau_j\}_{1 \leq j \leq n}$. Then V_t is clearly the increasing limit of $V_t^{\Delta_k}$ as $k \uparrow \infty$, and hence

$$\begin{aligned} E_f^x\{V_t\} &= \lim_{k \to \infty} E_f^x\{V_t^{\Delta_k}\} = \lim_{k \to \infty} \sum_{j=1}^n E_f^x\{|Y_{\tau_j} - Y_{\tau_{j-1}}|\} \\ &\leq \text{const. } (1 + |x|)t < \infty , \quad t \in Q_+ , \end{aligned}$$

and similarly $E_f^x\{U_t\} < \infty$, $t \in Q_+$. On the other hand, since V_t and U_t

are non-decreasing in $t \in Q_+$, $Y_t = Y_0 + V_t - U_t$, $t \in Q_+$, has right hand limits. Therefore the limit

$$\tilde{X}_t = \lim_{\substack{s \downarrow t \\ s \in Q_+}} X_s , \qquad t \in R_+$$

exists and gives a desired modification of X_t.

By virtue of Theorem 1, the probability measures P_f^x and P_f can be constructed on *the space of R^3-valued right continuous functions on R_+ having bounded variation on each finite t-interval*. From now on, Ω denotes this restricted space and so the σ-fields $\mathscr{B}$ and $\mathscr{B}_t$ are the ones on this new Ω. We call $\{\Omega, \mathscr{B}, X_t, P_f : f \in \mathscr{P}\}$ *a Markov process associated with* (1.2).

§ 3. Point process $\{Z_t\}$

3.1. In general suppose we are given a probability space $(\Omega, \mathscr{F}, P)$, a Borel subset S of R^d and an extra point ∂ not belonging to S. An $S \cup \{\partial\}$-valued process $\{Z_t(\omega), t > 0\}$ defined on $(\Omega, \mathscr{F}, P)$ is called *a point process on* S, if i) $Z_t(\omega)$ is jointly measurable in (t, ω), and ii) $\{t : Z_t(\omega) \in S\}$ is a countable set.

Given a point process $\{Z_t, t > 0\}$ on S, we put

$$p(A) = \sum_t \chi_A(t, Z_t) , \qquad A \in \mathscr{B}((0, \infty) \times S) ,$$

$$p_t(B) = \sum_{0 < s \le t} \chi_B(Z_s) , \qquad B \in \mathscr{B}(S) , \quad t \ge 0 .$$

Given also a σ-finite Borel measure λ on S, we introduce the following definition. A point process $\{Z_t, t > 0\}$ is called a *Poisson point process on S with characteristic measure* λ, if for any *disjoint* family $A_1, \cdots, A_n$ of Borel sets in $(0, \infty) \times S$ such that $\tilde{\lambda}(A_k) = \int_{A_k} dt d\lambda < \infty$ $(1 \le k \le n)$ we have

$$P\{p(A_k) = m_k, k = 1, \cdots, n\} = \prod_{k=1}^{n} \left\{ e^{-\tilde{\lambda}(A_k)} \frac{(\tilde{\lambda}(A_k))^{m_k}}{m_k !} \right\}$$
$$\text{for } m_1, \cdots, m_n \in N .$$

Then the following characterization of Poisson point processes is well-known.

Theorem 2. *Suppose we are given a point process $\{Z_t, t > 0\}$ on S, a σ-finite Borel measure λ on S and an increasing family $\{\mathscr{F}_t\}$ of sub-σ-fields of $\mathscr{F}$. If, for each $B \in \mathscr{B}(S)$ with $\lambda(B) < \infty$, $\{p_t(B) - \lambda(B)t,$*

$t \geq 0\}$ is an $\{\mathcal{F}_t\}$-martingale, then $\{Z_t, t > 0\}$ is a Poisson point process on S with characteristic measure λ.

In the above case, we also call $\{Z_t, t > 0\}$ an $\{\mathcal{F}_t\}$-adapted Poisson point process.

3.2. Suppose we are given a Markov process $\{\Omega, \mathcal{B}, X_t, P_f : f \in \mathcal{P}\}$ introduced in § 2, and put $Z_t = X_t - X_{t-}$. Then $\{Z_t, t > 0\}$ is a point process on $R_0^3 = R^3 - \{0\}$. The purpose of this subsection is to find the compensator of $\{Z_t\}$.

Fixing $f \in \mathcal{P}$, we introduce the following notations.

$$\mathcal{F} = \text{the completion of } \mathcal{B} \text{ with respect to } P_f \,,$$

$$\mathcal{F}_t^0 = \{A \in \mathcal{F} : P_f(A \ominus B) = 0 \text{ for some } B \in \mathcal{B}_t\} \,,$$

$$\mathcal{F}_t = \textstyle\bigcap_{\varepsilon>0} \mathcal{F}_{t+\varepsilon}^0 \,.$$

We now think of the unit interval $(0, 1)$ as a probability space by considering the Lebesgue measure on $\mathcal{B}((0, 1))$, and we take an R^3-valued stochastic process $\{Y_t(\alpha), t \geq 0\}$ defined on this probability space $(0, 1)$ with the following properties: (a) Sample paths of $\{Y_t\}$ are right continuous and have bounded variation on each finite t-interval, and (b) $\{Y_t, t \geq 0\}$ is equivalent in law to the process $\{X_t, t \geq 0, P_f\}$. Next, we put $S = (0, \pi) \times (0, 2\pi) \times (0, 1)$ and denote by $\sigma = (\theta, \varepsilon, \alpha)$ a generic element of S. On S we consider the Borel measure λ defined by $d\lambda = Q(d\theta) \otimes d\varepsilon \otimes d\alpha$. We also put

$$a(x, x_1, \theta, \theta, \varepsilon) = x' - x \,,$$

$$a(t, x, \sigma) = a(x, Y_t(\alpha), \theta, \varepsilon) \qquad \text{for } \sigma = (\theta, \varepsilon, \alpha) \,,$$

$$n(x, x_1, A) = \int_{(0,\pi)\times(0,2\pi)} \delta_{x'-x}(A) Q(d\theta) d\varepsilon \,, \qquad A \in \mathcal{B}(R_0^3) \,,$$

$$n_u(x, A) = \int_{R^3} n(x, x_1, A) u(dx_1) \,, \qquad A \in \mathcal{B}(R_0^3) \,.$$

Then, $\langle n_{u(t)}(y, \cdot), \psi \rangle = \langle \lambda, \psi(a(t, y, \cdot)) \rangle$ holds for any non-negative Borel function ψ on R^3 with $\psi(0) = 0$, where $u(t, \cdot) = \int_{R^3} f(dx) e_f(t, x, \cdot)$. Finally we put

$$p_t(\varphi) = \sum_{s \leq t} \varphi(s, X_{s-}, Z_s)$$

for a real valued Borel function φ on $R_+ \times R^3 \times R^3$ with $\varphi(t, y, 0) = 0$.

Theorem 3. *Let $\varphi : R_+ \times R^3 \times R^3 \to R$ be a Borel function with $\varphi(t, y, 0) = 0$ and assume that*

$$(3.1) \qquad \int_0^t ds \int_{R^3} u(s, dy) \int_{R_0^3} n_{u(s)}(y, dz)\, |\varphi(s, y, z)| < \infty$$

for each $t \in R_+$. Then

$$p_t(\varphi) - \int_0^t \langle \lambda, \varphi(s, X_s, a(s, X_s, \cdot)) \rangle ds$$

is an $\{\mathscr{F}_t\}$-martingale with respect to P_f. In other words

$$\int_0^t \langle \lambda, \varphi(s, X_s, a(s, X_s, \cdot)) \rangle ds$$

is the compensator of $p_t(\varphi)$.

Remark. The condition (3.1) is satisfied, if $|\varphi(t, y, z)| \leq$ const. $|z|$.

Before going to the proof of this theorem, we prepare some lemmas. First we assume that $\varphi \in C_0^\infty(R_+ \times R^3 \times R_0^3)$. For fixed $t > 0$ we consider a partition of $[0, t]$:

$$\Delta : 0 = t_0 < t_1 < \cdots < t_n = t\,,$$

and put

$$I_\Delta = E_f^x \Big\{ \sum_{k=1}^n \varphi(t_{k-1}, X_{t_{k-1}}, X_{t_k} - X_{t_{k-1}}) \Big\}\,.$$

Then, with the notation $\varphi^{k,y}(z) = \varphi(t_{k-1}, y, z - y)$ we have

$$\begin{aligned} I_\Delta &= \sum_{k=1}^n \int_{R^3} e_f(t_{k-1}, x, dy) \int_{R^3} e_{u(t_{k-1})}(t_k - t_{k-1}, y, dz) \varphi^{k,y}(z) \\ &= \sum_{k=1}^n \int_{R^3} e_f(t_{k-1}, x, dy) \int_0^{t_k - t_{k-1}} ds \int_{R^3} e_{u(t_{k-1})}(s, y, dz) K_{u(t_{k-1}+s)} \varphi^{k,y}(z)\,, \\ &= I'_\Delta + I''_\Delta\,, \end{aligned}$$

where

$$\begin{aligned} I'_\Delta = \sum_{k=1}^n \int_{R^3} e_f(t_{k-1}, x, dy) \int_0^{t_k - t_{k-1}} ds \int_{R^3} e_{u(t_{k-1})}(s, y, dz) \\ \times \chi_\varepsilon(z - y) K_{u(t_{k-1}+s)} \varphi^{k,y}(z)\,, \end{aligned}$$

and I''_Δ denotes a similar formula obtained from I'_Δ with the replacement of

$$\chi_\varepsilon(z - y) = \begin{cases} 1 & \text{for } |z - y| > \varepsilon \\ 0 & \text{for } |z - y| \leq \varepsilon \end{cases}$$

by $\bar{\chi}_\varepsilon(z-y) = 1 - \chi_\varepsilon(z-y)$.

Lemma 4. *If $\varphi \in C_0^\infty(R_+ \times R^3 \times R_0^3)$, then for any $\varepsilon > 0$*

$$I'_\Delta \to 0 \quad as \quad |\Delta| = \max_{1 \le k \le n} (t_k - t_{k-1}) \to 0 .$$

Proof. By Lemma 1 there exists a constant c independent of y, z, k and s such that

$$|K_{u(t_{k-1}+s)}\varphi^{k,y}(z)| \le c(1 + |z|) .$$

If we put

$$\rho(z) = \begin{cases} 0 & \text{for } |z| \le \varepsilon/2 \\ 2|z|/\varepsilon - 1 & \text{for } \varepsilon/2 < |z| < \varepsilon \\ 1 & \text{for } |z| \ge \varepsilon , \end{cases}$$

$$\xi^y(z) = c\rho(y-z)(1+|z|) ,$$

then we have

$$\chi_\varepsilon(z) \le \rho(z) ,$$

$$|I'_\Delta| \le \sum_{k=1}^n \int_{R^3} e_f(t_{k-1}, x, dy) \int_0^{t_k - t_{k-1}} ds \int_{R^3} e_{u(t_{k-1})}(s, y, dz)\xi^y(z) .$$

Since the support of φ is compact, the integral with respect to $e_{u(t_{k-1})}(s, y, dz)$ in the above may be performed only on some compact set B of R^3. Moreover, $\xi^y(z)$ is Lipschitz continuous as a function of z for each fixed y, and the Lipschitz constant is bounded as far as y is on B. Therefore, by Lemma 2 we have

$$\begin{aligned} \int_{R^3} e_{u(t_{k-1})}(s, y, dz)\xi^y(z) &= \int_0^s d\tau \int_{R^3} e_{u(t_{k-1})}(\tau, y, dz) K_{u(t_{k-1}+\tau)}\xi^y(z) \\ &\le \text{const.} \int_0^s d\tau \int_{R^3} e_{u(t_{k-1})}(\tau, y, dz)(1 + |z|) , \end{aligned}$$

and hence

$$\begin{aligned} |I'_\Delta| &\le \text{const.} \sum_{k=1}^n \int_{R^3} e_f(t_{k-1}, x, dy) \int_0^{t_k - t_{k-1}} ds \int_0^s d\tau \\ &\qquad \times \int_{R^3} e_{u(t_{k-1})}(\tau, y, dz)(1 + |z|) \\ &= \text{const.} \sum_{k=1}^n \int_0^{t_k - t_{k-1}} ds \int_0^s d\tau \int_{R^3} e_f(t_{k-1} + \tau, x, dz)(1 + |z|) \end{aligned}$$

$$\leq \text{const.} \sum_{k=1}^{n} (t_k - t_{k-1})^2 \to 0\,, \qquad \text{as } |\Delta| \to 0\,.$$

Lemma 5. *If $\varphi \in C_0^\infty(R_+ \times R^3 \times R_0^3)$, then*

$$\lim_{|\Delta|\to 0} I_\Delta = E_f^x\left\{\int_0^t \langle n_{u(s)}(X_s, \cdot), \varphi(s, X_s, \cdot)\rangle ds\right\}.$$

Proof. Define $\Delta(s)$, $0 \leq s \leq t$, by

$$\Delta(0) = 0\,, \quad \Delta(s) = t_{k-1} \qquad \text{for } t_{k-1} < s \leq t_k \ (1 \leq k \leq n)\,,$$

and for $\varepsilon > 0$ (small enough) put

$$\begin{aligned}\Phi(s, y, z, z_1) &= (K\varphi^{k,y})(z, z_1)\bar{\chi}_\varepsilon(z - y)\\ &= \int_{R_0^3} \varphi(s, y, z - y + w)\bar{\chi}_\varepsilon(z - y)n(z, z_1, dw)\,.\end{aligned}$$

Then

$$\begin{aligned}I'_\Delta &= \sum_{k=1}^{n} \int_{R^3} e_f(t_{k-1}, x, dy) \int_0^{t_k - t_{k-1}} ds \int_{R^3} e_{u(t_{k-1})}(s, y, dz)\\ &\qquad\qquad \times \int_{R^3} u(t_{k-1} + s, dz_1)\Phi(t_{k-1}, y, z, z_1)\\ &= \sum_{k=1}^{n} \int_0^{t_k - t_{k-1}} ds \int_0^1 d\alpha E_f^x\{\Phi(t_{k-1}, X_{t_{k-1}}, X_{t_{k-1}+s}, Y_{t_{k-1}+s}(\alpha))\}\\ &= \int_0^1 d\alpha E_f^x\left\{\int_0^t \Phi(\Delta(s), X_{\Delta(s)}, X_s, Y_s(\alpha))ds\right\}\\ &= \int_0^1 d\alpha E_f^x\left\{\int_{[0,t]\times R_0^3} dsn(x_s, Y_s(\alpha), dz)\right.\\ &\qquad\qquad \left.\times\, \varphi(\Delta(s), X_{\Delta(s)}, X_s - X_{\Delta(s)} + z)\bar{\chi}_\varepsilon(X_s - X_{\Delta(s)})\right\}\\ &\to \int_0^1 d\alpha E_f^x\left\{\int_{[0,t]\times R_0^3} dsn(X_{s-}, Y_s(\alpha), dz)\varphi(s, X_{s-}, z)\right\}\,, \ \text{as } |\Delta| \to 0\\ &= E_f^x\left\{\int_0^t \langle n_{u(s)}(X_s, \cdot), \varphi(s, X_s, \cdot)\rangle ds\right\}.\end{aligned}$$

This combined with Lemma 4 completes the proof.

Lemma 6. *Let $\varphi: R_+ \times R^3 \times R^3 \to R_+$ be a Borel function with $\varphi(t, y, 0) = 0$. Then*

$$E_f^x\{p_t(\varphi)\} = E_f^x\left\{\int_0^t \langle \lambda, \varphi(s, X_s, a(s, X_s, \cdot))\rangle ds\right\}. \tag{3.2}$$

Proof. We first consider the case $\varphi \in C_0^\infty(R_+ \times R^3 \times R_0^3)$. In this case we have

$$p_t^\Delta(\varphi) \equiv \sum_{k=1}^{n} \varphi(t_{k-1}, X_{t_{k-1}}, X_{t_k} - X_{t_{k-1}}) \to p_t(\varphi) ,$$

as $|\Delta| \to 0$. Since $|\varphi(s, y, z)| \leq \text{const.} |z|$, we have also

$$|p_t^\Delta(\varphi)| \leq \text{const.} \sum_{k=1}^{n} |X_{t_k} - X_{t_{k-1}}| \leq V ,$$

where V is some P_f^x-integrable random variable. Therefore by Lebesgue's dominated convergence theorem,

$$\begin{aligned} E_f^x\{p_t(\varphi)\} &= \lim_{|\Delta|\to 0} E_f^x\{p_t^\Delta(\varphi)\} \\ &= E_f^x\left\{\int_0^t \langle n_{u(s)}(X_s, \cdot), \varphi(s, X_s, \cdot)\rangle ds\right\} \\ &= E_f^x\left\{\int_0^t \langle \lambda, \varphi(s, X_s, a(s, X_s, \cdot))\rangle ds\right\} , \end{aligned}$$

and this proves (3.2) for smooth φ. Finally, it is easy to remove the smoothness condition of φ in (3.2).

We now come to the proof of Theorem 3. For $0 \leq s \leq t$ we have

$$p_t(\varphi) = p_s(\varphi) + p_{t-s}(\varphi^s) \circ \Theta_s ,$$

$$\begin{aligned} &\int_0^t \langle \lambda, \varphi(\tau, X_\tau, a(\tau, X_\tau, \cdot))\rangle d\tau \\ &\quad = \int_0^s \langle \lambda, \varphi(\tau, X_\tau, a(\tau, X_\tau, \cdot))\rangle d\tau \\ &\qquad + \int_0^{t-s} \langle n_{u(s+\tau)}(X_\tau(\Theta_s\omega), \cdot), \varphi^s(\tau, X_\tau(\Theta_s\omega), a(\tau, X_\tau(\Theta_s\omega), \cdot))\rangle d\tau , \end{aligned}$$

where $\varphi^s(\tau, y, z) = \varphi(s + \tau, y, z)$. Therefore, if we put

$$q_t(\varphi) = p_t(\varphi) - \int_0^t \langle \lambda, \varphi(\tau, X_\tau, a(\tau, X_\tau, \cdot))\rangle d\tau ,$$

then we have $q_t(\varphi) = q_s(\varphi) + q_{t-s}(\varphi^s) \circ \Theta_s$, and hence by using the Markov property (2.1a)

$$\begin{aligned} E_f\{q_t(\varphi) \,|\, \mathscr{B}_s\} &= q_s(\varphi) + E_f\{q_{t-s}(\varphi^s) \circ \Theta_s \,|\, \mathscr{B}_s\} \\ &= q_s(\varphi) + E_{u(s)}^{X_s}\{q_{t-s}(\varphi^s)\} \\ &= q_s(\varphi) , \qquad\qquad P_f\text{-a.s.} \end{aligned}$$

Thus $\{q_t(\varphi)\}$ is a $\{\mathscr{B}_t\}$-martingale, and hence an $\{\mathscr{F}_t\}$-martingale.

Corollary. $X_t = X_0 + \sum_{s\leq t} Z_s$, P_f-a.s. .

Proof. For $\varphi(z) = z$ we have $\sum_{s\leq t} Z_s = p_t(\varphi)$, and $E_f\{p_t(\varphi)\} = E_f\{X_t - X_0\}$ by making use of Theorem 3 and Lemma 2. Therefore $X_t - X_0 - \sum_{s\leq t} Z_s$ is a martingale with respect to P_f ; also it is continuous and has bounded variation on each finite t-interval. But, such a martingale must be 0, and so the corollary is proved.

§ 4. Derivation of stochastic differential equation

Notations are the same as in the subsection 3.2. We now know the compensator of the point process $\{Z_t\}$, and so the representation (1.3b) of $\{Z_t\}$ by means of certain Poisson point process $\{\hat{Z}_t\}$ might be a consequence of general works due to Grigelionis [1] and Karoui-Lepeltier [2]. However we give the construction of $\{\hat{Z}_t\}$ in detail, because we wish the whole proof to be self-contained. The construction given here seems to be simpler.

Theorem 4. *For each fixed $f \in \mathscr{P}$ we can find a probability space $\{\hat{\Omega}, \hat{\mathscr{F}}, \hat{P}\}$, an increasing family $\{\hat{\mathscr{F}}_t\}$, a mapping π from $\hat{\Omega}$ onto Ω and an $\{\hat{\mathscr{F}}_t\}$-adapted Poisson point process $\{\hat{Z}_t, t > 0\}$ on S with characteristic measure λ, having the following properties.*

(i) $\pi^{-1}(\mathscr{F}_t) \subset \hat{\mathscr{F}}_t$, $\pi^{-1}(\mathscr{F}) \subset \hat{\mathscr{F}}$ *and* $\hat{P}\{\pi^{-1}(A)\} = P_f\{A\}$ *for* $A \in \mathscr{F}$.

(ii) $X_t \circ \pi = X_0 \circ \pi + \sum_{s\leq t} a(s, X_{s-} \circ \pi, \hat{Z}_s)$, $\hat{P}$*-a.s.,*

or what is the same

$$X_t \circ \pi = X_0 \circ \pi + \int_{(0,t]\times S} a(X_s \circ \pi, Y_s(\alpha), \theta, \varepsilon)\hat{p}(dsd\sigma) , \qquad \hat{P}\text{-a.s.} ,$$

where $\hat{p}(A) = \sum \chi_A(t, \hat{Z}_t)$ *for* $A \in \mathscr{B}((0, \infty) \times S)$.

Lemma 7. *There exists $Q(t, y, z, A)$ such that*

(i) *for fixed $t \geq 0$, $y \in R^3$ and $z \in R^3_0$, $Q(t, y, z, \cdot)$ is a probability measure on S,*

(ii) *for fixed $A \in \mathscr{B}(S)$, $Q(t, y, z, A)$ is jointly measurable in (t, y, z),*

(iii) *for any nonnegative Borel function φ on $R^3 \times S$ with $\varphi(0, \sigma) = 0$ and for any $t \geq 0$, $y \in R^3$,*

$$\int_S \lambda(d\sigma)\varphi(a(t, y, \sigma), \sigma) = \int_{R^3_0\times S} n_{u(t)}(y, dz)Q(t, y, z, d\sigma)\varphi(z, \sigma) .$$

Proof of the lemma. For fixed $t \geq 0$, $y \in R^3$ and $A \in \mathscr{B}(S)$ we put

$$n^A_{u(t)}(y, B) = \int_A \chi_B(a(t, y, \sigma))\lambda(d\sigma) , \qquad B \in \mathscr{B}(R^3_0) .$$

Then $Q(t, y, z, A)$, as a function of z, should be the Radon-Nikodym derivative of $n^A_{u(t)}(y, \cdot)$ with respect to $n_{u(t)}(y, \cdot)$. A nice version of $Q(t, y, z, A)$ as stated in the lemma can be obtained by making use of the convergence theorem of martingales.

In order to prove Theorem 4 we must prepare some probability spaces together with various quantities defined on them.

1°. $\{\Omega, \mathscr{F}, P\}$: This is, of course, the basic probability space on which our Markov process $\{X_t\}$ has been given. Let

$$B_0 = \{z \in R^3 : |z| > 1\} , \qquad B_n = \left\{z \in R^3 : \frac{1}{n+1} < |z| \le \frac{1}{n}\right\}$$

$$n \ge 1 ,$$

and define $\{\mathscr{F}_t\}$-stopping times T_{nk}, $n \ge 0$, $k \ge 1$, by

$$T_{n0}(\omega) = \inf\{t > 0 : Z_t(\omega) \in B_n\} , \qquad n \ge 0$$

$$T_{nk}(\omega) = \inf\{t > T_{n,k-1}(\omega) : Z_t(\omega) \in B_n\} , \qquad n \ge 0 , k \ge 1 .$$

2°. $\{\Omega', \mathscr{F}', P'\}$: This is the probability space obtained by taking $\Omega' = (0, 1)$, $\mathscr{F}' = \mathscr{B}((0, 1))$ and P' = the Lebesgue measure (restricted on $\mathscr{F}'$). On this probability space we choose a sequence of independent random variables $\{\xi_{nk}(\omega') : n \ge 0,\ k \ge 1\}$, each being uniformly distributed on $(0, 1)$. Moreover we choose a jointly measurable function $Y(t, y, z, \omega')$ on $R_+ \times R^3 \times R^3_0 \times \Omega'$ such that, for each $t \ge 0$, $y \in R^3$ and $z \in R^3_0$, $Y(t, y, z, \cdot)$ is an S-valued random variable defined on $\{\Omega', \mathscr{F}', P'\}$ and with probability distribution $Q(t, y, z, \cdot)$.

3°. $\{\tilde{\Omega}, \tilde{\mathscr{F}}, \tilde{P}\}$: We choose an arbitrary probability space $\{\tilde{\Omega}, \tilde{\mathscr{F}}, \tilde{P}\}$ on which there is defined a Poisson point process $\{\tilde{Z}_t, t > 0\}$ on S with characteristic measure λ, and put $\tilde{\mathscr{F}}_t = \sigma\{\tilde{Z}_s : s \le t\}$.

Now we can construct all that we need.

(a) $\hat{\Omega} = \Omega \times \Omega' \times \tilde{\Omega}$, $\hat{\mathscr{F}} = \mathscr{F} \otimes \mathscr{F}' \otimes \tilde{\mathscr{F}}$, $\hat{P} = P_f \otimes P' \otimes \tilde{P}$.

(b) $\hat{\mathscr{F}}_t$ = the σ-field on $\hat{\Omega}$ generated by all sets of the form

$$(A \cap B) \times A' \times \tilde{A} , \tag{4.1}$$

where $B \in \mathscr{F}_t$, $\tilde{A} \in \tilde{\mathscr{F}}_t$ and

$$A = \{T_{nk} \le t \text{ for } \forall(n, k) \in M\} ,$$

$$A' = \{\xi_{nk} \in B_{nk} \text{ for } \forall(n, k) \in M\} ,$$

for some $M \subset N \times N$ and $B_{nk} \in \mathscr{B}((0,1))$.

(c)
$$Z'_t = \begin{cases} Y(T_{nk}, X_{nk}, Z_{nk}, \xi_{nk}) & \text{if } t = T_{nk} \\ 0 & \text{if } t \neq T_{nk} \quad \text{for } \forall(n,k)\,, \end{cases}$$

where $X_{nk} = X_{T_{nk}-}$ and $Z_{nk} = X_{T_{nk}} - X_{T_{nk}-}$.

(d)
$$Z''_t = \begin{cases} \tilde{Z}_t & \text{if } a(t, X_{t-}, \tilde{Z}_t) = 0 \\ 0 & \text{otherwise}\,. \end{cases}$$

(e)
$$\hat{Z}_t = Z'_t + Z''_t\,.$$

Then, (i) of Theorem 4 is obvious with the self-evident notation π (projection), and the rest will be proved in the following two lemmas.

Lemma 8. *$\{\hat{Z}_t, t > 0\}$ is an $\{\hat{\mathscr{F}}_t\}$-adapted Poisson point process on S with characteristic measure λ.*

Proof. Since the $\{\hat{\mathscr{F}}_t\}$-adaptedness of $\{\hat{Z}_t, t > 0\}$ is clear from the definition of $\hat{Z}_t$, it is enough to prove that

$$\hat{E}\{\hat{p}_t(\varphi) - \hat{p}_s(\varphi) \mid \hat{\mathscr{F}}_s\} = (t-s)\langle \lambda, \varphi \rangle\,, \qquad \hat{P}\text{-a.s.}$$

for any λ-integrable Borel function φ on S and $s \leq t$, where $\hat{p}_t(\varphi) = \sum_{s \leq t} \varphi(\hat{Z}_s)$. For this, it is also enough to prove that

$$\int_{\hat{A}} \{\hat{p}_t(\varphi) - \hat{p}_s(\varphi)\} d\hat{P} = (t-s)\langle \lambda, \varphi \rangle \hat{P}\{\hat{A}\}$$

for any set $\hat{A}$ of the form (4.1). Since

$$p'_t(\varphi) - p'_s(\varphi) = \sum_{n,k} \chi_{(s,t]}(T_{nk})\varphi(Y(T_{nk}, X_{nk}, Z_{nk}, \xi_{nk}))\,,$$

we have

$$\begin{aligned}
(4.2) \quad & \int_{(A \cap B) \times A'} \{p'_t(\varphi) - p_s(\varphi)\} d(P_f \otimes P') \\
&= \int_{(A \cap B) \times A'} \sum_{(n,k) \notin M} \chi_{(s,t]}(T_{nk})\varphi(Y(T_{nk}, X_{nk}, Z_{nk}, \xi_{nk})) d(P_f \otimes P') \\
&= P'\{A'\} \sum_{(n,k) \notin M} \int_{(A \cap B) \times \Omega'} \chi_{(s,t]}(T_{nk})\varphi(Y_{nk}, X_{nk}, Z_{nk}, \xi_{nk})) d(P_f \otimes P') \\
&= P'\{A'\} \int_0^1 d\alpha \int_{A \cap B} \sum_{s < \tau \leq t} \chi(Z_\tau)\varphi(Y(\tau, X_{\tau-}, Z_\tau, \alpha)) dP_f{}^{*)} \\
&= P'\{A'\} \int_0^1 d\alpha \int_{(A \cap B) \times (s,t] \times R_0^3} dP_f d\tau n_{u(\tau)}(X_{\tau-}, dz)\varphi(Y(\tau, X_{\tau-}, z, \alpha))
\end{aligned}$$

---

*) $\chi(z) \neq 1$ for $z \neq 0$, and $=0$ for $z = 0$.

$$= \int_{(A\cap B)\times A'} d(P_f \otimes P') \int_{(s,t]\times R_0^3 \times S} d\tau n_{u(\tau)}(X_{\tau-}, dz) Q(\tau, X_{\tau-}, z, d\sigma)\chi(z)\varphi(\sigma)$$

$$= \int_{(A\cap B)\times A'} d(P_f \otimes P') \int_{(s,t]\times S} d\tau\lambda(d\sigma)\chi(a(\tau, X_{\tau-}, \sigma))\varphi(\sigma);$$

in the above we used Theorem 3 and Lemma 7. We have also

$$(4.3) \qquad \int_{\hat{A}} \{p''_t(\varphi) - p''_s(\varphi)\} d\hat{P}$$

$$= \int_{(A\cap B)\times A'} d(P_f \otimes P') \int_{\tilde{A}} \{p''_t(\varphi) - p''_s(\varphi)\} d\tilde{P}$$

$$= \int_{(A\cap B)\times A'} d(P_f \otimes P') \int_{\tilde{A}} d\tilde{P} \int_{(s,t]\times S} d\tau\lambda(d\sigma) \times \{1 - \chi(a(\tau, X_{\tau-}, \sigma))\}\varphi(\sigma)$$

$$= \int_{\hat{A}} d\hat{P} \int_{(s,t]\times S} d\tau\lambda(d\sigma)\{1 - \chi(a(\tau, X_{\tau-}, \sigma))\}\varphi(\sigma).$$

Combining (4.2) with (4.3), we obtain

$$\int_{\hat{A}} \{\hat{p}_t(\varphi) - \hat{p}_s(\varphi)\} d\hat{P} = \tilde{P}\{\tilde{A}\} \int_{(A\cap B)\times A'} \{p'_t(\varphi) - p'_s(\varphi)\} d(P_f \otimes P') + \int_{\hat{A}} \{p''_t(\varphi) - p''_s(\varphi)\} d\hat{P} = (t-s)\langle\lambda, \varphi\rangle \hat{P}\{\hat{A}\},$$

as was to be proved.

Lemma 9. *$Z_t \circ \pi = a(t, X_{t-}, \hat{Z}_t)$, $\hat{P}$-a.s., in which we have put $a(t, y, 0) = 0$.*

Proof. Since the sets $\{t : Z'_t \in S\}$ and $\{t : Z''_t \in S\}$ are disjoint with $\hat{P}$-probability 1, we have $a(t, X_{t-}, \hat{Z}_t) = a(t, X_{t-}, Z'_t) + a(t, X_{t-}, Z''_t) = a(t, X_{t-}, Z'_t)$ with $\hat{P}$-probability 1, and hence it is enough to prove that for each fixed n and k

$$Z_{nk} = a(T_{nk}, X_{nk}, Y(T_{nk}, X_{nk}, Z_{nk}, \xi_{nk})), \qquad \hat{P}\text{-a.s.}.$$

Putting

$$\Phi^{\omega'}(t, y, z) = \chi_{B_n}(z)\,|\,z - a(t, y, Y(t, y, z, \xi_{nk}(\omega')))|,$$

and then applying Theorem 3 and Lemma 7, we have

$$
\begin{aligned}
&\hat{E}\{\chi_{(0,t]}(T_{nk})\,|Z_{nk} - a(T_{nk}, X_{nk}, Y(T_{nk}, X_{nk}, Z_{nk}, \xi_{nk}))|\} \\
&\quad \le \hat{E}\Big\{\sum_{s\le T_{nk}\wedge t} \Phi^{\omega'}(s, X_{s-}, Z_s)\Big\} \\
&\quad \le E'\Big\{\int_{[0,t]\times R^3\times R_0^3} ds u(s, dy) n_{u(s)}(y, dz)\Phi^{\omega'}(s, y, z)\Big\} \\
&\quad = \int_{[0,t]\times R^3\times R_0^3\times S} ds u(s, dy) n_{u(s)}(y, dz) Q(s, y, z, d\sigma)\chi_{B_n}(z)\,|z - a(s, y, \sigma)| \\
&\quad = \int_{[0,t]\times R^3\times S} ds u(s, dy)\lambda(d\sigma)\chi_{B_n}(a(s, y, \sigma))\,|a(s, y, \sigma) - a(s, y, \sigma)| = 0\ ,
\end{aligned}
$$

as required.

Combining the above two lemmas with the corollary to Theorem 3, we now complete the proof of Theorem 4.

§ 5. Concluding remark

The existence and uniqueness of solutions for the stochastic differential equation (1.3) were discussed by Tanaka [4]. The main results of [4] read as follows.

(A) Let $f \in \mathscr{P}$ be fixed. Then, on a suitable probability space $\{\hat{\Omega}, \hat{\mathscr{F}}, \hat{P}\}$ with an increasing family $\{\hat{\mathscr{F}}_t\}$ of sub-σ-fields we can construct an $\{\hat{\mathscr{F}}_t\}$-adapted Poisson point process $\{\hat{Z}_t, t > 0\}$ on S with characteristic measure λ and an $\{\hat{\mathscr{F}}_t\}$-adapted right continuous process $\{X_t, t \ge 0\}$ on R^3 with initial distribution f such that

(i) $\int_0^t \hat{E}\{|X_s|\}ds < \infty$ for each $t < \infty$,

(ii) (1.3b) holds with probability 1,

where $\{Y_t(\alpha), t \ge 0\}$ is some right continuous process defined on the probability space $\{(0, 1), d\alpha\}$ and is equivalent in law to the solution process $\{X_t, t \ge 0\}$. Moreover, the probability measure on the path space induced by $\{X_t, t \ge 0\}$ is uniquely determined from f.

(B) Let $f \in \mathscr{P}$ and $x \in R^3$ be fixed. Then, we can construct a Poisson point process $\{\hat{Z}_t, t > 0\}$ on S with characteristic measure λ and an $\{\hat{\mathscr{F}}_t\}$-adapted right continuous process $\{X_t^x, t \ge 0\}$ on R^3 such that

(iii) $\int_0^t \hat{E}\{|X_s^x|\}ds < \infty$ for each $t < \infty$,

(iv) $X_t^x = x + \sum_{s\le t} a(s, X_s^x, \hat{Z}_s)$, a.s.,

where $a(s, y, \sigma)$ is the same as used in (1.3b). Moreover, the probability measure on the path space induced by $\{X_t^x, t \ge 0\}$ is uniquely determined by f and x, and

$$
e_f(t, x, A) = \hat{P}\{X_t^x \in A\}\ , \qquad A \in \mathscr{B}(R^3)\ ,
$$

gives a transition function associated with (1.2) in the sense of § 2.

Combining these results with Theorem 4, we obtain the existence and *uniqueness* of Markov process associated with (1.2).

References

[1] Grigelionis, B., On representation of integer-valued random measures by means of stochastic integrals with respect to the Poisson measure, Liet. Mat. Sb., **XI** (1971), 93–108, (Russian).

[2] Karoui, N. E. and Lepeltier, J. P., Représentation des processus ponctuels multivariés à l'aide d'un processus de Poisson, Z. Wahrscheinlichkeitstheorie verw. Gebiete **39** (1977), 111–133.

[3] McKean, H. P., A class of Markov processes associated with non-linear parabolic equations, Proc. Nat. Acad. Sci. U.S.A., **56** (1966), 1907–1911.

[4] Tanaka, H., On Markov process corresponding to Boltzmann's equation of Maxwellian gas, Proc. 2nd Japan-USSR Symp. Prob. Th., Lecture Notes in Mathematics, **330**, Springer, 478–489.

[5] Uhlenbeck, G. E. and Ford, G. W., Lectures in statistical mechanics, Lectures in Appl. Math., vol. **1**, Amer. Math. Soc., Providence, 1963.

Department of Mathematics
Hiroshima University
Hiroshima 730, Japan

Proc. of Intern. Symp. SDE
Kyoto 1976, pp. 427–436

On the Stochastic Differential Equation for a Brownian Motion with Oblique Reflection on the Half Plane*)

Masaaki TSUCHIYA

§ 1. Introduction

Let $(B(t), b(t))$ be a two-dimensional Brownian motion starting from the origin and consider the stochastic differential equation:

$$(1) \qquad \begin{cases} dy(t) = db(t) + d\phi(t) \\ dx(t) = dB(t) + a(x(t))d\phi(t) , \end{cases}$$

where the first equation of (1) means the so-called Skorohod equation which determines a reflecting Brownian motion $y(t)$ on $[0, \infty)$ and its local time $\phi(t)$ at 0 (i.e., $y(t) = y(0) + b(t) + \phi(t)$ and $\phi(t) = -\min_{0 \leq s \leq t} \{(y(0) + b(s)) \wedge 0\}$**)).

The equation (1) is a special one of the stochastic equations introduced by N. Ikeda [4] and S. Watanabe [13], who solved the equations with Lipschitz continuous coefficients. By the results of Ikeda and Watanabe, it follows that if the coefficient $a(x)$ is Lipschitz continuous, the equation (1) has a unique solution $(x(t), y(t))$ and the process $\boldsymbol{P}^a$ defined by the solution $(x(t), y(t))$ becomes the Brownian motion with oblique reflection on the upper half plane $\bar{D} = \{(x, y); -\infty < x < \infty, y \geqq 0\}$ (i.e., it has the absorbing Brownian motion as a minimal process and satisfies the boundary condition $(\partial u/\partial y) + a(x)(\partial u/\partial x) = 0$).

In this paper, we shall treat the equation (1) with a *non-smooth coefficient* $a(x)$. For the equation (1) with a discontinuous coefficient $a(x)$, we need to generalize the concept of the solution. If a continuous process $x(t)$ satisfies

$$v \cdot (x(t_2) - x(t_1)) \leqq v \cdot (B(t_2) - B(t_1)) + \int_{t_1}^{t_2} v \cdot a(x(s))d\phi(s)$$

for $v = \pm 1$ and $t_1, t_2 \in [0, \infty)$, $t_1 < t_2$, then following E. D. Conway [2], we say that $(x(t), y(t))$ is a *relaxed solution* of the equation (1), where

*) A part of the results of the paper was announced in [11].
**) $a \wedge b = \min(a, b)$.

$$\overline{v \cdot a}(x) = \lim_{\delta \downarrow 0} \operatorname{ess\,sup}_{|y-x| \leqq \delta} v \cdot a(y) .$$

Our construction of the solutions of the equation (1) differs from the one of Ikeda and Watanabe and is carried out by piecing together the paths of the interior process and the paths of the boundary process (see Lemma 2.1). It should be noticed that such a piecing together procedure plays an important role in the construction of diffusion processes with boundary conditions by S. Watanabe [14].

The purpose of the paper is to show the existence of a relaxed solution such that it is a diffusion process on $\bar{D}$ and to give a relation between the relaxed solutions of the equation (1) and the processes constructed by M. Motoo (cf. [6]).

§ 2. Results

Assumption. In this note, we assume that the coefficient $a(x)$ is essentially bounded with respect to Lebesgue measure.

Definition 2.1. Let $(\Omega, \mathscr{F}, P; \mathscr{F}_t)$ be a probability space with an $(\mathscr{F}_t)$-adapted two-dimensional Brownian motion $(B(t), b(t))$ starting from the origin, i.e., $B(t)$ and $b(t)$ are continuous martingales with respect to $(\mathscr{F}_t)$ and for $s < t$ and $\lambda, \mu \in (-\infty, \infty)$,

$$E[\exp\{i\lambda(B(t) - B(s)) + i\mu(b(t) - b(s))\} | \mathscr{F}_s] \\ = \exp\{-\tfrac{1}{2}(t-s)(\lambda^2 + \mu^2)\} .$$

If a continuous process $x(t)$ which is defined on the probability space $(\Omega, \mathscr{F}, P; \mathscr{F}_t)$ and is adapted to $(\mathscr{F}_t)$ satisfies, with probability one, the inequality

$$v \cdot (x(t_2) - x(t_1)) \leqq v \cdot (B(t_2) - B(t_1)) + \int_{t_1}^{t_2} \overline{v \cdot a}(x(s)) d\phi(s)$$

for $v = \pm 1$ and all $t_1, t_2 \in [0, \infty)$, $t_1 < t_2$, then we say that $z(t) = (x(t), y(t))$ is a *relaxed solution* of the equation (1) on the probability space $(\Omega, \mathscr{F}, P; \mathscr{F}_t)$.

Remark 2.1. For an essentially bounded function $f(x)$,

$$\bar{f}(x) = \lim_{\delta \downarrow 0} \operatorname{ess\,sup}_{|y-x| \leqq \delta} f(y)$$

is a bounded Borel function, so that using Theorem 12, Chapter VII of P. A. Meyer [5], we have the equality

$$\int_{t_1}^{t_2} \overline{v \cdot a}(x(s))d\phi(s) = \int_{\phi(t_1)}^{\phi(t_2)} \overline{v \cdot a}(x(\phi^{-1}(s)))ds ,$$

where $\phi^{-1}(t) = \inf\{u ; \phi(u) > t\}$ (i.e., the inverse local time).

Then we have the following theorem.

Theorem 2.1. *On some probability space $(\Omega, \mathscr{F}, P ; \mathscr{F}_t)$, for each $z \in \bar{D}$ there exists a relaxed solution $z(t, z, \omega)$ of the equation* (1) *with initial value z such that $\{\Omega, z(t, z, \omega), \mathscr{F}_t, P\}$ is a diffusion process on the upper half plane $\bar{D}$.*

The proof will be given in [12].

We need the following stochastic differential equation on the boundary:

$$d\xi(t) = dl(t) + a(\xi(t))dt , \tag{2}$$

where $l(t)$ is a one-dimensional symmetric Cauchy process starting from the origin.

Definition 2.2. Let $(W, \mathscr{B}, Q ; \mathscr{B}_t)$ be a probability space with a $(\mathscr{B}_t)$-adapted one-dimensional symmetric Cauchy process $l(t)$ ($l(0) = 0$), i.e., $l(t)$ is a Lévy process defined on $(W, \mathscr{B}, Q ; \mathscr{B}_t)$, it is adapted to $(\mathscr{B}_t)$ and for $s < t$, $\lambda \in (-\infty, \infty)$,

$$E[\exp\{i\lambda(l(t) - l(s))\} | \mathscr{B}_s] = \exp\{-|\lambda| (t - s)\} .$$

By a *relaxed solution* of the equation (2) on the probability space $(W, \mathscr{B}, Q ; \mathscr{B}_t)$, we mean a process $\xi(t)$ defined on it such that

(i) with probability one, $\xi(t)$ is right continuous and has left-hand limits as a function of t,

(ii) it is adapted to $(\mathscr{B}_t)$ and

(iii) with probability one, it satisfies

$$v \cdot (\xi(t_2) - \xi(t_1)) \leq v \cdot (l(t_2) - l(t_1)) + \int_{t_1}^{t_2} \overline{v \cdot a}(\xi(s))ds$$

for $v = \pm 1$ and all $t_1, t_2 \in [0, \infty)$, $t_1 < t_2$.

The following lemma is necessary to prove Theorem 2.1 and gives our procedure of constructing solutions.

Lemma 2.1. (1°) *If on the probability space $(\Omega, \mathscr{F}, P ; \mathscr{F}_t)$ there exists a relaxed solution $z(t) = (x(t), y(t))$ of the equation* (1), *then on the probability space $(\Omega, \mathscr{F}, P ; \mathscr{F}_{\phi^{-1}(t)})$ with a Cauchy process $l(t) = B(\phi^{-1}(t)) - B(\phi^{-1}(0))$, $\xi(t) = x(\phi^{-1}(t))$ becomes a relaxed solution of the equation* (2) *with initial value $x(0) + B(\phi^{-1}(0))$. Moreover, let us put*

$$X(t) \equiv X(t,\omega) = \begin{cases} x(0) + B(t,\omega) & \textit{if } (t,\omega) \in \Omega^0 \\ \xi(\phi(t,\omega),\omega) & \textit{if } (t,\omega) \in \Omega^1 \\ \lim_{n\to\infty} \xi(\phi(t-1/n,\omega),\omega) & \textit{if } (t,\omega) \in \Omega^2 \\ \lim_{n\to\infty} \xi(\phi(\sigma-1/n,\omega),\omega) + B(t,\omega) - B(\sigma,\omega) & \\ & \textit{if } (t,\omega) \in \Omega^3 . \end{cases}$$

Then

$$P[x(t) = X(t) \textit{ for all } t \geqq 0] = 1 ,$$

where let us set

$$\Omega^0 = \{(t,\omega)\,;\, 0 \leqq t < \phi^{-1}(0,\omega)\} ,$$
$$\Omega^1 = \{(t,\omega)\,;\, \phi^{-1}(\phi(t,\omega),\omega) = t\} ,$$
$$\Omega^2 = \{(t,\omega)\,;\, \lim_{n\to\infty} \phi^{-1}(\phi(t-1/n,\omega),\omega) = t < \phi^{-1}(\phi(t,\omega),\omega)\} ,$$
$$\Omega^3 = \{(t,\omega)\,;\, \lim_{n\to\infty} \phi^{-1}(\phi(t-1/n,\omega),\omega) > t\}$$

and

$$\sigma \equiv \sigma(t,\omega) = \begin{cases} \lim_{n\to\infty} \phi^{-1}(\phi(t,\omega)-1/n,\omega) & \textit{if } (t,\omega) \in \Omega^3 \\ t & \textit{otherwise.} \end{cases}$$

(2°) *For the probability space $(\Omega, \mathscr{F}, P\,;\, \mathscr{F}_t)$, assume that $\mathscr{F}_t = \mathscr{F}_{t+}$ for all $t \geqq 0$ and $\mathscr{F}_0$ contains all null sets of $\mathscr{F}$ and further assume that for any $x \in (-\infty, \infty)$ there exists a relaxed solution $\xi(t,x,\omega)$ of the equation* (2) *with initial value x on the probability space $(\Omega, \mathscr{F}, P\,;\, \mathscr{F}_{\phi^{-1}(t)})$ such that $\xi(t,x,\omega)$ is $\mathscr{B}[0,\infty) \times \mathscr{B}(-\infty,\infty) \times \mathscr{F}^{*)}$-measurable in (t,x,ω). Define the process $X(t)$ as in* (1°) *from $\xi(t) = \xi(t, x(0) + B(\phi^{-1}(0)),\omega)$. Then $(X(t), y(t))$ becomes a relaxed solution of the equation* (1) *and $P[X(\phi^{-1}(t)) = \xi(t)$ for all $t \geqq 0] = 1$.*

For each $z = (x,y) \in \bar{D}$, we can construct the minimum relaxed solution $z_1(t,z,\omega) = (x_1(t\,;\,x,y\,;\,\omega), y(t,y,\omega))$ and the maximum relaxed solution $z_2(t,z,\omega) = (x_2(t\,;\,x,y\,;\,\omega), y(t,y,\omega))$ of the equation (1) with initial value z, i.e., for any relaxed solution $z(t,z,\omega) = (x(t\,;\,x,y\,;\,\omega), y(t,y,\omega))$,

$$P[x_1(t\,;\,x,y\,;\,\omega) \leqq x(t\,;\,x,y\,;\,\omega) \leqq x_2(t\,;\,x,y\,;\,\omega) \text{ for all } t \geqq 0] = 1$$

(cf. [11]). Hence, if

$$P[z_1(t,z,\omega) = z_2(t,z,\omega) \text{ for all } t \geqq 0] = 1$$

---

*) $\mathscr{B}[0,\infty)$ (resp. $\mathscr{B}(-\infty,\infty)$) is the σ-field consisting of Borel sets in $[0,\infty)$ (resp. $(-\infty,\infty)$).

for each $z \in \bar{D}$, we say that the *uniqueness* holds for the equation (1).

Definition 2.3. We say that a function $f(x)$ satisfies a *one-sided Lipschitz condition*, if one of the following conditions is satisfied;

(L-1) there exists a positive constant K such that

$$(x-y)(f(x)-f(y)) \leqq K(x-y)^2 \quad \text{for all } x, y \in (-\infty, \infty) \quad \text{or}$$

(L-2) there exists a positive constant K such that

$$-(x-y)(f(x)-f(y)) \leqq K(x-y)^2 \quad \text{for all } x, y \in (-\infty, \infty) .$$

Theorem 2.2. *Suppose that the coefficient $a(x)$ satisfies one of the conditions;*

(i) *$a(x)$ is bounded continuous,*

(ii) *$\|a\|_\infty < \sqrt{2\pi}$*) or*

(iii) *$a(x)$ satisfies a one-sided Lipschitz condition.*

Then the uniqueness holds for the equation (1) *and moreover if the condition* (i) *or* (ii) *holds, then the relaxed solutions coincide with the ordinary solutions.*

Proof. If $a(x)$ is bounded continuous, then by Theorem 3.1 in [9] the uniqueness holds.

Next assume $\|a\|_\infty < \sqrt{2\pi}$. Note that if x_0 is a Lebesgue point of $a(x)$, $\underline{a}(x_0) = \bar{a}(x_0) = a(x_0)$ and almost all real numbers are the Lebesgue point of $a(x)$, where

$$\underline{a}(x) = \lim_{\delta \downarrow 0} \operatorname*{ess\,inf}_{|y-x| \leqq \delta} a(y) .$$

For any process $\alpha(t, \omega)$ defined on the probability space $(\Omega, \mathscr{F}, P; \mathscr{F}_{\phi^{-1}(t)})$ such that it is adapted to $(\mathscr{F}_{\phi^{-1}(t)})$ and

$$P[|\alpha(t)| \leqq \|a\|_\infty \text{ a.e. } t] = 1 ,$$

let us set

$$\xi(t) = x + l(t) + \int_0^t \alpha(s) ds .$$

Then by the calculations in [10], we can show that there exists a positive constant K such that

$$E\left[\int_0^\infty e^{-t} f(\xi(t)) dt\right] \leqq K \|f\|_2 \qquad \text{for all } f \in L^2(-\infty, \infty),\ x \in (-\infty, \infty) .$$

*) $\|a\|_\infty = \text{ess sup}\, |a(x)|$.

Hence, for any null set N in $(-\infty, \infty)$,

$$P[\xi(t) \notin N \ a.e.\ t] = 1 .$$

Therefore, using Lemma 2.1 it follows that the relaxed solutions of the equation (1) coincide with the ordinary solutions of the equation (1) and moreover by Theorem 6.1 in [10], the uniqueness holds.

Finally, assume that $a(x)$ satisfies a one-sided Lipschitz condition. If the condition (L-1) holds, then the uniqueness is proved by the usual arguments in the theory of ordinary differential equations (cf. [3]) and if the condition (L-2) holds, then the similar arguments to the proof of Theorem 4.2 in [9] yield the uniqueness. Q.E.D.

The following proposition is used later.

Proposition 2.1. *For each $z = (x, y) \in \bar{D}$, there exists a sequence $\{a_n(x)\}$ of smooth functions such that $a_n(x) \to a(x)$ a.e. x as $n \to \infty$, $\|a_n\|_\infty \leqq \|a\|_\infty$ and the solution $z_n(t) = (x_n(t), y_n(t))$ of the equation (1) with the coefficient $a_n(x)$ and initial value z converges uniformly on each finite time interval to a relaxed solution $z(t) = (x(t), y(t))$ of the equation (1) with the coefficient $a(x)$ and initial value z.*

Proof. Let us set $a_n(x) = a(x) * \rho_n(x)$, where ρ_n is the usual mollifier with the support $\{x ; |x| \leqq 1/n\}$. On a probability space $(\Omega, \mathcal{F}, P ; \mathcal{F}_t)$, we consider the equation (1) with the coefficient $a_n(x)$ and initial value (x, y) and denote by $(x_n(t), y(t))$ the solution, i.e., with probability one $y(t) = y + b(t) + \phi(t)$ and

$$x_n(t) = x + B(t) + \int_0^{\phi(t)} a_n(x_n(\phi^{-1}(s)))ds \qquad \text{for all } t \geqq 0 .$$

It is easy to see that the distributions of the processes $(x_n(t), B(t), b(t), \phi(t))$ $(n = 1, 2, \cdots)$ are tight on the space $C[0, \infty) \times C[0, \infty) \times C[0, \infty) \times C[0, \infty)$*) and the processes satisfy the condition of the theorem in [8] page 9. So, we can construct a sequence**) of processes $(\tilde{x}_n(t), \tilde{B}_n(t), \tilde{b}_n(t), \tilde{\phi}_n(t))$ $(n = 0, 1, \cdots)$ defined on the probability space $([0, 1], \mathcal{B}[0, 1], dx)$***) such that with probability one $(\tilde{x}_n(t), \tilde{B}_n(t), \tilde{b}_n(t), \tilde{\phi}_n(t))$ converges to $(\tilde{x}_0(t), \tilde{B}_0(t), \tilde{b}_0(t), \tilde{\phi}_0(t))$ in the sense of the metric of $C[0, \infty) \times C[0, \infty) \times C[0, \infty) \times C[0, \infty)$ as $n \to \infty$ and moreover they satisfy

$$(3) \quad \tilde{x}_n(t) = x + \tilde{B}_n(t) + \int_0^{\tilde{\phi}_n(t)} a_n(\tilde{x}_n(\tilde{\phi}_n^{-1}(s)))ds \qquad (n = 1, 2, \cdots)$$

*) For the notation and the metric of $C[0, \infty)$ see [1].

**) For simplicity, we write a subsequence like a whole sequence.

***) dx denotes Lebesgue measure.

and $(\tilde{B}_n(t), \tilde{b}_n(t))$ $(n = 0, 1, \cdots)$ are two-dimensional Brownian motions $(\tilde{B}_n(0) = \tilde{b}_n(0) = 0)$, $\tilde{\phi}_n(t) = -\min_{0 \leq s \leq t} \{(y + \tilde{b}_n(s)) \wedge 0\}$ and $\tilde{y}_n(t) = y + \tilde{b}_n(t) + \tilde{\phi}_n(t)$ $(n = 0, 1, \cdots)$.

Using Theorem 12, Chapter VII in [5] and the fact that 0 is regular for $\{0\}$ with respect to $\tilde{y}_n(t)$ $(n = 0, 1, \cdots)$, we can show that with probability one, $\tilde{\phi}_n^{-1}(t)$ converges to $\tilde{\phi}_0^{-1}(t)$ at each continuity point t of the function $\tilde{\phi}_0^{-1}(\cdot)$.

If n tends to infinity in the equation (3), it follows that with probability one

$$\tilde{x}_0(t) = x + \tilde{B}_0(t) + \lim_{n\to\infty} \int_0^{\tilde{\phi}_n(t)} a_n(\tilde{x}_n(\tilde{\phi}_n^{-1}(s)))ds \qquad \text{for all } t \geqq 0 .$$

Let us set $\tilde{\xi}_n(t) = \tilde{x}_n(\tilde{\phi}_n^{-1}(t))$ $(n = 0, 1, \cdots)$. If $\tilde{\phi}_0^{-1}(t)$ is continuous at s_0, then $\tilde{\xi}_n(s_0) \to \tilde{\xi}_0(s_0)$ as $n \to \infty$, i.e., for any $\delta > 0$ there exists a positive integer n_1 such that $|\tilde{\xi}_n(s_0) - \tilde{\xi}_0(s_0)| \leqq \delta/4$ for all $n \geqq n_1$. Therefore, for all $n \geqq n_1$,

$$a_n(\tilde{\xi}_n(s_0)) \leqq \operatorname*{ess\,sup}_{|y-\tilde{\xi}_n(s_0)| \leqq \delta/4} a_n(y) \leqq \operatorname*{ess\,sup}_{|y-\tilde{\xi}_0(s_0)| \leqq \delta/2} a_n(y) .$$

On the other hand,

$$\operatorname*{ess\,sup}_{|y-\tilde{\xi}_0(s_0)| \leqq \delta/2} a_n(y) = \operatorname*{ess\,sup}_{|y-\tilde{\xi}_0(s_0)| \leqq \delta/2} \int_{|z| \leqq 1/n} \rho_n(z) a(y-z)dz .$$

So, if $n > 2/\delta$ and $|y - \tilde{\xi}_0(s_0)| \leqq \delta/2$, then $|y - z - \tilde{\xi}_0(s_0)| \leqq \delta$ for any $z : |z| \leqq 1/n$ and hence for all $n > 2/\delta$

$$\int_{|z| \leqq 1/n} \rho_n(z) a(y-z)dz \leqq \int_{|z| \leqq 1/n} \rho_n(z) \operatorname*{ess\,sup}_{|y-\tilde{\xi}_0(s_0)| \leqq \delta} a(y)dz = \operatorname*{ess\,sup}_{|y-\tilde{\xi}_0(s_0)| \leqq \delta} a(y) .$$

Setting $n_0 = \max(n_1, [2/\delta] + 1)$, for all $n \geqq n_0$

$$a_n(\tilde{\xi}_n(s_0)) \leqq \operatorname*{ess\,sup}_{|y-\tilde{\xi}_0(s_0)| \leqq \delta} a(y) .$$

Calculating similarly, for any $\delta > 0$ there exists a positive integer n_0 such that for all $n \geqq n_0$

$$a_n(\tilde{\xi}_n(s_0)) \geqq \operatorname*{ess\,inf}_{|y-\tilde{\xi}_0(s_0)| \leqq \delta} a(y) .$$

Thus if $\tilde{\phi}_0^{-1}(t)$ is continuous at s_0, then for $v = \pm 1$

$$\overline{\lim_{n\to\infty}}\, v \cdot a_n(\tilde{\xi}_n(s_0)) \leqq \overline{v \cdot a}(\tilde{\xi}_0(s_0)) .$$

Therefore using the uniform boundedness of $\{a_n(x)\}$,

$$\lim_{n\to\infty}\int_0^{\phi_0(t)} v\cdot a_n(\tilde{\xi}_n(s))ds \leqq \int_0^{\phi_0(t)} \overline{\lim_{n\to\infty}}\, v\cdot a_n(\tilde{\xi}_n(s))\,ds$$
$$\leqq \int_0^{\phi_0(t)} \overline{v\cdot a(\tilde{\xi}_0(s))}ds$$

and

$$\lim_{n\to\infty}\int_0^{\tilde{\phi}_n(t)} v\cdot a_n(\tilde{\xi}_n(s))ds = \lim_{n\to\infty}\int_0^{\phi_0(t)} v\cdot a_n(\tilde{\xi}_n(s))ds\,.$$

Consequently, with probability one

$$\begin{cases} \tilde{y}_0(t) = y + \tilde{b}_0(t) + \tilde{\phi}_0(t) & \text{for all } t \geqq 0 \\ v\cdot(\tilde{x}_0(t_2) - \tilde{x}_0(t_1)) \leqq v\cdot(\tilde{B}_0(t_2) - \tilde{B}_0(t_1)) \\ \qquad + \displaystyle\int_{\tilde{\phi}_0(t_1)}^{\tilde{\phi}_0(t_2)} \overline{v\cdot a(\tilde{x}_0(\tilde{\phi}_0^{-1}(s)))}ds \end{cases}$$

for $v = \pm 1$ and all $t_1, t_2 \in [0, \infty)$, $t_1 < t_2$, and this shows that $(\tilde{x}_0(t), \tilde{y}_0(t))$ is a relaxed solution of the equation (1) with the coefficient $a(x)$. Q.E.D.

Recently, M. Motoo (cf. [6]) has investigated in general set up the boundary problem for a Brownian motion in the upper half plane. Let $\mathscr{P}$ be the family of periodic processes with period 2π on D^* and $\mathscr{P}_0$ be the closure of smooth processes in $\mathscr{P}$, which are introduced by Motoo [6], where $D^* = D \cup \{\partial\}$ denotes the topological space obtained by identifying the boundary ∂D of $D = \{(x, y)\,;\, -\infty < x < \infty,\, y > 0\}$ in $\bar{D}$. Denote by Φ the natural mapping from $\bar{D}$ to D^*.

Motoo has shown that there exists a one to one correspondence between $\mathscr{P}_0$ and the family $\mathfrak{B}$ of the equivalence classes $[a(x)]$ for every periodic function $a(x)$ ($\not\equiv \pm\infty$ a.e. x) with period 2π (cf. Theorem 8.3 in [6]).

In the following theorem, we assume that the coefficient $a(x)$ of the equation (1) is periodic with period 2π.

Theorem 2.3. *Suppose that the uniqueness holds for the equation* (1). *Then the process $\Phi z(t)$ in $\mathscr{P}$ obtained from the relaxed solution $z(t)$ of the equation* (1) *with the coefficient $a(x)$ becomes Motoo's process $\mathbf{P}^a$ in $\mathscr{P}_0$ corresponding to* $[a(x)]$.

Proof. By virtue of the uniqueness, the approximating sequence $\{a_n(x)\}$ of the coefficient $a(x)$ in Proposition 2.1 can be chosen independently of the initial value $z = (x, y) \in \bar{D}$.

Therefore, using the uniform convergence of $(x_n(t), y_n(t))$, we can easily prove that $\boldsymbol{P}^{a_n} \rightarrow \boldsymbol{P}^a$ in $\mathscr{P}$ as $n \rightarrow \infty$, where $\boldsymbol{P}^{a_n}$ (resp. $\boldsymbol{P}^a$) denotes the process defined by $\Phi\tilde{z}_n(t) = \Phi(\tilde{x}_n(t), \tilde{y}_n(t))$ (resp. $\Phi\tilde{z}_0(t) = \Phi(\tilde{x}_0(t), \tilde{y}_0(t))$). Q.E.D.

As a consequence of the above theorems, we can use the results of M. Motoo [7]. For example,

(i) $$a(x) = \begin{cases} -1 & x < 0 \\ 1 & x > 0 , \end{cases}$$

then the uniqueness holds for the equation (1) with the coefficient $a(x)$ and with probability one, the path of the solution $z(t)$ cannot hit the origin.

(ii) $$a(x) = \begin{cases} 1 & x < 0 \\ -1 & x > 0 , \end{cases}$$

then the uniqueness holds for the equation (1) with the coefficient $a(x)$ and the path of the solution $z(t)$ hits the origin with positive probability. For the solution of the equation (1) with a smooth coefficient, such a fact does not occur.

References

[1] Billingsley, P., Convergence of probability measures, Wiley, 1968.

[2] Conway, E. D., Stochastic equations with discontinuous drift I, II, Trans. Amer. Math. Soc., **157** (1971), 235–245 and Indiana Univ. Math. J., **22** (1972), 91–99.

[3] Filippov, A. F., Differential equations with discontinuous right-hand side, Mat. Sb., **51-93** (1960), 99–168, (English transl.: Amer. Math. Soc. Transl., **2-42** (1964), 199–231).

[4] Ikeda, N., On the construction of two dimensional diffusion processes satisfying Wentzell's boundary conditions and its application to boundary value problems, Mem. Coll. Sci. Univ. Kyoto, Ser. A, **33** (1961), 367–427.

[5] Meyer, P. A., Probabilités et potentiel, Hermann, 1966.

[6] Motoo, M., Brownian motions in the half plane with singular inclined periodic boundary conditions, Topics in Prob. Theory, edited by D. W. Stroock and S. R. S. Varadhan, New York Univ. (1973), 163–179.

[7] ——, Hitting probability of a single point set with respect to a Brownian motion on the half plane with oblique reflection, Semi. on Prob., **40** (1973), 23–48, (in Japanese).

[8] Skorohod, A. V., Studies in the theory of random processes, Kiev, 1961, (English transl.: Addison Wesley, 1965).

[9] Tanaka, H., Tsuchiya, M. and Watanabe, S., Perturbation of drift-type for Lévy processes, J. Math. Kyoto Univ., **14** (1974), 73–92.

[10] Tsuchiya, M., On a small drift of Cauchy process, J. Math. Kyoto Univ., 10 (1970), 475-492.

[11] ——, On the relaxed solutions of a certain stochastic differential equation, III USSR-Japan Symp. on Prob. Theory II, Tashkent, 1975, 286–295.

[12] ——, On the stochastic differential equation for a two dimensional Brownian motion with boundary condition, in preparation.

[13] Watanabe, S., On stochastic differential equations for multi-dimensional diffusion processes with boundary conditions I, II, J. Math. Kyoto Univ., 11 (1971), 169–180 and 545–551.

[14] ——, Construction of diffusion processes with Wentzell's boundary conditions by means of Poisson point processes, III USSR-Japan Symp. on Prob. Theory II, Tashkent, 1975, 311–345.

College of Liberal Arts
Kanazawa University
Kanazawa 920, Japan

Proc. of Intern. Symp. SDE
Kyoto 1976, pp. 437–461

Excursion Point Process of Diffusion and Stochastic Integral

Shinzo WATANABE

Dedicated to Professor Kiyosi Itô on his 60th birthday

§ 1. Introduction

Let $g(t)$ be a locally summable function on $[0, \infty)$ and $e_\alpha = (l_\alpha, r_\alpha)$ be a family of disjoint open intervals in $(0, \infty)$ such that $[0, \infty)\backslash\bigcup_\alpha e_\alpha$ has the Lebesgue measure 0. Then, by a well known theorem of Lebesgue integration,

$$\int_0^t g(s)ds = \sum_\alpha \int_{l_\alpha\wedge t}^{r_\alpha\wedge t} g(s)ds . \tag{1.1}$$

In other words, setting $G(t) = \int_0^t g(s)ds$,

$$G(t) = \sum_\alpha [G(r_\alpha \wedge t) - G(l_\alpha \wedge t)] , \tag{1.1$'$}$$

the right-hand side being absolutely convergent. Now, let $B(t)$ be a one-dimensional Wiener process and $g(s)$ be a non-anticipative measurable process such that $\int_0^t g^2(s)ds < \infty$ a.s. for all $t > 0$. Then, by Itô's stochastic integral, $G(t) = \int_0^t g(s)dB(s)$ is defined as a continuous process. It is well known that, if $\int_0^t |g(s)|\, ds = 0$ a.s. then $G(t) = 0$ a.s.. But, for a family of disjoint random intervals e_α such that $[0, \infty)\backslash\bigcup_\alpha e_\alpha$ has the Lebesgue measure 0 a.s., (1.1)$'$ is no longer true even if the right-hand side is absolutely convergent, as is seen in the following example.

Let $(X(t), B(t), \psi(t))$ be a system of real stochastic processes such that, with probability one,

(i) $t \to X(t)$ is continuous, $X(0) = 0$ and $X(t) \geq 0$,

(ii) $t \to \psi(t)$ is continuous, non-decreasing, $\psi(0) = 0$ and $\psi(t) = \int_0^t I_{\{0\}}(X(s))d\psi(s)$,

(iii) $t \to B(t)$ is continuous, $B(0) = 0$ and it is a one-diemnsional Wiener process,

(iv) $X(t) = B(t) + \psi(t)$.

It is well known, by Lévy and Skorohod (cf. [9], [12]), that such a system exists, is unique in the law sense and $X(t)$ is a *reflecting Brownian motion* on $[0, \infty)$ (i.e., equivalent in law to the absolute value of a Wiener process): given $B(t)$, $X(t)$ and $\psi(t)$ are found as $\psi(t) = -\operatorname{Min}_{0 \le s \le t} B(s)$ and $X(t) = B(t) + \psi(t)$, and, given $X(t)$, $B(t)$ and $\psi(t)$ are found as

$$\psi(t) = \lim_{\varepsilon \downarrow 0} \frac{1}{2\varepsilon} \int_0^t I_{[0,\varepsilon)}(X(s))ds$$

and $B(t) = X(t) - \psi(t)$.

Now, let $Z = \{s; X(s) = 0\}$. It is well known that the Lebesgue measure of $Z = 0$ a.s. and $[0, \infty)\backslash Z$ is a countable disjoint union of open intervals $e_\alpha = (l_\alpha, r_\alpha)$. e_α is called an *excursion interval* of $X(t)$ and the part $\{X(t), t \in e_\alpha\}$ is called an *excursion*. Let $g(s) \equiv 1$. Then $G(t) = \int_0^t g(s)dB(s) = B(t)$ and, by (ii),

$$\begin{aligned} B(r_\alpha \wedge t) - B(l_\alpha \wedge t) &= X(r_\alpha \wedge t) - X(l_\alpha \wedge t) \\ &= \begin{cases} 0 & \text{if } r_\alpha \le t \text{ or } l_\alpha \ge t \\ X(t) & \text{if } l_\alpha < t < r_\alpha . \end{cases} \end{aligned}$$

Thus, we have $B(t)$ in the left-hand side of (1.1)′ and $X(t) = B(t) + \psi(t)$ in the right-hand side.

This fact suggests us that, generally, if $g(s)$ is $\sigma(X(u), u \le s)$-adapted and $G(t) = \int_0^t g(s)dB(s)$, then the following formula holds:

$$(1.2) \qquad \sum_\alpha [G(r_\alpha \wedge t) - G(l_\alpha \wedge t)] = G(t) + \int_0^t g(s)d\psi(s) .$$

In fact, we can show that (1.2) is true if $g(s)$ is right continuous with left-hand limits and if we interpret the sum in the left-hand side (which is usually not absolutely convergent) as the limit in probability of the finite sum $\sum^\alpha_{r_\alpha - l_\alpha > \varepsilon} [G(r_\alpha \wedge t) - G(l_\alpha \wedge t)]$ as $\varepsilon \downarrow 0$.

Purpose of this paper is to obtain some formulas like (1.2) in a general case of multi-dimensional diffusion processes with boundary conditions and discuss some of their applications.

§ 2. A class of diffusion processes with boundary conditions and main theorems

Let $D = \{x = (x_1, x_2, \cdots, x_n); x_n \ge 0\}$ be the upper half-space of

R^n, $\partial D = \{x \in D\,; x_n = 0\}$ be its boundary and $\mathring{D} = \{x \in D\,; x_n > 0\}$ be its interior. Suppose that the following functions are given:

$$\begin{aligned}
\sigma(x) &= \{\sigma_k^i(x)\}_{i,k=1}^n: && x \in D \longrightarrow R^n \otimes R^n \\
b(x) &= \{b^i(x)\}_{i=1}^n: && x \in D \longrightarrow R^n \\
\tau(x) &= \{\tau_l^i(x)\}_{i,l=1}^{n-1}: && x \in \partial D \longrightarrow R^{n-1} \otimes R^{n-1} \\
\beta(x) &= \{\beta^i(x)\}_{i=1}^{n-1}: && x \in \partial D \longrightarrow R^{n-1}.
\end{aligned}$$

Set

$$(2.1) \qquad a^{ij}(x) = \sum_{k=1}^n \sigma_k^i(x)\sigma_k^j(x)\,, \quad x \in D\,, \quad i, j = 1, 2, \cdots, n$$

$$(2.2) \qquad \alpha^{ij}(x) = \sum_{l=1}^{n-1} \tau_l^i(x)\tau_l^j(x)\,, \quad x \in \partial D\,, \quad i, j = 1, 2, \cdots, n-1$$

and we assume that

(2.3) *all σ, b, τ and β are bounded, Lipschitz continuous, $a^{ij}(x)$ are of class C^3 and $a^{nn}(x) \geq c$ for some positive constant c.*

Let $(\Omega, \mathscr{F}, P)$ be a complete probability space and $(\mathscr{F}_t)_{t\geq 0}$ be an right-continuous increasing family of sub σ-fields of $\mathscr{F}$ each containing all P-null sets. Let $(X(t), B(t), M(t), \psi(t))$ be a system of stochastic processes on $(\Omega, \mathscr{F}, P)$ *adapted to the family* $(\mathscr{F}_t)$ such that

(i) $t \to X(t)$ is a D-valued continuous process,
$t \to B(t)$ is an R^n-valued continuous process,
$t \to M(t)$ is an R^{n-1}-valued continuous process,
$t \to \psi(t)$ is an $R^+ = [0, \infty)$-valued continuous increasing process,

such that $B(0) = 0$, $M(0) = 0$, $\psi(0) = 0$ a.s.

(ii) $\psi(t)$ increases only when $X(t) \in \partial D$, i.e.,

$$\int_0^t I_{\partial D}(X(s))d\psi(s) = \psi(t) \qquad \text{a.s.}\,,$$

(iii) $(B(t), M(t))$ is a system of $\mathscr{F}_t$-martingales such that $\langle B^i, B^j\rangle_t = \delta_{ij}\cdot t$, $\langle B^i, M^l\rangle_t = 0$ and $\langle M^l, M^k\rangle_t = \delta_{lk}\cdot\psi(t)$, $i, j = 1, 2, \cdots, n$, $l, k = 1, 2, \cdots, n-1$,

(iv) with probability one,

$$X^i(t) - X^i(0) = \sum_{k=1}^n \int_0^t \sigma_k^i(X(s))dB^k(s) + \int_0^t b^i(X(s))ds$$

$$(2.4) \qquad + \sum_{l=1}^{n-1} \int_0^t \tau_l^i(X(s))dM^l(s) + \int_0^t \beta^i(X(s))d\psi(s)$$

$$i = 1, 2, \cdots, n-1$$

$$X^n(t) - X^n(0) = \sum_{k=1}^{n} \int_0^t \sigma_k^n(X(s))dB^k(s) + \int_0^t b^n(X(s))ds + \psi(t)$$

where $\int dB$ and $\int dM$ are understood in the sense of martingale stochastic integral, (cf. [10]).

It is shown in [15] that, for a given Borel probability measure μ on D, there exists such a system $\mathfrak{X} = (X(t), B(t), M(t), \psi(t))$ defined on a suitable $(\Omega, \mathcal{F}, P; \mathcal{F}_t)$ such that $P(X(0) \in dx) = \mu(dx)$ and furthermore, the law of any such system $\mathfrak{X}$ is uniquely determined. $X = (X(t))_{t\geq 0}$ defines a diffusion process on D and it satisfies

$$(2.5) \qquad \int_0^t I_{\partial D}(X(s))ds = 0 \qquad \text{a.s.} .$$

Roughly, X is the diffusion process corresponding to the differential operator

$$(2.6) \qquad Af(x) = \frac{1}{2} \sum_{i,j=1}^{n} a^{ij}(x)f''_{x_i x_j}(x) + \sum_{i=1}^{n} b^i(x)f'_{x_i}(x)$$

with Wentzell's boundary condition (cf. [17])

$$(2.7) \qquad Lf(x) \equiv \frac{1}{2} \sum_{i,j=1}^{n-1} \alpha^{ij}(x)f''_{x_i x_j}(x) + \sum_{i=1}^{n-1} \beta^i(x)f'_{x_i}(x) + f'_{x_n}(x) = 0 ,$$

$$x \in \partial D .$$

In the following, we fix the above system $\mathfrak{X} = (X(t), B(t), M(t), \psi(t))$. Set

$$(2.8) \qquad \boldsymbol{Z} = \{s \geq 0 : X(s) \in \partial D\} .$$

Then, by (2.5), $\boldsymbol{Z}$ has the Lebesgue measure 0, a.s., and $(0, \infty)\backslash \boldsymbol{Z} = \bigcup e_\alpha$ where $e_\alpha = (l_\alpha, r_\alpha)$'s are mutually disjoint open intervals. Each e_α is called an *excursion interval* of $X(t)$ and the part $(X(t), t \in e_\alpha)$ an *excursion of* $X(t)$. Let $\mathcal{F}_t^X$ be the sub σ-field generated by $\{X(s), s \leq t\}$ and P-null sets. Let $g(s)$ be an $(\mathcal{F}_t^X)$-adapted, well measurable process such that $s \to E(g(s)^2)$ is locally bounded and let us define the following stochastic integral:

$$(2.9) \qquad \int_0^t g(s)d\hat{X}^i(s) = \sum_{k=1}^{n} \int_0^t g(s)\sigma_k^i(X(s))dB^k(s) , \qquad i = 1, 2, \cdots, n$$

It is easy to see that for each i, $G^i(t) = \int_0^t g(s)d\hat{X}^i(s)$ is an $(\mathscr{F}_t^X)$-adapted continuous process. We define $\int_{e_\alpha \cap [0,t]} g(s)d\hat{X}^i(s)$ by $G^i(r_\alpha \wedge t) - G^i(l_\alpha \wedge t)$.

Theorem 1. *Let $g(s)$ be an $(\mathscr{F}_t^X)$-adapted, well measurvable process such that $E[g^2(s)]$ is locally bounded in s. Then*

$$E\Big\{\sum_\alpha \Big(\int_{e_\alpha \cap [0,t]} g(s)d\hat{X}^i(s)\Big)^2\Big\} = E\Big[\int_0^t g(s)^2 a^{ii}(X(s))ds\Big] \tag{2.10}$$

More generally, if $h(s)$ is a similar process as $g(s)$,

$$\begin{aligned} &E\Big\{\sum_\alpha \Big(\int_{e_\alpha \cap [0,t]} g(s)d\hat{X}^i(s)\Big)\Big(\int_{e_\alpha \cap [0,t]} h(s)d\hat{X}^j(s)\Big)\Big\} \\ &\quad = E\Big[\int_0^t g(s)h(s)a^{ij}(X(s))ds\Big] \qquad i, j = 1, 2, \cdots, n\,. \end{aligned} \tag{2.11}$$

Next, we consider the sum $\sum_\alpha \int_{e_\alpha \cap [0,t]} g(s)d\hat{X}^i(s)$. Generally, this sum is not absolutely convergent and we have to give a sense for it. *We define $\sum_\alpha^* \int_{e_\alpha \cap [0,t]} g(s)d\hat{X}^i(s)$ as the limit in probability of the finite sum*

$$\sum_{\substack{\alpha \\ r_\alpha - l_\alpha > \varepsilon}} \int_{e_\alpha \cap [0,t]} g(s)d\hat{X}^i(s)$$

as $\varepsilon \downarrow 0$ when and only when the limit exists.

Theorem 2. *Let $g(s)$ be an $(\mathscr{F}_t^X)$-adapted process such that $s \to g(s)$ is right continuous having left-hand limits and $s \to E[g(s)^2]$ is locally bounded. Then the following sum exists and the equality holds:*

$$\begin{aligned} &\sum_\alpha^* \int_{e_\alpha \cap [0,t]} g(s)d\hat{X}^i(s) = \int_0^t g(s)d\hat{X}^i(s) \\ &\quad + \int_0^t g(s)\frac{a^{ni}(X(s))}{a^{nn}(X(s))}d\psi(s)\,, \qquad i = 1, 2, \cdots, n\,. \end{aligned} \tag{2.12}$$

In proving these theorems,*) we formulate the collection of excursions of X as a point process on a function space and make use of some stochastic calculus on point processes. The idea of formulating excursions as a point process is due to K. Itô [8], (cf. Dynkin [4], Maisonneve [11]).

*) We rewrite these theorems in more convenient and general forms (cf. Theorems 1′ and 2′ of §5) in §5.

§ 3. Excursions as a point process

First, we recall some basic definitions of point processes on arbitrary state space, (cf. [6], [8], [16]). Let $(X, \mathscr{B}_X)$ be a measurable space. By a *point function p on X*, we mean a map $p: D_p \subset (0, \infty) \to X$ where the domain D_p is a countable subset of $(0, \infty)$. p defines a counting measure $N_p(dt, dx)$ on $(0, \infty) \times X$ by

$$(3.1) \quad N_p((0, t] \times U) = \sharp\{s \in D_p;\, s \leq t, p(s) \in U\} \qquad t > 0,\ U \in \mathscr{B}_X .$$

A point process is obtained by randomizing the notion of point functions. Let Π_X be the set of all point functions and $\mathscr{B}(\Pi_X)$ be the smallest σ-field on Π_X with respect to which all $p \to N_p((0, t] \times U)$, $t > 0$, $U \in \mathscr{B}_X$, are measurable. A *point process p* on X is, by definition, a $(\Pi_X, \mathscr{B}(\Pi_X))$-valued random variable: i.e., a measurable map $p: \Omega \ni \omega \to p(\omega) \in \Pi_X$ defined on a probability space Ω.

From now on, we fix a probability space $(\Omega, \mathscr{F}, P)$ with a family $(\mathscr{F}_t)_{t \geq 0}$ as in § 2. A point process on X is called *$(\mathscr{F}_t)$-adapted*, if $\{N_p((0, t] \times U)\}_{U \in \mathscr{B}_X}$ is $\mathscr{F}_t$-measurable for each $t \geq 0$.

Definition 3.1. A point process p on X is called *of the class* (QL)*) (with respect to $\mathscr{F}_t$) if

(i) it is $\mathscr{F}_t$-adapted,

(ii) it has a *continuous compensating measure*:

to be precise, there exists a σ-finite random measure $\phi_p(dt, dx)$ on $[(0, \infty) \times X, \mathscr{B}(0, \infty) \times \mathscr{B}_X]$ and a sequence $U_n \in \mathscr{B}_X$ such that $U_n \nearrow$ and $\bigcup_n U_n = X$, such that, if we set $\Gamma = \{U \in \mathscr{B}_X;\, {}^\exists n,\, U \subset U_n\}$, then for every $U \in \Gamma$, $t \to \phi_p((0, t] \times U)$ is a continuous, $(\mathscr{F}_t)$-adapted, integrable (i.e., $E(\phi_p((0, t] \times U)) < \infty$) increasing process and

$$t \longrightarrow N_p((0, t] \times U) - \phi_p((0, t] \times U)$$

is an $(\mathscr{F}_t)$-martingale.

ϕ_p is uniquely determined by p and is called the *compensator* of N_p. A stationary Poisson point process with the characteristic measure $n(dx)$ in the sense of Itô [8], where n is a σ-finite measure on $(X, \mathscr{B}_X)$, is characterized, in this context, as a point process of the class (QL) with the compensator $\phi_p(dtdx) = dt\, n(dx)$. Given a point process p of the class (QL), we can define the *stochastic integral* as follows. A real-function $f(t, x, \omega)$ defined on $[0, \infty) \times X \times \Omega$ is called *$(\mathscr{F}_t)$-predictable* if the mapping $(t, x, \omega) \to f(t, x, \omega)$ is $\mathscr{S}/\mathscr{B}(R)$-measurable, where $\mathscr{S}$ is the

---

*) (QL) denotes "quasi-left continuous".

smallest σ-field on $[0, \infty) \times X \times \Omega$ with respect to which all g with the following properties are measurable:

(i) for each $t > 0$, $(x, \omega) \to g(t, x, \omega)$ is $\mathscr{B}_X \times \mathscr{F}_t$-measurable

(ii) for each (x, ω), $t \to g(t, x, \omega)$ is left-continuous.

Let $$F_p = \Big\{f(t, x, \omega)\,;\, (\mathscr{F}_t)\text{-predictable and}$$

$$\int_0^{t+}\int_X |f(s, x, \cdot)| N_p(ds, dx) < \infty \qquad \text{a.s. } \forall t > 0\Big\}$$

$$F_p^1 = \Big\{f(t, x, \omega)\,;\, (\mathscr{F}_t)\text{-predictable and}$$

$$E\Big[\int_0^t\int_X |f(s, x, \cdot)|\, \phi_p(ds, dx)\Big] < \infty\,, \qquad \forall t > 0\Big\}$$

$$F_p^2 = \Big\{f(t, x, \omega)\,;\, (\mathscr{F}_t)\text{-predictable and}$$

$$E\Big[\int_0^t\int_X |f(s, x, \cdot)|^2\, \phi_p(ds, dx)\Big] < \infty\,, \qquad \forall t > 0\Big\}$$

and

$F_p^{2,\,\mathrm{loc}} = \{f(t, x, \omega)\,;\; (\mathscr{F}_t)$-predictable and $\exists T_n:\; \mathscr{F}_t$-stopping times such that $T_n \uparrow \infty$ a.s. and $I_{[0, T_n]}(t) \cdot f(t, x, \omega) \in F_p^2$, $n = 1, 2, \cdots\}$.

For $f \in F_p$, we set

$$P_{f,p}(t) = \int_0^{t+}\int_X f(s, x, \cdot) N_p(ds, dx) \tag{3.2}$$
$$\Big(= \sum_{\substack{s \le t\\ s \in D_p}} f(s, p(s), \cdot)\Big)\,.$$

Note that this is an a.s. absolutely convergent sum. For $f \in F_p^1 \cap F_p^2 (\subset F_p)$, we set

$$Q_{f,p}(t) = P_{f,p}(t) - \int_0^t\int_X f(s, x, \cdot)\phi_p(ds, dx) \tag{3.3}$$
$$= \sum_{\substack{s \le t\\ s \in D_p}} f(s, p(s), \cdot) - \int_0^t\int_X f(s, x, \cdot)\phi_p(ds, dx)\,.$$

It is easy to see that $Q_{f,p}(t)$ is a square-integrable, $(\mathscr{F}_t)$-martingale such that

$$(3.4) \qquad \langle Q_{f,p}\rangle(t) = \int_0^t \int_X f^2(s, x, \cdot)\phi_p(ds, dx) .$$

This definition is extented to $f \in \boldsymbol{F}_p^2$ by usual limiting procedure and $Q_{f,p}(t)$ is an $(\mathscr{F}_t)$-square integrable martingale. Also, it is extended, further, to $f \in \boldsymbol{F}_p^{2,\,\mathrm{loc}}$ and $Q_{f,p}$ is then an $(\mathscr{F}_t)$-locally square integrable martingale. Sometimes, we denote $Q_{f,p}(t)$ symbollically as

$$(3.5) \qquad Q_{f,p}(t) = \int_0^{t+} \int_X f(s, x, \cdot)[N_p(ds, dx) - \phi_p(ds, dx)] .$$

We refer to [3], [6], [13], [16] for the basic Itô's formula.

A typical point process in the theory of Markov processes is that defined by jumps of trajectories and the compensator is described by the Lévy system of the process (cf. [10]). As another example of point processes in the theory of Markov processes, we are now going to discuss the excursion point process of a diffusion process.

Let, as in § 2, a system $\mathfrak{X} = (X(t), B(t), M(t), \psi(t))$ be given on $(\Omega, \mathscr{F}, P; \mathscr{F}_t)$. Let

(3.6) $W = C([0, \infty) \to D)$ (=the set of all continuous function $w\colon [0, \infty) \ni t \to w(t) \in D$)

and let $\mathscr{W} \subset W$ be defined by

$$(3.7) \qquad \mathscr{W} = \{w \in W;\ w(0) \in \partial D, \quad {}^{\exists}\sigma(w) > 0 \quad \text{such that}$$
$$\text{(i)}\ t \in (0, \sigma(w)) \Rightarrow w(t) \in \mathring{D} \text{ and (ii) } t \geq \sigma(w) \Rightarrow w(t) = w(\sigma(w)) \in \partial D\}.$$

A point process p on $\mathscr{W}$, which we call the excursion point process of the diffusion $X = (X(t))$, is defined in the following way. Let $A(t)$ be the right-continuous inverse of $t \to \psi(t)$:

$$(3.8) \qquad A(t) = \inf\{u\colon \psi(u) > t\} .$$

Let

$$(3.9) \qquad D_p = \{s \in (0, \infty);\ A(s-) < A(s)\}$$

and, for $s \in D_p$, let $p(s) \in \mathscr{W}$ be defined by

$$(3.10) \qquad [p(s)](t) = \begin{cases} X(t + A(s-)) , & 0 \leq t \leq A(s) - A(s-) \\ X(A(s)) , & t \geq A(s) - A(s-) . \end{cases}$$

Clearly, $\sigma[p(s)] = A(s) - A(s-)$. This point process is clearly adapted to $(\tilde{\mathscr{F}}_t)$ where $\tilde{\mathscr{F}}_t = \mathscr{F}_{A(t)}$. It is a point process of the class (QL) with respect to $(\mathscr{F}_t)$ and its compensator $\phi_p(dt, dw)$ has the form $dt \cdot Q^{X(A(t))}(dw)$ where $Q^\xi(dw)$, $\xi \in \partial D$, is a system of σ-finite measures on $\mathscr{W}$ which is an X^0-Markovian measure corresponding to an entrance law $K^\xi(t, dx) = Q^\xi\{w; w(t) \in dx\}$ of X^0-process such that $w(0) = \xi$ a.s. $Q^\xi(dw)$, where $X^0 = (X^0(t))$ is the *absorbing barrier process*: $X^0(t) = X(t \wedge \sigma^*_{\partial D})$, $\sigma^*_{\partial D} = \inf\{t \geq 0; X(t) \in \partial D\}$. If X is the *reflecting Brownian motion* on D, i.e. the case

$$\begin{aligned} &\sigma^i_k(x) \equiv \delta^i_k\,, \quad b^i(x) \equiv 0\,, \quad i, k = 1, 2, \cdots, n \\ &\tau^i_k(x) \equiv 0\,, \quad \beta^i(x) \equiv 0\,, \quad i, k = 1, 2, \cdots, n-1\,, \end{aligned} \tag{3.11}$$

the Q^ξ is given explicitly as the unique σ-finite measure on $\mathscr{W}$ such that

$$\begin{aligned} &Q^\xi(w; w(t_1) \in E_1, w(t_2) \in E_2, \cdots, w(t_m) \in E_m, \sigma(w) > t_m) \\ &= \int_{E_1} K^\xi(t_1, x_1)dx_1 \int_{E_2} p^0(t_2 - t_1, x_1, x_2)dx_2 \\ &\times \cdots \int_{E_m} p^0(t_m - t_{m-1}, x_{m-1}, x_m)dx_m \\ &\qquad\qquad E_i \in \mathscr{B}(\mathring{D}),\ 0 < t_1 < t_2 \cdots < t_m \end{aligned} \tag{3.12}$$

where

$$K^\xi(t, x) = \prod_{i=1}^{n-1} \frac{1}{\sqrt{2\pi t}} \exp - \frac{(x_i - \xi_i)^2}{2t} \cdot \sqrt{\frac{2}{\pi t^3}}\, x_n \exp - \frac{x_n^2}{2t} \tag{3.13}$$

$$\begin{aligned} p^0(t, x, y) &= \prod_{i=1}^{n-1} \frac{1}{\sqrt{2\pi t}} \exp - \frac{(x_i - y_i)^2}{2t} \\ &\cdot \frac{1}{\sqrt{2\pi t}}\left(\exp - \frac{(x_n - y_n)^2}{2t} - \exp - \frac{(x_n + y_n)^2}{2t}\right) \\ &\qquad\qquad t > 0,\ x, y \in \mathring{D}\,. \end{aligned} \tag{3.14}$$

In this case, if we define another point process $\hat{p}$ on $\mathscr{W}$ by

$$D_{\hat{p}} = D_p \quad \text{and,} \quad \text{for } s \in D_{\hat{p}} \tag{3.15}$$

$$[\hat{p}(s)](t) = \begin{cases} X(t + A(s-)) - X(A(s-))\,, & 0 \leq t \leq A(s) - A(s-) \\ X(A(s)) - X(A(s-))\,, & t \geq A(s) - A(s-)\,, \end{cases}$$

then the compensator $\phi_p(dt, dw)$ has the form $dt \cdot Q^0(dw)$ and hence, it is a stationary Poisson point process with the characteristic measure Q^0. It is called the *Poisson point process of Brownian excursions*.

Fundamental formulas are as follows, (cf. [4], [5], [11], [16]). First, we introduce the following notations:

(3.16) $\rho_t : W \longrightarrow W$ defined by

$(\rho_t w)(s) = w(t \wedge s)$ (stopped path)

(3.17) $\theta_t : W \longrightarrow W$ defined by

$(\theta_t w)(s) = w(t + s)$ (shifted path)

(3.18) $\rho_{\partial D} : W \longrightarrow W$ defined by

$$(\rho_{\partial D} w)(t) = w(t \wedge \sigma_{\partial D}(w))$$

where

$$\sigma_{\partial D}(w) = \inf\{t > 0 \,;\, w(t) \in \partial D\} \tag{3.19}$$

(stopped path on reaching the boundary).

Let, by X, be denoted the path $t \to X(t)$ which is clearly a W-valued random variable.

Excursion formula I. Let $Z(s)$ be $\tilde{\mathscr{F}}_t := \mathscr{F}_{A(t)}$-predictable non-negative process and $f(s, w, w')$ be a non-negative Borel function on $(0, \infty) \times W \times \mathscr{W}$. Then,

$$\begin{aligned} &E\Big\{ \sum_{\substack{s \le t \\ s \in D_p}} Z(s) \cdot f(A(s-), \rho_{A(s-)}X, \rho_{\partial D}[\theta_{A(s-)}X]) \Big\} \\ &\qquad = E\Big\{ \int_0^t Z(s) \Big[\int_{\mathscr{W}} f(A(s), \rho_{A(s)}X, w') Q^{X(A(s))}(dw') \Big] ds \Big\} \end{aligned} \tag{3.20}$$

Excursion formula II. Let $Z(s)$ be $(\mathscr{F}_t)$-well measurable non-negative process and $f(s, w, w')$ be as above. Then,

$$\begin{aligned} &E\Big\{ \sum_{\substack{s \le \psi(t) \\ s \in D_p}} Z(A(s-)) \cdot f(A(s-), \rho_{A(s-)}X, \rho_{\partial D}[\theta_{A(s-)}X]) \Big\} \\ &\qquad = E\Big\{ \int_0^t Z(s) \Big[\int_{\mathscr{W}} f(s, \rho_s X, w') Q^{X(s)}(dw') \Big] d\psi(s) \Big\} . \end{aligned} \tag{3.21}$$

Let $t > 0$ be fixed and set

$$l(t) = \begin{cases} \sup\{s < t \,;\, X(s) \in \partial D\} \\ 0 \qquad \text{if } \{\ \} = \phi . \end{cases} \tag{3.22}$$

$l(t)$ is the *last exit time* from ∂D before t. Clearly $l(t) = A(\psi(t)-)$.

Last exit formula. Let $Z(s)$ be $\mathscr{F}_s$-well measurable non-negative process and $g(s, s', w, w')$ be a non-negative Borel function on $(0, \infty) \times (0, \infty) \times W \times \mathscr{W}$. Then

$$E\{Z(l(t))\cdot g(t-l(t), l(t), \rho_{l(t)}X, \rho_{\partial D}[\theta_{l(t)}X])I_{\{l(t)>0\}}\}$$

(3.23)
$$= E\left\{\int_0^t Z(s)\left[\int_{\mathscr{W}} g(t-s, s, \rho_s X, w')I_{\{\sigma(w')>t-s\}}Q^{X(s)}(dw')\right]d\psi(s)\right\}$$

Note that (3.20) is just rephrasing the fact that the compensator of p is $ds\,Q^{X(A(s))}(dw)$ and (3.21) is obtained from (3.20) by a time substitution. (3.23) is obtained from (3.21) by setting $f(s, w, w') = g(t-s, s, w, w')I_{\{\sigma(w')>t-s\}}$ (cf. Maisonneuve [11]).

Finally, let us note that the integrand $\sum_\alpha \left(\int_{e_\alpha \cap [0,t]} g(s)d\hat{X}^i(s)\right)^2$ in the left-hand side of (2.10) is written, in the context of this section, as $\sum_{\substack{s\in D_p\cap\{0\}\\ s\leq\psi(t)}} \left(\int_{A(s-)}^{A(s)\wedge t} g(s)d\hat{X}^i(s)\right)^2$ and the left-hand side of (2.12) is written as $\sum^*_{\substack{s\in D_p\cap\{0\}\\ s\leq\psi(t)}} \int_{A(s-)}^{A(s)\wedge t} g(s)d\hat{X}^i(s)$.

§ 4. Proof of theorems

We first prove the special case of the reflecting Brownian motion and reduce the general case to this special case.

(i) **The case of the reflecting Brownian motion**: i.e. the case $\sigma^i_k(x) \equiv \delta^i_k$, $b^i(x) \equiv 0$, $\tau^i_l(x) \equiv 0$, $\beta^i(x) \equiv 0$.

In this case, the system $(X(t), B(t), \psi(t))$ satisfies

(4.1)
$$\begin{aligned} X_i(t) - X_i(0) &= B_i(t), \qquad i = 1, 2, \cdots, n-1 \\ X_n(t) - X_n(0) &= B_n(t) + \psi(t). \end{aligned}$$

Without loss of generality, we may suppose that the system is given in the canonical way: $\Omega = W$ $(=C([0, \infty) \to D))$, $X(t, w) = w(t)$ $\mathscr{F}_t =$ *the completion of* $\sigma\{w(s); s \leq t\}$.

Let $\xi \in \partial D$. On the measure space $(\mathscr{W}, \mathscr{B}(\mathscr{W}), Q^\xi)$ where $\mathscr{B}(\mathscr{W})$ is the completion of the σ-field generated by Borel cylinder sets of $\mathscr{W}$ and Q^ξ is a σ-finite measure on $(\mathscr{W}, \mathscr{B}(\mathscr{W}))$ determined by (3.12), we can define the stochastic integrals as follows. Let $\mathscr{B}_t(\mathscr{W})$ be the completion of the σ-field generated by Borel cylinder sets up to the time t and $\Phi(s, w)$ be $\mathscr{B}_t(\mathscr{W})$-adapted, well-measurable process such that

$$\int_{\mathscr{W}}\left[\int_0^{\sigma(w)\wedge t} |\Phi(s, w)|^2\, ds\right] Q^\xi(dw) < \infty$$

for every $t > 0$.

Since, for $t_0 > 0$, $w_i(t + t_0) - w_i(t_0)$ $(i = 1, 2, \cdots, n)$ is a continuous

$\mathscr{B}_{t+t_0}(\mathscr{W})$-martingale with respect to $Q^\xi(\cdot\,|\,\sigma(w) > t_0)$, the stochastic integral $\int_{t_0}^t \Phi(s, w)dw_i(s)$ is defined and as is easily seen $\lim_{t_0 \downarrow 0} \int_{t_0}^t \Phi(s, w)dw_i(s)$ exists in $L^2(Q^\xi)$ which we denote as $\int_0^t \Phi(s, w)dw_i(s)$. Clearly,

$$\int_{\mathscr{W}} \left[\left(\int_0^t \Phi(s, w)dw_i(s)\right)^2\right] Q^\xi(dw) = \int_{\mathscr{W}} \left[\int_0^{\sigma(w)\wedge t} \Phi^2(s, w)ds\right] Q^\xi(dw), \tag{4.2}$$

and, for every $\mathscr{B}_{t_0}(\mathscr{W})$-measurable $H(w) \in L^2(Q^\xi)$,

$$\begin{aligned} &\int_{\mathscr{W}} \left[\int_0^t \Phi(s, w)dw_i(s)\right] \cdot H(w)Q^\xi(dw) \\ &= \int_{\mathscr{W}} \left[\int_0^{t_0} \Phi(s, w)dw_i(s)\right] \cdot H(w)Q^\xi(dw), \qquad \text{if } 0 < t_0 < t. \end{aligned} \tag{4.3}$$

Now, let $g(s) = g(s, w)$ be $(\mathscr{F}_t)$-adapted well-measurable process such that $s \to E(g^2(s))$ is locally bounded. For given $s > 0$, $w \in W$, $w' \in \mathscr{W}$, we set

$$\Phi_g^s(u, w, w') = \begin{cases} g(s + u, [w, w']_s) & \text{if } w(s) = w'(0) \\ 0 & \text{otherwise} \end{cases} \tag{4.4}$$

where $[w, w']_s \in W$ is defined by

$$[w, w']_s(u) = \begin{cases} w(u), & 0 \le u \le s \\ w'(u - s), & u \ge s. \end{cases} \tag{4.5}$$

Let $t > 0$ be given and fixed. Let

$$f_2^t(s, w, w') = \left[\int_0^{(t-s)\wedge\sigma(w')} \Phi_g^s(u, w, w')dw_i'(u)\right]^2 \tag{4.6}$$

in the sense of the stochastic integral on the measure space $(\mathscr{W}, \mathscr{B}(\mathscr{W}), Q^{w(s)})$ explained above. By (4.2),

$$\begin{aligned} &\int_{\mathscr{W}} f_2^t(s, w, w')Q^{w(s)}(dw') \\ &= \begin{cases} \int_{\mathscr{W}} \left[\int_s^{(\sigma(w)+s)\wedge t} g(u, [w, w']_s)^2 du\right] Q^{w(s)}(dw') & \text{if } w(s) = w'(0) \\ 0 & \text{otherwise.} \end{cases} \end{aligned} \tag{4.7}$$

It is clear from the definition that

$$(4.8)\qquad \begin{aligned} f_2^t(A(s-), \rho_{A(s-)}X, \rho_{\partial D}[\theta_{A(s-)}X]) &= \left(\int_{A(s-)}^{A(s)\wedge t} g(u)dX_i(u)\right)^2 \\ &= \left(\int_{A(s-)}^{A(s)\wedge t} g(u)dB_i(u)\right)^2 . \end{aligned}$$

Then, by the excursion formula II and (4.7),

$$\begin{aligned} E&\left\{\sum_{\substack{s\in D_p\cup\{0\}\\ s\le\psi(t)}} \left(\int_{A(s-)}^{A(s)\wedge t} g(u)dB_i(u)\right)^2\right\} \\ &= E\left\{\left(\int_0^{\sigma_{\partial D}\wedge t} g(u)dB_i(u)\right)^2\right\} + E\left\{\sum_{\substack{s\in D_p\\ s\le\psi(t)}} \left(\int_{A(s-)}^{A(s)\wedge t} g(u)dB_i(u)\right)^2\right\} \\ &= E\left\{\int_0^{\sigma_{\partial D}\wedge t} g(u)^2 du\right\} \\ &\quad + E\left\{\int_0^t d\psi(s)\int_{\mathscr{W}}\left[\int_s^{(\sigma(w')+s)\wedge t} g(u, [w, w']_s)^2 du\right] Q^{w(s)}(dw')\right\} \\ &= E\left\{\int_0^{\sigma_{\partial D}\wedge t} g(u)^2 du\right\} + E\left\{\sum_{\substack{s\in D_p\\ s\le\psi(t)}} \int_{A(s-)}^{A(s)\wedge t} g(u)^2 du\right\} \\ &= E\left\{\int_0^t g(u)^2 du\right\} . \end{aligned}$$

This proves (2.10) in this case and (2.11) is proved similarly.

Now, we will prove Theorem 2. In this case, (2.12) is given in the form:

$$(4.9)\qquad \sum_{\substack{s\in D_p\cup\{0\}\\ s\le\psi(t)}}^{*} \int_{A(s-)}^{A(s)\wedge t} g(s)dB_i(s) = \int_0^t g(s)dB_i(s) \qquad i = 1, 2, \cdots, n-1 ,$$

$$(4.10)\qquad \sum_{\substack{s\in D_p\cup\{0\}\\ s\le\psi(t)}}^{*} \int_{A(s-)}^{A(s)\wedge t} g(s)dB_n(s) = \int_0^t g(s)dB_n(s) + \int_0^t g(s)d\psi(s) .$$

We will prove (4.10) only: (4.9) is proved similarly and, in fact, much easier. In the following, we call $g(s)$ a *step process* if there exists a sequence of $(\mathscr{F}_t)$-stopping times $\sigma_0 \equiv 0 < \sigma_1 < \sigma_2 < \cdots < \sigma_n < \cdots \to \infty$ and $\mathscr{F}_{\sigma_i}$-measurable random variable $g_i (i = 0, 1, 2, \cdots)$ such that $g(s) = g_i$ if $s \in [\sigma_i, \sigma_{i+1})$, $i = 0, 1, 2, \cdots$.

Lemma 1. *Let $g(s)$ be a step process. Then* (4.10) *holds.*

Proof. If, for example, $g(s) \equiv 1$, then (4.10) is trivially true: we have

$$\int_{A(s-)}^{A(s)\wedge t} dB_n(s) = \int_{A(s-)}^{A(s)\wedge t} dX_n(s) = X_n(A(s)\wedge t) - X_n(A(s-))$$

$$= \begin{cases} 0\,, & s\in D_p\,, \quad A(s)\leq t \\ X_n(t)\,, & s\in D_p\,, \quad A(s-)\leq t < A(s) \\ X_n(\sigma_{\partial D\wedge t}) - X_n(0)\,, & s=0\,, \end{cases}$$

and hence the left-hand side of (4.10) is equal to $X_n(t) - X_n(0) = B_n(t) + \psi(t)$. A similar argument applies if $g(s)$ is a step process.

Lemma 2. *Let $g(s)$ be an $\mathscr{F}_t$-adapted process such that $s \to g(s)$ is right continuous having left-hand limits. Then, for every $\varepsilon > 0$, there exists a step process $g_\varepsilon(s)$ such that*

(4.11) $$|g_\varepsilon(s) - g(s)| \leq \varepsilon \qquad \textit{for every } s.$$

Proof. Let a sequence $\{\sigma_n\}$ of stopping times be defined by $\sigma_0 = 0$ and

$$\sigma_n = \inf\{t > \sigma_{n-1}\,; |g(t) - g(\sigma_{n-1})| \geq \varepsilon\} \wedge n\,, \qquad n = 1, 2, \cdots.$$

Then, $\sigma_n \uparrow \infty$ and we set $g_\varepsilon(s) = g(\sigma_n)$, $s \in [\sigma_n, \sigma_{n+1})$, $n = 1, 2, \cdots$. (4.11) clearly holds.

Lemma 3. *Let, for $\xi \in \partial D$ and $t > 0$, we set, for $B \in \mathscr{B}(\mathscr{W})$,*

(4.12) $$\mu^{\xi,t}(B) = Q^\xi(B\,|\,\sigma > t) = \frac{Q^\xi(B\,;\sigma > t)}{Q^\xi(\sigma > t)}\,.$$

Then $\mu^{\xi,t}$ is a Markovian measure on $\mathscr{W}$ concentrated on $\{w \in \mathscr{W}\,; w(0) = \xi, \sigma(w) > t\}$ such that

(4.13) $$\begin{aligned} &\mu^{\xi,t}\{w\,; w(t_1)\in E_1, w(t_2)\in E_2\cdots, w(t_n)\in E_n\} \\ &= \int_{E_1} dx_1 \int_{E_2} dx_2 \cdots \int_{E_n} dx_n k^\xi(t_1, x_1) \prod_{i=1}^{n-1} p(t_i, x_i\,; t_{i+1}, x_{i+1}) \\ &\qquad 0 < t_1 < t_2 \cdots < t_n < t\,, \qquad E_i \in \mathscr{B}(\mathring{D}) \end{aligned}$$

where

$$k^\xi(s, x) = \frac{K^\xi(s, x)h(t-s, x)}{K(t)} \qquad s > 0,\ x \in \mathring{D}$$

$$p(s, x\,; u, y) = \frac{h(t-u, y)}{h(t-s, x)}\cdot p^0(u-s, x, y) \quad 0 < s < u < t,\ x, y \in \mathring{D}$$

and

$$h(s, x) = \int_D p^0(s, x, y)\,dy = \frac{2}{\sqrt{2\pi s}} \int_0^{x_n} \exp\left\{-\frac{\eta^2}{2s} d\eta\right\}$$

$$K(t) = \int_D K^\xi(t, x)dx = \sqrt{\frac{2}{\pi t}}\,.$$

This lemma is proved easily by (3.12), (3.13) and (3.14).

Corollary. *The process* $b_n(s) = w_n(s) - \int_0^s A(t, u, w(u))du$ $(0 \leq u < t)$ *is a one-dimensional Wiener process with respect to the probability measure* $\mu^{\xi,t}$ *on* $\mathscr{W}$ *where*

$$\text{(4.14)} \quad A(t, u, x) = \frac{(\partial/\partial x_n)h(t-u, x)}{h(t-u, x)} = \frac{\exp\{-x_n^2/2(t-u)\}}{\int_0^{x_n} \exp\{-\mu^2/2(t-u)\}d\mu} \qquad t > u > 0,\ x \in \mathring{D}$$

Lemma 4. *Let* $g(s)$ *be a "bounded"* $\mathscr{F}_t$*-well measurable process. Then for fixed s and, for any* $\varepsilon \geq \varepsilon' > 0$,

$$\text{(4.15)} \quad \int_{\mathscr{W}} \left| \int_0^{\varepsilon'} \Phi_g^s(u, w, w')dw_n'(u) \right| I_{\{\sigma(w') > \varepsilon\}} Q^{w(s)}(dw') \leq \left(\sqrt{\frac{2}{\pi}} + 1\right) \|g\|_\infty\,,$$

where $\|g\|_\infty = \sup_{u,w} |g(u, w)|$ *and* Φ_g^s *is defined by* (4.4).

Proof. By the above corollary,

$$\begin{aligned}
&\int_{\mathscr{W}} \left| \int_0^{\varepsilon'} \Phi_g^s(u, w, w')dw_n'(u) \right| I_{\{\sigma(w') > \varepsilon\}} Q^{w(s)}(dw') \\
&\leq \int_{\mathscr{W}} \left\{ \left| \int_0^{\varepsilon'} \Phi_g^s(u, w, w')db_n(u) \right| \right\} \mu^{w(s),\varepsilon}(dw') \cdot Q^{w(s)}\{w' : \sigma(w') > \varepsilon\} \\
&\quad + \int_{\mathscr{W}} \left\{ \int_0^{\varepsilon'} |\Phi_g^s(u, w, w')A(\varepsilon, u, w'(u))|\,du \right\} \mu^{w(s),\varepsilon}(dw') \\
&\qquad\qquad \cdot Q^{w(s)}\{w' : \sigma(w') > \varepsilon\} \\
&\leq \|g\|_\infty \cdot \varepsilon'^{1/2} \cdot \sqrt{\frac{2}{\pi\varepsilon}} + \|g\|_\infty \leq \left(\sqrt{\frac{2}{\pi}} + 1\right) \|g\|_\infty\,.
\end{aligned}$$

Remark. $Q^\xi(w\,;\sigma(w) > \varepsilon) = \int_D K^\xi(\varepsilon, x)dx = \sqrt{\frac{2}{\pi\varepsilon}}$

$$\int_D w_n(t)Q^\xi(dw) = \int_D x_n K^\xi(t,x)dx = 1\,, \qquad t>0\,.$$

Now, let $g(s)$ be an $\mathscr{F}_t$-well measurable process such that $s \to E(g^2(s))$ is locally bounded and let us introduce the following notations:

$$(4.16) \qquad S_\varepsilon(t) = \sum_{\substack{s\leq\psi(t)\\ A(s)-A(s-)>\varepsilon}} \int_{A(s-)}^{A(s)\wedge t} g(s)dB_n(s)$$

$$(4.17) \quad Y_\varepsilon(t) = \int_0^t d\psi(s)\Big[\int_{\mathscr{W}}\Big\{\int_0^{\sigma(w')\wedge(t-s)} \Phi_g^s(u,w,w')dw'_n(u) \cdot I_{\{\sigma(w')>\varepsilon\}}\Big\}Q^{x(s)}(dw')\Big]$$

$$(4.18) \qquad M_\varepsilon(t) = S_\varepsilon(t) - Y_\varepsilon(t)\,.$$

Lemma 5. *Let $g(s)$ be an $\mathscr{F}_t$-well measurable process such that $s \to E(g^2(s))$ is locally bounded. Then $M_\varepsilon(t) \to \int_0^t g(s)dB_n(s)$ in $L^2(\Omega,P)$ as $\varepsilon \to 0$.*

Proof. Set, for $t \geq s \geq 0$,

$$(4.19) \qquad f_t(s,w,w') = \int_0^{(t-s)\wedge\sigma(w')} \Phi_g^s(u,w,w')dw'_n(u)\,.$$

Then, clearly

$$M_\varepsilon(t) = \int_0^{\sigma_{\partial D}\wedge t} g(u)dB_n(u)\cdot I_{\{\sigma_{\partial D}>\varepsilon\}}$$

$$+ \int_0^{\psi(t)}\int_{\mathscr{W}} f_t(A(s-),\rho_{A(s-)}X,w')I_{\{\sigma(w')>\varepsilon\}}[N_p(dsdw') - Q^{X(A(s))}(dw')ds]$$

and hence, it is easy to see that $M_\varepsilon(t) \to M(t)$ in $L^2(\Omega,P)$ as $\varepsilon \to 0$ where

$$M(t) = \int_0^{\sigma_{\partial D}\wedge t} g(u)dB_n(u)$$

$$+ \int_0^{\psi(t)}\int_{\mathscr{W}} f_t(A(s-),\rho_{A(s-)}X,w')[N_p(ds,dw') - Q^{X(A(s))}(dw')ds]\,.$$

Note also

$$E(M^2(t)) = E\Big[\int_0^{\sigma_{\partial D}\wedge t} g^2(u)du\Big] + E\Big[\int_0^t d\psi(s)\int_{\mathscr{W}} f_t^2(s,\rho_s X,w')Q^{X(s)}(dw')\Big]$$

$$= E\Big[\int_0^t g^2(u)du\Big]\,.$$

Next, we assume that $g(s)$ is bounded and prove that $M(t)$ is an $\mathscr{F}_t$-martingale. It is sufficient to show that, for any bounded Borel measurable functions $F_1(w), F_2(w)$ on W and $0 < t_1 < t_2$, by setting

$$H(w) = F_1(\rho_{A(\psi(t_1)-)}w)\cdot F_2(\rho_{t_1 - A(\psi(t_1)-)}[\theta_{A(\psi(t_1)-)}w]) ,$$

$$E(M(t_2)H) = E(M(t_1)\cdot H) . \tag{4.20}$$

We prove the following estimate

$$E(M_\varepsilon(t_2)\cdot H) = E(M_\varepsilon(t_1)\cdot H) + o(1) \qquad (\varepsilon \downarrow 0) \tag{4.21}$$

from which (4.20) follows. Now

$$\begin{aligned} M_\varepsilon(t_2) &= \int_0^{\sigma_{\partial D}\wedge t_2} g(u)dB_n(u)\cdot I_{\{\sigma_{\partial D}>\varepsilon\}} \\ &\quad + \sum_{\substack{s\le\psi(t_2),\, s\in D_p \\ \sigma(\theta_{A(s-)}X)>\varepsilon}} f_{t_2}(A(s-), \rho_{A(s-)}X, \rho_{\partial D}[\theta_{A(s-)}X]) \\ &\quad - \int_0^{t_2} d\psi(s)\int_{\mathscr{W}} f_{t_2}(s, \rho_s w, w')I_{\{\sigma(w')>\varepsilon\}}Q^{X(s)}(dw') \\ &= I_1^\varepsilon + I_2^\varepsilon - I_3^\varepsilon , \end{aligned}$$

say and let a similar expression for $M_\varepsilon(t_1)$ be as

$$M_\varepsilon(t_1) = J_1^\varepsilon + J_2^\varepsilon - J_3^\varepsilon .$$

Then, it is easy to see that

$$E(I_1^\varepsilon\cdot H) = E(J_1^\varepsilon\cdot H) + o(1) .$$

By the excursion formula II, we have

$$E[(I_2^\varepsilon - I_3^\varepsilon)\cdot H] = E[(I_{21}^\varepsilon - I_{31}^\varepsilon)\cdot H]$$

where

$$I_{21}^\varepsilon = \sum_{\substack{s\le\psi(t_1),\, s\in D_p \\ \sigma(\theta_{A(s-)}X)>\varepsilon}} f_{t_2}(A(s-), \rho_{A(s-)}X, \rho_{\partial D}[\theta_{A(s-)}X]) ,$$

$$I_{31}^\varepsilon = \int_0^{t_1} d\psi(s)\int_{\mathscr{W}} f_{t_2}(s, \rho_s w, w')I_{\{\sigma(w')>\varepsilon\}}Q^{X(s)}(dw') .$$

If $\varepsilon < t_2 - t_1$ and $s \le t_1$,

$$\begin{aligned} &\int_{\mathscr{W}} f_{t_2}(s, \rho_s w, w')I_{\{\sigma(w')>\varepsilon\}}Q^{X(s)}(dw') \\ &\quad = \int_{\mathscr{W}} \left[\int_0^\varepsilon \Phi_g^s(u, w, w')dw_n'(u)\right]\cdot I_{\{\sigma(w')>\varepsilon\}}Q^{X(s)}(dw') \end{aligned}$$

and hence,

$$I_{31}^{\varepsilon} = J_3^{\varepsilon} + \int_{t_1-\varepsilon}^{t_1} d\psi(s)\Big[\int_{\mathscr{W}} \Big\{\int_0^{\varepsilon} \Phi_g^s(u, w, w')dw'_n(u)$$

$$- \int_0^{(t_1-s)} \Phi_g^s(u, w, w')dw'_n(u)\Big\}\cdot I_{\{\sigma(w')>\varepsilon\}} Q^{X(s)}(dw')\Big] \equiv J_3^{\varepsilon} + \delta(\varepsilon)$$

By Lemma 4, $E|\delta(\varepsilon)| = o(1)$ and hence

$$E(I_{31}^{\varepsilon}\cdot H) = E(J_3^{\varepsilon}\cdot H) + o(1) .$$

Next,

$$\begin{aligned} I_{21}^{\varepsilon} = & \sum_{\substack{s<\psi(t_1),\, s\in D_p \\ \sigma(\theta_{A(s-)}X)>\varepsilon}} f_{t_1}(A(s-), \rho_{A(s-)}X, \rho_{\partial D}[\theta_{A(s-)}X]) \\ & + f_{t_2}(l(t_1), \rho_{l(t_1)}X, \rho_{\partial D}[\theta_{l(t_1)}X]) \\ \equiv & \, I_{211}^{\varepsilon} + I_{212}^{\varepsilon} \end{aligned}$$

where $l(t_1) = A(\psi(t_1)-)$. By the last exit formula,

$$\begin{aligned} E(I_{212}^{\varepsilon}\cdot H) = & E\Big[\int_0^{t_1} F_1(\rho_s X)d\psi(s)\Big[\int_{\mathscr{W}} \Big\{\int_0^{\sigma(w')\wedge(t_2-s)} \Phi_g^s(u, w, w')dw'_n(u)\Big\} \\ & \times I_{\{\sigma(w')>\varepsilon\vee(t_1-s)\}}\cdot F_2(\rho_{t_1-s}w')\Big] Q^{X(s)}(dw')\Big] \\ = & E\Big[\int_0^{t_1} F_1(\rho_s X)d\psi(s)\Big[\int_{\mathscr{W}} \Big\{\int_0^{\varepsilon\vee(t_1-s)} \Phi_g^s(u, w, w')dw'_n(u)\Big\} \\ & \times I_{\{\sigma(w')>\varepsilon\vee(t_1-s)\}}\cdot F_2(\rho_{t_1-s}w')\Big] Q^{X(s)}(dw')\Big] \\ = & E\Big[\int_0^{t_1} F_1(\rho_s X)d\psi(s)\Big[\int_{\mathscr{W}} \Big\{\int_0^{(t_1-s)} \Phi_g^s(u, w, w')dw'_n(u)\Big\} \\ & \times I_{\{\sigma(w')>\varepsilon\vee(t_1-s)\}}\cdot F_2(\rho_{t_1-s}w')\Big] Q^{X(s)}(dw')\Big] \\ & + E\Big[\int_{t_1-\varepsilon}^{t_1} F_1(\rho_s X)d\psi(s)\Big[\int_{\mathscr{W}} \Big\{\int_0^{\varepsilon} \Phi_g^s(u, w, w')dw'_n(u) \\ & - \int_0^{t_1-s} \Phi_g^s(u, w, w')dw'_n(u)\Big\} I_{\{\sigma(w')>\varepsilon\}}\cdot F_2(\rho_{t_1-s}w')\cdot Q^{X(s)}(dw')\Big]\Big] \\ \equiv & \, a_1(\varepsilon) + a_2(\varepsilon) . \end{aligned}$$

Then, clearly

$$E(I_{211}^{\varepsilon}\cdot H) + a_1(\varepsilon) = E(J_2^{\varepsilon}\cdot H)$$

and by Lemma 4, $a_2(\varepsilon) = o(1)$. Putting all the above estimates together, we have (4.21).

Let $g(s)$ be a *bounded step process*. By Lemma 1,

$$S_\varepsilon(t) \longrightarrow \int_0^t g(s)dB_n(s) + \int_0^t g(s)d\psi(s)$$

a.s. and hence

$$Y_\varepsilon(t) = S_\varepsilon(t) - M_\varepsilon(t) \rightarrow \int_0^t g(s)dB_n(s) + \int_0^t g(s)d\psi(s) - M(t)$$

in probability. By Lemma 4, it is clear that this limit has the form $\int_0^t h(s)d\psi(s)$ for some bounded adapted process $h(s)$. Thus we have

$$\int_0^t h(s)d\psi(s) = \int_0^t g(s)dB_n(s) + \int_0^t g(s)d\psi(s) - M(t) .$$

From this, we must have

$$\int_0^t h(s)d\psi(s) = \int_0^t g(s)d\psi(s)$$

and

$$\int_0^t g(s)dB_n(s) = M(t) .$$

Let $g(s)$ be general. Then, we take a sequence $\{g_k(s)\}$ of bounded step processes such that

$$E\left[\int_0^t |g_k(t) - g(s)|^2\, ds\right] \rightarrow 0 \qquad (k \rightarrow \infty) .$$

It is easy to see that $M_k(t)$ corresponding to $g_k(s)$ converges to $M(t)$ and hence $M(t) = \int_0^t g(s)dB_n(s)$.

Lemma 6. *Let $g(s)$ be an $\mathscr{F}_t$-adapted process such that $s \rightarrow g(s)$ is right-continuous having left-hand limits and $s \rightarrow E[g^2(s)]$ is locally bounded. Let $Y^\varepsilon(t)$ be defined by* (4.17). *Then*

$$\lim_{\varepsilon \downarrow 0} Y_\varepsilon(t) = \int_0^t g(u)d\psi(u) \tag{4.22}$$

in the sense of limit in probability.

Proof. First, let $g(s)$ be a step process. We have

$$Y_\varepsilon(t) = -M_\varepsilon(t) + S_\varepsilon(t) .$$

and by Lemma 1, $S_\varepsilon(t) \to \int_0^t g(u)dB_n(u) + \int_0^t g(u)d\psi(u)$ a.s. and by Lemma 5, $M_\varepsilon(t) \to \int_0^t g(u)dB_n(u)$ in L^2-sense. Thus (4.22) holds. Let $g(s)$ be general. By Lemma 2, there exists a sequence $\{g_k(s)\}$ of step processes such that

$$g_k(s) - \frac{1}{k} \le g(s) \le g_k(s) + \frac{1}{k} .$$

Since $Y_\varepsilon(t)$ is expressed also as

$$Y_\varepsilon(t) = \int_0^t d\psi(s) \int_{\mathscr{W}} \left[\int_0^{\sigma(w')\wedge(t-s)\wedge\varepsilon} \Phi_g^s(u, w, w')A(\varepsilon, u, w'(u))du \right. \\ \left. \times I_{\{\sigma(w')>\varepsilon\}}\right] Q^{X(s)}(dw) ,$$

we have

$$Y_\varepsilon^{g_k}(t) - \frac{1}{k}\psi(t) \le Y_\varepsilon(t) \le Y_\varepsilon^{g_k}(t) + \frac{1}{k}\psi(t)$$

where $Y_\varepsilon^{g_k}$ corresponds to g_k. As is already shown, $Y_\varepsilon^{g_k} \to \int_0^t g_k(u)d\psi(u)$ as $\varepsilon \to 0$ and hence, letting $\varepsilon \to 0$ and then $k \to \infty$, we see easily that (4.22) holds.

Now, we are ready to prove (4.10). By Lemma 5 and 6,

$$S_\varepsilon(t) = M_\varepsilon(t) + Y_\varepsilon(t)$$

converges in probability, as $\varepsilon \to 0$, to

$$\int_0^t g(s)dB_n(s) + \int_0^t g(s)d\psi(s)$$

and this proves (4.10).

(ii) **General case.**

The proof of Theorem 1 is similar as in the case of reflecting Brownian motion. As for Theorem 2, first we consider the case $\sigma_k^n(x) \equiv \delta_k^n$, $b_n(x) \equiv 0$. In this case,

$$[B_1(t), B_2(t), \cdots, B_{n-1}(t), X_n(t) = X_n(0) + B_n(t) + \psi(t)]$$

is a reflecting Brownian motion and the proof of the case (i) applies term by term. Next, the case $\sigma_k^n(x) \equiv \delta_k^n$ is reduced to the former case by a change of drift (cf. [15]). Finally, the general case is reduced to the former case by the following change of coordinates and the change of Brownian motion: let $a^{ij}(x)$ be defined by (2.1). By the assumption, $a^{ij}(x)$ are of C^3-class and it is not difficult to see there exists a C^2-function $f(x)$ on D such that $f(x) \geq 0$, $f(x) = 0$ if and only if $x \in \partial D$ and $\sum_{i,j} a^{ij}(\partial f/\partial x_i)\cdot(\partial f/\partial x_j) \equiv 1$ on ∂D. Let $\mathfrak{X} = (X(t), B(t), M(t), \psi(t))$ be a stochastic system corresponding to the coefficients $[\sigma, b, \tau, \beta]$. Then $\hat{\mathfrak{X}} = (\hat{X}(t), B(t), M(t), \psi(t))$, where

$$\hat{X}_i(t) = X_i(t)\,, \qquad i = 1, 2, \cdots, n-1$$
$$\hat{X}_n(t) = f(X(t))$$

is the system corresponding to $[\hat{\sigma}, \hat{b}, \hat{\tau}, \hat{\beta}]$ with $\hat{a}^{nn}(x) \equiv 1$. By a change of Brownian motion $B(t)$ to $\hat{B}(t)$, (cf. [15]), we may assume $\hat{\sigma}_k^n(x) = \delta_k^n$.

§ 5. Applications

1. Variation of the diffusion along the boundary

Let a system $\mathfrak{X} = (X(t), B(t), M(t), \psi(t))$ be given as in § 2. Then, by Itô's formula and Theorem 2, we see that, for $t > 0$,

$$V_{\mathring{D}}(f) := \sum^*_{\substack{s \in D_p \cup \{0\} \\ s \leq \psi(t)}} [f(X_{t\wedge A(s)}) - f(X_{A(s-)})]\,,$$

where $\sum^*_{\substack{s \in D_p \cup \{0\} \\ s \leq \psi(t)}}$ is understood as the limit in probability of the finite sum $\sum_{\substack{s \leq \psi(t) \\ A(s) - A(s-) > \varepsilon}}$, exists for $f \in C_b^2(D)$ and

$$V_{\partial D}(f) := f(X_t) - f(X_0) - V_{\mathring{D}}(f)\,,$$

which may be called naturally as the *variation of $t \to f(X_t)$ along the boundary*, is given by

$$\begin{aligned} V_{\partial D}(f) = & \sum_{i=1}^{n-1}\sum_{l=1}^{n-1} \int_0^t \frac{\partial f}{\partial x_i}(X_s)\tau_l^i(X_s)dM_s^l \\ & + \int_0^t \Bigg[\sum_{i=1}^{n-1} \left\{\beta^i(X_s) - \frac{a^{ni}(X_s)}{a^{nn}(X_s)}\right\} \frac{\partial f}{\partial x_i}(X_s) \\ & + \frac{1}{2}\sum_{i,j=1}^{n-1} \alpha^{ij}(X_s)\frac{\partial^2 f}{\partial x_i \partial x_j}(X_s)\Bigg] d\psi(s)\,. \end{aligned} \tag{5.1}$$

In particular, $V_{\partial D}(f) = 0$ for all $f \in C_b^2(D)$ if and only if we have identically on ∂D

$$(5.2)\qquad \begin{aligned} \alpha^{ij}(x) &= 0\,, \qquad i, j = 1, 2, \cdots, n-1 \\ \beta^i(x) &= \frac{a^{ni}(x)}{a^{nn}(x)}\,, \qquad i, j = 1, 2, \cdots, n-1\,. \end{aligned}$$

In this case, it is natural to call the diffusion X_t as *normally reflecting diffusion process*. Indeed, if $\{a^{ij}(x)\}$ is non-degenerate, the vector field $\{\beta^i(x)\}_{i=1}^n$ on ∂D satisfying (5.2) ($\beta^n(x) \equiv 1$) is in the direction of the inner normal determined by the metric tensor $a_{ij}(x) = (a^{ij}(x))^{-1}$. Thus, a characteristic feature of a normally reflecting diffusion process is that, for any $\mathscr{F}_t^X$-adapted process $g(s)$ such that $s \to g(s)$ is right-continuous with left-hand limits and $s \to E[g^2(s)]$ is locally bounded, if we define $\int_0^t g(s)dX^i(s)$ ($i = 1, 2, \cdots, n$) as a usual quasi-martingale (or semi-martingale) stochastic integral, the formula

$$(5.3)\qquad \sum_{\substack{s \le \psi(t) \\ s \in D_p \cup \{0\}}}^{*} \int_{A(s-)}^{A(s)\wedge t} g(s)dX^i(s) = \int_0^t g(s)dX^i(s)$$

holds for every $i = 1, 2, \cdots, n$. In other words, if $f \in C_b^2(D)$, $t \to f(X_t)$ is a quasi-martingale (or semi-martingale, i.e. a sum of a martingale and a process with bounded variation) and if $\int_0^t g(s)df(X_s)$ is defined as a quasi-martingale stochastic integral, a *normally reflecting diffusion is characterized by the fact that*

$$(5.4)\qquad \sum_{\substack{s \le \psi(t) \\ s \in D_p \cup \{0\}}}^{*} \int_{A(s-)}^{A(s)\wedge t} g(s)df(X_s) = \int_0^t g(s)df(X_s)$$

holds for every $f \in C_b^2(D)$.

2. Multiplicative operator functionals of diffusion X

Let a system $\mathfrak{X} = (X(t), B(t), M(t), \psi(t))$ be given as in § 2. Let $A_{ij}^k(x)$, $i, j = 1, 2$, $k = 0, 1, 2, \cdots, n$ be bounded continuous functions on D.

First, we remark that Theorem 1 and Theorem 2 of § 2 can be rewritten in the following form:

Theorem 1′. *Let* $\mathscr{F}_t^0 =$ *the completion of* $\sigma(X(u), B(u), u \le t)$. *Let* $g(s)$ *be an* $\mathscr{F}_t^0$*-adapted well-measurable process such that* $s \to E[g^2(s)]$ *is locally bounded. Then*

$$(5.5)\quad E\Big(\sum_{\substack{s\in D_p\cup\{0\}\\ s\leq\psi(t)}}\Big[\int_{A(s-)}^{A(s)\wedge t} g(s)dB^k(s)\Big]^2\Big) = E\Big[\int_0^t g^2(s)ds\Big],$$

$$k = 1, 2, \cdots, n$$

Theorem 2′. *Let $g(s)$ be an $\mathscr{F}_t^0$-adapted process such that $s\to g(s)$ is right-continuous with left-hand limits and $s\to E[g^2(s)]$ is locally bounded. Then*

$$(5.6)\quad \begin{aligned}\sum_{\substack{s\in D_p\cup\{0\}\\ s\leq\psi(t)}}^{*}\int_{A(s-)}^{A(s)\wedge t} g(s)dB^k(s) &= \int_0^t g(s)dB^k(s)\\ &+\int_0^t g(s)\frac{\sigma_k^n(X(s))}{a^{nn}(X(s))}d\psi(s)\,, \qquad k = 1, 2, \cdots, n\,.\end{aligned}$$

In fact, we replace $X(t)$ by $X(t) = (B^1(t), \cdots, B^n(t), X^1(t), \cdots, X^n(t))$ and apply Th. 1 and Th. 2 for this system. Then we obtain Th. 1′ and Th. 2′ at once. Let

$$(5.7)\quad l(t) = \begin{cases}\sup\{s\leq t\,;\, X(s)\in\partial D\}\\ 0 \qquad \text{if}\ \{\quad\} = \phi\,.\end{cases}$$

Then, for fixed t, $l(t) = A(\phi(t)-)$ a.s. and by (5.5),

$$(5.8)\quad E\Big[\Big\{\int_{l(t)}^t g(u)dB^i(u)\Big\}^2\Big] \leq E\Big[\int_0^t g^2(u)du\Big]\,.$$

We consider the following stochastic differential equation for $M(t) = (M_{ij}(t))_{i,j=1}^2$:

$$(5.9)\quad \begin{aligned} M_{i1}(t) &= \delta_{i1} + \sum_{k=1}^n\sum_{l=1}^2\int_0^t M_{il}(s)A_{l2}^k(X(s))dB^k(s)\\ &\qquad + \sum_{l=1}^2\int_0^t M_{il}(s)A_{1l}^0(X(s))ds\,,\\ M_{i2}(t) &= I_{\{\sigma_{\partial D}>t\}}\cdot\Big(\delta_{i2} + \sum_{k=1}^n\sum_{l=1}^2\int_0^t M_{il}(s)\cdot A_{l2}^k(X(s))dB^k(s)\\ &\qquad + \sum_{l=1}^2\int_0^t M_{il}(s)A_{l2}^0(X(s))ds\Big)\\ &\quad + I_{\{\sigma_{\partial D}\leq t\}}\Big(\sum_{k=1}^n\sum_{l=1}^2\int_{l(t)}^t M_{il}(s)A_{l2}^k(X(s))dB^k(s)\\ &\qquad + \sum_{l=1}^2\int_{l(t)}^t M_{il}(s)A_{l2}^0(X(s))ds\Big)\,.\end{aligned}$$

By the estimate (5.8), we can easily show that there exists a unique adapted solution $M(t)$ and it is a multiplicative 2×2-matrix (cf. [14]). Using

(5.6), we see that*) if $f_1(x)$ and $f_2(x)$ are in $C_b^2(D)$ such that $Lf_1 = 0$ and $f_2 = 0$ on ∂D, then

$$M(t)f(X(t)) - M(0)f(X(0)) = \text{a martingale} + \int_0^t M(s)(Af)(X(s))ds \tag{5.10}$$

where

$$(Af)_i(x) = \frac{1}{2}Af_i(x) + \sum_{j,k=1}^{n}\sum_{l=1}^{2} A_{il}^k(x)\sigma_k^j(x)\frac{\partial f^l}{\partial x_j}(x) + \sum_{j=1}^{2} A_{ij}^0(x)f_j(x), \quad i = 1, 2. \tag{5.11}$$

Thus, $u(t, x) = E[M(t)f(X(t))]$ gives us a stochastic solution of the following system of heat equations with boundary conditions:

$$\begin{aligned} \frac{\partial u_i}{\partial t} &= (Au)_i \qquad i = 1, 2 \\ u_i|_{t=0} &= f_i \\ Lu_1|_{\partial D} &= 0, \qquad u_2|_{\partial D} = 0 \end{aligned} \tag{5.12}$$

cf. [1], [2], [7].

References

[1] Airault, H., Resolution stochastique d'un probleme de Dirichlet-Neumann, C. R. Acad. Sci. Paris, t. **280**, 781–784.

[2] ——, Perturbations singulieres et solutions stochastiques de problemes de D. Neumann-Spencer, J. Math. pures Appl., **54** (1976), 233–268.

[3] Dolean-Dade, C. et Meyer, P. A., Integrales stochastique par rapport aux martingales locales, Seminaire de Prob. IV Lecture Notes in Math., **124**, Springer (1970), 77–107.

[4] Dynkin, E. B., Wanderings of a Markov process, Theor. Probability Appl., **16** (1971), 401–428.

[5] Getoor, R. K. and Sharpe, M. J., Last exit decompositions and distributions, Indiana Univ. Math. J., **23** (1973), 377–404.

[6] Grigelionis, B., Stochastic point processes and martingales, Liet. matem. ring., **XV-3** (1975), 101–114 (in Russian).

[7] Ikeda, N. and Watanabe, S., Heat equation and diffusion on Riemannian manifold with boundary, These Proceedings, 75–94.

---

*) L and A are defined in §2.

[8] Itô, K., Poisson point processes attached to Markov processes, Proc. 6-th Berkeley Symp., III (1972), 225–239.
[9] Itô, K. and McKean, H. P. Jr., Diffusion processes and their sample paths, Springer, 1965.
[10] Kunita, J. and Watanabe, S., On square integrable martingales, Nagoya Math. J., **30** (1967), 209–245.
[11] Maisonneuve, B., Exit systems, Ann. Probability, **3** (1975), 399–411.
[12] McKean, H. P. Jr., Stochastic integrals, Academic Press, 1969.
[13] Meyer, P. A., Un cours sur les integrales stochastiques, Seminaire de Prob., X, Lecture Notes in Math., Springer, 1976, 245–400.
[14] Pinsky, M. A., Multiplicative operator functionals and their asymptotic properties, Advances in Appl. Probability, **3** (1974), 1–100.
[15] Watanabe, S., On stochastic differential equations for multi-dimensional diffusion processes with boundary conditions, J. Math. Kyoto Univ., **11** (1971), 169–180.
[16] ——, Poisson point process of Brownian excursions and its applications to diffusion processes, Proc. Symp. in Pure Math. AMS., **31** (1977), 153–164.
[17] Wentzell, A. D., On boundary conditions for multi-dimensional diffusion processes, Theor. Probability Appl., **4** (1959), 164–177.

Department of Mathematics
Kyoto University
Kyoto 606, Japan

Proc. of Intern. Symp. SDE
Kyoto 1976, pp. 463–491

Approximation of Markovian Control Systems by Discrete Control Policies

Keigo YAMADA

Abstract

Given Markovian control systems described by stochastic differential equations with discounted costs, we consider the problem of finding optimal controls by discretizing in time the original continuous systems. Some sufficient conditions are given under which the discretized systems approximate the original one. Noting that the discretized problem is essentially a discrete-time Markovian decision process, by using a conventional dynamic programming technique we get a sequence of stationary Markov controls which are optimal for the discretized systems. We show that under appropriate conditions a weak limit of this sequence of Markov controls is optimal for the original system. A sufficient condition is also given for the existence of such a weak limit. These results are applied to a continuous time inventory control problem and are used to obtain the characterization of optimal ordering policy.

Contents

§ 1. Introduction

This paper concerns a system described by stochastic differential equations of the form

$$dx_1(t) = g_1(x(t))dt$$
$$(1) \qquad dx_2(t) = g_2(x(t))dt + f(x(t), u_t)dt + \sigma(x(t))dw(t)$$

where $x(t) = (x_1(t), x_2(t))$ is a state vector, $w(t)$ is a standard Wiener process, and the control u_t is generally a functional of the past history of the state process $x(t)$. If u_t is a function of the present state $x(t)$, then $x(t) = (x_1(t), x_2(t))$ is a homogeneous Markov process with degenerate (non-uniformly elliptic) differential operator. The objective is to choose the control u_t so as to minimize the discounted cost:

$$(2) \qquad J_u = E\int_0^\infty e^{-\beta t}h(x(t), u_t)dt\ , \qquad \beta > 0\ .$$

This type of problem was treated in Kushner [1] where a Hamilton-Jacobi type non-linear partial differential equation was introduced. The difficulty of using this equation for the analysis of optimal controls lies in establishing the existence of the solution of the equation. In [1] the existence of such a solution is shown under strong assumptions for system equations. Furthermore, even if the existence of the solution is assured, it is often difficult to compute or characterize optimal controls from the Hamilton-Jacobi equation.

The approach taken here is, in a sense, quite direct. First we approximate the original problem (1) and (2) by the control system where decisions are made at discrete times with an equal decision interval and then try to find an optimal control for the original system as a limit of optimal policies for discretized (or approximated) problems. The reason for taking such an approach is that such discretized problems are essentially discrete-time control problems and in some cases discrete-time problems are easier to analyze than continuous-time problems. For example, the application of dynamic programming to discrete-time systems is rather straightforward, and for analyzing the structure of optimal controls the discrete-time dynamic programming equation is often more convenient than Hamilton-Jacobi equations mentioned above. Since such discretized problems belong to so called discrete-time Markovian decision problems, we can also expect to utilize the results of such relatively well developed area. From this point our approach may be viewed as a trial to bridge the gap between the discrete and continuous time decision problems which have been studied separately.

The problem for such an approach of discretizing the continuous-time control problems is that whether the discretization can really approximate the original problem or not, i.e., as the decision interval tend to zero, do optimal values and controls for the discretized problems converge to those of the original continuous-time problem? This paper treats this question for the control system (1–2), and gives conditions for the justification of our procedure.

For an arbitrary $\varDelta > 0$, the discretization of the systems (1–2) by the interval $\varDelta$ is meant by making decisions at only discrete times $k\varDelta$,

$k = 0, 1, 2, \cdots$, and using these decision values during the intervals. The set of such controls is denoted by $\mathscr{U}_\Delta^d$ (see the next section for the exact definition). Under appropriate assumptions, we have

$$\inf_{u \in \mathscr{U}_\Delta^d} J_u \to \inf_{u \in \mathscr{U}} J_u$$

as $\Delta \downarrow 0$, where $\mathscr{U}$ is the set of all non-anticipative controls. This is easily obtained with the aid of techniques in Davis and Varaiya [5]. Noting that the problem of finding u^* in $\mathscr{U}_\Delta^d$ such that

$$\inf_{u \in \mathscr{U}_\Delta^d} J_u = J_{u^*}$$

is essentially a discrete-time Markovian decision problem, we can show that u^* can be taken as a stationary Markov control, i.e., a policy depending only on the state at decision times. These facts are shown in Theorem 1. Writing $u_\Delta^* = u^*$, where u_Δ^* is a stationary Markov control, our main result is in Theorem 2 of Section 4 where a weak limit of u_Δ^* (if it exists) as $\Delta \downarrow 0$ is shown to be an optimal stationary Markov control for the system (1–2) under appropriate assumptions. These assumptions are not so general, but cover some important classes in applications. In Corollary 1 of this theorem, a sufficient condition is given for the existence of a subsequence which has a weak limit. This corollary also shows the existence of an optimal stationary Markov control which does not depend on the initial state of the system. As long as the optimality of stationary Markov controls for the system (1–2) is concerned, this will be obtained under much weaker conditions by using the new type Hamilton-Jacobi equation in Davis and Varaiya [2] just as in Davis [4].

In Section 5, to show how Theorem 2 does work, the foregoing results are applied to a typical continuous-time inventory control problem where the order rate has an upperlimit A. We can show that the optimal ordering policy is of bang-bang type, i.e., there exists a constant $\bar{z}$ such that the optimal policy $v^*(x)$, where x is the present inventory level, is given by

$$v^*(x) = \begin{cases} 0: & x \geq \bar{z} \\ \mathrm{A}: & x < \bar{z} \end{cases}$$

which can be obtained as a weak limit of optimal ordering policies for discretized inventory problems. The characterization of optimal policies for such discretized inventory systems in well studied. The result itself may be obtained via Hamilton-Jacobi equation method.

Finally a remark on the cost functional. We have chosen a discounted cost functional in (2) since it yields generally stationary Markov controls as an optimal policy and this fact enables us to discuss the convergence problem of optimal stationary Markov controls for the discretized problem. Hence from this point of view we could have chosen other cost criteria as long as they yield stationary Markov controls as optimal policies. The cost functional

$$J_u = E\int_0^\tau h(x(t), u_t)dt ,$$

where τ is the escape time of a bounded region, is such an example.

§ 2. Preliminaries

Various conditions on systems equations are described here. These assumptions are assumed to hold in this and next sections. In Section 4 we shall need further conditions and they will be presented there. We assume the following.

$(A_g)g_i : R^d \to R^{d_i} (i = 1, 2)$ satisfies a Lipshitz condition and a linear growth condition of the form:

$$|g_i(x) - g_i(y)| \leqq K|x - y|$$
$$|g_i(x)| \leqq K(1 + |x|) .$$

$(A_f)f : R^d \times U \to R^{d_2}$ (here U, the control set, is a compact subset of R^m) is bounded measurable, continuous in the second argument for each fixed first argument, and uniformly equi-continuous, i.e., for an arbitrary $\varepsilon > 0$, there exists a $\delta(\varepsilon) > 0$ such that

$$|u - u'| \leqq \delta(\varepsilon) \Rightarrow |f(x, u) - f(x, u')| \leqq \varepsilon \quad \text{for all } x \in R^d .$$

$(A_\sigma)\sigma : R^d \to R^{d_2 \times d_2}$ is a bounded matrix function which satisfies the same type of Lipshitz and linear growth conditions as in (A_g). It also has a bounded inverse $\sigma^{-1}(\cdot)$.

$(A_h)h : R^d \times U \to R^1$ is bounded measurable and uniformly equi-continuous in the second argument as in (A_f).

Let C be the space of continuous functions from $[0, \infty)$ to R^d with the topology of uniform convergence on compact sets, $\mathscr{B}(C)$ the Borel σ-field of C, and $\{\mathscr{B}_t\}$ the increasing family of sub σ-fields of $\mathscr{B}(C)$ generated by $\{z(s), s \leqq t, z \in C\}$.

Now various admissible classes of control policies will be defined. Let $\mathscr{U}$ (the non-anticipative control policies) denote the set of measurable functions $u: [0, \infty) \times C \to U$ such that $u(t, \cdot)$ is $\mathscr{B}_t$-measurable for each t, $\mathscr{M}$ (the stationary Markov control policies) the set of $u \in \mathscr{U}$ such that $u(t, z) = v(z(t))$ for a measurable function $v: R^d \to U$. A control $u \in \mathscr{U}$

will be called a discrete non-anticipative control policy if there exist decision times $k\Delta$, $k = 0, 1, \cdots, \Delta$ a positive number, such that $u(t, \cdot) = u(k\Delta, \cdot)$ for $t \in [k\Delta, (k+1)\Delta)$. The set of such control policies is denoted by $\mathscr{U}^d$. The subset of $\mathscr{U}^d$, where the decision interval Δ is fixed, is denoted by $\mathscr{U}^d_\Delta$. Similarly a discrete stationary Markov control policy $u \in \mathscr{U}$ is defined as one where decisions are made at only discrete times $k\Delta$, $k = 0, 1, \cdots$ and there exists a measurable function $v: R^d \to U$ such that $u(t, z) = v(z(k\Delta))$ for $t \in [k\Delta, (k+1)\Delta)$. Such a class is denoted by $\mathscr{M}^d$, and when the decision interval Δ is fixed, it is denoted by $\mathscr{M}^d_\Delta$.

As for the existence and uniqueness of solution of the system equation (1), we have the following result. Let $w(t)$ be a standard Wiener process on a probability space $(\Omega, \mathscr{B}, P)$. Then by assumptions (A_g) and (A_σ), the equation

$$(3)\qquad \begin{aligned} dx_1(t) &= g_1(x(t))dt \\ dx_2(t) &= g_2(x(t))dt + \sigma(x(t))dw(t)\,, \qquad x(0) = x \end{aligned}$$

has a unique continuous solution $y(t) = (y_1(t), y_2(t))$. For a control $u \in \mathscr{U}$, let us define a random variable

$$\zeta_s^t(u) = \int_s^t (\sigma^{-1}(y(\eta))f(y(\eta), u(\eta, y)))'dw(\eta) - \frac{1}{2}\int_s^t |\sigma^{-1}(y(\eta))f(y(\eta), u(\eta, y))|^2\, d\eta$$

and a measure

$$P_u^t(d\omega) = \exp \zeta_0^t(u)P(d\omega)$$

on $(\Omega, \mathscr{B})$. Then by Girsanov [6], assumptions (A_σ) and (A_f) guarantees that P_u^t is a probability measure on $(\Omega, \mathscr{F}_t)$, where $\mathscr{F}_t$ is a sub σ-field of $\mathscr{B}$ generated by the processes $y(s)$ and $w(s)$, $s \leqq t$, and P_u^t can be uniquely extended to a probability measure P_u on $(\Omega, \vee \mathscr{F}_t)$, where, $\vee \mathscr{F}_t$ is the minimal σ-field containing all $\mathscr{F}_t$. Note that for $s \leqq t$, $P_u^t(A) = P_u^s(A)$ for all $A \in \mathscr{F}_s$ by the martiagale property of $\exp \zeta_0^t(u)$, and hence we can use a unique extension argument of measures (see, for example, [12, Theorem 3.2, p. 139]).

Then the process

$$\tilde{w}(t) = w(t) - \int_0^t \sigma^{-1}(y(s))f(y(s), u(s, y))ds$$

is a Wiener process under the measure P_u, and the pair of processes $(y(t), \tilde{w}(t))$ on the space $(\Omega, \mathscr{B}, P_u)$ is a solution of the equation (1) (see

Girsanov [6]). Let μ_u be the probability measure on $(C, \mathscr{B}(C))$ induced by the solution $y(t)$ under the measure P_u. Then we can show that the solution of (1) is unique in the sense that any solutions with different Wiener processes induce the unique probability measure μ_u on the space $(C, \mathscr{B}(C))$ (Kushner [10]). When necessary the notation P_u^x and μ_u^x will be used to indicate the initial point x of the process $y(t)$. Hereafter we shall confine ourselves to the solution $(y(t), \tilde{w}(t))$ defined on the space $(\Omega, \mathscr{B}, P_u)$. The cost functional $J_u(x)$ is defined as

$$J_u(x) = E^{P_u^x} \int_0^\infty e^{-\beta t} h(y(t), u(t, y))dt$$

$$= E^{\mu_u^x} \int_0^\infty e^{-\beta t} h(z(t), u(t, z))dt$$

where $E^{P_u^x}(\mu_u^x)$ denotes the expectation under the measure $P_u^x(\mu_u^x)$. Finally the following inequality will be used often later:

$$E^P \exp 2\zeta_0^t(u) \leqq \exp Kt .$$

§ 3. Approximation by discrete stationary Markov controls

In this section we shall show that under our assumptions the infimum costs for discretized problems converge to the infimum cost for the original problem and the optimal policy for each discretized problem can be a discrete stationary Morkov control. Let

$$V(x) = \inf_{u \in \mathscr{U}} J_u(x) .$$

Then we have:

Theorem 1. *Under assumptions* (A_g), (A_σ), (A_f), *and* (A_h),

$$\inf_{u \in \mathscr{U}_\Delta^d} J_u(x) \to V(x) \qquad (\Delta \to 0)$$

and, for an arbitrary $\Delta > 0$, *there exists a stationary Markov control* $u_\Delta \in \mathscr{M}_\Delta^d$, *such that*

$$\inf_{u \in \mathscr{U}_\Delta^d} J_u(x) = J_{u_\Delta}(x) .$$

Proof. To see the first assertion, since

$$E^{P_u} \int_T^\infty e^{-\beta t} h(y(t), u(t, y))dt \leqq K \int_T^\infty e^{-\beta t} dt \to 0$$

as $T \to \infty$ uniformly in $u \in \mathscr{U}$, it is sufficient to prove that for an arbitrary $u \in \mathscr{U}$ and an arbitrary sequence $\Delta_n \downarrow 0$ there exists a sequence of controls $u_n \in \mathscr{U}^d_{\Delta_n}$ such that

$$E^{P^T_n} \int_0^T e^{-\beta t} h(y(t), u_n(t, y))dt \to E^{P^T_u} \int_0^T e^{-\beta t} h(y(t), u(t, y))dt$$

where $P^T_n = P^T_{u_n}$. Note that for any $\mathscr{F}_T$-measurable random variable ξ, $E^{P_u}\xi = E^{P^T_u}\xi$. Now since u is bounded (U is a compact set), there exists a sequence of $\{u_n\} \subset \mathscr{U}^d_{\Delta_n}$ such that $\Delta_n \downarrow 0$ as $n \to \infty$ and

$$\int_0^T E^P |u_n(t, y(\cdot, \omega)) - u(t, y(\cdot, \omega))|^2 dt \to 0 .$$

Hence, as in [5], by equi-continuity of f and boundedness of σ^{-1} and f, we can show that

$$\int_0^T E^P |\sigma^{-1}(y(t))f(y(t), u_n(t, y)) - \sigma^{-1}(y(t))f(y(t), u(t, y))|^2 dt \to 0$$

(see pp. 397–398, [5]). Thus

$$\int_0^T \sigma^{-1}(y(t))f(y(t), u_n(t, y))dw(t) \to \int_0^T \sigma^{-1}(y(t))f(y(t), u(t, y))dw(t)$$

in $L_2(P)$ (the space of square-integrable random variables on $(\Omega, \mathscr{B}, P)$). Similarly by boundedness of σ^{-1} and f,

$$\int_0^T |\sigma^{-1}(y(t))f(y(t), u_n(t, y))|^2 dt \to \int_0^T |\sigma^{-1}(y(t))f(y(t), u(t, y))|^2 dt$$

in $L_2(P)$.

These imply that

$$\zeta^T_0(u_n) \to \zeta^T_0(u)$$

in probability and hence

$$\exp \zeta^T_0(u_n) \to \exp \zeta^T_0(u) \tag{4}$$

in $L_1(P)$ (see p. 390 [5] or Girsanov [6]). Thus we get the conclusion by the same argument as in [5]. Q.E.D.

To see the second assertion, let

$$V_\Delta(x) = \inf_{u \in \mathscr{U}^d_\Delta} J_u(x) .$$

We should note that to find a control which attains the value $V_\Delta(x)$ is essentially a discrete-time Markovian decision problem with a discounted cost and hence we can use the result and technique developed in such a field. The discussion given below follows the method in Ross [7, p. 122]. The only essentially different point of our model from that given in Ross [7, p. 122] is that the action space U is not finite but a compact set.

Let us define $\bar{h}(x, a) : R^d \times U \to R^1$ by

$$\bar{h}(x, a) = E^{P_a^x} \int_0^\Delta e^{-\beta s} h(y(s), a) ds$$

where P_a^x is defined for any $a \in U$ by

$$\begin{aligned} P_a^x(d\omega) &= \exp \zeta_0^\Delta(a) P(d\omega) \\ &= \exp \left[\int_0^\Delta \sigma^{-1}(y(s)) f(y(s), a) d\omega(s) \right. \\ &\qquad \left. - \frac{1}{2} \int_0^\Delta |\sigma^{-1}(y(s)) f(y(s), a)|^2 \, ds \right] P(d\omega) \end{aligned}$$

and $y(0) = x$.

Let us also define a distribution function $F(y \mid x, a)$ by

$$F(y \mid x, a) = P_a^x(y(\Delta) \leqq y) .$$

Then we can show the following dynamic programming equation:

$$V_\Delta(x) = \inf_{a \in U} \left\{ \bar{h}(x, a) + e^{-\beta\Delta} \int V_\Delta(y) dF(y \mid x, a) \right\} . \tag{5}$$

The inequality

$$V_\Delta(x) \geqq \inf_{a \in U} \left\{ \bar{h}(x, a) + e^{-\beta\Delta} \int V_\Delta(y) dF(y \mid x, a) \right\}$$

can be shown by the exactly same argument as in [7, p. 122]. To go the other way, first note that $\bar{h}(x, a)$ and $\int V_\Delta(y) dF(y \mid x, a)$ are both continuous in a for an arbitrary fixed x. This follows from the already proved fact that $a_n \to a$ implies

$$\exp \zeta_0^\Delta(a_n) \to \exp \zeta_0^\Delta(a) \qquad \text{in } L_1(P)$$

and the fact that $V_\Delta(\cdot)$ is bounded. Then since U is compact, there exists a measurable selection $u_\Delta(x)$ such that

$$\inf_{a \in U} \left\{ \bar{h}(x, a) + e^{-\beta\Delta} \int V_\Delta(y) dF(y \mid x, a) \right\}$$

$$= \bar{h}(x, u_\Delta(x)) + e^{-\beta\Delta} \int V_\Delta(y) dF(y \mid x, u_\Delta(x)) \ .$$

(See, for example, Benes [8]).

Now the rest of the proof of the converse inequality is just like that given in Ross [7, p. 122] and hence it is omitted.

Once we have the dynamic programming equation (5), then by following the same argument as in [7, p. 124, Theorem 6.3] we can show that $u_n(\cdot)$ is optimal for an arbitrary initial point i.e., $V_\Delta(x) = J_{u_\Delta}(x)$ for all x. Q.E.D.

Remark 1. In a later application, we shall treat the problem where the cost function h is not bounded and has a form

$$h(x, u) = c(u) + k(x)$$

where $c(\cdot)$ is an equi-continuous bounded function and $k(\cdot)$ is not necessarily bounded, but satisfies the conditions:

$$E^P k^2(y(t)) < \infty \qquad \forall t > 0$$

$|E^{P_u} k(y(t))| \leqq g(t)$ for some $g(t)$ and for all $u \in \mathscr{U}$, and

$$\int_0^\infty e^{-\beta t} g(t) dt < \infty \ . \tag{6}$$

For this form of the cost functional, the first assertion of Theorem 1 is still true under (A_g), (A_f) and (A_σ) as long as $J_u(x)$ is well defined for each $u \in \mathscr{U}$. To see this, let u_n and u be the same as in the proof of the first assertion in Theorem 1. Then it is sufficient to show that

$$E^{P_n} \int_0^\infty e^{-\beta t} c(u_n(t, y)) dt \to E^{P_u} \int_0^\infty e^{-\beta t} c(u(t, y)) dt$$

and

$$E^{P_n} \int_0^\infty e^{-\beta t} k(y(t))\, dt \to E^{P_u} \int_0^\infty e^{-\beta t} k(y(t)) dt$$

where $P_n = P_{u_n}$.

Since $C(\cdot)$ satisfies (A_h), the first convergence comes from the first assertion in Theorem 1. To see the second convergence, it is sufficient to prove that

$$E^{P^t_n} k(y(t)) \to E^{P^t_u} k(y(t)) \tag{7}$$

for each t.

Let us define

$$k_N(x) = \begin{cases} k(x) & \text{if } |k(x)| < N \\ 0 & \text{otherwise.} \end{cases}$$

Then

$$\begin{aligned} &|E^{P^t_n}[k(y(t)) - k_N(y(t))]| \\ &\leq (E^P \exp 2\zeta_0^t(u_n))^{1/2} \{E^P[k_N(y(t)) - k(y(t))]^2\}^{1/2} \\ &\leq K\{E^P[k_N(y(t)) - k(y(t))]^2\}^{1/2} . \end{aligned}$$

Since $[k_N(y(t)) - k(y(t))]^2 \leq k(y(t))^2$ and $E^P k(y(t))^2 < \infty$, we have, by Lebesgue's bounded convergence theorem, the last term goes to zero as $N \to \infty$ uniformly in n. For a fixed N,

$$E^{P^t_n} k_N(y(t)) \to E^{P^t_u} k_N(y(t))$$

since $k_N(\cdot)$ is bounded and by the fact of (4).

Similarly

$$E^{P^t_u} k_N(y(t)) \to E^{P^t_u} k(y(t)) .$$

Hence we have (7), and by (6) Lebesgue's bounded convergence theorem yields the desired result.

We also should note that the second assertion in Theorem 1 holds as long as $h(x, a)$ and $\int V_A(y) dF(y \mid x, a)$ is continuous in a even if h is not bounded, and condition (6) is a sufficient condition for such continuity. Note that all the conditions in Remark 1 are satisfied in the later application of Section 5.

§ 4. Convergence of discrete stationary Markov controls $u_n(\cdot)$ to an optimal control

In the preceding section, we have shown that discrete stationary Markov controls $u_n(\cdot)$ approximate the optimal value $V(\cdot)$, *i.e.*, $J_{u_n}(x) \to V(x)$ for all x. In this section we shall consider whether a limit function $\bar{u}(\cdot)$ of $u_n(\cdot)$, if it exists in some sense, is an optimal control or not. For example, suppose $u_n(x) \to \bar{u}(x)$ for each x, then is $\bar{u}(\cdot)$ an optimal control? When $u_n(x) \to \bar{u}(x)$ for each x, then $f(x, u_n(x)) \to f(x, \bar{u}(x))$ and $h(x, u_n(x)) \to h(x, \bar{u}(x))$ for each x under the assumptions (A_f) and (A_h). Then this implies that $f(\cdot, u_n(\cdot))$ and $h(\cdot, u_n(\cdot))$ converge weakly to $f(\cdot, \bar{u}(\cdot))$ and $h(\cdot, \bar{u}(\cdot))$ respectively on each compact set in R^d, *i.e.*, on each compact set A, $f(\cdot, u_n(\cdot))$ and $h(\cdot, u_n(\cdot))$ converge weakly to $f(\cdot, \bar{u}(\cdot))$ and $h(\cdot, \bar{u}(\cdot))$ respectively in $L_1(A)$ where $L_1(A)$ is the space of integrable functions on A. We shall denote the weak convergence in above by $f(\cdot, u_n(\cdot)) \Rightarrow f(\cdot, \bar{u}(\cdot))$ and $h(\cdot, u_n(\cdot)) \Rightarrow h(\cdot, \bar{u}(\cdot))$.

Now we shall consider the following problem: Suppose $u^*(\cdot)$ is a measurable function from R^d to U such that $u_n(\cdot) \Rightarrow u^*(\cdot)$, $f(\cdot, u_n(\cdot)) \Rightarrow f(\cdot, u^*(\cdot))$ and $h(\cdot, u_n(\cdot)) \Rightarrow h(\cdot, u^*(\cdot))$, then is $u^*(\cdot)$ be an optimal stationary Markov control? If, in the above problem $u^*(\cdot)$ is optimal, then from the above discussion we see that the limit function $\bar{u}(\cdot)$ of $u_n(\cdot)$, *i.e.*, $\bar{u}(x) = \lim u_n(x)$ for each x, is an optimal control provided $\bar{u}(\cdot)$ exists. In the problem formulated just now, for $u^*(\cdot)$ to be optimal we shall need the existence of a density for the process $y(t)$ of the equation (3) and some properties on it.

In Corollary 1 of Theorem 2, a sufficient condition is given for the existence of a subsequence $\{u_n\}$ and a measurable control $u^*(\cdot)$ such that $u_n \Rightarrow u^*$, $f(\cdot, u_n(\cdot)) \Rightarrow f(\cdot, u^*(\cdot))$ and $h(\cdot, u_n(\cdot)) \Rightarrow h(\cdot, u^*(\cdot))$. Let us impose the following assumptions:

(B_p) For the unique solution $y(t)$ of (3), there exists a density $p(t, x, y)$ such that (1) $p(\cdot, x, y)$ is continuous (2) for any $f_n(\cdot)$ and $f(\cdot)$ such that $f_n \Rightarrow f$,

$$\int_D f_n(y)p(t, x, y)dy \to \int_D f(y)p(t, x, y)dy$$

for each t and x and each compact set D. (If, for example, $p(t, x, \cdot)$ is continuous, then this is satisfied.)
(3) there exists a β' such that for an arbitrary $t > 0$

$$\int_0^t \left(\int_{R^d} p(t, x, y)^{\beta'} dy \right)^{1/\beta'} dt \leqq M(t)$$

and

$$\int_{R^d} p(t, x, y)^{\beta'} dy \leqq M'(t_0, t_1)$$

uniformly on each compact t-set $[t_0, t_1]$, $t_0 > 0$.

(4) $$\int_{D_m^c} p(t, x, y)dy \to 0 \qquad (m \to \infty)$$

uniformly in any compact t-set where $D_m = \{x : |x| \leqq m,\ x \in R^d\}$ and D_m^c is the complementary set of D_m.

(B_f) $f(\cdot, \cdot)$ is bounded and satisfies a Lipshitz condition:

$$|f(x, u) - f(x', u')| \leqq K(|x - x'| + |u - u'|) .$$

Remark 2. The above assumptions on the density of the process are generally not easy to check. Here as an example, we can show the density of the Wiener process satisfies the assumption (B_p). This fact will be used later in Section 5. We also note that if the equation (3) is non-degenerate and bounded coefficient, *i.e.*, $g_1 \equiv 0$ and $|g_2| \leqq K$, then the solution $y(t)$ of (3) has a density which satisfies (B_p). As a case where the equation (3) is degenerate and the density of the solution $y(t)$ satisfies (B_p), some of the following type equations:

$$dx(t) = Ax(t)dt + Bdw(t) , \qquad x(o) = x$$

where the pair (A, B) is controllable, satisfy this condition. For example, we can check that the case where

$$A = \begin{pmatrix} 0 & 1 \\ 0 & 0 \end{pmatrix}, \qquad B = \begin{pmatrix} 0 \\ 1 \end{pmatrix}$$

satisfies (B_p).

Theorem 2. *Let us assume* $(A_g), (B_f), (A_\sigma), (A_h)$ *and* (B_p). *Then* u^* *is an optimal stationary Markov control, i.e.,*

$$V(x) = \inf_{u \in \mathscr{U}} J_u(x) = \inf_{u \in \mathscr{M}} J_u(x) = J_{u^*}(x)$$

for all x.

Proof. Since $J_{u_n}(x) \to V(x)$ for all x, it is sufficient to prove that $J_{u_n}(x) \to J_{u^*}(x)$ for all x. As in Section 2, for $u_n \in \mathscr{M}^d_{\Delta_n}$ let $P_{u_n} = P_n$ be the transformed probability measure on $(\Omega, \mathscr{B})$. Similarly P_0 corresponds to the control u^*. Let μ_n and μ_0 be the probability measures on $(C, \mathscr{B}(C))$ induced by the unique solution $y(t)(y(0) = x)$ under the measure P_n and P_0 respectively. Then by denoting $t_\Delta = [t/\Delta]\Delta$ where $[a]$ is the maximal integer not greater than a, we have

$$\begin{aligned} J_{u_n}(x) &= E^{P_n} \int_0^\infty e^{-\beta t} h(y(t), u_n(y(t_{\Delta_n})))dt \\ &= E^{\mu_n} \int_0^\infty e^{-\beta t} h(z(t), u_n(z(t_{\Delta_n})))dt . \end{aligned}$$

Hence if we prove that

$$(8)\qquad E^{\mu_n}\int_0^\infty e^{-\beta t}h(z(t), u_n(z(t_{\Delta_n})))dt - E^{\mu_n}\int_0^\infty e^{-\beta t}h(z(t), u^*(z(t)))dt \to 0$$

and that

$$(9)\qquad E^{\mu_n}\int_0^\infty e^{-\beta t}h(z(t), u^*(z(t)))dt \to E^{\mu_0}\int_0^\infty e^{-\beta t}h(z(t), u^*(z(t)))dt\ ,$$

then $J_{u_n}(x) \to J_{u^*}(x)$ for all x and this implies $J_{u^*}(x) = V(x)$ for all x. The desired results (8) and (9) will be proved in the following lemmas. Noting that

$$E^{\mu_n}\int_T^\infty e^{-\beta t}h(z(t), u(t, z))dt \to 0$$

as $T \to \infty$ uniformly in $u \in \mathscr{U}$ since $h(\cdot, \cdot)$ is bounded, it is sufficient to show that, for an arbitrary $T > 0$,

$$(10)\qquad E^{\mu_n}\int_0^T e^{-\beta t}h(z(t), u_n(z(t_{\Delta_n})))dt - E^{\mu_n}\int_0^T e^{-\beta t}h(z(t), u^*(z(t)))dt \to 0$$

and

$$(11)\qquad E^{\mu_n}\int_0^T e^{-\beta t}h(z(t), u^*(z(t)))dt \to E^{\mu_0}\int_0^T e^{-\beta t}h(z(t), u^*(z(t)))dt\ .$$

In the following Lemma 1 and Lemma 2, the measures μ_n and μ_0 will be confined on the space $(C_T, \mathscr{B}(C_T))$ where C_T is the space of continuous functions from $[0, T]$ to R^d with sup-norm. Furthermore, although u_n is a function from $R^d \to U$, the notation u_n is also used as a discrete stationary Markov control.

Lemma 1. *For an arbitrary $t > 0$,*

$$E^{\mu_n}\left\{\int_0^t [f(z(s), u_n(z(s_{\Delta_n}))) - f(z(s), u^*(z(s)))]ds\right\}^2 \to 0\ .$$

Remark. The same thing does hold for h instead of f.

Proof. The proof is quite technical and complicated. Let $g_n(z_s)$ be the term in the blaket [ ] in the above equation. Then

$$E^{\mu_n}\left[\int_0^t g_n(z_s)ds\right]^2 = E^{\mu}\left[\int_0^t g_n(z_s)ds\right]^2 \exp\zeta_0^t(u_n)$$
$$\le \left(E^{\mu}\exp 2\zeta_0^t(u_n)\right)^{1/2}E^{\mu}\left[\int_0^t g_n(z_s)ds\right]^4\Big)^{1/2}$$

$$\leqq K(t)E^{\mu}\left[\int_0^t g_n(z_s)ds\right]^2 .$$

The last inequality is due to the boundedness of g_n. Hence to prove Lemma 1, it is sufficient to show

$$E^{\mu}\left[\int_0^t g_n(z_s)ds\right]^2 \to 0 .$$

Now

$$\left[\int_0^t g_n(z_s)ds\right]^2 = [J_1 + J_2 + J_3]^2 \leqq 8[J_1^2 + J_2^2 + J_3^2]$$

where

$$J_1^2 = \left[\int_0^t [f(z(s), u_n(z(s_{\Delta_n}))) - f(z(s_{\Delta_n}), u_n(z(s_{\Delta_n})))]ds\right]^2$$

$$J_2^2 = \left[\int_0^t [f(z(s_{\Delta_n}), u_n(z(s_{\Delta_n}))) - f(z(s_{\Delta_n}), u^*(z(s_{\Delta_n})))]ds\right]^2$$

and

$$J_3^2 = \left[\int_0^t [f(z(s_{\Delta_n}), u^*(z(s_{\Delta_n}))) - f(z(s), u^*(z(s)))]ds\right]^2 .$$

Then by the Lipshitz condition on $f(B_f)$,

$$J_1^2 \leqq K \int_0^t |z(s) - z(s_{\Delta_n})|^2\, ds$$

and

$$E^{\mu}J_1^2 \leqq K \int_0^t E^{\mu}\,|z(s) - z(s_{\Delta_n})|^2\, ds .$$

Since $s_{\Delta_n} \to s$ as $\Delta_n \to 0$ and $z(\cdot)$ is continuous,

$$|z(s) - z(s_{\Delta_n})|^2 \to 0$$

as $\Delta_n \to 0$ for all $z \in C$. Since $|z(s) - z(s_{\Delta_n})|^2 \leqq 2 \sup_{0\leqq s\leqq t} z(s)^2$ and $2E^{\mu} \sup_{0\leqq s\leqq t} z(s)^2 < \infty$, $E^{\mu}\,|z(s) - z(s_{\Delta_n})|^2 \to 0$ by Lebesgue's bounded convergence theorem and hence

$$\int_0^t E^{\mu}\,|z(s) - z(s_{\Delta_n})|^2\, ds \to 0$$

again by Lebesgue's bounded convergence theorem. This leads to the result

$$E^{\mu}J_1^2 \to 0 \quad \text{as} \quad \Delta_n \to 0 .$$

Next let us prove

$$E^{\mu}J_2^2 \to 0 \quad \text{as} \quad \Delta_n \to 0 .$$

Defining

$$f_n(x) = f(x, u_n(x)) - f(x, u^*(x)) ,$$

$$E^{\mu}J_2^2 = E^{\mu}\left[\int_0^t f_n(z(s_{\Delta_n}))ds\right]^2$$

$$= \int_0^t \int_0^t E^{\mu} f_n(z(s_{\Delta_n})) f_n(z(u_{\Delta_n})) ds du .$$

Let $s < u$. Then noting that the unique solution $y(t)$ of (3) with $y(0) = x$ is a Markov process having a density $p(t, x, y)$ which satisfies the condition (B_p) and μ is the probability measure induced by $y(t)$ under the measure P,

$$E^{\mu} f_n(z(s_{\Delta_n})) f_n(z(u_{\Delta_n}))$$

$$= \int f_n(y)\left(\int f_n(v) p(u_{\Delta_n} - s_{\Delta_n}, y, v) dv\right) p(s_{\Delta_n}, x, y) dy .$$

We shall show that

$$A_n(y) \equiv \int f_n(v) p(u_{\Delta_n} - s_{\Delta_n}, y, v) dv \to 0$$

as $\Delta_n \to 0$. $A_n(y)$ can be written as

$$A_n(y) = \left[\int f_n(v) p(u - s, y, v) dv\right]$$

$$+ \left[\int f_n(v) p(u_{\Delta_n} - s_{\Delta_n}, y, v) dv - \int f_n(v) p(u - s, y, v) dv\right] .$$

The first term in the above equation is

$$\int_{D_m} f_n(v) p(u - s, y, v) dv + \int_{D_m^c} f_n(v) p(u - s, y, v) dv$$

where D_m is a compact set defined by $D_m = \{x : x \in R^d, |x| \leq m\}$. Since $f_n(\cdot)$ is uniformly bounded in n,

$$\int_{D_m^c} f_n(v) p(u - s, y, v) dv \to 0$$

as $m \to \infty$ uniformly in n. On the other hand since $f_n \Rightarrow 0$,

$$\int_{D_m} f_n(v)p(u-s, y, v)dv \to 0$$

as $n \to \infty$ for an arbitrarily fixed m by the assumption $(B_p - 2)$, and the first term in $A_n(y)$ converges to zero. The absolute value of the second term in $A_n(y)$ is less than

$$(*) \qquad K \int |p(u_{\Delta_n} - s_{\Delta_n}, y, v) - p(u-s, y, v)|\, dv\, .$$

And this is written as

$$K \int_{D_m} |p(u_{\Delta_n} - s_{\Delta_n}, y, v) - p(u-s, y, v)|\, dv + K \int_{D_m^c} |p(u_{\Delta_n} - s_{\Delta_n}, y, v) - p(u-s, y, v)|\, dv\, .$$

The last term tends to zero uniformly in n as $m \to \infty$ by the assumption $(B_p - 4)$. For a fixed m, the first term satisfies

$$\int_{D_m} |p(u_{\Delta_n} - s_{\Delta_n}, y, v) - p(u-s, y, v)|^{\beta'}\, dv \leqq \left[\left(\int_{D_m} p(u_{\Delta_n} - s_{\Delta_n}, y, v)^{\beta'} dv\right)^{1/\beta'} + \left(\int_{D_m} p(u-s, y, v)^{\beta'}\right)^{1/\beta'}\right]^{\beta'} \leqq K$$

uniformly in n due to $(B_p - 3)$ since $s < u$ and for sufficiently large n $u_{\Delta_n} - s_{\Delta_n}$ is contained in a compact set $[t_0, t]$, $t_0 > 0$. Then since

$$p(u_{\Delta_n} - s_{\Delta_n}, y, v) \to p(u-s, y, v)$$

by $(B_p - 1)$ we have

$$\int_{D_m} |p(u_{\Delta_n} - s_{\Delta_n}, y, v) - p(u-s, y, v)|\, dv \to 0$$

as $n \to \infty$ (see [14, Theorem 6, p. 72]). From these, the term $(*)$ converges to zero as $\Delta_n \to 0$. Hence we have shown that $A_n(y) \to 0$ as $\Delta_n \to 0$. Now noting $|f_n| \leqq K$, $|A_n| \leqq K$ and

$$\int f_n(y)A_n(y)p(s_{\Delta_n}, x, y)dy = \int f_n(y)A_n(y)(p(s_{\Delta_n}, x, y) - p(s, x, y))dy$$

$$+ \int f_n(y) A_n(y) p(s, x, y) dy ,$$

by the similar argument as above we get

$$E^\mu f_n(z(s_{\Delta_n})) f_n(z(u_{\Delta_n})) \to 0$$

as $\Delta_n \to 0$. This implies $E^\mu J_2^2 \to 0$ as $\Delta_n \to 0$.

Finally assuming that $u^*(\cdot)$ is continuous, we shall show $E^\mu J_3^2 \to 0$ as $\Delta_n \to 0$. By the assumption (B_f),

$$E^\mu J_3^2 \leqq E^\mu \left[\int_0^t K[|z(s_{\Delta_n}) - z(s)|^2 + |u^*(z(s_{\Delta_n})) - u(z(s))|^2] ds \right] .$$

Now the rightside converges to zero by the assumption that u^* is continuous.

Thus we have shown that

$$E^\mu \left[\int_0^t g_n(z_s) ds \right]^2 \to 0$$

under the continuity assumption of u^*. Now we shall prove the lemma where u^* is not necessarily continuous. Let $\varepsilon > 0$ be arbitrary. Then, since u^* is bounded, we can find a bounded continuous function $\bar{u}^*(\cdot)$ such that

$$\int_{R^d} (u^*(x) - \bar{u}^*(x))^2 dx \leqq \varepsilon . \tag{12}$$

Let us define

$$g(x, u) = f(x, u) - f(x, u^*(x)) + f(x, \bar{u}^*(x)) .$$

Then

$$g(\cdot, u_n(\cdot)) \Rightarrow f(\cdot, \bar{u}^*(\cdot)) . \tag{13}$$

Now

$$\begin{aligned} E^\mu \left[\int_0^t f_n(z_s) ds \right]^2 &= E^\mu \left[\int_0^t \{ f(z(s), u_n(z(s_{\Delta_n}))) - g(z(s), u_n(z(s_{\Delta_n}))) \} ds \right. \\ &\quad + \int_0^t \{ g(z(s), u_n(z(s_{\Delta_n}))) - f(z(s), \bar{u}^*(z(s))) \} ds \\ &\quad \left. + \int_0^t \{ f(z(s), \bar{u}^*(z(s))) - f(z(s), u^*(z(s))) \} ds \right]^2 \\ &\leqq 8(E^\mu I_1^2 + E^\mu I_2^2 + E^\mu I_3^2) \end{aligned}$$

where I_1, I_2 and I_3 are obviously defined.

Since $\bar{u}^*(\cdot)$ is continuous, we have $E^\mu I_2^2 \to 0$ as $\Delta_n \to 0$ by noting (13) and the fact proven already. Let us estimate $E^\mu I_1^2$.

$$E^\mu I_1^2 \leqq t \int_0^t E^\mu[f(z(s), u_n(z(s_{\Delta_n}))) - g(z(s), u_n(z(s_{\Delta_n})))]^2 ds$$

$$= t \int_0^t E^\mu[f(z(s), u^*(z(s))) - f(z(s), \bar{u}^*(z(s)))]^2 ds$$

$$\leqq Kt \int_0^t E^\mu[u^*(z(s)) - \bar{u}^*(z(s))]^2 ds .$$

And

$$E^\mu[u^*(z(s)) - \bar{u}^*(z(s))]^2$$

$$= \int_{R^d} [u^*(y) - \bar{u}^*(y)] p(s, x, y) dy$$

$$\leqq K\Big(\int_{R^d} [u^*(y) - \bar{u}^*(y)]^2 dy\Big)^{1/\beta}\Big(\int_{R^d} p(s, x, y)^{\beta'} dy\Big)^{1/\beta'}$$

$$\leqq K\varepsilon^{1/\beta} M^{1/\beta'} \qquad (1/\beta + 1/\beta' = 1)$$

by (12) and $(B_p - 3)$. Hence we have

$$\lim_{\Delta_n \to 0} E^\mu I_1^2 \leqq K t^2 \varepsilon^{1/\beta} M^{1/\beta'}$$

and, since $\varepsilon > 0$ was arbitrary, $\lim_{\Delta_n \to 0} E^\mu I_1^2 = 0$.

We have at the same time shown that $\lim_{\Delta_n \to 0} E^\mu I_3^2 = 0$. Q.E.D.

The following lemma is essentially due to Stroock and Varadhan [9]. Since we are dealing with the cases of unbounded coefficients in system equation (1) and with the discrete stationary Markov controls, the proof is somewhat involved but essentially follows the martingale theoretical method given in [9].

Lemma 2. *$\mu_n \Rightarrow \mu_0$, i.e., μ_n converges weakly to μ_0.*

Proof. First we should note that the solution of the system equation (1) is unique in the sense of probability law, *i.e.*, any solutions with different Wiener processes $w(t)$ induce the same probability measure on $(C, \mathscr{B}(C))$ (for the proof, see, for example, Kushner [10]). Secondly, writing the system equation (1) in a simpler form:

$$x(t) = x + \int_0^t \bar{f}(x(t), u(t)) dt + \int_0^t \bar{\sigma}(x(t)) d\bar{w}(t)$$

where

$$\bar{f}(x(t), u(t)) = \begin{pmatrix} g_1(x(t)) \\ g_2(x(t)) + f(x(t), u(t)) \end{pmatrix}$$

$$\bar{\sigma}(x) = \begin{pmatrix} 0 & 0 \\ 0 & \sigma(x) \end{pmatrix}$$

$\bar{w}(t)$: d-dimensional Wiener process,

we note that, for any $\theta \in R^d$,

$$X_u^\theta(t) = \left(\theta, z(t) - x - \int_0^t \bar{f}(z(s), u(s, z))ds\right)$$

is a μ_u-martingale with respect to $\mathscr{B}_t$ with its increasing process

$$A_u^\theta(t) = \int_0^t \theta' a(z(s))\theta ds \qquad \text{where } a(x) = \bar{\sigma}(x)\bar{\sigma}'(x) .$$

Furthermore the uniqueness of the solution of the system equation (1) implies that μ_u is the unique probability measure under which $X_u^\theta(t)$ is a martingale with its increasing process $A_u^\theta(t)$ (for the proof, see Priouret [11, Theorem 6, Chap. III, p. 71]). Although, in [11], diffusion processes are treated, the proof applies directly to our case).

Now from the fact just mentioned, μ_n is the unique probability measure such that

$$X_n^\theta(t) = \left(\theta, z(t) - x - \int_0^t \bar{f}(z(s), u_n(z(s_{\Delta_n})))ds\right)$$

is a martingale with the increasing process

$$A_n^\theta(t) = \int_0^t \theta' a(z(s))\theta ds .$$

Similarly

$$X_0^\theta(t) = \left(\theta, z(t) - x - \int_0^t \bar{f}(z(s), u^*(z(s)))\, ds\right)$$

is a μ_0-martingale where μ_0 is the uniqune such probability measure. The process for showing $\mu_n \Rightarrow \mu_0$ consists of the following two steps. The first of these is quite standard and consists of showing that the family $\{\mu_n\}$ is weakly compact. The second step involves identifying any limit of $\{\mu_n\}$ as μ_0. That $\{\mu_n\}$ is weakly compact is easily established by noting the following inequality: for $0 \leqq t_1 \leqq t_2$,

$$
\begin{aligned}
E^{\mu_n}\{|z(t_2) - z(t_1)|^3\} \\
&= E^{P_n^T}|y(t_2) - y(t_1)|^3 = E^P|y(t_2) - y(t_1)|^3 \exp \zeta_0^T(u_n) \\
&\leqq (E^P(|y(t_2) - y(t_1)|^3)^{4/3})^{3/4}(E^P \exp 4\zeta_0^T(u_n))^{1/4} \\
&\leqq H(t_2 - t_1)^{3/2}
\end{aligned}
$$

where $y(t)$ is the unique solution of (3) and we have used the facts that

$$E^P|y(t_2) - y(t_1)|^4 \leqq K(t_2 - t_1)^2$$

and

$$E^P \exp 4\zeta_0^T(u_n) \leqq K(T) .$$

Thus we have shown that $\{\mu_n\}$ is weakly compact. From this, we have a subsequence $\{\mu_n\}$, indexed by n, such that $\mu_n \Rightarrow \mu'$. We shall show that $\mu' = \mu_0$. To do this, it is sufficient to prove that $X_0^\theta(t)$ defined above is a μ'-martingale, since then $\mu' = \mu_0$ from the uniqueness of such measures as was mentioned above. Noting that $X_n^\theta(t)$ is a μ_n-martingale, it suffices to show that

$$\lim_{n\to\infty} E^{\mu_n}[\xi \cdot X_n^\theta(t)] = E^{\mu'}[\xi \cdot X_0^\theta(t)] \tag{14}$$

for any bounded continuous $\mathscr{B}_s$-measurable function ξ $(s \leqq t)$ defined on the space $(C_T, \mathscr{B}(C_T))$. Now

$$
\begin{aligned}
E^{\mu_n}\xi \cdot X_n^\theta(t) = E^{\mu_n}\xi\Big(\theta' z(t) - \theta' x - \int_0^t \theta_1' g_1(z(s))ds \\
- \int_0^t \theta_2' g_2(z(s))\,ds - \int_0^t \theta_2' f(z(s), u_n(z(s_{\Delta_n})))ds\Big)
\end{aligned}
$$

where $\theta = (\theta_1, \theta_2)'$, $\theta_1 \in R^{d_1}$, $\theta_2 \in R^{d_2}$. We shall calculate the limit of each term in the above. Let us define, for an arbitray $N > 0$,

$$
z_N(t) = \begin{cases} \theta' z(t) & \text{if } |\theta' z(t)| \leqq N \\ N & \text{if } \theta' z(t) > N \\ -N & \text{if } \theta' z(t) < -N . \end{cases}
$$

Then since $z_N(t)$ is bounded and continuous in z, we have

$$E^{\mu_n}\xi z_N(t) \to E^{\mu'}\xi z_N(t)$$

as $n \to \infty$. On the other hand, since ξ is bounded,

$$|E^{\mu_n}\xi(\theta' z(t) - z_N(t))| \leqq KE^\mu|\theta' z(t) - z_N(t)| \exp \zeta_0^t(u_n)$$

$$\leqq K(E^{\mu}|\theta' z(t) - z_N(t)|^2)^{1/2}(E^{\mu} \exp 2\zeta_0^t(u_n))^{1/2}$$
$$\leqq K(t)E^{\mu}|\theta' z(t) - z_N(t)|^2 \to 0$$

as $N \to \infty$ uniformly in n.

The last convergence is due to Lebesgue's bounded convergence theorem, since we have

$$|\theta' z(t) - z_N(t)|^2 \leqq 4|\theta' z(t)|^2$$

and

$$\int |\theta' z(t)|^2 \, d\mu(z) = \int |\theta' y(t)|^2 \, dP(\omega) < \infty$$

where $y(t)$ is the unique solution of (3). Combining these results, we have

$$E^{\mu_n}\xi\theta' z(t) \to E^{\mu'}\xi\theta' z(t) .$$

Similarly, although g_1 and g_2 are not bounded, using the same technique, we have

$$E^{\mu_n} \int_0^t \theta_i' g_i(z(s))ds \to E^{\mu'} \int_0^t \theta_i' g_i(z(s))ds$$

($i = 1, 2$) on the basis of assumption (A_g). Finally we shall show

$$E^{\mu_n}\xi \int_0^t \theta_2' f(z(s), u_n(z(s_{\Delta_n})))ds \to E^{\mu'}\xi \int_0^t \theta_2' f(z(s), u^*(z(s)))ds .$$

Since

$$E^{\mu_n}\xi \int_0^t \theta_2' f(z(s), u_n(z(s_{\Delta_n})))ds \to E^{\mu_n}\xi \int_0^t \theta_2' f(z(s), u^*(z(s)))ds \to 0$$

by Lemma 1, it suffices to show that

$$E^{\mu_n}\xi \int_0^t \theta_2' f(z(s), u^*(z(s)))ds \to E^{\mu'}\xi \int_0^t \theta_2' f(z(s), u^*(z(s)))ds .$$

But this can be proved again by the similar technique as above ***i.e.***, by noting that there exists a sequence of bounded continuous functions $h_N(z)$ such that

$$E^{\mu}\left(h_N(z) - \int_0^t \theta_2' f(z(s), u^*(z(s)))\,ds\right)^2 \to 0$$

as $N \to \infty$, and by the fact that $E^{\mu} \exp 2\zeta_0^t(u_n) \leqq K(t)$. Now we have shown (18) and this implies $\mu' = \mu_0$ and $\mu_n \Rightarrow \mu_0$.

Proof of Theorem 2. Now the desired result (10) is a direct consequence of Lemma 1. For the result (11), it is sufficient to show that

$$E^{\mu_n}h(z(t), u^*(z(t))) \to E^{\mu_0}h(z(t), u^*(z(t))) .$$

Noting that there exists a sequence $h_m(z)$ of bounded continuous functions such that

$$E^{\mu}[h(z(t), u^*(z(t))) \to h_m(z)]^2 \to 0 ,$$

we have

$$\begin{aligned} E^{\mu_n}h(z(t), u^*(z(t))) = & [E^{\mu_n}h(z(t), u^*(z(t))) - E^{\mu_n}h_m(z)] \\ & + [E^{\mu_n}h_m(z) - E^{\mu_0}h_m(z)] \\ & + [E^{\mu_0}h_m(z) - E^{\mu_0}h(z(t), u^*(z(t)))] . \end{aligned}$$

The first and third terms go to zero as $m \to \infty$ uniformly in μ_n and μ_0 since

$$\begin{aligned} & |E^{\mu_n}h(z(t), u^*(z(t))) - E^{\mu_n}h_m(z)| \\ & \leqq \{E^{\mu}[h(z(t), u^*(z(t))) - h_m(z)]^2\}^{1/2}(\exp 2\zeta_0^T(u_n))^{1/2} \\ & \leqq K(T)\{E^{\mu}[h(z(t), u^*(z(t))) - h_m(z)]^2\}^{1/2} \to 0 . \end{aligned}$$

The second term goes to zero as $n \to \infty$ for an arbitrarily fixed m since $\mu_n \Rightarrow \mu_0$ and $h_m(z)$ is bounded and continuous. Q.E.D.

Corollary 1. *Suppose that, for each x, the set*

$$\begin{pmatrix} f(x, U) \\ h(x, U) \end{pmatrix}$$

is convex, then under (A_g), (B_f), (A_σ), (A_h) *and* (B_p), *there exists an optimal stationary Markov control* $u^*(\cdot)$ *which does not depend on the initial point of the process, i.e.,*

$$V(x) = \inf_{u \in \mathscr{U}} J_u(x) = \inf_{u \in \mathscr{M}} J_u(x) = J_{u^*}(x)$$

for all x, and $u_n \Rightarrow u^*$ *for a subsequence* $\{u_n\}$.

Proof. By a standard argument (see Yamada [13]), under the assumption we can show that there exists a measurable function $u^*(\cdot): R^d \to U$ and a subsequence $\{u_n(\cdot)\}$ for controls such that

$$u_n \Rightarrow u^*, \; f(\cdot, u_n(\cdot)) = f(\cdot, u^*(\cdot)) \quad \text{and} \quad h(\cdot, u_n(\cdot)) \Rightarrow h(\cdot, u^*(\cdot)) .$$

Then by Theorem 2 u^* is optimal. Q.E.D.

Corollary 2. *Suppose $u_n(x) \to u^*(x)$ for each x. Then under (A_g), (B_f), (A_σ), (A_h) and (B_p), u^* is optimal.*

Remark 3. Although the cost function h was assumed to be bounded in Theorem 2 and Corollary 1, as in Remark 1 h need not be bounded if h has the same form and the same conditions as in Remark 1. The proof of this fact can be done via the same techniques used in Remark 1 and Lemma 1. For the sake of completeness, the outline will be stated here.

As for (8),

$$E^{\mu_n}\int_0^\infty e^{-\beta t}h(z(t), u_n(z(t_{\Delta_n})))\,dt - E^{\mu_n}\int_0^\infty e^{-\beta t}h(z(t), u^*(z(t)))dt$$
$$= E^{\mu_n}\int_0^\infty e^{-\beta t}[c(u_n(z(t_{\Delta_n}))) - c(u^*(z(t)))]dt \to 0$$

as $n \to \infty$ by Lemma 2 since $c(u_n) \Rightarrow c(u^*)$ (Note that instead of assuming $h(\cdot, u_n(\cdot)) \Rightarrow h(\cdot, u^*(\cdot))$, we assume that $c(u_n(\cdot)) \Rightarrow c(u^*(\cdot))$. As for (9), it is sufficient to show that

$$(*1) \qquad E^{\mu_n}k(z(t)) \to E^{\mu_0}k(z(t))$$

and

$$(*2) \qquad E^{\mu_n}c(u^*(z(t))) \to E^{\mu_0}c(u^*(z(t)))$$

according to condition (6). By condition (6),

$$E^{\mu_0}k^2(z(t)) < \infty\ .$$

Hence there exists a sequence of bounded continuous functions $k_N(z)$ such that

$$E^{\mu_0}(k_N(z) - k(z(t)))^2 \to 0 \qquad (N \to \infty)\ .$$

Then by the same technique as in the proof of Theorem 2 we can show (*1). Similar discussion applies to (*2).

§ 5. Application: Continuous-time inventory problem

Let us apply the preceding result to determine the optimal ordering policy for a continuous-time inventory problem. Let $x(t)$ be the inventory level of a commodity at time t, $u(t)$ the ordering rate, $m(t)$ the total demand until time t, then allowing the negative inventory level (backlog case), we have

$$x(t) = x + \int_0^t u(s)ds - m(t) .$$

The usual assumption for the demand process $m(t)$ is that it has stationary independent increments. Here we shall assume that $m(t)$ can be approximately represented as

$$m(t) = m \cdot t + w(t)$$

where $w(t)$ is a standard Wiener process and m is the average demand rate. We could do the analysis without this approximation once in the preceding sections Wiener processes $w(t)$ can be replaced by other more general processes such as semimartingales. This replacement is certainly possible for some cases including processes with stationary independent increments. Here we do not go into the detail with respect to this point since it involves much and needs some recent results on martingale theory. Now under our assumption, the inventory level $x(t)$ can be represented as

$$x(t) = x + \int_0^t u(s)\,ds - \int_0^t mds - w(t) .$$

The discounted cost incurred for the infinite horizon is

$$J_u(x) = E \int_0^\infty e^{-\beta s}(c(u(s)) + k(x(s)))ds$$

where it is assumed that

$$c(u(s)) = c \cdot u(s)$$

and

$$k(x(s)) = \begin{cases} h \cdot x(s) & x(s) \geqq 0 \\ -p \cdot x(s) & x(s) \leqq 0 . \end{cases}$$

Let us assume that the control value set U is compact, *i.e.*, $U = [0, A]$. The model considered here is a continuous-time version of typical discrete-time inventory models treated in the literature. By using a usual technique in discrete-time inventory models, it can be shown that the optimal ordering policy for the discretized problem with a modified cost of this inventory model has a simple form. As a result of limiting procedure, we can show that the optimal ordering policy for the original inventory model is of bang-bang type, *i.e.*, there exists a number $\bar{z}$ and the ordering rate v^* is given by

$$v^*(x) = \begin{cases} 0: & x \geqq \bar{z} \\ A: & x < \bar{z} \end{cases}$$

where x is the present level of the inventory.

To have a complete correspondence to discrete-time inventory models, let us define for a $u \in \mathscr{U}^d_{\Delta_n}$ a modified cost given by

$$\bar{J}_u(x) = E^{P_u} \int_0^\infty e^{-\beta s} cu(s) ds + E^{P_u} \sum_{l=0}^\infty \Delta_n e^{-\beta l \Delta_n} k(y(l\Delta_n))$$

where $y(t)$ is the unique solution of

$$x(t) = x - \int_0^t m ds - w(t) \tag{15}$$

and $w(t)$ is a standard Wiener process defined on $(\Omega, \mathscr{B}, P)$. Let us define

$$\bar{V}_n(x) = \inf_{u \in \mathscr{U}^d_{\Delta_n}} \bar{J}_u(x) ,$$

$$V_n(x) = \inf_{u \in \mathscr{U}^d_{\Delta_n}} J_u(x)$$

and

$$V(x) = \inf_{u \in \mathscr{U}} J_u(x) .$$

Let $\bar{v}_n(\cdot)$ and $v_n(\cdot)$ be optimal stationary Markov controls for $\bar{V}_n$ and V_n respectively. The existence of such controls is guaranteed by the result in Section 3, Remark 1. Now once we can show that

$$J_{\bar{v}_n}(x) \to V(x) \tag{16}$$

for all x, then the limiting function $v(\cdot)$ of $\bar{v}_n(\cdot)$, if it exists, is an optimal control (ordering policy) for the original problem, *i.e.*,

$$J_v(x) = V(x) \qquad \text{for all } x .$$

Since the problem of finding $\bar{v}_n(\cdot)$ for the discretized problem with the modified cost $\bar{J}_u(x)$ has a complete analogy to the usual discrete-time inventory model, we can characterize the form of $\bar{v}_n(\cdot)$ by the conventional technique in inventory theory, and the limiting form of $\bar{v}_n(\cdot)$ can be determined.

Now let us prove (16). Since

$$\bar{J}_{v_n}(x) \geqq \bar{J}_{\bar{v}_n}(x) \quad \text{and} \quad J_{v_n}(x) \leqq J_{\bar{v}_n}(x)$$

for all x and $J_{v_n}(x) \to V(x)$, it is sufficient to show that

$$\lim_{n\to\infty} |\bar{J}_{v_n}(x) - J_{v_n}(x)| = 0 \tag{17}$$

and

$$\lim_{n\to\infty} |J_{v_n}(x) - J_{v_n}(x)| = 0\,. \tag{18}$$

Let us prove (17).

$$\begin{aligned} J_{v_n}(x) - \bar{J}_{v_n}(x) &= E^{P_{v_n}} \sum_{l=0}^{\infty} \int_{l\Delta_n}^{(l+1)\Delta_n} [e^{-\beta s} k(y(s)) - e^{-\beta s} k(y(l\Delta_n))]\, ds \\ &= E^{P_{v_n}} \sum_{l=0}^{T_n} A_l + E^{P_{v_n}} \sum_{T_n+1}^{\infty} A_l \qquad \left(T_n \equiv \left[\frac{T}{\Delta_n}\right]\right) \end{aligned} \tag{19}$$

where $[a]$ denotes the largest integer not greater than a and A_l is obviously defined. As in Remark 1,

$$\begin{aligned} &E^{P_{v_n}} \sum_{T_n+1}^{\infty} A_l \\ &\quad = E^{P_{v_n}} \sum \int_{l\Delta_n}^{(l+1)\Delta_n} [e^{-\beta s} k(y(s)) - e^{-\beta l\Delta_n} k(y(l\Delta_n))]\, ds \to 0 \end{aligned}$$

as $T \to \infty$ uniformly in n. For the first term in (19),

$$\begin{aligned} \left| E^{P_{v_n}} \sum_{l=0}^{T_n} A_l \right| &= |E^P(\textstyle\sum A_l) \exp \zeta_0^T(v_n)| \\ &\leqq (E^P(\textstyle\sum A_l)^2)^{1/2} (E^P \exp 2\zeta_0^T(v_n))^{1/2} \\ &\leqq K(T) \left(E^P \left(\sum_{l=0}^{T_n} A_l \right)^2 \right)^{1/2} . \end{aligned}$$

Since $k(\cdot)$ is continuous and the path $y(t)$ is continuous with probability one,

$$\sum_{l=0}^{T_n} A_l \to 0 \qquad (n \to \infty)$$

with probability one. To apply Lebesgue's bounded convergence theorm, let us estimate

$$\begin{aligned} \left(\sum_{l=0}^{T_n} A_l \right)^2 &\leqq 2 \left(\sum_{l=0}^{T_n} \int_{l\Delta_n}^{(l+1)\Delta_n} e^{-\beta s} k(y(s))\, ds \right)^2 \\ &\quad + 2 \left(\sum_{l=0}^{T_n} \int_{l\Delta_n}^{(l+1)\Delta_n} e^{-\beta l\Delta_n} k(y(l\Delta_n))\, ds \right)^2 . \end{aligned}$$

Since $|k(y(s))| \leqq K(1 + |y(s)|) \leqq K + K \sup_{0 \leqq s \leqq T} |y(s)|$, the last term in the above equation is less than

$$K(T)\Big(K + K \sup_{0\leqq s \leqq T} |y(s)|\Big)^2 \leqq K'(T)\Big(1 + \sup_{0\leqq s \leqq T} |y(s)|^2\Big)$$

with probability one. Since $E^P \sup_{s \leqq T} y(s)^2 < \infty$, we can apply Lebesgue's bounded convergence theorem and

$$E^P\Big(\sum_{l=0}^{T_n} A_l\Big)^2 \to 0 \, .$$

Thus we have shown (17). By the similar argument, we have the result (18). The next thing is to determine the form of $\bar{v}_n(\cdot)$. This can be done by the conventional technique used in discrete-time inventory problems. As in Section 3, Remark 1, we have the following dynamic programming equation for $\bar{V}_n(x)$:

$$\text{(20)} \qquad \bar{V}_n(x) = \inf_{a \in U} \Big\{\bar{h}_n(x, a) + e^{-\beta \Delta_n} \int \bar{V}_n(y) dF_n(y \,|\, x, a)\Big\}$$

where

$$\bar{h}_n(x, a) = E^P \int_0^{\Delta_n} e^{-\beta s} c \cdot a ds + E^P e^{-\beta \Delta_n} k(x(\Delta_n))$$

$$= K(\Delta_n) \cdot a + e^{-\beta \Delta_n} E^P k(x(\Delta_n))$$

$$K(\Delta_n) = c \int_0^{\Delta_n} e^{-\beta s} ds$$

$$x(\Delta_n) = x - m\Delta_n + a\Delta_n - w(\Delta_n)$$

and

$$F_n(y \,|\, x, a) = P\{\omega : x(\Delta_n) \leqq y\} \, .$$

Let $\varphi_n(\cdot)$ be the density of the random variable $w(\Delta_n)$. Then

$$\bar{h}_n(x, a) = K(\Delta_n) \cdot a + e^{-\beta \Delta_n} \int_{z - m\Delta_n}^{\infty} p \cdot (\xi - z + m\Delta_n) \varphi_n(\xi) d\xi$$

$$+ e^{-\beta \Delta_n} \int_{-\infty}^{z - m\Delta_n} h \cdot (z - m\Delta_n - \xi) \varphi_n(\xi) d\xi$$

$$= K(\Delta_n) a + L_n(z)$$

where $L_n(z)$ is obviously defined and $z = x + a\Delta_n$. Hence (20) can be written as

$$\overline{V}_n(x) = \inf_{z\in[x,x+A\Delta_n]} \{K(\Delta_n)(1/\Delta_n)(z-x) + L_n(z)$$
$$+ e^{-\beta\Delta_n}\int \overline{V}_n(y)\varphi(y + m\Delta_n - z)dy\}$$
$$= \inf_{z\in[x,x+A\Delta_n]} \{K(\Delta_n)(1/\Delta_n)(z-x) + L_n(z)$$
$$+ e^{-\beta\Delta_n}\int \overline{V}_n(z - m\Delta_n + \xi)\varphi(\xi)d\xi\}.$$

By the contraction mapping argument we can show that $\overline{V}_n(\cdot)$ is convex (see, for example, [7, p. 126]).

Hence

$$G_n(z) = K(\Delta_n)(1/\Delta_n)(z-x) + L_n(z)$$
$$+ e^{-\beta\Delta_n}\int \overline{V}_n(z - m\Delta_n + \xi)\varphi(\xi)d\xi$$

is again a convex function. Since $\lim_{z\to\infty} G_n(z) = \lim_{z\to-\infty} G_n(z) = \infty$, there exists a finite minimizing point $\bar{z}_n$ of $G_n(z)$ which does not depend on x. Thus the optimal ordering policy $v_n(\cdot)$ is given by

$$\overline{v}_n(x) = \begin{cases} 0: & \bar{z}_n \leqq x \\ (1/\Delta_n)(\bar{z}_n - x): & x \leqq \bar{z}_n \leqq x + \Delta_n A \\ A: & \bar{z}_n \geqq x + \Delta_n A\,. \end{cases}$$

There exists a subsequence of $\{\bar{z}_n\}$, indexed by n, such that $\bar{z}_n \to \bar{z}$ where $\bar{z}$ might be ∞ or $-\infty$. For such a subsequence $\overline{v}_n(x) \Rightarrow v^*(x)$ where

$$v^*(x) = \begin{cases} 0: & x \geqq \bar{z} \\ A: & x < \bar{z}\,. \end{cases}$$

Then by the result in Section 4, Remark 3, $v^*(\cdot)$ is an optimal ordering policy and we see that v^* is of bang-bang type. Note that all the assumptions made in Section 4 are satisfied in this example.

References

[1] Kushner, H. J., Optimal discounted stochastic control for diffusion processes, SIAM J. Control, 5 (1967), 520–531.

[2] Davis, M. H. A. and Varaiya, P., Dynamic programming conditions for partially observable stochastic systems, SIAM J. Control, 11 (1973), 226–261.

[3] Davis, M. H. A., On the existence of optimal policies in stochastic control, SIAM J. Control, **11** (1973), 587–594.
[4] ——, Optimal control of a degenerate Markovian system (in Lecent Mathematical Developments in Control), Academic Press, London/New York, 1973, 255–266.
[5] Davis, M. H. A. and Varaiya, P. P., Information states for linear stochastic systems, J. Math. Anal. Appl., **37** (1972), 384–402.
[6] Girsanov, I. V., On transforming a certain class of stochastic processes by absolutely continuous substitution of measures, Theor. Probability Appl., **5** (1960), 285–301.
[7] Ross, S. M., Applied probability models with optimization applications, Holden Day, San Francisco, 1970.
[8] Benes, V. E., Existence of optimal stochastic control laws, SIAM J. Control, **9** (1971), 446–472.
[9] Stroock, D. W. and Varadhan, S. R. S., Diffusion processes with continuous coefficients, I, II, Comm. Pure Appl. Math., **22** (1969), 345–400, 479–530.
[10] Kushner, H. J., Approximations to and local properties of diffusions with discontinuous controls, J. Optimization Theory Appl., **14** (1974), 131–150.
[11] Priouret, P., Processus de diffusion et equations differentiells stochastiques, Ecole d'Eté de Probabilités de Saint-Flour III-1973, Lecture Notes in Math., **390**, Springer, (1974), 37–113.
[12] Parthasarathy, K. R., Probability measures on metric spaces, Academic Press, 1967.
[13] Yamada, K., Optimal control of diffusion processes with discontinuous coefficients, Technical Report No. 45 (1972), Department of Statistics, Stanford University.
[14] Gikhman, I. I. and Skorokhod, A. V., Introduction to the theory of random processes, W.B. Saunders Company, 1965.

Institute of Information Sciences
University of Tsukuba
Sakuramura Ibaraki-ken 300-31
Japan

Proc. of Intern. Symp. SDE
Kyoto 1976, pp. 493-507

Localization of Conditions on the Coefficients of Diffusion Type Equations in Existence Theorems

M. P. YERSHOV

§ 0. Introduction

The aim of this paper is to obtain theorems concerning "localization" of conditions on the coefficients of a stochastic differential equation under which this equation is solvable. The localization means that the conditions on the coefficients themselves are replaced by those on the "truncated" coefficients obtained from the original ones by stopping trajectories.

We will consider even a more general problem than that of solving a stochastic differential equation: the problem of constructing a continuous quasimartingale with given drift and diffusion functionals which determine, respectively, the part of bounded variation and the squared variation.

The quasimartingale problem is formulated in Section 1.

Section 2 contains a very short survey on known results in the quasimartingale problem.

Our main result is Theorem 3.2. Its more constructive corollaries are given in Section 4.

In Section 5, a general statement of the explosion problem is proposed.

§ 1. The quasimartingale problem

1.1. Notation

C the space of continuous functions $x=\{x_t\}_{t\in T}$ on $T=[0,\infty)$, starting from zero: $x_0=0$, with the topology of uniform convergence on compacta.

$\mathscr{C}_t$ the σ-algebra in C generated by cylinder sets $\{x : x_s \in \Gamma\}$, $s\in[0,t]$, Γ being Borel sets of the real line $\boldsymbol{R}^1$.

$\mathscr{C}=\mathscr{C}_\infty$ coincides with the Borel σ-algebra in C with respect to the topology introduced.

1.2. Definition

A measure μ in $(C, \mathscr{C})$ is called a *local* (wide sense) *quasimartingale* if there exists a causal transform

$$g : (C, \{\mathscr{C}_t\}) \to (C, \{\mathscr{C}_t\})$$

(i.e., for any t, $g^{-1} \circ \mathscr{C}_t \subset \mathscr{C}_t$) with the following properties:

(1) $\forall \boldsymbol{x} \in C$, $g \circ \boldsymbol{x}$ has bounded variation on each segment $[0, t]$, $t < \infty$;

(2) for some sequence $\{\tau_n\}_{n=1,2,\ldots}$ of stopping times with respect to $\{\mathscr{C}_t\}$ ($\{\boldsymbol{x} : \tau_n(\boldsymbol{x}) \leq t\} \in \mathscr{C}_t\ \forall t$) such that

$$\mu(\{\boldsymbol{x} : \tau_n(\boldsymbol{x}) \leq N\}) \xrightarrow[n]{} 0 \qquad \forall N > 0 \,,$$

$$\forall t \in [0, \infty) \,, \qquad \forall n = 1, 2, \ldots$$

$$\int |x_{t \wedge \tau_n(x)} - (g \circ \boldsymbol{x})_{t \wedge \tau_n(x)}| \, \mu(dx) < \infty \,,$$

we have ($s \leqq t$):

$$E_\mu((x_{t \wedge \tau_n(x)} - (g \circ \boldsymbol{x})_{t \wedge \tau_n(x)}) - (x_{s \wedge \tau_n(x)} - (g \circ \boldsymbol{x})_{s \wedge \tau_n(x)}) \,|\, \mathscr{C}_{s \wedge \tau_n(x)}) = 0$$

where E_μ is the conditional expectation on the probability space $(C, \mathscr{C}, \mu)$.

1.3. Remark

Condition 1.2 (2) means, of course, that the stochastic process $\xi(\boldsymbol{x}) = \{\xi_t(\boldsymbol{x})\}_{t \in [0, \infty)}$:

$$\xi_t(x) = x_t - (g \circ \boldsymbol{x})_t$$

on $(C, \mathscr{C}, \mu)$ is a continuous local (narrow sense) martingale with respect to $\{\mathscr{C}_t\}$. It is now well-known that, in this case, there exists a unique (mod μ) causal transformation

$$A : (C, \{\mathscr{C}_t\}) \to (C, \{\mathscr{C}_t\})$$

with the following properties:

(1) $\forall \boldsymbol{x} \in C$, $A \circ \boldsymbol{x}$ is a non-decreasing function on $[0, \infty)$;

(2) $A \circ \boldsymbol{x} = \langle \xi(\boldsymbol{x}) \rangle = \int [d\xi_t(\boldsymbol{x})]^2$ (the squared variation of ξ).

1.4. Definition

The mappings g and A with the above properties will be called respectively the *drift* and the *diffusion* of the local quasimartingale μ.

1.5. The quasimartingale problem

Suppose that we are given causal transformations g and A from $(C, \{\mathscr{C}_t\})$ into itself, satisfying conditions 1.2 (1) and 1.3 (1) respectively. *Under what additional restrictions are they the drift and the diffusion, respectively, of some local quasimartingale?*

If g and A are of the form:

$$(1)\qquad \begin{aligned}(g \circ x)_t &= \int_0^t b(s, x)ds\,,\\ (A \circ x)_t &= \int_0^t a^2(s, x)ds\,,\end{aligned}$$

where the functionals $b(s, x)$ and $a(s, x)$ depend on x_u $(u \leq s)$ only, we arrive at the problem of solving the diffusion type equation with the drift and diffusion coefficients b and a respectively.

§ 2. Known results in the quasimartingale problem

The classical existence conditions, in the case 1.5 (1), are Lipschitz continuity of b and a (in a certain integral sense) in x and at most linear growth of $b(t, x)$ and $a(t, x)$ with respect to

$$\|x\|_t = \sup_{0 \leq s \leq t} |x_s|\,.$$

Various modifications of such conditions were obtained by K. Itô, M. Nisio, W. H. Fleming, A. V. Skorokhod and many others.

All these conditions were connected with the method of successive approximations. However, *within the scope of the quasimartingale problem*, the compactness-of-measures method of solving stochastic differential equations, developed by A. V. Skorokhod [2], turns out to be more fruitful. By using this method, one replaces the Lipschitz continuity by the simple one (however, the growth restrictions remain). For $b(s, x)$ and $a(s, x)$ in 1.5 (1) depending essentially on the whole past of x up to time s, this was first done by D. W. Stroock and S. R. S. Varadhan [5]: the coefficient a was allowed to be continuous in both arguments. As for coefficient b, it is easy even to get rid of continuity in x by using the Girsanov theorem on the absolute continuity of measures; therefore, one can often restrict oneself to the case $b \equiv 0$.

The author has been interested in the quasimartingale problem, in its present formulation, since 1972. In the author's note [7], the case 1.5 (1) was considered with $b \equiv 0$ and $a(s, x)$ depending only on the present of x: $a(s, x) = \tilde{a}(s, x_s)$. The method used was that of time change and allowed $\delta(s, x)$ to be Lipschitz continuous in the first argument, while its dependence on x could be arbitrary.

S. Watanabe showed [6] that the time-change method is applicable under weaker conditions on the diffusion coefficient.

In [9], the case of $g \equiv 0$ and general A, i.e. not only of the form 1.5 (1), was considered. By using a combination of time-change, compactness-of-measures and extension-of-measures arguments, it was shown that for the quasimartingale problem to have a solution it is sufficient that A should be continuous and such that $A \circ C$ is relatively compact in C. For A of the form 1.5 (1), it reduces to the condition that $a(s, \boldsymbol{x})$ should be, in some sense, bounded and continuous in $\boldsymbol{x}$ only, thus generalizing the result of D. W. Stroock and S. R. S. Varadhan [5] in the one-dimensional case.

A. V. Skorokhod [3] showed that a result, essentially the same as that of [9], can be obtained, by some approximation technique, from the result of D. W. Stroock and S. R. S. Varadhan [5]. Moreover A. V. Skorokhod, as well as D. W. Stroock and S. R. S. Varadhan, considered the case of multidimensional martingales in which the time-change construction is inapplicable. However, in the one-dimensional case, the proof of [9] seems to be shorter than that of D. W. Stroock and S. R. S. Varadhan—A. V. Skorokhod.

A. V. Skorokhod [3] proves even a stronger result than that of [9] but only because he uses, in contrast to [5] and [9], the localization which, in fact, is absolutely independent of the method the "unlocalized" result is obtained by. This will be illustrated by the present paper.

For the sake of completeness we formulate the following "unlocalized" result of [8].

Theorem. *Let g and A be causal transformations from $(C, \{\mathscr{C}_t\})$ into $(C, \{\mathscr{C}_t\})$ and, for any $\boldsymbol{x} \in C$, $g \circ \boldsymbol{x}$ has bounded variation on finite intervals**) *and $A \circ \boldsymbol{x}$ is non-decreasing. Let, in addition, g and A be continuous and the sets $g \circ C$ and $A \circ C$ be relatively compact in C.*

Then g and A are the drift and the diffusion respectively of some local quasimartingale.

This result can not be obtained directly from the corresponding one for $g \equiv 0$ by using the Girsanov type theorem.

In conclusion of our short survey, we mention a result due to A. V. Skorokhod: he proves in [4] that the diffusion A of a martingale (i.e. the diffusion of a quasimartingale with zero drift) of the form 1.5 (1) can

*) This theorem, without any changes in the proof of [8] remains true if we allow the drift to have unbounded variation (with the obvious modification in the formulation of the quasimartingale problem).

be chosen so that $(A \circ \boldsymbol{x})_t$ would be continuous in both the variables t and $\boldsymbol{x}$ on a fixed system of compact subsets of $[0, \infty) \times C$.

§ 3. Localization of conditions for the quasimartingale problem to have a solution

3.1. Notation

$$\forall x \in C\,, \qquad \forall n = 1, 2, \cdots$$
$$\lambda_n(\boldsymbol{x}) = \inf\{t : \|\boldsymbol{x}\|_t = n - t\}$$
$$\left(\|\boldsymbol{x}\|_t = \sup_{0 \leqq s \leqq t} |x_s|\right).$$

Clearly, for each n, λ_n is a stopping time with respect to $\{\mathscr{C}_t\}$ ($\{x : \lambda_n(x) \leqq t\} \in \mathscr{C}_t \ \forall t$) and

$$\lambda_n(\boldsymbol{x}) < n \qquad \forall \boldsymbol{x} \in C\,.$$

For given g and A from $(C, \{\mathscr{C}_t\})$ into itself, put

$$g_n : \quad (g_n \circ \boldsymbol{x})_t = (g \circ \boldsymbol{x})_{t \wedge \lambda_n(x)}\,,$$
$$A_n : \quad (A_n \circ \boldsymbol{x})_t = (A \circ \boldsymbol{x})_{t \wedge \lambda_n(x)}\,.$$

Note that, if g and A are continuous, so are g_n and A_n for each n.

3.2. Theorem on localization

Let g and A be causal transformations from $(C, \{\mathscr{C}_t\})$ into itself such that, for any $\boldsymbol{x} \in C$, $g \circ \boldsymbol{x}$ has bounded variation on finite intervals and $A \circ \boldsymbol{x}$ is non-decreasing. Let, in addition, g and A be continuous and, for each $n = 1, 2, \cdots$, the sets $g_n \circ C$ and $A_n \circ C$ be relatively compact in C, and let μ_n be a corresponding to g_n and A_n, by Theorem of Section 2, quasimartingale.

If

$$(1) \qquad \forall t \ \lim_n \sup_{n' \geq n} \mu_{n'}(\{\boldsymbol{x} : \lambda_n(\boldsymbol{x}) \leq t\}) = 0\,,$$

then g and A are the drift and the diffusion respectively of some local quasimartingale.

This theorem enables extending the time-domain of a solution of the quasimartingale problem (of the stochastic differential equation, in the case 1.5 (1)) despite the fact that the "truncated" problems can have non-unique solutions.

Note that, if the "truncated" problems do have unique solutions, it is easy to see that condition (1) is equivalent to

$$\forall t \lim_n \mu_n(\{x : \lambda_n(x) \le t\}) = 0 .$$

3.3. Auxiliary results

Since C is a Polish space, according to Yu. V. Prokhorov's criterion, a sequence of probability measures $\{\mu_n\}$ on $(C, \mathscr{C})$ is relatively weakly compact iff the μ_n's are uniformly tight.

Lemma 1. *For $\{\mu_n\}$ to be relatively weakly compact, it is necessary and sufficient that, for any $\varepsilon > 0$ and $t \in [0, \infty)$, there existed a compact $K = K(\varepsilon, t) \subset C$ such that*

$$\inf_n \mu_n(\Theta_t \circ K) \ge 1 - \varepsilon$$

where $\Theta_t = \theta_t^{-1} \circ \theta_t$ and $\theta_t : C \to C$ is the stopping operator:

$$(\theta_t \circ x)_s = x_{s \wedge t} .$$

Proof. Necessity follows directly from Yu. V. Prokhorov's criterion since, for any $B \subset C$ and $t \in [0, \infty)$,

$$\Theta_t \circ B \supset B .$$

Sufficiency. For a given $\varepsilon > 0$ and any $N = 1, 2, \cdots$, put $K_N = K(\varepsilon/2^N, N)$.

The set

$$K = \bigcap_{N=1}^{\infty} \Theta_N \circ K_N$$

is compact in C since it is closed and, for any $t \in [0, \infty)$,

$$\theta_t \circ K \subset \theta_t \circ \Theta_{[t]+1} \circ K_{[t]+1} \subset \theta_t \circ K_{[t]+1}$$

is relatively compact in C ($[t]$ is the integer part of t).

By construction,

$$\inf \mu_n(K) \ge 1 - \sum_{N=1}^{\infty} \frac{\varepsilon}{2^N} = 1 - \varepsilon . \quad \blacksquare$$

For any stopping time $\tau = \tau(x)$ with respect to $\{\mathscr{C}_t\}$, let $\theta_\tau : C \to C$ be defined as follows:

$$(\theta_\tau \circ x)_t = x_{t \wedge \tau(x)} .$$

Let, for any $x \in C$,

$$\Theta_\tau \circ x = \{x' \in C : \theta_{\tau(x)} \circ x' = \theta_\tau \circ x\} .$$

(Note that the transformations $\theta_{\tau(x)}$ and θ_τ are different here:

$$(\theta_{\tau(x)} \circ x')_t = x'_{t\wedge\tau(x)} ,$$

$$(\theta_\tau \circ x')_t = x'_{t\wedge\tau(x')} ;$$

but it is easy to check that, for any $x' \in \Theta_\tau \circ x$, $\tau(x') = \tau(x)$.)

Lemma 2. *Let a sequence $\{\mu_n\}$ of measures in $(C, \mathscr{C})$ satisfy condition* 3.2 (1).

If, for any $\varepsilon > 0$ and $n = 1, 2, \cdots$, there exists a compact $\tilde{K} = \tilde{K}(\varepsilon, n) \subset C$ such that

$$\inf_{n' : n' \geq n} \mu_{n'}(\Theta_{\lambda_n} \circ \tilde{K}) \geq 1 - \varepsilon ,$$

then $\{\mu_n\}$ is relatively weakly compact.

Proof. Let $\varepsilon > 0$ and $t \in [0, \infty)$ be fixed, and let n be such that

$$\sup_{n' : n' \geq n} \mu_{n'}(\{x : \lambda_n(x) \leq t\}) \leq \frac{\varepsilon}{2} .$$

Take $K = \tilde{K}(\varepsilon/2, n)$. Then, for $n' \geq n$,

$$\begin{aligned} \mu_{n'}(\Theta_t \circ K) &\geq \mu_{n'}(\{x : \theta_t \circ x \in \theta_t \circ K,\ \lambda_n(x) > t\}) \\ &= \mu_{n'}(\{x : \theta_{t\wedge\lambda_n} \circ x \in \theta_{t\wedge\lambda_n} \circ K,\ \lambda_n(x) > t\}) \\ &\geq \mu_{n'}(\Theta_{t\wedge\lambda_n} \circ K) - \mu_{n'}(\{x : \lambda_n(x) \leq t\}) \\ &\geq \mu_{n'}(\Theta_{\lambda_n} \circ K) - \frac{\varepsilon}{2} \\ &\geq 1 - \frac{\varepsilon}{2} - \frac{\varepsilon}{2} = 1 - \varepsilon . \end{aligned}$$

Since the family $\{\mu_{n'}\}_{n'=1,\cdots,n-1}$ is trivially uniformly tight, we arrive, by Lemma 1, at the desired. ∎

3.4. Proof of Theorem on localization

Let us first prove that the sequence $\{\mu_n\}$ is relatively weakly compact.

Let $\varepsilon > 0$ and n be fixed. Let us show that there exists a compact $\tilde{K} \subset C$ such that, for any $n' \geq n$,

$$\mu_{n'}(\{x : \theta_{\lambda_n(x)} \circ G_n \circ x \in \tilde{K}\}) \geq 1 - \varepsilon$$

where $G_n \circ x = x - g_n \circ x$.

On the probability space $(C, \mathscr{C}, \mu_{n'})$, the stochastic process

$$\zeta^{n'}(x) = G_{n'} \circ x$$

is a martingale with respect to $\{\mathscr{C}_t\}$ with the squared variation

$$\langle \zeta^{n'}(x) \rangle = A_{n'} \circ x .$$

Hence the process, obtained from it by stopping at time λ_n, is also a martingale with respect to $\{\mathscr{C}_t\}$; this process coincides with $\theta_{\lambda_n(x)} \circ \zeta^n(x)$:

$$\theta_{\lambda_n(x)} \circ \zeta^{n'}(x) = \theta_{\lambda_n(x)} \circ G_{n'} \circ x = \theta_{\lambda_n(x)} \circ G_n \circ x .$$

Its squared variation

$$\langle \theta_{\lambda_n(x)} \circ \zeta^n(x) \rangle = \theta_{\lambda_n(x)} \circ A_{n'} \circ x = A_n \circ x .$$

It is well-known (cf. e.g. F. B. Knight [1]) that there exists a Wiener process $W = \{W_t\}$ on the probability space $(C, \mathscr{C}, \mu_{n'}) \times (C, \mathscr{C}, \nu_{n'})$ (obtained from $(C, \mathscr{C}, \mu_{n'})$ by a suitable "randomization") with respect to a flow of σ-algebras in it such that the family $\{(A_n \circ x)_t\}_{t \in [0,\infty)}$ is a time change with respect to this flow and

$$(\theta_{\lambda_n(x)} \circ \zeta^n(x))_t = W_{(A_n \circ x)_t}(x, y) , \qquad (x, y) \in C \times C .$$

Let $K \subset C$ be such a compact that

$$[\mu_{n'} \times \nu_{n'}](W^{-1} \circ \tilde{\tilde{K}}) \geq 1 - \varepsilon .$$

The set

$$\tilde{K}_0 = \{z : z_t = W_{(A_n \circ x)_t}(x, y),\ (x, y) \in W^{-1} \circ \tilde{\tilde{K}}\}$$

is relatively compact in C since $A_n \circ C$ is relatively compact and

$$\|z\|_t \leq \sup_{(x,y) \in W^{-1} \circ \tilde{\tilde{K}}} \|W(x, y)\|_{\sup_{x \in C} (A_n \circ x)_t} ,$$

$$\delta_{z;\,t}(h) \leq \sup_{(x,y) \in W^{-1} \circ \tilde{\tilde{K}}} \delta_{W(x,y);\ \sup_{x \in C} (A_n \circ x)_t} \Big(\sup_{x \in C} \delta_{A_n \circ x;\,t}(h) \Big)$$

where, for any $z \in C$, $h > 0$ and $t \in [0, \infty)$,

$$\delta_{z;\,t}(h) = \sup \{|z_s - z_{s'}|;\ 0 \leq s \,\&\, s' \leq t,\ |s - s'| \leq h\} .$$

For its closure, $\tilde{K}$, we have:

$$\mu_{n'}(\{x\colon \theta_{\lambda_n(x)} \circ G_n \circ x \in \tilde{K}\}) = \mu_{n'}(\{x\colon \theta_{\lambda_n(x)} \circ \zeta^n(x) \in \tilde{K}\})$$
$$\geq [\mu_{n'} \times \nu_{n'}](W^{-1} \circ \tilde{\tilde{K}}) \geq 1 - \varepsilon .$$

Thus a possibility of finding a $\tilde{K}$ with the desired property is guaranteed.

Now let $\hat{K}$ be the closure of $g_n \circ C$ and

$$K = \hat{K} + \tilde{K} = \{x + x'\colon x \in \hat{K},\ x' \in \tilde{K}\} .$$

For any $n' \geq n$,

$$\begin{aligned}
&\mu_{n'}(\Theta_{\lambda_n} \circ K) \\
&\quad \geq \mu_{n'}(\{x\colon \theta_{\lambda_n(x)} \circ g_{n'} \circ x \in \theta_{\lambda_n(x)} \circ \hat{K}, \theta_{\lambda_n(x)} \circ G_{n'} \circ x \in \theta_{\lambda_n(x)} \circ \tilde{K}\}) \\
&\quad = \mu_{n'}(\{x\colon g_n \circ x \in g_n \circ C, \theta_{\lambda_n(x)} \circ G_n \circ x \in \theta_{\lambda_n(x)} \circ \tilde{K}\}) \\
&\quad = \mu_{n'}(\{x\colon \theta_{\lambda_n(x)} \circ G_n \circ x \in \theta_{\lambda_n(x)} \circ \tilde{K}\}) \\
&\quad \geq \mu_{n'}(\{x\colon \theta_{\lambda_n(x)} \circ G_n \circ x \in \tilde{K}\}) \\
&\quad \geq 1 - \varepsilon .
\end{aligned}$$

Thus we have proved that the sequence $\{\mu_n\}$ is relatively weakly compact.

Let μ be a limiting point of $\{\mu_n\}$: $\mu_{n'} \to \mu$ weakly. Since $\mu_{n'}$ is a solution of the quasimartingale problem for $(g_{n'}, A_{n'})$ and, for fixed t and n, the functions

$$x \mapsto x_{t\wedge\lambda_n(x)} ,$$
$$x \mapsto (g \circ x)_{t\wedge\lambda_n(x)} ,$$
$$x \mapsto (A \circ x)_{t\wedge\lambda_n(x)}$$

are continuous and bounded, we obtain:

$$\begin{aligned}
&E_\mu((x_{t\wedge\lambda_n(x)} - (g \circ x)_{t\wedge\lambda_n(x)}) - (x_{s\wedge\lambda_n(x)} - (g \circ x)_{s\wedge\lambda_n(x)}) \mid \mathscr{C}_{s\wedge\lambda_n}) \\
&\qquad = \lim_{n'} E_{\mu_{n'}}(\cdots) \\
&\qquad = \lim_{n'} E_{\mu_{n'}}((G_{n'} \circ x)_{t\wedge\lambda_n(x)} - (G_{n'} \circ x)_{s\wedge\lambda_n(x)} \mid \mathscr{C}_{s\wedge\lambda_n}) \\
&\qquad = 0 ,
\end{aligned}$$

$$\begin{aligned}
&E_\mu([(x_{t\wedge\lambda_n(x)} - (g \circ x)_{t\wedge\lambda_n(x)})^2 - (A \circ x)_{t\wedge\lambda_n(x)}] \\
&\qquad\qquad - [(x_{s\wedge\lambda_n(x)} - (g \circ x)_{s\wedge\lambda_n(x)})^2 - (A \circ x)_{s\wedge\lambda_n(x)}] \mid \mathscr{C}_{s\wedge\lambda_n}) \\
&\qquad = \lim_{n'} E_{\mu_{n'}}(\cdots) \\
&\qquad = \lim_{n'} E_{\mu_{n'}}([(G_{n'} \circ x)^2_{t\wedge\lambda_n(x)} - (A_{n'} \circ x)_{t\wedge\lambda_n(x)}] \\
&\qquad\qquad - [(G_{n'} \circ x)^2_{s\wedge\lambda_n(x)} - (A_{n'} \circ x)_{s\wedge\lambda_n(x)}] \mid \mathscr{C}_{s\wedge\lambda_n}) \\
&\qquad = 0 .
\end{aligned}$$

(Here we made use of the fact that $\mathscr{C}_{s\wedge\lambda_n} = \theta^{-1}_{s\wedge\lambda_n} \circ \mathscr{C}$.) Hence μ is a local quasimartingale with the drift g (it follows directly from 1.2 (2) with τ_n replaced by λ_n) and diffusion A (the latter equality is another definition of the squared variation).

§ 4. Constructive conditions

The following lemma, a modification of the Gronwall-Bellman lemma, is usually used to obtain constructive conditions of localization for the coefficients of stochastic differential equations to ensure the existence of a solution.

4.1. Lemma. *Let $H(s)$, $s \leq t$, be a non-decreasing function, $H(0) = 0$. Let $\varphi(s)$, $s \leq t$, be an integrable on $[0, t]$ with respect to dH function satisfying, for each $s \leq t$, the inequality*

$$\varphi(s) \leq k \int_0^s \varphi(u)dH(u) + \psi(s)$$

where $k \geq 0$ is a constant and ψ is an integrable on $[0, t]$ with respect to dH function.

Then

$$\varphi(t) \leq \psi(t) + k \int_0^t e^{k(H(t)-H(s))}\psi(s)dH(s) .$$

For the proof, one uses a trivial iterative process.

Remark. If ψ is a non-decreasing function, we obtain, from this lemma, the following coarser but simpler estimate:

$$\varphi(t) \leq \psi(t)(1 + e^{kH(t)}) . \tag{1}$$

4.2. Theorem. *In Theorem 3.2, for 3.2 (1) to be satisfied, it is sufficient that*

$$\forall m = 1, 2, \cdots , \quad \forall s \in [0, \infty) , \quad \forall \boldsymbol{x} \in C ,$$

$$\|g_m \circ \boldsymbol{x}\|_s^2 + 4(A_m \circ \boldsymbol{x})_s \leq k_1(s ; m) \int_0^s (1 + \|\boldsymbol{x}\|_u^2)dH(u ; m) + k_2(s ; m) \|\boldsymbol{x}\|_s^2$$

where $k_1(s ; m)$, $k_2(s ; m)$ and $H(s ; m)$ are non-decreasing in s for each fixed m functions such that, for any m, $k_2(s ; m) < 1/2$ for all $s \in [0, \infty)$, for any m, $H(0 ; m) = 0$ and, for each $s \in [0, \infty)$,

(1) $$\frac{k(s\,;m)}{m^2}e^{k(s;\,m)} \xrightarrow[m]{} 0$$

where

$$k(s\,;m) = \frac{2k_1(s\,;m)H(s\,;m)}{1 - 2k_2(s\,;m)}.$$

4.3. Examples

To make condition (1) more transparent we give a simpler sufficient condition.

Assume that*)

$$k(m) \lesssim 2\ln\frac{m}{\ln m}$$

(here we write $k(m)$ instead of $k(s\,;m)$ with a fixed s). Then

$$\frac{k(m)e^{k(m)}}{m^2} \lesssim \frac{1}{m^2}\left[(2\ln m - 2\ln\ln m)\frac{m^2}{(\ln m)^2}\right]$$
$$= \frac{2}{\ln m} - \frac{2\ln\ln m}{(\ln m)^2} \xrightarrow[m]{} 0$$

thus implying (1).

In terms of the transformations g and A, one gets the following condition implying that of Theorem 4.2:

$$\forall s \in [0,\infty)\,,\qquad \forall \boldsymbol{x} \in C\,,$$

$$\|g\circ\boldsymbol{x}\|_s^2 + 4(A\circ\boldsymbol{x})_s \le k_0(\ln(1+\|\boldsymbol{x}\|_s))^{1-\varepsilon}\int_0^s(1+\|\boldsymbol{x}\|_u^2)dH(u) + \alpha\|\boldsymbol{x}\|_s^2$$

where $\varepsilon > 0$, $k_0 > 0$ and $\alpha < 1/2$ are constants, $H(s)$ is a non-decreasing function. (Compare with the localization condition in A. V. Skorokhod [3] where the quasimartingale problem with $g \equiv 0$ was considered:

$$(A\circ\boldsymbol{x})_s \le K_s\left(1 + \int_0^s x_u^2\,du\right) + q\,\|\boldsymbol{x}\|_s^2$$

where K_s is a non-decreasing function and $q < 9/32\sqrt{78}$.)

---

*) The writing $\alpha(m) \lesssim \beta(m)$ ($\alpha(m) \ge 0$, $\beta(m) \ge 0$) means that

$$\limsup_m \frac{\alpha(m)}{\beta(m)} \le 1\,.$$

In fact, under this condition, the condition of Theorem 4.2 is satisfied with:

$$k_1(s\,;m) \equiv k_0(\ln(1+m))^{1-\varepsilon}\,,$$
$$k_2(s\,;m) \equiv \alpha\,,$$
$$H(s\,;m) \equiv H(s) - H(0)$$

since

$$k(s\,;m) = \frac{2k_0(\ln(1+m))^{1-\varepsilon}(H(s)-H(0))}{1-2\alpha} = o\left(\ln\frac{m}{\ln m}\right).$$

Let us further assume that g and A are of the form 1.5 (1); in other words, that we deal with a stochastic differential equation. Then the above condition is implied by the following:

$$\left(\int_0^s |b(u,\boldsymbol{x})|\,du\right)^2 + 4\int_0^s a^2(u,\boldsymbol{x})du$$
$$\leq k_0(\ln(1+\|\boldsymbol{x}\|_s))^{1-\varepsilon}\int_0^s (1+\|\boldsymbol{x}\|_u^2)dH(u) + \alpha\,\|\boldsymbol{x}\|_s^2$$

where ε, k_0, α and H are the same as just above.

If we add to the latter condition the following two:

(i) b and a are continuous in $\boldsymbol{x}$;

(ii) $|b(u,\boldsymbol{x})|$ and $a^2(u,\boldsymbol{x})$ are bounded by an $h(\|\boldsymbol{x}\|_u)$ where $h(\xi)$, $\xi \in \boldsymbol{R}$, is an increasing function,

we obtain an example of a complete set of conditions sufficient for the existence of a solution of the stochastic differential equation with the coefficients b and a. In fact, (ii) implies then that the sets $g_n \circ C$ and $A_n \circ C$ are relatively compact.

4.4. Proof of Theorem 4.2

Let G_n be defined as in No. 3.4. For each $m = 1, 2, \cdots$, and $s \in [0, \infty)$,

$$\|\boldsymbol{x}\|_{s\wedge\lambda_m(x)}^2 \leq 2\,\|g_m \circ \boldsymbol{x}\|_{s\wedge\lambda_m(x)}^2 + 2\,\|G_m \circ \boldsymbol{x}\|_{s\wedge\lambda_m(x)}^2$$

and, for each $n \geq m$, by a well-known martingale inequality, taking into account that

$$(G_m \circ \boldsymbol{x})_{s\wedge\lambda_m(x)} = (G_n \circ \boldsymbol{x})_{s\wedge\lambda_m(x)}\,, \qquad (A_n \circ \boldsymbol{x})_{s\wedge\lambda_m(x)} = (A_m \circ \boldsymbol{x})_s\,,$$

we obtain:

$$\int_C \|\boldsymbol{x}\|^2_{s\wedge\lambda_m(x)}\mu_n(dx)$$

$$\le 2\int_C \|g_m\circ\boldsymbol{x}\|^2_{s\wedge\lambda_m(x)}\mu_n(dx) + 2\int_C \|G_n\circ\boldsymbol{x}\|^2_{s\wedge\lambda_m(x)}\mu_n(dx)$$

$$\le 2\int_C \|g_m\circ\boldsymbol{x}\|^2_{s\wedge\lambda_m(x)}\mu_n(dx) + 8\int_C (C_n\circ\boldsymbol{x})^2_{s\wedge\lambda_m(x)}\mu_n(dx)$$

$$= 2\int_C \|g_m\circ\boldsymbol{x}\|^2_{s\wedge\lambda_m(x)}\mu_n(dx) + 8\int_C (A_n\circ\boldsymbol{x})_{s\wedge\lambda_m(x)}\mu_n(dx)$$

$$= \int_C (2\|g_m\circ\boldsymbol{x}\|^2_{s\wedge\lambda_m(x)} + 8(A_m\circ\boldsymbol{x})_{s\wedge\lambda_m(x)})\mu_n(dx)$$

$$\le 2k_1(s;m)\int_0^s\left(1+\int_C \|\boldsymbol{x}\|^2_{u\wedge\lambda_m(x)}\mu_n(dx)\right)dH(u;m)$$

$$+ 2k_2(s;m)\int_C \|\boldsymbol{x}\|^2_{s\wedge\lambda_m(x)}\mu_n(dx)\,.$$

Thus we have the estimate:

$$\int_C \|\boldsymbol{x}\|^2_{s\wedge\lambda_m(x)}\mu_n(dx)$$

$$\le \frac{2k_1(s;m)}{1-2k_2(s;m)}\int_0^s\int_C \|\boldsymbol{x}\|^2_{u\wedge\lambda_m(x)}\mu_n(dx)dH(u;m) + \frac{2k_1(s;m)H(s;m)}{1-2k_2(s;m)}$$

and, by 4.1 (1),

$$\int_C \|\boldsymbol{x}\|^2_{t\wedge\lambda_m(x)}\mu_n(dx) \le k(t;m)[1+e^{k(t;\,m)}]\,.$$

By applying the Chebyshev inequality, we arrive at the following relation:

$$\sup_n \mu_n(\{\boldsymbol{x}\colon \lambda_m(x)\le t\}) = \sup_n \mu_n(\{\boldsymbol{x}\colon \|\boldsymbol{x}\|_{t\wedge\lambda_m(x)} \ge m-t\})$$

$$\le \frac{k(t;m)}{(m-t)^2}[1+e^{k(t;\,m)}] \xrightarrow[m]{} 0$$

by 4.2 (1). ■

§ 5. A general explosion problem

Let Ω be a set of points ω with a flow of σ-algebras $\not f = \{\mathscr{F}_t\}_{t\in R_+}$ of its subsets.

The set of stopping times with respect to $\not{f}$, i.e. such mappings $\tau : \Omega \to \bar{\boldsymbol{R}}_+$ that

$$\{\omega : \tau(\omega) \leq t\} \in \mathscr{F}_t \qquad \forall t \in \boldsymbol{R}_+$$

forms a lattice with respect to the usual partial ordering.

We call two (probability) measures μ_σ and μ_τ on $\mathscr{F}_\sigma$ and $\mathscr{F}_\tau$ ($\mathscr{F}_\tau = \{A \in \bigvee_t \mathscr{F}_t : A \cap \{\tau \leq t\} \in \mathscr{F}_t \ \forall t\}$ respectively, where σ and τ are $\not{f}$-stopping times, *compatible* if their restrictions $\mu_\sigma | \mathscr{F}_\sigma \cap \mathscr{F}_\tau$ and $\mu_\tau | \mathscr{F}_\sigma \cap \mathscr{F}_\tau$ coincide. (Note that $\mathscr{F}_\sigma \cap \mathscr{F}_\tau = \mathscr{F}_{\sigma \wedge \tau}$.)

Given two compatible measures μ_σ and μ_τ on $\mathscr{F}_\sigma$ and $\mathscr{F}_\tau$ respectively, there exists one and only one measure $\mu_{\sigma \vee \tau}$ on $\mathscr{F}_{\sigma \vee \tau}$ $(= \mathscr{F}_\sigma \vee \mathscr{F}_\tau)$ compatible with μ_σ and μ_τ. To show this one can make use of the fact that every set A from $\mathscr{F}_{\sigma \vee \tau}$ can be represented as a disjoint union of a set A_1 from $\mathscr{F}_\sigma$, a set A_2 from $\mathscr{F}_\tau$ and a set A_3 from $\mathscr{F}_{\sigma \wedge \tau}$:

$$A_1 = [A \cap \{\sigma < \tau\}] \in \mathscr{F}_\tau \,,$$
$$A_2 = [A \cap \{\tau < \sigma\}] \in \mathscr{F}_\sigma \,,$$
$$A_3 = [A \cap \{\tau = \sigma\}] \in \mathscr{F}_{\sigma \wedge \tau} \,.$$

Let $\boldsymbol{T}$ be a set of $\not{f}$-stopping times. We say that a family of measures $\{\mu_\tau\}_{\tau \in \boldsymbol{T}}$ is *compatible* if any two measures of this family are compatible.

Given any compatible family of measures, we may imbed it, taking into account the above remark, into a larger compatible family so that the index set $\boldsymbol{T}$ would become a *hereditary lattice*:

$$\sigma \leq \tau \in \boldsymbol{T} \Rightarrow \sigma \in \boldsymbol{T} \,,$$
$$\sigma \in \boldsymbol{T} \;\&\; \tau \in \boldsymbol{T} \Rightarrow \sigma \vee \tau \in \boldsymbol{T} \,.$$

Let $\{\mu_\tau\}_{\tau \in \boldsymbol{T}}$ be a compatible family of measures with $\boldsymbol{T} \neq \emptyset$ being a *proper* hereditary sublattice of the lattice of all $\not{f}$-stopping times. We call this family *unextendable* if, for no stopping time $\tau' \notin \boldsymbol{T}$, there exists a measure $\mu_{\tau'}$ on $\mathscr{F}_{\tau'}$ such that the family $\{\{\mu_\tau\}_{\tau \in \boldsymbol{T}}, \mu_{\tau'}\}$ is compatible.

Hypothesis. *For any unextendable family* $\{\mu_\tau\}_{\tau \in \boldsymbol{T}}$, *there exists a unique $\not{f}$-stopping time κ such that either*

(i) $\kappa \notin \boldsymbol{T}$, *and* $\sigma < \kappa \Rightarrow \sigma \in \boldsymbol{T}$

or

(ii) $\kappa \in \boldsymbol{T}$, *and* $\mu_\kappa(\{\omega : \sigma > \kappa\}) > 0 \Rightarrow \sigma \notin \boldsymbol{T}$.

(Here σ stands for $\not{f}$-stopping times.)

It would be natural to call H the *explosion time for* $\{\mu_\tau\}_{\tau \in \boldsymbol{T}}$.

Condition 3.2 (1) was imposed to guarantee that our process (local quasimartingale with the drift g and diffusion A) would not be explosive.

References

[1] Knight, F. B., A reduction of continuous square integrable martingales to Brownian motion, Martingales, (A Report on a Meeting at Oberwolfach, May 17–23, 1970), Lecture Notes in Math., **190** (1971), 19–31, Springer-Verlag.

[2] Skorokhod, A. V., Studies in the theory of stochastic processes, Kiev, 1961, (Russian).

[3] ——, A remark on square integrable martingales, Theor. Probability Appl., **20**-1 (1975), 195–198.

[4] ——, K-martingales and stochastic equations, Proc. of School-seminar on the Theory of Stochastic Processes, Druskininkai, November 25–30, 1974, 195–234, Vilnius 1975, (Russian).

[5] Stroock, D. W. and Varadhan, S. R. S., Diffusion processes with continuous coefficients, I, Comm. Pure Appl. Math., **22**-3 (1969), 345–400.

[6] Watanabe, S., Solution of stochastic differential equations by a time change, Appl. Math. and Optim., **2** (1975), 90–96.

[7] Yershov, M. P., On stochastic equations, Proc. 2nd Japan-USSR Sympos. Probability Theory, August 1972, Kyoto, Lecture Notes in Math., **330** (1973), 527–530, Springer-Verlag.

[8] ——, Studies in the theory of stochastic equations, a dissertation, Moscow, 1974, (Russian).

[9] Yershov (Ershov), M. P., The existence of a martingale with given diffusion functional, Theor. Probability Appl., **19**-4 (1974), 633–655.

Steklov Mathematical Institute
Vavilova 42, Moscow
V-333, U.S.S.R.